PUBLICATIONS OF THE NEWTON INSTITUTE

Idempotency

T0269278

Publications of the Newton Institute

Edited by H.P.F. Swinnerton-Dyer
Executive Director, Isaac Newton Institute for Mathematical Sciences

The Isaac Newton Institute for Mathematical Sciences of the University of Cambridge exists to stimulate research in all branches of the mathematical sciences, including pure mathematics, statistics, applied mathematics, theoretical physics, theoretical computer science, mathematical biology and economics. The four six-month-long research programmes it runs each year bring together leading mathematical scientists from all over the world to exchange ideas through seminars, teaching and informal interaction.

Associated with the programmes are two types of publication. The first contains lecture courses, aimed at making the latest developments accessible to a wider audience and providing an entry to the area. The second contains proceedings of workshops and conferences focusing on the most topical aspects of the subjects.

Idempotency

Edited by

Jeremy Gunawardena
BRIMS, Hewlett–Packard Research Laboratories, Bristol

CAMBRIDGE
UNIVERSITY PRESS

CAMBRIDGE UNIVERSITY PRESS
Cambridge, New York, Melbourne, Madrid, Cape Town, Singapore, São Paulo

Cambridge University Press
The Edinburgh Building, Cambridge CB2 8RU, UK

Published in the United States of America by Cambridge University Press, New York

www.cambridge.org
Information on this title: www.cambridge.org/9780521553445

First published 1998
This digitally printed version 2008

A catalogue record for this publication is available from the British Library

ISBN 978-0-521-55344-5 hardback
ISBN 978-0-521-05538-3 paperback

Contents

Foreword

Hewlett-Packard's *Basic Research Institute in the Mathematical Sciences*[1] (BRIMS) was set up in 1994 as a joint undertaking between Hewlett-Packard Laboratories in Bristol and the Isaac Newton Institute for Mathematical Sciences[2] in Cambridge.

BRIMS represents a pioneering attempt to create an environment for fundamental research, which draws upon the best aspects of both academia and industry. It exemplifies our belief that science and technology should stimulate and influence each other.

Proceedings of workshops and conferences at BRIMS are eligible for publication in this series of *Publications of the Newton Institute* by virtue of a special arrangement with Cambridge University Press. We welcome such volumes as evidence of the quality and vitality of scientific activity at BRIMS.

Dr John M. Taylor, OBE, FEng, FIEE
Director
Hewlett-Packard Laboratories Europe

Sir Michael Atiyah, OM, FRS
Director
Isaac Newton Institute

---

[1]http://www-uk.hpl.hp.com/brims/
[2]http://www.newton.cam.ac.uk/

Preface

This volume arose out of a workshop on *Idempotency* that was held at Hewlett-Packard's Basic Research Institute in the Mathematical Sciences (BRIMS) from 3–7 October 1994.

The word idempotency signifies the study of semirings in which the addition operation is idempotent: $a + a = a$. The best-known example is the *max-plus* semiring, $\mathbb{R} \cup \{-\infty\}$, in which addition is defined as $\max\{a, b\}$ and multiplication as $a + b$, the latter being distributive over the former. Interest in such structures arose in the late 1950s through the observation that certain problems of discrete optimisation could be linearised over suitable idempotent semirings. More recently the subject has established rich connections with automata theory, discrete event systems, nonexpansive mappings, nonlinear partial differential equations, optimisation theory and large deviations. These new developments are the focus of this volume.

The papers brought together here consist of expanded contributions to the BRIMS workshop as well as some invited contributions. The papers were all reviewed, in each case by at least one person who was not associated with the workshop. Although the level of acceptance was more relaxed than for a journal, the reviewing process was conducted seriously. Not all the contributions were accepted and many of those that were have been substantially improved as a result of the reviews. I am grateful to all the reviewers for their time and effort and I regret that, for reasons of confidentiality, it does not seem appropriate to mention their names here.

The opening paper is a survey of the subject written specially for this volume. It has an extensive bibliography and tries to fill in some of the background to the remaining papers.

It is a pleasure to thank several people: Kate Elenor, of Hewlett-Packard, for her invaluable assistance in organising the workshop; Peter Goddard, at that time Deputy Director of the Newton Institute and currently Master of St. John's College, Cambridge, for his support in setting up the publication arrangement; Margaret Powell, of BRIMS, for so efficently keeping track of the reviewing process; Tom Rokicki, of Hewlett-Packard, for installing LaTeX2$_\epsilon$ on my workstation, remotely from Palo Alto, at very short notice—always a pleasure to watch an expert at work; Imre Simon, of the Universidade de São Paulo, whose tropical participation made such a difference to the workshop itself; David Tranah, the Publishing Director for Mathematical Sciences at CUP, for his composure in the face of repeatedly missed deadlines; and, finally, all the contributors and participants for their patience with the editor, despite the inordinate delay in getting the volume to press.

Jeremy Gunawardena, Bristol

List of Participants

The list below includes the authors of papers in this volume and those who attended the BRIMS workshop.

AKIAN, Marianne .... marianne.akian@inria.fr
        INRIA, BP 105, 78153 Le Chesnay, FRANCE
D'ALESSANDRO, Flavio ....
        IBP, 4 Place Jussieu, 75252 Paris, FRANCE
BACCELLI, François .... baccelli@sophia.inria.fr
        INRIA, BP 93, 06902 Sophia-Antipolis, FRANCE
BORRIELLO, Gaetano .... gaetano@cs.washington.edu
        University of Washington, Seattle WA 98195, USA
COFER, Darren D. .... cofer_darren@htc.honeywell.com
        Honeywell Technology Center, Minneapolis MN 55418 USA
COHEN, Guy .... cohen@cas.ensmp.fr
        CAS, École des Mines de Paris, 77305 Fontainebleau, FRANCE
CUNINGHAME-GREEN, Raymond A. ....
        University of Birmingham, Birmingham B15 2TT, UK
DEL MORAL, Pierre .... delmoral@cict.fr
        LSP-CNRS, Université Paul Sabatier, 31062 Toulouse, FRANCE
DUDNIKOV, Peter ....
        2555 10 Boyarka, Kievskoi obl., ul. Lenina 42-1, UKRAINE
GARG, Vijay K. .... garg@ece.utexas.edu
        University of Texas, Austin TX 78712, USA
GAUBERT, Stéphane .... stephane.gaubert@inria.fr
        INRIA, BP 105, 78153 Le Chesnay, FRANCE
GAUJAL, Bruno .... gaujal@sophia.inria.fr
        INRIA, BP 93, 06902 Sophia Antipolis, FRANCE
GUNAWARDENA, Jeremy .... jhcg@hplb.hpl.hp.com
        BRIMS, Hewlett-Packard Labs, Bristol BS12 6QZ, UK
GÜREL, Ayla .... gurel@eenet.ee.emu.edu.tr
        Eastern Mediterranean University, Famagusta, via Mersin 10, TURKEY
JEAN-MARIE, Alain .... ajm@sophia.inria.fr
        INRIA, BP 93, 06902 Sophia Antipolis, FRANCE
KOLOKOLTSOV, Vassili N. .... v.kolokoltsov@maths.ntu.ac.uk
        Nottingham Trent University, Nottingham NG1 4BU, UK
KROB, Daniel .... daniel.krob@litp.ibp.fr
        LITP/IBP, CNRS-Université Paris 7, 75251 Paris, FRANCE
LEUNG, Hing .... hleung@cs.nmsu.edu
        New Mexico State University, Las Cruces NM 88003, USA

LEWIS, Frank L. .... flewis@controls.uta.edu
    University of Texas at Arlington, Fort Worth TX 76118, USA
LITVINOV, Grigori .... litvinov@islc.msk.su
    Nagornaya 27-4-72, Moscow 113186, RUSSIA
MAIRESSE, Jean .... Jean.Mairesse@sophia.inria.fr
    INRIA, BP 93, 06902 Sophia-Antipolis, FRANCE
MASCARI, Gianfranco .... mascari@nextiac.iac.rm.cnr.it
    CNR-IAC, 137 Viale del Policilinico, 00161 Rome, ITALY
MASLOV, Victor P. .... maslov@amath.msk.ru
    Inst of Elec & Math, Moscow 109028, RUSSIA
NUSSBAUM, Roger D...................... nussbaum@math.rutgers.edu
    Rutgers University, New Brunswick NJ 08903, USA
OLSDER, Geert-Jan .... g.j.olsder@math.tudelft.nl
    Delft University of Technology, 2600GA Delft, HOLLAND
PASTRAVANU, Octavian C. .... opastrav@ac.tuiasi.ro
    Technical University of Iasi, 6600 Iasi, ROMANIA
PEDICINI, Marco .... marco@sparc1.iac.rm.cnr.it
    CNR-IAC, 137 Viale del Policilinico, 00161 Rome, ITALY
PIN, Jean-Eric .... Jean-Eric.Pin@litp.ibp.fr
    LITP/IBP, CNRS-Université Paris 7, 75251 Paris, FRANCE
QUADRAT, Jean-Pierre .... Jean-Pierre.Quadrat@inria.fr
    INRIA, BP 105, 78153 Le Chesnay, FRANCE
SAKAROVITCH, Jacques .... jacques.sakarovitch@ibp.fr
    IBP, 4 Place Jussieu, 75252 Paris, FRANCE
SALUT, Gérard ....
    LAAS du CNRS, 31077 Toulouse, FRANCE
SAMBORSKIĬ, Serguei .... samborsk@matin.math.unicaen.fr
    Université de Caen, 14032 Caen, FRANCE
SIMON, Imre .... is@ime.usp.br
    Universidade de São Paulo, 05508-900 São Paulo, SP, BRASIL
VIOT, Michel ....
    01300 IZIEU, FRANCE
WAGNEUR, Edouard .... wagneur@auto.emn.fr
    École des Mines de Nantes, 44049 Nantes, FRANCE
WALKUP, Elizabeth A. .... ewalkup@ichips.intel.com
    Intel Corporation, Hillsboro OR 97124, USA
WEBER, Andreas .... weber@tiger.psc.informatik.uni-frankfurt.de
    Martin Luther Universität, D-06099 Halle a.d. Saale, GERMANY

An Introduction to Idempotency

Jeremy Gunawardena

1   Introduction

The word *idempotency* signifies the study of semirings in which the addition operation is idempotent: $a + a = a$. The best-known example is the *max-plus* semiring, $\mathbb{R} \cup \{-\infty\}$, in which addition is defined as $\max\{a, b\}$ and multiplication as $a + b$, the latter being distributive over the former. Interest in such structures arose in the 1950s through the observation that certain problems of discrete optimisation could be linearised over suitable idempotent semirings. Cuninghame-Green's pioneering book, [CG79], should be consulted for some of the early references. More recently, intriguing new connections have emerged with automata theory, discrete event systems, nonexpansive mappings, nonlinear partial differential equations, optimisation theory and large deviations, and these topics are discussed further in the subsequent sections of this paper. The phrase *idempotent analysis* first appears in the work of Kolokoltsov and Maslov, [KM89].

Idempotency has arisen from a variety of sources and the different strands have not always paid much attention to each other's existence. This has led to a rather parochial view of the subject and its place within mathematics; it is not as well-known nor as widely utilised as perhaps it should be. The workshop on which this volume is based was organised, in part, to address this issue. With this in mind, we have tried to present here a coherent account of the subject from a mathematical perspective while at the same time providing some background to the other papers in this volume. We have said rather little about what are now standard topics treated in the main books in the field, [CG79, Zim81, CKR84, BCOQ92, MK94, KMa] and [Mas87a, Chapter VIII]. We have tried instead to direct the reader towards the open problems and the newer developments, while pointing out the links with other areas of mathematics. However, we make no pretence at completeness. A different view of the subject is put forward by Litvinov and Maslov in their survey paper in this volume, [LM].

Stéphane Gaubert and Jean Mairesse provided the author with extensive and detailed suggestions on the first draft of this paper. It is a pleasure to thank them, as well as François Baccelli, Grigori Litvinov, Pierre Del Moral and Jacques Sakarovitch, for their comments, which resulted in many corrections and improvements. Any errors or omissions that remain are entirely the responsibility of the author.

2   Dioids

2.1   Introduction

In this section we introduce idempotent semirings, or dioids, and study some
of the main examples. As we shall see, dioids occur more widely than one
might suspect and certain classes of dioids have been extensively studied
under other names. The intention is to give the reader a sense of the scope
of the subject before looking at more specialised topics in later sections.

2.2   Semirings and dioids

We recall that a semigroup is a set with an associative binary operation while
a monoid is a semigroup with a distinguished identity element, [KS86].

Definition 2.1 *A semiring is a set, S, with two binary operations—denoted
with the usual conventions for addition ("+") and multiplication ("×" or
".")—and two distinguished elements, $0, 1 \in S$, such that $0 \neq 1$ and*

- *$(S, +, 0)$ is a commutative monoid with identity element 0,*

- *$(S, \times, 1)$ is a monoid with identity element 1,*

- *$a.0 = 0.a = 0$ for all $a \in S$,*

- *$a.(b + c) = a.b + a.c$, $(b + c).a = b.a + c.a$, for all $a, b, c \in S$.*

A dioid, or idempotent semiring, is a semiring, D, such that:

- *$a + a = a$ for all $a \in D$.*

A commutative dioid is one in which $a.b = b.a$ for all $a, b \in D$. A dioid is an
idempotent semi-skewfield (respectively, idempotent semifield) if its nonzero
elements form a group (respectively, commutative group) under multiplication.

The word dioid (or dioïde) is used by Kuntzmann, [Kun72], to mean simply
a semiring or *double monoid*, a usage followed by others, [GM84]. There seems
little point in wasting a new word on something well-known and we follow the
modern custom, [BCOQ92, §4.2], of using dioid as a synonym for idempotent
semiring.

It is less customary to use + and × to denote the operations in a dioid
and many authors use $\oplus$ and $\otimes$. We shall not do so here because, in our
view, formulae involing the latter operations are difficult to read and fail to
exploit the analogy with classical algebra. However, our choice does lead to

difficulties because some of the best-known examples of dioids are based on the classical number systems and it can be unclear whether, for instance, + refers to the dioid addition or to classical addition. To resolve confusion arising from this, we shall use the symbol := between formulae. This will imply that the symbols on the left hand side refer to the dioid under consideration while the symbols on the right hand side have their classical, or customary, meanings. Hence $a \times b := a + b$ means that dioid mulitipliation is equal to the customary addition, where *customary* should be clear from the context in which the formula appears.

2.3   Examples of dioids

Before going any further, it is best to see some examples.

1. The Boolean dioid: $\mathbb{B} = \{0, 1\}$, with $0, 1$ thought of as integers and addition and multiplication defined as maximum and minimum respectively.

2. The max-plus dioid: $\mathbb{R}_{\max} = \mathbb{R} \cup \{-\infty\}$ with $a + b := \max\{a, b\}$ and $a \times b := a + b$. Here, $0 := -\infty$ and $1 := 0$. $\mathbb{R}_{\max}$ is an idempotent semifield. (So too is $\mathbb{B}$ but for rather trivial reasons!) $\mathbb{R}_{\max}$ sometimes appears in its isomorphic form as the min-plus dioid: $\mathbb{R}_{\min} = \mathbb{R} \cup \{+\infty\}$, with $a + b := \min\{a, b\}$ and $a \times b := a + b$. Here, $0 := +\infty$ and $1 := 0$.

3. The tropical dioid: $\mathbb{N}_{\min} = \mathbb{N} \cup \{+\infty\}$, with $a + b := \min\{a, b\}$ and $a \times b := a + b$. Here, $0 := +\infty$ and $1 := 0$.

4. Let S be a set. The set of subsets of S, $\mathcal{P}(S)$, or the set of finite subsets of S, $\mathcal{P}^{Fin}(S)$, with addition and multiplication defined as union and intersection respectively, are both dioids. Here, $0 := \emptyset$ and $1 := S$. Similarly, let T be the set of open sets of any topology with addition and multiplication defined as union and intersection, respectively. Then T is a commutative dioid. We see that dioids are at least as common as topological spaces!

5. Suppose that $(M, ., 1_M)$ is a monoid. The monoid operation can be used to give $\mathcal{P}(M)$ or $\mathcal{P}^{Fin}(M)$ a different dioid structure from that in the previous example. Addition is still given by union of subsets but multiplication is defined by $U \times V := \{u.v \mid u \in U, v \in V\}$ where $U, V \subseteq M$. Here $0 := \emptyset$ and $1 := \{1_M\}$. There are several interesting examples of such dioids. Take $M = \mathbb{R}^d$ with the standard vector space addition as the monoid operation. The dioid multiplication in $\mathcal{P}(\mathbb{R}^d)$ then corresponds to Minkowski addition of subsets, [Hei95]. Another example is the dioid of formal languages. If A is a set, let A^* be the set of finite strings (words) over A with the monoid operation defined by

juxtaposition of strings and the empty string, ϵ, as the identity element. The subsets of A^* are called (formal) languages over the alphabet A. The dioid $\mathcal{P}(A^*)$ is noncommutative if A has more than one element.

6. Suppose that $(G, +, 0)$ is an abelian group, not necessarily finite. For instance, $G = \mathbb{R}^n$. Let $(\mathbb{R}_{\min})_b(G)$ denote the set of functions $u : G \to \mathbb{R}_{\min}$ which are bounded below. If $u, v \in (\mathbb{R}_{\min})_b(G)$, define addition pointwise, $(u + v)(x) := u(x) + v(x)$, and multiplication as convolution: $(u.v)(x) := \inf_{y \in G}\{u(y) + v(x - y)\}$. Here $x - y$ is calculated in the group G. The boundedness of u and v ensures that this is well-defined. $(\mathbb{R}_{\min})_b(G)$ is a convolution dioid, [ST]. The operation $u.v$ is called *inf convolution* in convex analysis, [Aub93, Exercise 1.4].

7. Let R be a commutative ring and let $Spec(R)$ denote the set of ideals of R. $Spec(R)$ becomes a dioid when addition is defined by the sum of ideals, $I + J = \{a + b \mid a \in I,\ b \in J\}$, and multiplication by the product, $I.J = \{a_1.b_1 + \cdots + a_n.b_n \mid a_i \in I,\ b_j \in J\}$, [Vic89, Chapter 12].

8. If D is a dioid, then $M_n(D)$ will denote the dioid of $n \times n$ matrices with entries in D with matrix addition and mulitiplication as the operations.

9. Let D be a dioid and A a set. The dioid of formal power series over D with (noncommuting) variables in A, denoted $D\langle\langle A \rangle\rangle$, is the set of D-valued functions on A^* with addition defined pointwise and multiplication as convolution: if $f, g \in D(A^*)$, then $(f.g)(w) = \sum_{u.v=w} f(u).g(v)$. By restricting to the sub-dioid of functions which are nonzero at only finitely many points, we recover the polynomials over D with (non-commuting) variables in A, denoted $D\langle A \rangle$.

2.4 Homomorphisms and semimodules

Many of the basic concepts in the theory of rings can be defined in a similar way for semirings and can hence be specialised to dioids. For the most part, the idempotency plays no special role. We run through some of the most important definitions here; for more details see [Gol92]. The reader who is familiar with these elementary concepts might still want to absorb some of the notation.

Let R, S be semirings. A homomorphism of semirings from R to S is a function $f : R \to S$ which is a homomorphism of monoids for both addition and multiplication:

$$f(a + b) = f(a) + f(b) \quad f(0_R) = 0_S$$
$$f(a.b) = f(a).f(b) \quad f(1_R) = 1_S .$$

For example, we can define a function $\alpha : \mathcal{P}(A^*) \to \mathbb{B}\langle\langle A \rangle\rangle$ by $\alpha(U)(w) = 1$ if $w \in U$, $\alpha(U)(w) = 0$ if $w \notin U$. The reader can check that this is an isomorphism of semirings. That is, there exists a homomorphism of semirings, $\beta : \mathbb{B}\langle\langle A \rangle\rangle \to \mathcal{P}(A^*)$, which is inverse to α: $\alpha\beta = 1_{\mathbb{B}\langle\langle A \rangle\rangle}$ and $\beta\alpha = 1_{\mathcal{P}(A^*)}$. By restriction, β also induces an isomorphism between $\mathbb{B}\langle A \rangle$ and $\mathcal{P}^{Fin}(A^*)$.

Let $(M, +)$ be a commutative monoid and R a semiring. M is said to be a left R-semimodule if there is an action of R on M, $R \times M \to M$, called scalar multiplication and denoted $r.a$, such that, for all $r, s \in R$ and $a, b \in M$,

$$
\begin{array}{ll}
r.(a + b) = r.(a) + r.(b) & r.(0_M) = 0_M \\
(r + s).a = r.a + s.a & 0_R.a = 0_M \\
r.(s.a) = (r.s).a & 1_R.a = a \; .
\end{array}
$$

It is worth noting at this point that if R happens to be a dioid, then M must necessarily have an idempotent addition:

$$ a + a = 1.a + 1.a = (1 + 1).a = 1.a = a \; . $$

A useful source of R-semimodules is provided by the following construction. Let A be any set and let $R(A)$ denote the set of all functions $u : A \to R$. This has a natural structure as a left R-semimodule: addition is defined pointwise, $(u + v)(a) = u(a) + v(a)$, and scalar multiplication by $(r.u)(a) = r.u(a)$. When A is infinite, $R(A)$ contains inside itself another useful R-semimodule. Define the support of u by $supp(u) = \{a \in A \mid u(a) \neq 0\}$ and let $R^{Fin}(A)$ denote the set of functions u with finite support. It is clear that $R^{Fin}(A)$ is a sub-semimodule of $R(A)$.

If M, N are R-semimodules, a homomorphism of R-semimodules from M to N is a homomorphism of monoids $f : M \to N$ which respects the scalar multiplication: $f(r.a) = r.f(a)$. We can construct homomorphisms between the R-semimodules $R^{Fin}(A)$ as follows. If $f : A \to B$ is any set function, then f can be extended to a function, also denoted f for convenience, $f : R^{Fin}(A) \to R^{Fin}(B)$ where $f(u)(b) = \sum_{f(a)=b} u(a)$. The finiteness of $supp(u)$ ensures that this is well-defined. The reader can check that this is a homomorphism of R-semimodules.

At this point, it will be convenient to make use of the language of category theory, essentially to clarify the nature of $R^{Fin}(A)$. The reader who is unfamiliar with this language, but who nevertheless has an intuitive understanding of what it means for an object to be *freely generated*, will not lose much by ignoring the details below. The canonical reference for those wishing to know more is MacLane's book, [Mac71].

The constructions given above define a functor from the category of sets, Set, to the category of R-semimodules, SMod_R, which is left adjoint to the forgetful functor which forgets the R-semimodule structure. That is, if A is a set and M is an R-semimodule, then there is a natural one-to-one correspondence between the respective sets of homomorphisms:

$$ \mathsf{Hom}(A, M)_{\mathsf{Set}} \longleftrightarrow \mathsf{Hom}(R^{Fin}(A), M)_{\mathsf{SMod}_R} \; . \qquad (2.1) $$

This expresses the fact that $R^{Fin}(A)$ is the free R-semimodule generated by A: any set function $f : A \to M$ can be extended to a unique homomorphism of R-semimodules $f : R^{Fin}(A) \to M$. If $u \in R^{Fin}(A)$, the extension is given by $f(u) = \sum_{a \in A} u(a).f(a)$, the finiteness of $supp(u)$ again ensuring that this is well-defined.

If M is an R-semimodule then the set $\mathsf{Hom}(M, M)_{\mathsf{SMod}_R}$ has both an addition, $(f + g)(a) = f(a) + g(a)$, and a multiplication given by composition of functions, $(f.g)(a) = f(g(a))$. This defines a semiring, over which M is a left semimodule under $f.a = f(a)$. For finitely generated free modules, $\mathsf{Hom}(M, M)$ is well known. Let A be a finite set, so that $R^{Fin}(A) = R(A)$. By (2.1), an element of $\mathsf{Hom}(R(A), R(A))$ is uniquely determined by a function $A \to R(A)$. This, in turn, may be uniquely specified by a function $A \times A \to R$. In other words, by an element of $R(A \times A)$, or, after choosing an ordering $A = \{a_1, \ldots, a_n\}$, by an $n \times n$ matrix over R. It is easy to show that this gives an isomorphism of semirings, $\mathsf{Hom}(R(A), R(A)) \longrightarrow M_n(R)$.

2.5   The partial order in a dioid

Our constructions have, so far, been valid for arbitrary semirings. The reader might be forgiven for thinking that the theory of dioids has no special character of its own. This is not the case. The idempotency gives rise to a natural partial order in a dioid which differentiaties the theory from that of more general semirings.

Proposition 2.1 ([Joh82, §I.1.3]) *Let A be a commutative semigroup with an idempotent addition. Define $a \preceq b$ whenever $a + b = b$. Then $(A, \preceq)$ is a sup-semilattice (a partially ordered set in which any two elements have a least upper bound). Furthermore,*

$$\max\{a, b\} = a + b. \tag{2.2}$$

Conversely, if $(A, \preceq)$ is a sup-semilattice and $+$ is defined to satisfy (2.2), then $(A, +)$ is an idempotent semigroup. These two constructions are inverse to each other.

It follows that any dioid, D, has a natural ordering, denoted $\preceq$ or $\preceq_D$. Addition is a monotonic operation and 0 is the least element. Distributivity implies that left and right multiplication are semilattice homomorphisms; in particular, they are monotonic.

We see from Proposition 2.1 that a dioid may be thought of as a semilattice-ordered semigroup, [Fuc63]. There is a continuing tradition of work on idempotency from this perspective, [CG79, Zim81, But94].

The reader might note that the dioids introduced in §2.3 fall into two main classes: those whose partial order is derived from the order on $\mathbb{R}$ and those derived from inclusion of subsets.

2.6   Free dioids

We turn now to an examination of the dioids arising as sets of subsets (Examples 4 and 5 in §2.3). Once again, it is convenient to use the language of category theory.

The constructions $\mathcal{P}(M)$ and $\mathcal{P}^{Fin}(M)$ extend to functors from the category of monoids, Monoid, to the category of dioids, Dioid: if $f : M \to N$ is a homomorphism of monoids, then define $f : \mathcal{P}(M) \to \mathcal{P}(N)$ by $f(U) = \{f(u) \mid u \in U\}$ and similarly for $\mathcal{P}^{Fin}(-)$.

Proposition 2.2 *The functor* $\mathcal{P}^{Fin}$: Monoid $\to$ Dioid *is left adjoint to the forgetful functor from* Dioid *to* Monoid *which forgets the dioid addition.*

Proof Let M be a monoid and D a dioid. It is required to show that there is a natural one-to-one correspondence between the respective sets of homomorphisms:

$$\mathsf{Hom}(M, D)_{\mathsf{Monoid}} \longleftrightarrow \mathsf{Hom}(\mathcal{P}^{Fin}(M), D)_{\mathsf{Dioid}} .$$

Let $q : M \to D$ be a homomorphism of monoids. Define $\theta(q) : \mathcal{P}^{Fin}(M) \to D$ by $\theta(q)(U) = \sum_{u \in U} q(u)$. It is an exercise to check that θ establishes the required correspondence. $\square$

In other words, $\mathcal{P}^{Fin}(M)$ is the free dioid generated by the monoid M. The construction A^* can also be extended to a functor from Set to Monoid which is left adjoint to the forgetful functor which forgets the monoid multiplication. A^* is hence the free monoid generated by A. These constructions can also be specialised to the commutative case. The free commutative monoid generated by A is simply $\mathbb{N}^{Fin}(A)$. By putting these remarks together, we obtain the following characterisations.

Proposition 2.3 ([Gau92, Proposition 1.0.7]) *If A is a set, then $\mathcal{P}^{Fin}(A^*)$ is the free dioid, and $\mathcal{P}^{Fin}(\mathbb{N}^{Fin}(A))$ the free commutative dioid ([Shu, Theorem 4.1]), generated by A.*

2.7   Quantales

As we saw in §2.4, the difference between $\mathcal{P}(M)$ and $\mathcal{P}^{Fin}(M)$ is similar to the difference between power series and polynomials. In the former, an infinite number of additions can be meaningfully performed. Dioids of this type have been extensively studied under another name.

Definition 2.2 *A quantale, or complete dioid ([BCOQ92, Definition 4.32]), Q, is a dioid in which the underlying sup-semilattice has arbitrary suprema over which the multiplication distributes: $\forall S \subseteq Q$ and $\forall a \in Q$,*

- *there exists a least upper bound* $\sup_{x \in S} x$;

- $a.(\sup_{x \in S} x) = \sup_{x \in S}(a.x)$, $(\sup_{x \in S} x).a = \sup_{x \in S}(x.a)$.

The word *quantale* was coined by Mulvey in his study of the constructive foundations of quantum theory, [Mul86]. His definition does not require an identity element for multiplication.

If $(V, \leq)$ is any partial order and $S \subseteq V$ any subset of V, then it is easy to see that

$$\inf_{x \in S} x = \sup\{y \in V \mid y \leq x, \ \forall x \in S\} , \qquad (2.3)$$

in the sense that, if either side exists, so does the other, and both are equal. Hence, every quantale has a greatest element $\mathsf{T} = \inf \emptyset$, and any subset has an infimum. However, multiplication does not necessarily distribute over infima, even though it does over suprema.

We regard a quantale Q as endowed with an infinitary addition given by $\sum_{x \in S} x = \sup_{x \in S} x$, following the identification in (2.2). Homomorphisms of quantales are required to preserve the multiplication and the infinitary addition. The category of quantales, **Quant**, forms a subcategory of **Dioid**. $\mathcal{P}(-)$ defines a functor from **Monoid** to **Quant** which is left adjoint to the forgetful functor which forgets the quantale multiplication. Hence, we obtain the following characterisations.

Proposition 2.4 *If A is a set, then $\mathcal{P}(A^*)$ is the free quantale generated by A ([AV93, Theorem 2.1]) and $\mathcal{P}(\mathbb{N}^{Fin}(A))$ is the free commutative quantale generated by A.*

Topological spaces form another source of quantales, as pointed out in Example 4 in §2.3. These quantales are special because multiplication corresponds to minimum. Quantales of this type are called frames, [Joh82, Chapter 2] and are characterised by having 1 as the greatest element and an idempotent multiplication, $a^2 = a$, [JT84, Proposition III.1]. Because they provide an extended concept of topological space, frames have proved important in formulating an appropriate abstract setting—Grothendieck's notion of *topos*—for modern algebraic geometry. In the course of proving structure theorems for topoi, Joyal and Tierney have developed the theory of complete semimodules over frames, [JT84].

Quantales have been studied for a number of reasons. On the one hand the infinitary addition allows the definition of the binary operation $a \rightarrow b$, characterised by $c \preceq a \rightarrow b$ if, and only if, $c.a \preceq b$. (This is best understood as an application of the Adjoint Functor Theorem, [Joh82, §I.4.2].) This operation is referred to variously as a residuation, implication or pseudo-complement, [DJLC53, BJ72, Joh82, BCOQ92]. When Q is a frame, the

operation $\neg x = x \to 0$ has many of the features of a negation, except that
the law of the excluded middle, $x + \neg x = \mathsf{T}$, may fail. Logicians have consequently studied frames as models for intuitionistic logic, [Vic89]. More
recently, quantales have served the same purpose for Girard's linear logic.
They are also used as semantic models in computer science. For details and
references, see the paper by Mascari and Pedicini in this volume, [MP], or
[AV93].

On the other hand, despite the infinitary operation, quantales and frames
are still algebraic theories, in the sense that they can be defined in terms of
generators and relations, [Joh82, §II.1.2]. For frames, this leads to the study
of topological spaces from an algebraic viewpoint, sometimes called *pointless
topology*, [Joh83].

For the purposes of the present paper, quantales are important for the
following reason. The discrete dynamical system $x_{i+1} = a.x_i + b$ appears in
numerous applications. In general, a may be an element of a dioid D and
x_i and b elements of some left D-semimodule. The problem is to understand
the asymptotic behaviour of the sequence $x_0, x_1, \ldots$. A useful first step is to
determine the equilibrium points of the system, where

$$x = a.x + b . \tag{2.4}$$

In a quantale, appropriate equilibrium solutions can always be constructed.
(In this volume, Walkup and Borriello consider the more general problem of
solving $a.x + b = c.x + d$ when $a, c \in M_n(\mathbb{R}_{\mathsf{max}})$, [WB].)

Definition 2.3 *If Q is a quantale and $a \in Q$, the Kleene star of a, denoted
a^*, is defined by*

$$a^* = 1 + a + a^2 + a^3 + \cdots = \sup_{0 \le i} a^i . \tag{2.5}$$

It is sometimes helpful to use also $a^+ = a + a^2 + \cdots$. By the infinitary
distributive law, $a.a^* = a^+$.

Proposition 2.5 ([Con71, Theorem III.2]) *Let Q be a quantale and $a, b \in Q$.
Then $a^*.b$ is the least solution of (2.4).*

Proof $a.(a^*.b) + b = a^+.b + b = a^*.b$. Hence $a^*.b$ is a solution. Since multiplication and addition are both monotonic, if $x \preceq y$ then $a.x + b \preceq a.y + b$. Now
let s be any other solution. Since $0 \preceq s$, it follows that $b \preceq s$. By induction,
$(1 + a + a^2 + \cdots + a^n).b \preceq s$. It is now not difficult to show that $a^*.b \preceq s$, as
required. □

We see from this that a^* has a "rational" flavour: it is analogous to $(1-a)^{-1}$ in customary algebra. The star operation satisfies many identities, [Con71, Chapter 3], of which we mention only one:

$$(a + b)^* = (a^*.b)^*.a^* . \tag{2.6}$$

The left hand side is a sup of terms of the form $a^{i_1}b^{j_1}\cdots a^{i_m}b^{j_m}$. On the right hand side, $(a^*.b)^*$ is a sup of similar terms except for those ending in powers of a. The product with a^* supplies the missing terms. The problem of finding a complete set of identities for $+$, $\times$ and * is very subtle, [Con71, Chapter 12].

The uniqueness of the solution given in Proposition 2.5 appears to be an open problem. It is a classical result in automata theory, Arden's Lemma, [HU79, Exercise 2.22], that, in the free quantale $\mathcal{P}(A^*)$, if $1 \not\preceq a$ then $a^*.b$ is the unique solution of (2.4). This is not true in general. In any frame, $a^2 = a$ and 1 is the greatest element. Hence the function $f(x) = a.x + b$ is itself idempotent, $f^2 = f$, implying that any element of the form $a.x + b$ is a solution of (2.4). On the other hand, in any quantale, if $1 \preceq a$, it is easy to see that $a^*.(u + b)$ is a solution of (2.4) for any $u \in Q$. Another relevant result is [BCOQ92, Theorem 4.76].

It is convenient to mention here a technical trick which is well known in certain quarters. It is used by Walkup and Borriello in their paper in this volume, [WB], and it will be the key ingredient in the proof of Theorem 3.1. The proof is left as an exercise for the reader, who will need to make use of (2.6) and Proposition 2.5. There are several equivalent formulations: [Kui87, Theorem 2.5], [CMQV89, Lemma 2].

Lemma 2.1 ([Con71, Theorem III.4]) *Let Q be a quantale. The following identity holds for any block representation of a matrix over Q.*

$$\begin{pmatrix} A & B \\ C & D \end{pmatrix}^* = \begin{pmatrix} (A + BD^*C)^* & A^*B(D + CA^*B)^* \\ D^*C(A + BD^*C)^* & (D + CA^*B)^* \end{pmatrix}$$

2.8   Matrix dioids and graphs

If the dioid R is not a quantale, it may still be the case that a^* exists and that Proposition 2.5 continues to hold. Indeed, in any dioid, if $a \preceq 1$ then $a^* = 1$. In a matrix dioid, the existence of A^* is related to the eigenvalues of A, in analogy with classical results. For dioids such as $M_n(\mathbb{R}_{\mathsf{max}})$, this has been completely worked out, [BCOQ92, Theorem 3.17].

The case of matrix dioids is of particular interest for problems of discrete optimisation. This arises through the intimate relationship between matrices and graphs, which becomes particularly attractive in the idempotent context. Let R be a semiring and $T \in M_n(R)$ a matrix.

Definition 2.4 *The graph of T, $\mathcal{G}(T)$, is a directed graph on the vertices $\{1,\ldots,n\}$ with edges labelled by elements of R. There is an edge from i to j if, and only if, $T_{ij} \neq 0$. If there is such an edge then its label is T_{ij}.*

This gives a one-to-one correspondence between $n \times n$ matrices over R and directed graphs on $\{1,\ldots,n\}$ with edge labels in R.

Taking powers of A corresponds to building longer paths. A path p of length m from i to j is a sequence of vertices $i = v_0,\ldots,v_m = j$ such that $A_{v_i v_{i+1}} \neq 0$. We can write

$$A_{ij}^m = \sum_p |p|_{\mathsf{w}} \,,$$

where p runs over all paths of length m from i to j and $|p|_{\mathsf{w}}$ is the weight of the path: the product of the labels on the edges in the path, $|p|_{\mathsf{w}} = A_{v_0 v_1} \cdots \cdots A_{v_{m-1} v_m}$. For a general semiring, it is hard to deduce much from this. If R is a dioid, however, the sum corresponds to taking a maximum with respect to $\preceq_R$ and A_{ij}^m is the maximum weight among paths of length m from i to j. By choosing the dioid appropriately, a variety of discrete optimisation problems can be formulated in terms of matrices.

For instance, we can answer questions about the existence of paths in a directed (unlabelled) graph, G. Assume the vertices are labelled $\{1,\ldots,n\}$. Take $R = \mathbb{B}$ and label each edge of the graph with $1 \in \mathbb{B}$. Let A be the matrix in $M_n(\mathbb{B})$ corresponding to G. Then A^* (which exists since $\mathbb{B}$ and $M_n(\mathbb{B})$ are both quantales) gives the transitive closure of the edge relation in G: $A_{ij}^* = 1$ if, and only if, there is some path in G from i to j. Problems of enumeration, shortest paths, critical paths, reliability, etc. in graphs or networks can be formulated using other dioids, often as solutions to (2.4), [GM84, Chapter 2].

To calculate A^* efficiently, classical algorithms for computing the inverse matrix—Jacobi, Gauss–Seidel, Jordan—can be adapted to the idempotent setting, [Car71], [GM84, Chapter2]. For finding longest paths (for which the appropriate dioid is $\mathbb{R}_{\mathsf{max}}$) these correspond to well-known algorithms such as those of Bellman, Ford–Fulkerson and Floyd–Warshall, [Car71].

It is interesting to ask if properties of periodic graphs, which are infinite graphs with $\mathbb{Z}^n$-symmetry, can be studied by similar methods. Backes, in his thesis, has given a formula for longest paths in a periodic graph, [Bac94]. It seems likely that, at least when $n = 1$, this can be interpreted in idempotent terms using the matrix methods described above. Can a similiar interpretation be found when $n > 1$?

2.9 The max-plus dioid and lattice ordered groups

The max-plus dioid, $\mathbb{R}_{\mathsf{max}}$, has been of great importance for idempotency and deserves special mention. For a general reference see [BCOQ92, Chapter 3].

Cuninghame-Green's survey paper, [CG95], discuses several topics that we omit.

As remarked in §2.3, $\mathbb{R}_{\mathsf{max}}$ is an idempotent semifield. The group operation endows such structures with natural symmetry. The reader should have no difficulty in proving the following result by using the remarks in §2.5.

Lemma 2.2 *If D is an idempotent semi-skewfield then $(D, \preceq_D)$ is a lattice and, for $a, b \in D \backslash \{-\infty\}$, $\min\{a, b\} = (a^{-1} + b^{-1})^{-1}$.*

In the language of ordered algebraic structures, idempotent semi-skewfields are therefore lattice ordered groups and these sometimes form a convenient generalisation of $\mathbb{R}_{\mathsf{max}}$. They correspond to the *blogs* of [CG79], a name which has, understandably, not survived.

The most extensively studied aspect of max-plus is linear algebra: finitely generated free semimodules and their endomorphisms, [Bap95, BSvdD95, But94, GMb, OR88]. The spectral theory of max-plus matrices has been of particular interest because of the role of eigenvalues as a performance measure for discrete event systems; see §4.4. The basic observation is that eigenvalues correspond to maximum mean circuit weights in the associated graph.

Let us consider elements of $(\mathbb{R}_{\mathsf{max}})(\{1, \ldots, n\})$ as column vectors. Recall that u is an eigenvector of a matrix $A \in M_n(\mathbb{R}_{\mathsf{max}})$, with eigenvalue $\lambda \in \mathbb{R}_{\mathsf{max}}$, if $Au = \lambda u$. If g is a circuit in $\mathcal{G}(A)$—a path that returns to its starting vertex—let $|g|_\ell$ be its length: the number of edges in the circuit. Note that in $\mathbb{R}_{\mathsf{max}}$, the polynomial equation $x^k = a$ has a unique solution: $x = a^{1/k} := a/k$. (Dioids with this property are *radicable* in [CG79] or *algebraically complete* in [MS92].) Hence, the mean weight of a circuit, $\mu(g) := |g|_{\mathsf{w}}/|g|_\ell$, is a *bona fide* element of $\mathbb{R}_{\mathsf{max}}$.

Proposition 2.6 *Suppose that $A \in M_n(\mathbb{R}_{\mathsf{max}})$ and $Au = \lambda u$. Then, in $\mathbb{R}_{\mathsf{max}}$, $\lambda = \sum_g \mu(g)$, where g ranges over circuits of $\mathcal{G}(A)$ whose vertices lie in $supp(u)$. In particular, any two eigenvectors with the same support have the same eigenvalue.*

Results of this type are part of the folklore of idempotency and go back to [CG62], [Rom67] and [Vor67]; the formulation above is taken from [Gun94c, Lemma 4.5]. A great deal more is known, particularly over $\mathbb{R}_{\mathsf{max}}$, about the existence of eigenvectors, the structure of the set of eigenvectors, spectral projectors, etc, [BCOQ92, Chapter 3]. When A is an irreducible matrix (or, more generally, when $supp(u) = \{1, \cdots, n\}$), a description of the eigenvectors, valid for radicable idempotent semi-skewfields, appears in [CG79]. The general case for $\mathbb{R}_{\mathsf{max}}$ is discussed in [WXS90], and later in [Gau92], and for dioids more general than idempotent semi-skewfields in [DSb].

The behaviour of A^k as $k \to \infty$ is beautifully described by the following Cyclicity Theorem, which asserts that A^k is *asymptotically cyclic*.

Theorem 2.1 ([BCOQ92, Theorem 3.112]) *For any matrix $A \in M_n(\mathbb{R}_{max})$, there exists $d \in \mathbb{N}$ such that $A^{k+d} - A^k \to 0$ as $k \to \infty$, where subtraction should be taken in the customary sense.*

The convergence is with respect to the topology which arises by identifying $\mathbb{R}_{max}$ with the positive reals, $\{x \in \mathbb{R} \mid x > 0\}$ under the exponential map: $x \to \exp(x)$, [BCOQ92, 3.7.4]. (We shall use this identification again in §4.2 and will discuss the topology further in §5.4.3.) The asymptotic cyclicity of A (i.e. the least d) can be calculated in terms of the lengths of circuits in $\mathcal{G}(A)$, [BCOQ92, §3.7.1]. The Cyclicity Theorem is one of the main results in the linear algebra of $\mathbb{R}_{max}$.

The reader may notice here some analogy between the theory of max-plus matrices and that of nonnegative matrices, as described by Perron–Frobenius theory, [Min88]. This is one of the most intruiging puzzles in the whole subject. The analogy seems too close to be simply an accident but a satisfactory explanation of the relationship has not yet been found. We will return to this point in §4.2 and §6.5.

Over a field, every finitely generated module is free but this is not the case for semimodules over an idempotent semifield. Nonfree semimodules have been much less well-studied than their free counterparts. Moller, in his thesis, and Wagneur have independently found results which shed light on the structure of nonfree semimodules over dioids like $\mathbb{R}_{max}$, [Mol88, Wag91]. (See also [JT84].) Further discussion of this appears in Wagneur's paper in this volume, [Wag].

2.10   Finite dioids

As we have seen, dioids are very plentiful. However, as remarked in §2.5, the examples considered in 2.3 fall into two broad families. It is interesting to speculate on whether this reflects some fundamental underlying classification or is merely an accident resulting from our ignorance. Shubin, who seems to have been one of the few to take a systematic approach, has constructed all finite commutative dioids having at most 4 elements, [Shu, §2], and has found 14 pairwise nonisomorphic commutative dioids with exactly 4 elements. (Conway enumerates certain specialised dioids in [Con71, Chapter 12].) This suggests that dioids are too numerous to expect a simple classification. It would, nevertheless, be useful to have a structure theory for dioids to bring some order to this profusion.

In some respects the profusion can be misleading. The following observation, well known in the theory of lattice ordered groups, [Fuc63, page 89], shows that there are no idempotent analogues of the Galois fields.

Proposition 2.7 *The only dioid which is both a quantale and an idempotent*

semi-skewfield is the Boolean dioid, $\mathbb{B}$*. In particular,* ([Shu, Theorem 3.1])
there are no finite idempotent semi-skewfields other than $\mathbb{B}$*.*

Proof Let D be a dioid satisfying the hypotheses and let g be the greatest
element of D. Since $1 \preceq g$, it follows that $g \preceq g^2$. Hence $g = g^2$ and so $g = 1$.
Now suppose that $s \neq 0$. Since $s^{-1} \preceq 1$, it follows that $1 \preceq s$ and so $s = 1$.
Hence $D = \mathbb{B}$.   □

This negative result is an appropriate place to bring this section to a close.
We hope to have given some idea of the scope of the dioid concept. It would be
fair to say that most of the interesting questions about general dioids remain
unanswered, if not unasked. In the subsequent sections we shall be concerned
with specific dioids and will not touch on such general matters again.

3   Automata and Idempotency

Automata may be thought of as language recognisers, that is, as rules for
recognising strings of symbols from some (finite) alphabet. By defining appropriate rule schemes, computer scientists have defined classes of automata—
finite automata, push-down automata, stack automata, Turing machines, etc.,
[HU79]—and corresponding classes of languages, [Sal73]. Because languages
are elements of the free quantale, $\mathcal{P}(A^*)$, it should not come as a surprise
that automata theory has something to tell us about idempotent semirings.
In this paper, we shall only consider finite automata, where the connections
have been most studied. For a good overview see [Per90].

Our main objective in this section is to prove a generalisation of Kleene's
Theorem, [Kle]. This is the starting point of finite automata theory and it
appears in several papers in this volume, [Kro, Pin]. We then discuss briefly
the way in which the tropical dioid has been used to solve certain decision
problems of finite automata.

3.1   Quantales and Kleene's Theorem

It will be convenient to identify an element $u \in \mathbb{B}(\{1, \ldots, n\})$ with a column vector, as in §2.9, and to make use of the customary operations on
vectors and matrices. For instance, u^t will denote the transpose of u: $u^t =
(u(1), \ldots, u(n))$. Since the Boolean dioid, $\mathbb{B}$, can be regarded as a subdioid of
any dioid, D, vectors and matrices over $\mathbb{B}$ can always be regarded as vectors
and matrices over D.

Definition 3.1 *Let* Q *be a quantale and* $S \subseteq Q$*. An element* $q \in Q$ *is recognisable over* S *if, for some* n*, there exists a matrix* $T \in M_n(S)$ *and vectors*

$\iota, \phi \in \mathbb{B}(\{1, \ldots, n\})$ such that $q = \iota^{t} T^{*} \phi$. The set of elements recognisable over S will be denoted $\mathrm{Rec}(S)$.

Let A be a finite set and let $Q = \mathcal{P}(A^{*})$, the free quantale generated by A. The elements of Q are the languages over A. Let $S = \{q \in Q \mid q \preceq A\}$, where A is treated as an element of Q by identifying each $a \in A$ with the corresponding string of length 1. The elements of S can then be identified with the subsets of A. The data (ι, T, ϕ) over S correspond to a nondeterministic finite automaton, [Pin, §5.1]. The states of the automaton are $\{1, \ldots, n\}$ and the subsets ι and ϕ are the initial states and final states, respectively. T encodes the transitions of the automaton: if $T_{ij} = q$, where $q \subseteq A$, then if the automaton is in state i and encounters any symbol from the subset q, it may make a transition to state j. The possibility that the same symbol could occur in both T_{ij} and T_{ik}, where $j \neq k$, or that $T_{ij} = \emptyset$, allows for nondeterminism.

In the usual definition of a finite automaton, the effect of individual symbols is separated out. Let $\mu : A \to M_{n}(\mathbb{B})$ be given by $\mu(a)_{ij} = 1$ if, and only if, $a \in T_{ij}$. The matrix $\mu(a)$ identifies those transitions which recognise the symbol a. It follows that, in $M_{n}(\mathcal{P}(A^{*}))$,

$$T = \sum_{a \in A} (aI)\mu(a),$$

where I is the identity matrix. Since $M_{n}(\mathbb{B})$ is a multiplicative monoid, the function μ can be extended to a homomorphism of monoids $\mu : A^{*} \to M_{n}(\mathbb{B})$. Hence, ignoring initial and final states, we can think of an automaton in different ways: as a representation of the free monoid over finitely generated free $\mathbb{B}$ modules, or as a finitely generated semigroup (or monoid) of matrices generated by the subset $\{\mu(a) \mid a \in A\} \subseteq M_{n}(\mathbb{B})$. The slogan *automata are semigroups of matrices* is helpful to keep in mind, particularly when it comes to defining more general types of automata, as in §3.2 and §4.1.4.

What does the element $q = \iota^{t} T^{*} \phi$ correspond to in terms of automata? Recalling the discussion in §2.8 the reader can check that T_{ij}^{m} is the set of strings of length m which lead from state i to state j. It follows that q is the set of strings of any length which lead from some initial state to some final state. Hence, q is the language recognised by the automaton, [Pin, §5.1].

Kleene's original result, [Kle], which is the starting point of formal language theory, characterised the languages of finite automata; it was stated, in effect, for the free quantale, $\mathcal{P}(A^{*})$. Following Conway, [Con71], we state and prove it for an arbitrary quantale.

Definition 3.2 *Let Q be a quantale and $S \subseteq Q$. The rational closure of S, S^{*}, is the smallest subset of Q which contains S and is closed under $+$, $\times$ and $*$.*

Theorem 3.1 *Let Q be a quantale and $\{0,1\} \subseteq S \subseteq Q$. Then $\mathrm{Rec}(S) = S^*$.*

Proof We first show that $S^* \subseteq \mathrm{Rec}(S)$. Choose $s \in S$ and form the data below, which we may do since $0 \in S$.

$$\iota = \begin{pmatrix} 1 \\ 0 \end{pmatrix} \quad T = \begin{pmatrix} 0 & s \\ 0 & 0 \end{pmatrix} \quad \phi = \begin{pmatrix} 0 \\ 1 \end{pmatrix}.$$

Evidently, $\iota^t T^* \phi = s$ and so $S \subseteq \mathrm{Rec}(S)$. Now suppose that $q_1 = \iota_1^t T_1^* \phi_1$ and $q_2 = \iota_2^t T_2^* \phi_2$ and form the three sets of block matrix data below, which we may do since $\{0,1\} \subseteq S$.

$$\iota = \begin{pmatrix} \iota_1 \\ \iota_2 \end{pmatrix} \quad T = \begin{pmatrix} T_1 & 0 \\ 0 & T_2 \end{pmatrix} \quad \phi = \begin{pmatrix} \phi_1 \\ \phi_2 \end{pmatrix},$$

$$\iota = \begin{pmatrix} \iota_1 \\ 0 \end{pmatrix} \quad T = \begin{pmatrix} T_1 & \phi_1 \iota_2^t \\ 0 & T_2 \end{pmatrix} \quad \phi = \begin{pmatrix} 0 \\ \phi_2 \end{pmatrix},$$

$$\iota = \begin{pmatrix} 0 \\ 1 \end{pmatrix} \quad T = \begin{pmatrix} T_1 & \phi_1 \\ \iota_1^t & 0 \end{pmatrix} \quad \phi = \begin{pmatrix} 0 \\ 1 \end{pmatrix}.$$

(It is instructive to picture these constructions for finite automata.) The reader can now check, using Lemma 2.1, that $\iota^t T^* \phi$ for each set of data is, respectively, $q_1 + q_2$, $q_1 . q_2$ and q_1^*. It follows by induction that $\mathrm{Rec}(S)$ is closed under $+$, $\times$ and $*$. Hence, $S^* \subseteq \mathrm{Rec}(S)$, since S^* is the smallest set containing S with that property.

For the other way round, if $q = \iota^t T^* \phi$, where $T \in M_n(S)$, then, by Lemma 2.1, $T^* \in M_n(S^*)$. Hence $q \in S^*$. It follows that $\mathrm{Rec}(S) = S^*$, as required.   □

There is a mild embarrassment with this result: in the case of finite automata, where $Q = \mathcal{P}(A^*)$ and $S = \{q \preceq A\}$, $1 \notin S$! However, in this case, S has additional properties that allow the same result to go through.

Corollary 3.1 *Let Q be a quantale and $S \subseteq Q$. If $0 \in S$ and S is closed under $+$, then $\mathrm{Rec}(S) = S^*$.*

Proof By the Theorem, $\mathrm{Rec}(S \cup \{1\}) = (S \cup \{1\})^*$. Choose $q \in S \cup \{1\}$ and suppose that $q = \iota^t T^* \phi$ where $T \in M_n(S \cup \{1\})$. We can write $T = C + P$ where $C \in M_n(\mathbb{B})$ and $P \in M_n(S)$. By (2.6),

$$q = \iota^t (C + P)^* \phi = \iota^t (C^* P)^* (C^* \phi).$$

Since S is closed under $+$, $C^*P \in M_n(S)$ and, evidently, $C^*\phi \in \mathbb{B}(\{1, \ldots, n\})$. Hence, $q \in \text{Rec}(S)$ and so $\text{Rec}(S \cup \{1\}) = \text{Rec}(S)$. Furthermore, since $0 \in S$, it is easy to see that $(S \cup \{1\})^* = S^*$. Hence, $\text{Rec}(S) = S^*$. $\square$

The treatment we have given follows Conway, [Con71], and was also influenced by Kuich, [Kui87], who states the result in even greater generality. The essential ideas, but not the theorem itself, can be found in [CMQV89], which inexplicably fails to make any reference to the automata theory literature.

As we saw in §2.4, the free quantale, $\mathcal{P}(A^*)$, can also be thought of as the power series ring $\mathbb{B}\langle\langle A \rangle\rangle$. Schutzenberger has extended Kleene's result to power series rings, $R\langle\langle A \rangle\rangle$, where R is a semiring which is not necessarily idempotent. The * operation cannot now be defined on all of $R\langle\langle A \rangle\rangle$ but does exist on those series whose constant term is 0. This result, known as the Kleene–Schutzenberger Theorem, is discussed further in Krob's paper in this volume, [Kro, §2.2].

It is rather peculiar that there are two generalisations of Kleene's original result, in one of which idempotency plays a crucial role while in the other it is the free monoid structure of A^*. It would be very interesting to have a single formulation which includes both contexts and yet retains the clarity of Kleene's original result.

3.2 The tropical dioid

The star operation is not a simple one because of its infinitary nature. It is interesting to ask when it can be defined in a finite way. That is, whether

$$a^* = 1 + a + \cdots + a^m, \text{ for some } m. \tag{3.1}$$

In a dioid which is not a quantale, this gives a way of constructing a^*; Gondran and Minoux introduced the notion of m-regularity, $a^{m+1} = a^m$, for just this reason, [GMb, Definition 1]. In automata theory, (3.1) is known as the finite power property. In 1966, Brzozowski raised the question of whether it was decidable if a given rational set (i.e. the language of a finite automaton) had this property.

This celebrated problem was solved in the affirmative independently by Simon, [Sim78], and Hashiguchi, [Has79]. Simon's proof introduced the tropical dioid, $\mathbb{N}_{\min}$, into automata theory and initiated a deep exploration of decision problems related to the * operation. For an early survey, see [Sim88]. The basic idea is to use automata with multiplicities in $\mathbb{N}_{\min}$ or, equivalently, semigroups of matrices in $M_n(\mathbb{N}_{\min})$, to reformulate the finite power property as a Burnside problem. Recall that the original Burnside problem asks if a finitely generated group must necessarily be finite if each element has finite order. This is true for groups of matrices over a commutative ring but is false in general. An essential step in Simon's proof is to show that it is also true for semigroups of matrices over $\mathbb{N}_{min}$, [Sim78, Theorem C].

For further details, the reader cannot do better than turn to the papers in this volume devoted to this subject. The tutorial by Pin, [Pin], explains in more depth how the tropical dioid enters the picture, while that of Krob, [Kro], surveys a number of decision problems related to (3.1). D'Alessandro and Sakarovitch show that the finite power property also holds for rational subsets of the free group, [dSa]. Leung, in his thesis, introduced topological ideas into the study of * problems over the tropical dioid. His paper in this volume gives a new and simplified treatment of this approach, [Leu].

Gaubert has initiated the study of automata with multiplicities in $\mathbb{R}_{max}$ and has shown that the Burnside problem has a positive answer for semigroups of matrices in $M_n(\mathbb{R}_{max})$, [Gau]. The rationale for introducing such automata is that, just as the tropical dioid can be used to estimate frequencies (how often?), max-plus can be used to estimate durations (how long?). This leads naturally to the next section which studies problems of performance analysis. We move, correspondingly, from $\mathcal{P}(A^*)$ and $\mathbb{N}_{min}$ to $\mathbb{R}_{max}$.

4   Discrete Event Systems

4.1   Introduction

4.1.1   Examples and general questions

A discrete event system is one whose behaviour consists of the repeated occurrence of events, [Ho89, Scanning the Issue]. For example: a distributed computing system, in which an event might be the receipt of a message; a digital circuit, in which an event might be a voltage change on a wire; or a manufacturing process, in which an event might be the delivery of a part to a machine.

Our main interest will be in the long-term behaviour of the system. If a denotes an event and $t_i(a)$ denotes the time at which the i-th occurrence of this event takes place then we shall study the asymptotic behaviour of the sequence $t_1(a), t_2(a), \ldots$. Depending on the nature of the system, this question may have to be formulated stochastically: the $t_i(a)$ would then be random variables over some suitable measure space. This form of randomness is that of a random environment, as distinct from the additive noise that is customary in signal processing or linear systems theory.

It may not be immediately obvious that idempotency has anything to contribute to this. In fact, discrete event systems which can be modelled by max-plus matrices have been studied repeatedly, [CG62, Rei68, RH80, CDQV85, Bur90, RS94, ER95], although not all these authors have explicitly used the idempotency. In this volume, Gürel, Pastravanu and Lewis use max-plus matrices to study manufacturing systems, [GPL]; Cofer and Garg exploit the partial order structure of $\mathbb{R}_{max}$ to study supervisory control, [CGa]; and Cuninghame-Green uses polynomials over max-plus to study the realisability problem, [CGb].

There are many questions that can be asked about the long-term behaviour of discrete event systems. We shall consider only a few. Do there exist steady or periodic regimes? What form do they take? On average, how quickly does the next occurrence of a take place?

4.1.2 Mathematical models for discrete event systems

To answer questions such as those above, it is necessary to have a mathematical description of the system, which specifies a mechanism for determining when events occur. A variety of models have been proposed, [Ho89]. The automata of the previous section are convenient for modelling systems with state. They allow easy specification of how different choices of event can occur, depending on the current state of the system. To specify temporal behaviour, they can be augmented with information on the duration of events, as in the model proposed by Glasserman and Yao, [GY94]. A different approach is to specify how events are causally related to each other. An example of this is the classical Gantt chart, or PERT diagram. A somewhat related model is the task-resource model, studied by Gaubert and Mairesse in this volume, [GMa], in which tasks are specified by their durations and the resources they require. This is sometimes called a tetris model by analogy with the computer game of that name.

In the tetris model, resources are renewable and can be used repeatedly, like a machine in a factory. For consumable resources, a more complex model is required, which combines both state and causal aspects. Petri nets have proved popular in this respect, [Rei85]. A Petri net can be defined in terms of a finite set of tasks, Q, and a finite set of resource types, P. Each task, q, consumes and produces a basket of resources, which are specified as elements ${}^{\bullet}q, q^{\bullet}$, respectively, of $\mathbb{N}(P)$. The state of the system is given by an element, $c \in \mathbb{N}(P)$, which specifies the current availability of resources of each type. The only tasks, q, which can proceed in state c are those for which there are sufficient resources, i.e. ${}^{\bullet}q \leq c$; if q does proceed, the state of the system changes to $c - {}^{\bullet}q + q^{\bullet}$. (Inequality and addition are defined, as usual, pointwise in the semimodule $\mathbb{N}(P)$.) Temporal behaviour can be incorporated by, for instance, giving each resource a *holding time* for which it must be kept, before becoming available for consumption by a task. Timed Petri nets are the subject of the paper by Cohen, Gaubert and Quadrat in this volume, [CGQ].

Suppose that in a timed Petri net, for each resource, there is exactly one task which produces it and exactly one task which consumes it. Such a net can be described by a directed graph in which the vertices correspond to the tasks and the edges correspond to the resources. For obvious reasons, this is known as a timed event graph. Its importance lies in the fact that its temporal behaviour can be completely described as a linear system over $\mathbb{R}_{\max}$, [BCOQ92, §2.5]. That is, if $\vec{x}(k)$ denotes the vector of k-th occurrence times of the different tasks in the net, then, under reasonable conditions, there exist matrices $A_0, \dots, A_p \in M_n(\mathbb{R}_{\max})$, such that, for sufficiently large k,

$$\vec{x}(k) = A_0\vec{x}(k) + A_1\vec{x}(k-1) + \cdots + A_p\vec{x}(k-p) .$$

This result, that *timed event graphs are linear systems*, has been the key to an algebraic approach, based on idempotency, to the study of discete event systems, [CDQV85, CMQV89]. The spectral theory of max-plus matrices, as discussed in §2.9, and such results as the Cyclicity Theorem, Theorem 2.1, enable one to draw important and useful conclusions about the long-term behaviour of the timed event graph.

The linear theory also has implications for efficient simulation of discrete event systems. The discrete event simulators currently used for performance modelling in industry take no account of the underlying algebraic structure. However, as shown in [BC93], this can be used to develop more efficient simulation algorithms.

The main limitation of event graphs is their inability to capture conflict, or competition for resources. Each resource in the event graph is consumed by only one task. Tasks may proceed in parallel but they cannot pre-empt each other. (A finite automaton, in contrast, allows a choice among different outcomes.) Free choice nets are a class of Petri nets which include event graphs but allow some conflict, [DE95]. If there is a hierarchy of Petri net models then free choice nets are the obvious candidate for the next level of complexity beyond event graphs. In the untimed case they are known to have many interesting properties, [DE95]. In the timed case, the results of [BFG94] confirm that, at least from a mathematical standpoint, free choice nets are a good class to study.

It is clear from this discussion that there are a variety of different models for studying discrete event systems. There is not much consensus on a single fundamental model, [Ho89, Scanning the Issue]. Our approach in the rest of this section will be to extract certain features which appear both conceptually reasonable and common to many systems, and to study a basic model with just these features. This basic model, of topical functions, will include as a special case the linear theory based on max-plus. It can also be extended in several ways and used as a building block to model a wide class of discrete event systems. As we shall see, there are many unanswered questions about the basic model itself.

4.1.3   The basic model

Let $\mathbb{R}^n$ denote the space of functions $\mathbb{R}(1, \ldots, n)$, which we think of as column vectors. We use x, y, z for vectors and x_i instead of $x(i)$ for the i-th component of x. $\mathbb{R}^n$ acquires the usual pointwise ordering from the ordering on $\mathbb{R}$: $x \leq y$ if, and only if, $x_i \leq y_i$ for $1 \leq i \leq n$. If $x \in \mathbb{R}^n$ and $h \in \mathbb{R}$ then $x + h$ will denote the vector y for which $y_i = x_i + h$: the operation is performed on each component of the vector. This *vector-scalar* convention extends to equations and inequalities: $x = h$ will mean that $x_i = h$ for $1 \leq i \leq n$.

Definition 4.1 ([GK95]) *A topical function is a function, $F : \mathbb{R}^n \to \mathbb{R}^n$, such that if $x, y \in \mathbb{R}^n$ and $h \in \mathbb{R}$ then the following properties hold:*

- (monotonicity) $x \leq y \implies F(x) \leq F(y)$ M

- (homogeneity) $F(x + h) = F(x) + h$.   H

A topical function can be thought of as modelling a system with n events in which $x \in \mathbb{R}^n$ specifies the time of occurrence of each event and $F(x)$ specifies the times of next occurrences. The sequence $x, F(x), F^2(x), \ldots$ is then the sequence of occurrence times of events, whose asymptotic behaviour is the object of study. Definition 4.1 can be understood as follows: if the times of occurrences of some events are increased, then the times of next occurrences of all events cannot decrease; if the times of occurrences are all increased, or decreased, by exactly the same amount for each event, then so too, respectively, are the times of next occurrences. These properties appear very reasonable and can be seen to hold in some form for most discrete event systems.

Functions satisfying the conditions of Definition 4.1 have been introduced and studied independently by other authors, [Kola, Vin], but the material discussed below, as well as the name *topical*, is based on joint work of this author with Keane and Sparrow, [GK95, GKS96].

4.1.4   Extensions of the basic model

There are practical systems which can be modelled directly as topical functions, [Gun93, SS92], but such systems are rather restricted. They must usually be closed, or autonomous, in that they require only an initial condition to generate a sequence of occurrence times. Open systems, in contrast, require input to be regularly provided. To model this, it is more convenient to work, not with vectors in $\mathbb{R}^n$, but with appropriate functions $\mathbb{R} \to \mathbb{R}^n$, which represent input, or output, histories. This requires an extension of the theory of topical functions to infinite dimensional spaces, a problem studied in [Kola]. Another extension is to study semigroups of topical functions, which allows one to model the choice or conflict described in §4.1.2. As remarked at the end of §3.2, Gaubert has made initial investigations in this direction, [Gau]. Finally, stochastic systems can be modelled by random variables taking values in the space of topical functions. In this volume, Baccelli and Mairesse give an extensive discussion of ergodic theorems for random topical functions and for more general systems, [BM], while Gaujal and Jean-Marie study the problem of computing asymptotic quantities in stochastic systems, [GJM].

These three extensions bring a much wider class of discrete event systems within our scope. It remains an open problem to understand exactly how much wider. For instance, how can Petri nets be incorporated in this framework? What about the model of Glasserman and Yao?

Before going further, we need to see some concrete examples. We shall then discuss the properties of topical functions, with reference to the questions raised in §4.1.1.

4.2   Examples of topical functions

Let $\mathsf{Top}(n, n)$ denote the set of topical functions, $F : \mathbb{R}^n \to \mathbb{R}^n$. This set has a rich structure, which we need some notation to explain. If a, b are elements of some partially ordered set, let $a \vee b$ and $a \wedge b$ denote the least upper bound and greatest lower bound, respectively, when these exist. The set of functions $\mathbb{R}^n \to \mathbb{R}^n$ has a natural partial order defined pointwise from that on $\mathbb{R}^n$. Finally, let F^- denote the function $-F(-x)$.

Lemma 4.1 ([GK95, Lemma 1.1]) *Let $F, G \in \mathsf{Top}(n, n)$. Let $\lambda, \mu \in \mathbb{R}$ satisfy $\lambda, \mu \geq 0$ and $\lambda + \mu = 1$. Let $c \in \mathbb{R}^n$. Then $FG, F \vee G, F \wedge G, F + c, F^-$ and $\lambda F + \mu G \in \mathsf{Top}(n, n)$.*

This result is an immediate consequence of Definition 4.1. $\mathsf{Top}(n, n)$ is a distributive lattice under $\vee$ and $\wedge$. It is almost a Boolean algebra with F^- as complement but lacks top and bottom elements.

Definition 4.2 *A function $F : \mathbb{R}^n \to \mathbb{R}^n$ is said to be simple if each component, $F_i : \mathbb{R}^n \to \mathbb{R}$, can be written as $F_i(x) = x_j + a$, for some j and some $a \in \mathbb{R}$.*

Simple functions are clearly topical. Lemma 4.1 now provides a mechanism for building topical functions which are not simple.

Definition 4.3 ([Gun94c]) *A function $F : \mathbb{R}^n \to \mathbb{R}^n$ is said to be min-max if it can be constructed from simple functions by using only $\vee$ and $\wedge$. Such a function is max-only if it can be built using only $\vee$ and min-only if it can be built using only $\wedge$.*

Max-only functions provide the link to idempotency. Any max-only function can be placed in normal form:

$$F_i(x) := (x_1 + A_{i1}) \vee \cdots \vee (x_n + A_{in})$$

where $A \in M_n(\mathbb{R}_{\max})$. We may then write $F(x) = Ax$. Hence, max-only functions correspond to matrices over $\mathbb{R}_{\max}$ which satisfy the non-degeneracy condition:

$$\forall i, \exists j, \text{ such that } A_{ij} \neq -\infty . \tag{4.1}$$

A min-only function, dually, corresponds to a matrix over $\mathbb{R}_{\min}$. Min-max functions are nonlinear in the idempotent sense.

Lemma 4.2 ([GKS96]) *Choose $F \in \mathsf{Top}(n,n)$ and $a \in \mathbb{R}^n$. There exists a max-only function F_a such that $F(a) = F_a(a)$ and $F \leq F_a$. A dual statement holds for min-only functions.*

Corollary 4.1 ([GKS96]) *Let $F \in \mathsf{Top}(n,n)$. Then F can be written in the two forms:*

$$\bigwedge_{i \in I} G_i = F = \bigvee_{j \in J} H_j \tag{4.2}$$

where G_i are max-only, H_j are min-only and I and J may be uncountably infinite.

Min-max functions are, in a strict sense, finite topical functions. If either of the index sets, I or J, is finite, then F is a min-max function. In this case, the representations in (4.2) can be reduced to essentially unique normal forms, [Gun94c, Theorem 2.1]. Min-max functions capture some of the dynamical features of topical functions; see Corollary 4.2.

The infinite topical functions include some well-known functions in disguise. Let $\mathbb{R}^+$ denote the positive reals and $\mathbb{R}^{+0}$ the nonnegative reals. Let $\exp : \mathbb{R}^n \to (\mathbb{R}^+)^n$ and $\log : (\mathbb{R}^+)^n \to \mathbb{R}^n$ be defined componenentwise: $\exp(x)_i = \exp(x_i)$ and $\log(x)_i = \log(x_i)$. These are mutually inverse bijections between $\mathbb{R}^n$ and $(\mathbb{R}^+)^n$. Let $A : (\mathbb{R}^+)^n \to (\mathbb{R}^+)^n$ be any function on the positive cone and let $\mathcal{E}(A) : \mathbb{R}^n \to \mathbb{R}^n$ denote the function $\log(A(\exp))$. The functional $\mathcal{E}$ allows us to transport functions on the positive cone to functions on $\mathbb{R}^n$. Moreover, $\mathcal{E}(AB) = \mathcal{E}(A)\mathcal{E}(B)$, so A and $\mathcal{E}(A)$ have equivalent dynamic behaviour.

Now suppose that A is a nonnegative matrix, $A \in M_n(\mathbb{R}^{+0})$, which satisifes a similiar nondegeneracy condition to (4.1): $\forall i, \exists j$, such that $A_{ij} \neq 0$. Then A preserves the positive cone, when elements of $(\mathbb{R}^+)^n$ are interpreted as column vectors in the usual way. The reader can easily check that $\mathcal{E}(A)$ is a topical function. It follows that the theory of topical functions is a generalisation of Perron–Frobenius theory, [Min88]. It can further be shown, using Lemma 4.1, that a number of problems considered in the optimisation theory literature also fall within the theory of topical functions, [GKS96]. These include problems of deterministic optimal control, Markov decision processes and Leontief substitution systems.

4.3   Nonexpansiveness and perioidicity

If $x \in \mathbb{R}^n$, let $\| x \|$ denote the ℓ^∞ norm of x: $\| x \| = \vee_{1 \leq i \leq n} |x_i|$. A function $F : \mathbb{R}^n \to \mathbb{R}^n$ is nonexpansive in the ℓ^∞ norm if

- $\| F(x) - F(y) \| \leq \| x - y \|$.   N

The following observation was first made by Crandall and Tartar, [CT80]; see also [GK95, Proposition 1.1] for a proof adapted to the present context.

Proposition 4.1 *If $F : \mathbb{R}^n \to \mathbb{R}^n$ satisfies* H, *then* M *is equivalent to* N.

In particular, topical functions are nonexpansive. This constrains their dynamics in ways that are still not understood. We draw the reader's attention here to one result which has significance for discrete event systems.

One of the general questions raised in §4.1.1 concerned the existence of a periodic regime. In the light of the suggested interpretion of topical functions in §4.1.3, a periodic regime can be reasonably formulated as a generalised periodic point: a vector x such that $F^p(x) = x + h$ for some $p > 0$ and some $h \in \mathbb{R}$. (Recall that we are using the vector-scalar convention of §4.1.3.) The system returns after p occurrences with a shift of h in each occurrence time. Because of property H, this behaviour persists. We can, without loss of generality, consider only ordinary periodic points, because $F^p(x) = x + h$ if, and only if, $(F - h/p)^p(x) = x$ and, by Lemma 4.1, $F - h/p$ is topical.

The least p for which $F^p(x) = x$ is the period of F at x. What periods are possible for discrete event systems modelled by topical functions? It turns out, surprisingly, that there is a universal bound on the size of periods which depends only on the dimension of the ambient space.

Theorem 4.1 ([BW92]) *If $F : \mathbb{R}^n \to \mathbb{R}^n$ is nonexpansive in the ℓ^∞ norm and if p is the period of a periodic point of F, then $p \leq (2n)^n$.*

Results of this form originate in the work of Sine, [Sin90]. An up-to-date discussion, as well as complete references, can be found in the survey paper by Nussbaum in this volume, [Nus]. The bound in Theorem 4.1 is not tight: Nussbaum has conjectured that $p \leq 2^n$, and this can be shown to be best possible. The Nussbaum Conjecture has been proved only for $n \leq 3$ and remains the outstanding open problem in this area.

For topical functions, more can be said. The following is an immediate consequence of Lemma 4.2.

Corollary 4.2 ([GKS96]) *Let F be a topical function and $S \subseteq \mathbb{R}^n$ any finite set of vectors. There exists a min-max function G such that $F(s) = G(s)$ for all $s \in S$. In particular, any period of a topical function must be the period of a min-max function.*

It follows that it is sufficient to consider only min-max functions in determining the best upper bound for the periods of topical functions. By augmenting min-max functions with $F(-x)$, where F is min-max, it is possible to give a similar reduction for general nonexpansive functions, [GKS96]. Gunawardena and Sparrow, in unpublished work, have shown that there are min-max functions in dimension n with period ${}^n C_{[n/2]}$ and conjecture that this is the best upper bound for topical functions.

Fixed points, where $F(x) = x$, are of particular importance for discrete event systems. They represent equilibria, as in (2.4). The existence of fixed points for nonexpansive functions is a classical problem, [GK90]. For topical functions, Kolokoltsov has given a sufficient condition in terms of a game theoretic representation of topical functions, [Kola, Theorem 9]. For min-max functions, much stronger results are thought to hold, as in Theorem 4.3. This turns out to be related to the other question raised in §4.1.1, which forms the subject of the next section.

4.4 Cycle times

The rate at which events occur in a discrete event system is an important measure of its performance. The average elapsed time between occurrences, starting from the initial condition $x \in \mathbb{R}^n$, is given by $(F^k(x) - F^{k-1}(x) + \cdots + F(x) - x)/k$. The asymptotic average, as $k \to \infty$, is then

$$\lim_{k \to \infty} F^k(x)/k \, . \tag{4.3}$$

It is not at all clear that this limit exists in general. Suppose, however, that it does exist for a given F at some initial condition x and that y is some other initial condition. Since F is nonexpansive, $\| F^k(x) - F^k(y) \| \leq \| x - y \|$. Hence, if the limit (4.3) exists anywhere, it must exist everywhere, and must have the same value.

Definition 4.4 *Let $F \in \mathsf{Top}(n, n)$. The cycle time vector of F, $\chi(F) \in \mathbb{R}^n$, is defined as the value of (4.3) if that limit exists for some $x \in \mathbb{R}^n$, and is undefined otherwise.*

If F has a generalised fixed point, $F(x) = x + h$, then, by property H, $F^k(x) = x + kh$. Hence, $\chi(F) = h$. If F is a max-only function, and $A \in M_n(\mathbb{R}_{\mathsf{max}})$ is the corresponding max-plus matrix, then a generalised fixed point of F is simply an eigenvector of A and h is the corresponding eigenvalue. By Proposition 2.6, h is the largest mean circuit weight in $\mathcal{G}(A)$. Similarly, if $F = \mathcal{E}(A)$, where $A \in M_n(\mathbb{R}^{+0})$, then $\exp(x) \in (\mathbb{R}^+)^n$ is an eigenvector of A with eigenvalue $\exp(h)$. In this case $\exp(h)$ is the classical spectral radius of A. We see from this that $\chi(F)$ is a vector generalisation of the notion of eigenvalue, suitable for an arbitrary topical function.

However, if $A \in M_n(\mathbb{R}_{\max})$ is the max-plus matrix corresponding to the max-only function F, then not all eigenvectors of A can be generalised fixed points of F. They must lie in $\mathbb{R}^n$ and hence have no component equal to $-\infty$. To bring the other eigenvectors into the picture requires some form of compactification of $\mathbb{R}^n$, analogous to putting the boundary on the positive cone $(\mathbb{R}^+)^n$. It also suggests the existence of other cycle time vectors which give information on fixed points lying in different parts of this boundary. To make sense of this for topical functions remains an open problem.

When does $\chi(F)$ exist? The first indication that this is a difficult question came from the following result of Gunawardena and Keane.

Theorem 4.2 ([GK95]) *Let* $\{a_i\}$, $i \geq 1$, *be any sequence of real numbers drawn from the unit interval* $[0, 1]$. *There exists a function* $F \in \mathsf{Top}(3, 3)$, *such that* $F^i(0, 0, 0)_2 = a_1 + \cdots + a_i$.

In particular, $\chi(F)$ does not always exist and the result indicates the extent of the departure from convergence.

On the other hand, there is strong evidence that $\chi(F)$ does exist for min-max functions. We can formulate this by asking how $\chi(F)$ should behave with respect to the operations which preserve min-max functions. Let $\mathsf{MM}(n, n) \subseteq \mathsf{Top}(n, n)$ denote the set of min-max functions $\mathbb{R}^n \to \mathbb{R}^n$. It is easy to see that $\mathsf{MM}(n, n)$ is closed under all the operations of Lemma 4.1, with the exception of convex combination: $\lambda F + \mu G$. In particular, $\mathsf{MM}(n, n)$ is a distributive lattice. $\mathsf{MM}(n, n)$ also has a Cartesian product structure in which each $F \in \mathsf{MM}(n, n)$ is decomposed into its separate components $(F_1, \cdots, F_n)$. If $S \subseteq A_1 \times \ldots \times A_n$ is a subset of some such product, let $r(S)$ denote its rectangularisation:

$$r(S) = \{u \in A_1 \times \cdots \times A_n \mid \pi_k(u) \in S\},$$

where $\pi_k : A_1 \times \cdots \times A_n \to A_k$ is the projection on the k-th factor. It is always the case that $S \subseteq r(S)$ and if $S = r(S)$ we say that S is a rectangular subset. For example, if $F, G \in \mathsf{MM}(2, 2)$ then

$$r\{F, G\} = \{(F_1, F_2), (F_1, G_2), (G_1, F_2), (G_1, G_2)\}.$$

The lattice operations on $\mathsf{MM}(n, n)$ behave well with respect to the Cartesian product structure: if $S \subseteq \mathsf{MM}(n, n)$, then

$$\bigvee_{F \in S} F = \bigvee_{G \in r(S)} G \quad \text{and} \quad \bigwedge_{F \in S} F = \bigwedge_{G \in r(S)} G. \tag{4.4}$$

Conjecture 4.1 (The duality conjecture) $X : \mathsf{MM}(n,n) \to \mathbb{R}^n$ *always exists and is a homomorphism of lattices on rectangular subsets: if $S \subseteq \mathsf{MM}(n,n)$ is a finite, rectangular subset, then*

$$X\left(\bigvee_{F \in S} F \right) = \bigvee_{F \in S} X(F)$$

$$X\left(\bigwedge_{F \in S} F \right) = \bigwedge_{F \in S} X(F).$$

Theorem 4.3 ([Gun94a]) *Let $F \in \mathsf{MM}(n,n)$. If the duality conjecture is true in dimension n then $F(x) = x + h$ for some x, if, and only if, $X(F) = h$.*

Conjecture 4.1 was first stated in a different but equivalent form in [Gun94a]. By virtue of (4.4), it gives an algorithm for computing $X(F)$ for any min-max function in terms of simple functions, for which X can be trivally calculated. However, because of the rectangularisation required in (4.4), this algorithm is very infeasible. This raises issues of complexity about which little is known.

The conjecture is known to be true when S consists only of simple functions, that is, when $\bigvee_{F \in S} F$ is a max-only function and $\bigwedge_{F \in S} F$ is a min-only function. The conjecture has also been proved in dimension 2, where the argument is already non-trivial, [Gun94b]. Sparrow has shown that X exists for min-max functions in dimension 3 but his methods do not establish the full conjecture, [Spa96]. The duality conjecture remains the fundamental open problem in this area and a major roadblock to further progress in understanding topical functions.

The existence of X does not depend on the finiteness of min-max functions: it also exists for functions $\mathcal{E}(A)$ where $A \in M_n(\mathbb{R}^{+0})$, [GKS96]. At present, we lack even a conjecture as to which topical functions have a cycle time.

We have sketched some of the main results and open problems for topical functions, which we believe are fundamental building blocks for discrete event systems. Topical functions also provide a setting in which max-plus linearity, $A \in M_n(\mathbb{R}_{\mathsf{max}})$, and classical linearity, $A \in M_n(\mathbb{R}^{+0})$, coexist. As we mentioned in §2.9, the relationship between these is a very interesting problem. We shall to return to it in §6.5. Before that, we must venture into infinite dimensions.

5   Nonlinear Partial Differential Equations

5.1   Introduction

In this section we shall be concerned with scalar nonlinear first order partial differential equations of the form

$$F(x, u(x), Du(x)) = 0 \qquad (5.1)$$

where $F : \mathbb{R}^n \times \mathbb{R} \times \mathbb{R}^n \to \mathbb{R}$. Here, u is a real-valued function on some open subset $X \subseteq \mathbb{R}^n$, $u : X \to \mathbb{R}$, and $Du(x) \in \mathbb{R}^n$ can be thought of as the derivative of u at x, $Du = (\partial u/\partial x_1, \ldots, \partial u/\partial x_n)$.

It may seem implausible that idempotency has anything to say about differential equations: operations like $\max(u, v)$ do not preserve differentiable functions. However, remarkable advances have taken place in our understanding of nonlinear partial differential equations which enable us to give meaning to solutions of (5.1) which may not be differentiable anywhere. From one perspective, this can be viewed as borrowing ideas from convex analysis. Indeed, the origins of the differential calculus itself go back to Fermat's observation that the maxima of a function $f : \mathbb{R} \to \mathbb{R}$ occur at points where $df/dx = 0$. Efforts to generalise this to nondifferentiable functions have led to new notions of differentiability, [Aub93, Chapter 4], which are closely related to the ideas discussed below, [CEL84, Definition 1]. For ease of exposition, we take a different approach, but convex analysis forms a backdrop to much of what we say and its relationship to idempotency is badly in need of further investigation. Aubin's book provides an excellent foundation, [Aub93].

5.2   Viscosity solutions

The advances mentioned above centre around the concept of *viscosity solutions*, introduced by Crandall and Lions in a seminal paper in 1983 following related work of Kruzkov in the late 1960s. For references, see the more recent *User's Guide* of Crandall, Ishii and Lions, [CIL92], which discusses extensions of the theory to second order equations. The earlier survey by Crandall, Evans and Lions, [CEL84], is a model of lucid exposition and can be read, even by non-users, with pleasure and insight.

The word *viscosity* refers to a method of obtaining solutions to (5.1) as limits of solutions of a second order equation with a small parameter—the viscosity—as that parameter goes to zero. We discuss asymptotics further in §6.1. Solutions obtained by this method of vanishing viscosity can be shown to be viscosity solutions in the sense of Crandall, Evans and Lions, [CEL84, Theorem 3.1].

Nondifferentiable solutions to (5.1) arise in a number of applications. For instance, the initial value problem

$$u_t + H(Du) = 0$$
$$u(x,0) = u_0(x) \tag{5.2}$$

where $u : \mathbb{R}^{n-1} \times [0,\infty) \to \mathbb{R}$, arises in the context of optimisation. In mechanics it is known as the Hamilton–Jacobi equation, in optimal control as the Bellman equation and in the theory of differential games as the Isaacs equation. The function u represents the optimal value under the appropriate optimisation requirement. In mechanics, the Hamiltonian H and the initial conditions u_0 are often sufficiently smooth to give uniqueness and existence of smooth solutions, at least for small values of t. However, in the other contexts, neither the Hamiltonian nor the initial conditions need be differentiable and are sometimes not even continuous. See, for instance, the Hamiltonian found by Kolokoltsov and Maslov in their paper in this volume, [KMb], for multicriterial optimisation problems. It becomes important, therefore, to give a convincing account of what it means for u to be a solution of (5.2) or (5.1), when u is not differentiable.

An obvious approach is to restrict solutions to be nondifferentiable only on a set of measure zero. Unfortunately, there are usually lots of these. Consider the very simple example of (5.1) with $n = 1$ and $F(x,u,p) = p^2 - 1$, in other words, the equation $(du/dx)^2 = 1$. Suppose that we seek solutions on the interval $[-1,1]$ which are zero on the boundary. Then $u_1(x) = 1 - |x|$ is differentiable except at $x = 0$ and satisfies the boundary conditions. However, so too does $u_2(x) = -u_1(x)$ and the reader will see that there are infinitely many piecewise differentiable solutions of this form.

The viscosity method cuts through this problem by specifying conditions on the local behaviour of u which isolate one solution out of the many possibilities. The idea is beautifully simple. Let X be an open subset of $\mathbb{R}^n$ and suppose that $\phi : X \to \mathbb{R}$ is a C^1 function such that $u - \phi$ has a local maximum at x_0. If u is differentiable at x_0, then $Du(x_0) = D\phi(x_0)$. If u is not differentiable, why not use $D\phi(x_0)$ as a surrogate for $Du(x_0)$?

Definition 5.1 *An upper semi-continuous (respectively, lower semi-continuous) function $u : X \to \mathbb{R}$ is said to be a viscosity subsolution (respectively, supersolution) of (5.1) if, for all $\phi \in C^1(X)$, whenever $u - \phi$ has a local maximum (respectively, minimum) at $x_0 \in X$, then $F(x_0, u(x_0), D\phi(x_0)) \leq 0$ (respectively, ≥ 0). A viscosity solution is both a subsolution and a supersolution.*

We recall that $u : X \to \mathbb{R}$ is upper semi-continuous at $x \in X$ if $\inf_{U \ni x} \sup_{y \in U} f(y) \leq f(x)$ and lower semi-continuous if $\sup_{U \ni x} \inf_{y \in U} f(y) \geq f(x)$, where U runs through neighbourhoods of x, [Aub93, Chapter 1]. Definition 5.1 interleaves Definition 2 of [CEL84], for first order equations, with Definition 2.2 of [CIL92], for semi-continuous solutions.

Note that if u is not upper semi-continuous at x_0 then there is no $\phi \in C^1(X)$ such that $u - \phi$ has a local maximum at x_0. Without the restriction to upper or lower semi-continuous functions, the characteristic function of the rationals, for instance, would be a viscosity solution of any equation!

Definition 5.1 conceals many subtleties, [CIL92, §2]. For instance, in the example above, u_1 is both a subsolution and a supersolution. There are, in fact, no C^1 functions ϕ such that $u_1 - \phi$ has a local minimum at 0, and so the supersolution condition is vacuously satisfied at 0. However, u_2, while it is a subsolution for the same reason, is not a supersolution: the function $u_2 + 1$ has a minimum at 0 but $p^2 - 1 \not\geq 0$ when $p = -1$.

Definition 5.1 leads to elegant and powerful existence, uniqueness and comparison theorems which form the heart of the theory of viscosity solutions, [CIL92]. We mention only the following point, which is relevant to the discussion in §5.5. It is easy to see that for $a \in \mathbb{R}$ and $b \in \mathbb{R}^n$, $u(x,t) = a + b.x - H(b)t$ is a smooth solution of (5.2). Here $b.x$ denotes the standard inner product in $\mathbb{R}^n$. Hopf showed how these linear solutions could be combined to give a generalised global solution of (5.2).

Theorem 5.1 ([Hop65, Theorem 5a]) *Suppose that H is strictly convex and that $H(p)/|p| \to \infty$ as $|p| \to \infty$. Suppose further that u_0 is Lipschitz. Then,*

$$u(x,t) = \inf_{y \in \mathbb{R}^{n-1}} \left(u_0(y) + \sup_{z \in \mathbb{R}^{n-1}} \{z.(x-y) - tH(z)\} \right) \qquad (5.3)$$

satisfies (5.2) almost everywere in $\mathbb{R}^{n-1} \times [0,\infty)$. ([Eva84, Theorem 6.1]) If u_0 is also bounded then (5.3) is a viscosity solution of (5.2).

The reader familiar with convex analysis will note that the innermost term in (5.3) can be rewritten as $tL((x-y)/t)$, where $L : \mathbb{R}^{n-1} \to \mathbb{R}_{\min}$ is the Legendre–Fenchel transform of H, [Aub93, Definition 3.1]. L is called the Lagrangian in mechanics. Hence,

$$u(x,t) = \inf_{y \in \mathbb{R}^{n-1}} \{u_0(y) + tL((x-y)/t)\} . \qquad (5.4)$$

Kolokoltsov and Maslov have shown that this formula gives a C^1 solution of (5.2) throughout $\mathbb{R}^{n-1} \times [0,\infty)$ under weaker hypothses on H but stronger hypotheses on u_0, [KM89, §4].

5.3   The role of idempotency

What, then, are the contributions of idempotency to this area? A key intuition has been that equations of the form (5.2) should be considered as linear equations over $\mathbb{R}_{\max}$ or $\mathbb{R}_{\min}$. This idea was first put forward by Maslov in the Russian literature in 1984 and later in English translation in [Mas87b]. There is a hint of it already in Hopf's *basic lemma*, [Hop65, §2], and in the following idempotent superposition principle.

Proposition 5.1 ([CEL84, Proposition 1.3(a)]) *Let u, v be viscosity subsolutions (respectively, supersolutions) of (5.1). Then $\max(u, v)$ (respectively, $\min(u, v)$) is also a viscosity subsolution (respectively, supersolution).*

It follows that for equations in which F is independent of the second variable u—such as (5.2)—the space of viscosity subsolutions is almost a semimodule over $\mathbb{R}_{\max}$; it lacks only a zero element, $u(x) = -\infty$. Notice that supersolutions are linear over $\mathbb{R}_{\min}$ and to get access to viscosity solutions one must work with both dioids.

This linearity is very appealing. It suggests that the Hopf formula (5.4), from an idempotent viewpoint over $\mathbb{R}_{\min}$, is a Green's function representation with kernel $tL((x-y)/t)$. The kernel should arise when the initial condition is a Dirac function at 0 and the solution for general u_0 should then be obtained as a convolution with the kernel, in the usual way. Of course, the notions of Dirac function and convolution must be understood in the idempotent sense. As we shall see, formula (5.4) has exactly the right form for this to make sense. While this intuition is very suggestive, it has yet to be fully realised in a convincing manner. We try to unravel what has been done in this direction in §5.5. In the next sub-section we develop some of the language needed to discuss these ideas.

5.4 Functional analysis over dioids

5.4.1 Introduction

Let D be a dioid. The aim of this section is to study spaces of functions, $X \to D$, and their linear transformations, when X is an arbitrary set. In infinite dimensions, limiting conditions must be imposed, as in classical functional analysis.

5.4.2 Boundedly complete dioids

To begin with, let us use the partial order of the dioid to control the infinite behaviour. We have already had a taste of this with quantales in §2.7 and it is in keeping with the order theoretic nature of idempotency. We compare this with more conventional topological methods in §5.4.3.

Definition 5.2 *A boundedly complete (BC) dioid, D, is a dioid in which every subset which is bounded above has a least upper bound, over which the multiplication distributes: $\forall S \subseteq Q$ such that $\exists d \in Q$, for which $x \preceq d$ for all $x \in S$,*

- *there exists a least upper bound $\sup_{x \in S} x$;*

- *$a.(\sup_{x \in S} x) = \sup_{x \in S}(a.x)$, $(\sup_{x \in S} x).a = \sup_{x \in S}(x.a)$.*

As with quantales, we write $\sum_{x \in S} x = \sup_{x \in S} x$. Let D be a BC dioid and X an arbitrary set. Let $D_b(X)$ denote the set of functions to D which are bounded above: $D_b(X) = \{f : X \to D \mid \exists d \in D, \ f(x) \preceq d, \ \forall x \in X\}$. $D_b(X)$ is a sub-semimodule of $D(X)$. If D is a BC dioid, then $D_b(X)$ inherits bounded completeness from D, which allows us to write any element $u \in D_b(X)$ as a linear combination of characteristic functions. Let $\delta_A \in D_b(X)$ denote the characteristic function of $A \subseteq X$:

$$\delta_A(x) = \begin{cases} 1 & \text{if } x \in A \\ 0 & \text{otherwise}. \end{cases}$$

For $a \in X$, $\delta_{\{a\}}$ can be thought of as the idempotent Dirac function at a. When $D = \mathbb{R}_{\min}$, it corresponds to the indicator function of a point as used in convex analysis, [Aub93, Definition 1.2].

Choose $u \in D_b(X)$. Since $u(x) \preceq d$ for some $d \in D$, the functions $u(a).\delta_{\{a\}}(x)$ all lie in $D_b(X)$ and the set of such functions is bounded above by the constant function d. Hence we can write, in $D_b(X)$,

$$u = \sum_{a \in X} u(a).\delta_{\{a\}} . \tag{5.5}$$

We are interested in the linear transformations on $D_b(X)$ which preserve infinite sums of this form.

Definition 5.3 *Let D be a BC dioid and M a semimodule over D. M is a boundedly complete (BC) semimodule if every subset of M which is bounded above has a least upper bound over which the scalar multiplication distributes. A homomorphism of BC semimodules, $f : M \to N$, is a homomorphism of semimodules which preserves the least upper bound of any bounded subset.*

As remarked above, $D_b(X)$ is a BC semimodule over D. It would seem from (5.5) that $D_b(X)$ is the free BC semimodule generated by X but this is not quite right. To clarify its properties, consider the following more general construction. Let M be any BC semimodule and let $D_b(X, M) = \{f : X \to M \mid f(x) \preceq m, \text{ for some } m \in M\}$. $D_b(X, M)$ is also a BC semimodule over D and $D_b(X) = D_b(X, D)$. For fixed X, this defines a functor from the category BC-SMod$_D$ to itself.

Let $\lambda : D_b(X) \to M$ be a homomorphism of BC semimodules. For each $a \in X$, the value of λ on $\delta_{\{a\}}$ defines a function $\theta_M(\lambda) : X \to M$. Since $\delta_{\{a\}} \preceq \delta_X$, it follows that $\lambda(\delta_{\{a\}}) \preceq \lambda(\delta_X)$, since λ is a homomorphism of semimodules. By definition, $\theta_M(\lambda)(a) = \lambda(\delta_{\{a\}})$ and so $\theta_M(\lambda)$ is bounded above. Hence we have a function $\theta_M : \mathrm{Hom}(D_b(X), M)_{\text{BC-SMod}_D} \to D_b(X, M)$.

Proposition 5.2 *θ_M is an isomorphism of BC semimodules.*

This follows by what category theorists call *general nonsense* and can safely be left to the reader, as can the proofs of the corollaries below.

Corollary 5.1 *Choose $f, g \in D_b(X)$. The pairing $\langle f, g \rangle = \theta_D^{-1}(g)(f)$ defines a nonsingular, boundedly complete inner product $D_b(X) \times D_b(X) \to D$ whose value is given by $\langle f, g \rangle = \sum_{a \in X} f(a).g(a)$.*

The implication is that $D_b(X)$ is self-dual as a BC semimodule. We can also describe the structure of homomorphisms between function spaces. Let $\Lambda : D_b(X) \to D_b(Y)$ be a homomorphism of BC semimodules and choose $(x, y) \in X \times Y$. Let $\eta : \mathrm{Hom}(D_b(X), D_b(Y)) \to D_b(X \times Y)$ be defined by $\eta(\Lambda)(x, y) = \theta_{D_b(Y)}(\Lambda)(x)(y)$.

Corollary 5.2 *η is an isomorphism of BC semimodules. Furthermore, for $f \in D_b(X)$ and $y \in Y$, $\Lambda(f)(y) = \sum_{x \in X} f(x).\eta(\Lambda)(x, y)$.*

If $D = \mathbb{R}_{\mathsf{min}}$ and $X = Y = \mathbb{R}^n$, then the Hopf formula (5.4) has exactly this structure, emphasising once again the linearity of the solutions of (5.2).

Corollary 5.2 is essentially Shubin's kernel representation theorem, [Shu, Theorem 7.1]. We have reformulated his treatment to bring out its algebraic nature and to highlight BC dioids and BC semimodules. Shubin refers to BC dioids as *regular* and homomorphisms of BC semimodules as *normal*. Akian refers to *locally complete* dioids but does not imply by this that multiplication distributes over infinite sums, [Aki95].

5.4.3   Topologies on dioids

Consider any BC dioid or semimodule. If follows from (2.3) that any *nonempty* subset has an infimum. Hence, the upper and lower limits of a bounded sequence can be defined in the usual way and we can say what it means for a sequence to converge. This defines a topology, which we call the BC topology. (For technical reasons, one must work with generalised sequences indexed over a directed set and not just with sequences indexed over $\mathbb{N}$, [KMa].)

Lemma 5.1 ([KMa]) *Let $\rho : \mathbb{R}_{\mathsf{min}} \times \mathbb{R}_{\mathsf{min}} \to [0, \infty)$ be defined by $\rho(x, y) := |\exp(-x) - \exp(-y)|$ with $\exp(-\infty) := 0$. Then ρ is a metric whose underlying topology coincides with the BC topology on $\mathbb{R}_{\mathsf{min}}$.*

The metric can extended to function spaces over $\mathbb{R}_{\mathsf{min}}$ in the usual way. Let $f, g \in (\mathbb{R}_{\mathsf{min}})_b(X)$ and define $\rho_X(f, g) := \sup_{x \in X} \rho(f(x), g(x))$. The boundedness of f and g ensures that $\rho_X(f, g) \in [0, \infty)$ and gives a valid metric. We now have two topologies on $(\mathbb{R}_{\mathsf{min}})_b(X)$: the BC topology, corresponding to pointwise convergence, and the metric topology of ρ_X, corresponding to uniform convergence. These are undoubtedly different but the author knows of no results or examples comparing them in the literature.

It appears that homomorphisms of BC semimodules are not continuous in the BC topology. (Nor presumably in the metric topology.) Once again, this issue has not been studied in the literature. However, kernel representations as in Corollary 5.2 have been found for semimodule homomorphisms which are continuous in the metric topology, [KM89, §2]. The marked discrepancy in hypothesis between results like this and Corollary 5.2 indicates that we are still some way from a full understanding of function spaces and their linear transformation. We can formulate the kernel problem as follows. Let V, W be sub-semimodules of $(\mathbb{R}_{\min})_b(X)$ and $\lambda : V \to W$ a homomorphism of $\mathbb{R}_{\min}$ semimodules. Under what conditions on V, W and λ does λ have a kernel representation as in Corollary 5.2?

5.5   Idempotency and viscosity solutions

From now on we shall work entirely with $\mathbb{R}_{\min}$, equipped with the BC topology. For the remainder of this section, **we shall only use the customary notations in $\mathbb{R}_{\min}$.**

The pairing introduced in Corollary 5.1, which we may write as $\langle f, g \rangle = \inf_{x \in X}(f(x) + g(x))$, provides an elegant way of comparing functions $X \to \mathbb{R}_{\min}$. Let $\Phi \subseteq (\mathbb{R}_{\min})_b(X)$ be a set of test functions. If $f, g \in (\mathbb{R}_{\min})_b(X)$, we say that f and g are equivalent, $f \approx g$, if $\langle f, \phi \rangle = \langle g, \phi \rangle$ for all $\phi \in \Phi$. The equivalence class of f will be denoted $[f]$ and the set of equivalence classes Φ^*.

Each equivalence class contains a unique smallest element, which can be identified as follows. Let $P_\Phi : (\mathbb{R}_{\min})_b(X) \to (\mathbb{R}_{\min})_b(X)$ be given by

$$P_\Phi(f)(x) = \sup_{\phi \in \Phi}\{\langle f, \phi \rangle - \phi(x)\} \qquad (5.6)$$

which is well-defined provided that, for any $x \in X$, there is some $\phi \in \Phi$, such that $\phi(x) \neq +\infty$.

Lemma 5.2 ([Gonb, Theorem 1]) *Suppose that $\Phi \neq \emptyset$. Then $P_\Phi(f) \approx f$, and if $f \approx g$ then $P_\Phi(f) \leq g$ in $(\mathbb{R}_{\min})_b(X)$.*

Proof Suppose that $g \approx f$. Choose $\phi \in \Phi$ and $x \in X$. Then, by definition of the pairing, $\langle g, \phi \rangle \leq g(x) + \phi(x)$. Hence, $\langle g, \phi \rangle - \phi(x) \leq g(x)$ and so $\langle f, \phi \rangle - \phi(x) \leq g(x)$, since $g \approx f$. This holds for any ϕ and any $x \in X$ and so $P_\Phi(f) \leq g$. Now choose $\phi \in \Phi$ and $x \in X$ as before. Then, by definition of P_Φ, $\langle f, \phi \rangle - \phi(x) \leq P_\Phi(f)(x)$. Hence, $\langle f, \phi \rangle \leq P_\Phi(f)(x) + \phi(x)$. Since this holds for all $x \in X$, it follws that $\langle f, \phi \rangle \leq \langle P_\Phi(f), \phi \rangle$. But, by the second assertion proved above, $P_\Phi(f) \leq f$, and so, by linearity of the pairing, $\langle P_\Phi(f), \phi \rangle \le \langle f, \phi \rangle$. Hence, $\langle P_\Phi(f), \phi \rangle = \langle f, \phi \rangle$. Since this holds for any $\phi \in \Phi$ it follws that $\mathcal{P}_\Phi(f) \approx f$, as required. $\qquad\square$

Let $\lambda : \Phi \to \Phi$ be any function. The pairing $\langle f, g \rangle$ enables us to dualise λ to a function $\lambda^* : \Phi^* \to \Phi^*$ as follows. For any $\phi \in \Phi$ and any $f \in (\mathbb{R}_{\min})_b(X)$, let $\langle \lambda^*[f], \phi \rangle = \langle f, \lambda(\phi) \rangle$. This defines λ^* unambiguously. Note that if $\lambda(\Phi) \not\subseteq \Phi$, then λ^* is not well-defined on Φ^*, and may be multiple-valued. λ^* is analogous to the adjoint or transpose in classical linear analysis. If $\Phi = (\mathbb{R}_{\min})_b(X)$ and λ has a kernel representation with kernel $k(x, y)$, then λ^* has kernel $k(y, x)$.

Kolokoltsov and Maslov have introduced a notion of *generalised weak solution* of (5.2). To describe this, we first need to explain the test functions used by them. Let X be a locally compact topological space and let $c^\infty(X) \subseteq (\mathbb{R}_{\min})_b(X)$ denote the set of functions $f : X \to \mathbb{R}_{\min}$ which are bounded below, continuous with respect to the BC topology on $\mathbb{R}_{\min}$ and which tend to $+\infty$ at infinity. This last condition means simply that, for any $0 < d$, no matter how large, there exists a compact subset $K \subseteq X$, such that $d < f(x)$ whenever $x \notin K$. This is equivalent to f being lower semi-compact (or inf compact as in [Gonb]) in the sense of convex analysis, [Aub93, Definition 1.6].

Lemma 5.3 ([KM89, §2]) $P_{c^\infty(X)}(f)$ *is the lower semi-continuous closure of* f: $P_{c^\infty(X)}(f) = \sup\{g \in (\mathbb{R}_{\min})_b(X) \mid g \text{ continuous and } g \le f\}$. *Furthermore, if* f, g *are both lower semi-continuous and* $f \approx g$ *then* $f = g$.

Consider now the initial value problem (5.2). For $0 < t$, let $\lambda_t : (\mathbb{R}_{\min})_b(\mathbb{R}^{n-1}) \to (\mathbb{R}_{\min})_b(\mathbb{R}^{n-1})$ be the time evolution defined by the Hopf formula (5.4):

$$\lambda_t(u)(x) = \inf_{y \in \mathbb{R}^{n-1}} \{u(y) + tL((x-y)/t)\}. \tag{5.7}$$

According to Kolokoltsov and Maslov, a generalised weak solution of (5.2), with initial condition $u_0 \in (\mathbb{R}_{\min})_b(\mathbb{R}^{n-1})$, is $\lambda_t([u_0]) \in c^\infty(X)^*$, [KM89, §4]. For this to make sense, it is necessary that $\lambda_t^*(c^\infty(X)) \subseteq c^\infty(X)$ where λ_t^* is the transposed operator with kernel $tL((y-x)/t)$. Curiously, this is neither explicitly stated nor proved in [KM89].

Leaving this issue aside—it is a property of H and not of equation (5.2)—the implication here is that initial conditions with the same lower semi-continuous closure should have the same evolution under (5.2). This does appear to capture an essential feature of (5.2) but the definition itself is unconvincing. The rationale behind it appears to be an analogy with generalised solutions in the classical linear theory. The difficulty is that the operator $\partial_t(-) + H(D(-))$ is not itself defined on Φ^*; it is only the evolution operator (5.7) which is defined there. This differs from the classical theory where differential operators are themselves dualised to act on spaces of distributions, [ES92, §2.1.6], and one is left in no doubt as to what it means for a distribution to be a solution. It is also not clear why the set of test functions $c^{\infty}(X)$ should play such a fundamental role with respect to solutions of (5.2).

Maslov and Samborskiĭ have made important progress towards resolving these difficulties for equations of the form $H(Du) = 0$, [MSa]. They define a notion of derivative on elements of $c^{\infty}(X)^*$ and prove existence and uniqueness results for an appropriate boundary value problem, [MSa, Theorems 1 and 2]. In this volume, Samborskiĭ clarifies the structure of the space on which $H(D(-))$ can be considered to act and states further uniqueness results, [Sam], while Kolokoltsov uses the idea of generalised weak solutions to study a stochastic version of (5.2), [Kolb].

Maslov and Samborskiĭ remark that *"viscosity solutions coincide"* with those given by their results, [MSa, Remark 4], but fail to state the conditions under which such a comparison can be proved. Very recently, Gondran has announced a characterisation of viscosity solutions, [Gonb, Theorem 6], based on the ideas in [KM89] and [MSa]. (See also the more recent [Gona].) This is an important development which finally suggests a precise connection between viscosity solutions and idempotent methods. Unfortunately, the announcement contains no proofs! We limit ourselves to a brief statement of Gondran's result, as our final comment on this fascinating subject.

Definition 5.4 *Let X be a set and suppose that $\Phi \subseteq (\mathbb{R}_{\min})_b(X)$ is nonempty. The sequence $u_k \in (\mathbb{R}_{\min})_b(X)$ is said to converge weakly from below to $u \in (\mathbb{R}_{\min})_b(X)$ with respect to Φ, denoted $\liminf^w_{k \to \infty} u_k = u$, if $\liminf_{k \to \infty} \langle u_k, \phi \rangle = \langle u, \phi \rangle$, for all $\phi \in \Phi$.*

Weak convergence from above, $\limsup^w_{k \to \infty} u_k$, and weak convergence, $\lim^w_{k \to \infty} u_k$, can be defined in a similar way. Weak convergence was introduced by Maslov, [Mas87b].

Theorem 5.2 ([Gonb, Theorem 6]) *Let $X = \mathbb{R}^n$ and $\Phi = c^{\infty}(\mathbb{R}^n)$. Suppose that, $u_k \in (\mathbb{R}_{\min})_b(\mathbb{R}^n)$ is a sequence of C^1 functions such that $\lim^w_{k \to \infty} u_k = u$, for some $u \in (\mathbb{R}_{\min})_b(\mathbb{R}^n)$. Suppose further that the image sequence $F(x, u_k, Du_k)$ has the property that $\liminf^w_{k \to \infty} F(x, u_k, Du_k) = 0$. Then u is a viscosity supersolution of (5.1). A dual statement over $\mathbb{R}_{\max}$ characterises subsolutions.*

6 Optimisation and large deviations

6.1 Asymptotics

In the previous section we studied nonlinear equations in their own right. We touched on asymptotics in explaining the origins of the word viscosity in §5.2. In fact, asymptotics was a fundamental motivation behind Maslov's work and throws up some intriguing questions about the nature of idempotency.

The following example, which is taken from [Mas87b], gives a good illustration of how asymptotics enters the picture. Consider the heat equation in one dimension, with a small parameter, $0 < h$: $\partial_t = h\partial_x^2 u$. This is a linear equation, and if u_1, u_2 are solutions then so is $a_1 u_1 + a_2 u_2$ for $a_1, a_2 \in \mathbb{R}$. We can, however, apply a nonlinear transformation, such as $u = \exp(-w/h)$, and thereby arrive at a nonlinear equation

$$\partial_t w + (\partial_x w)^2 - h\partial_x^2 w = 0 \,,$$

which still satisfies a superposition principle. If w_1, w_2 are solutions of this then so is $-h \log(\exp(-(a_1 + w_1)/h) + \exp(-(a_2 + w_2)/h))$. Now let $h \to 0$. We obtain the nonlinear first order equation

$$\partial_t w + (\partial_x w)^2 = 0 \,,$$

which the reader will recognise as a Hamilton–Jacobi equation of the form (5.2) with $H(p) = p^2$. We see further that

$$\lim_{h \to 0^+} -h \log(\exp(-a/h) + \exp(-b/h)) = \min(a, b) \,, \tag{6.1}$$

so that the superposition principle reduces to $\min(a_1 + w_1, a_2 + w_2)$. Once again we have stumbled across the intuition that solutions of (5.2) are linear over $\mathbb{R}_{\min}$.

Limits of the form $\lim_{h \to 0} h \log(\exp(F(h)))$ are studied in a number of areas: large deviations, exponential asymptotics, etc. Calculations like that above have given rise to another intuition, arising out of asymptotics, that *idempotency appears in the large deviation limit*, at least over $\mathbb{R}_{\max}$ and $\mathbb{R}_{\min}$. How do we make sense of this?

Maslov, in an influential book published in 1987 in French translation, [Mas87a], put forward the idea of constructing an idempotent measure theory with measures taking values in $\mathbb{R}_{\min}$. This provides a foundation for a theory of optimisation which runs parallel to the theory of probability, based on classical measures. Recent work suggests that this idempotent optimisation theory is, in fact, the large deviation limit of classical probability theory. In the remaining sections we briefly discuss this circle of ideas.

6.2   Idempotent measures and integration

Let X be a set and choose $c \in (\mathbb{R}_{\min})_b(X)$. If $A \subseteq X$, let $c(A) = \langle \delta_A, c \rangle :=$ $\inf_{a \in A} c(a)$. This is the canonical example of an idempotent measure. If $\{A_i \subseteq X \mid i \in I\}$ is any family of subsets, then it follows easily that

$$c\left(\bigcup_{i \in I} A_i\right) = \sum_{i \in I} c(A_i) \,. \tag{6.2}$$

Unlike conventional measures, which are based on geometric concepts of area, idempotent measures are based on economic concepts of cost and the subsets A_i need not be pairwise disjoint. It is also not necessary, in order for (6.2) to hold, that the index set I be countable, although countable additivity always seems to be assumed. Idempotent measures are closely related to Choquet capacities and to so-called nonadditive set functions, [O'B96, Pap91].

More generally, one can seek an idempotent measure, k, on a restricted family, $\mathcal{F}$, of subsets, corresponding to a σ-algebra in measure theory. There is no obvious need for $\mathcal{F}$ to be closed under either complementation or intersection, although these properties are often assumed. It is now no longer necessary that a function $c \in (\mathbb{R}_{\min})_b(X)$ exists such that, for each $A \in \mathcal{F}$, $k(A) = \langle \delta_A, c \rangle$. If c does exist, it is called a *density* for k. By Lemma 5.2, provided $\cup_{A \in \mathcal{F}} A = X$, there is a unique least such density, given by $k_* = P_\Phi(c)$ where $\Phi = \{\delta_A \mid A \in \mathcal{F}\}$. Formula (5.6) shows that $k_*(x) = \sup_{A \ni x} k(A)$, which gives us a candidate for a density, should one exist. For measures defined on the open sets of a topology, densities always exist if the topological space is reasonable: for instance, a separable metric space, [MK94, Aki95].

Another way to think about measure is through integration. If $(X, \mathcal{F}, k)$ is a reasonable idempotent measure space and $f \in (\mathbb{R}_{\min})_b(X)$, then the idempotent integral is given by $\int f(x)dk = \langle f, k_* \rangle$. When k is the uniform measure, whose density is the characteristic function of X, this is written $\int f(x)dx$. This integral notation is widely used. It suggests analogues of various classical theorems on integration, such as those of Riesz, Fubini, etc. For the most part, these are not deep. Fubini's Theorem, for instance, is equivalent to the triviality that $\inf_{a \in A} \inf_{b \in B} f(a, b) = \inf_{b \in B} \inf_{a \in A} f(a, b)$, where $f \in (\mathbb{R}_{\min})_b(X \times Y)$. Akian's result, mentioned in the previous paragraph, tells us that idempotent measures on sufficiently nice topological spaces are always absolutely continuous with respect to the uniform measure. This is different from the classical Radon–Nikodým theorem.

6.3   Optimisation theory

The idea of an optimisation theory built on idempotent measures has been developed in the PhD theses of Bellalouna and Del Moral and by the Max-Plus Working Group. We will not discuss it in detail here; the reader should refer to the papers in this volume of Akian, Quadrat and Viot, [AQV], Del Moral, [Morb], and Del Moral and Salut, [MSc]. Quadrat's presentation to the International Congress in Zürich in 1994, [Gro95], includes a useful summary. The theory is in a state of vigorous development; there is not always agreement on terminology and many of the foundational definitions have not stabilised. To what extent it provides a true foundation for optimisation, as probability theory does for the study of random phenomena, is a question that must be addressed elsewhere.

Optimisation theory provides a conceptual framework which runs parallel to probability theory: cost variable, as opposed to random variable; value, as opposed to mean; independence of cost variables; conditional cost; Bellman chains, as opposed to Markov chains; etc. The analogue of the Gaussian distribution turns out to be $(x-a)^2/2b$, which is stable under inf convolution, as the customary Gaussian is under ordinary convolution, [Gro95]. Analogues of the laws of large numbers can be formulated, based on various notions of convergence of cost variables, [AQV, Morb].

Weak convergence of cost measures, as in [AQV, Definition 5.1] or [Morb, Definition 3], is defined to be weak convergence of their densities in the sense of Definition 5.4 but with respect to the set of continuous functions $f \in (\mathbb{R}_{\min})_b(X)$. This is larger than the set of test functions used for viscosity solutions in Theorem 5.2, which had prescribed behaviour at infinity. The differences between various sets of test functions is another foundational question which has not been adequately studied. Weak convergence of cost measures is exactly analogous to weak convergence of probability measures, [Wil91, Chapter 17].

6.4   Large deviations

Let $\mathbf{r}_i$ be a sequence of independent and identically distributed random variables. One of the basic questions in probability theory concerns the behaviour of the average $\mathbf{s}_k = (\mathbf{r}_1 + \cdots + \mathbf{r}_k)/k$. Under reasonable conditions, $\mathbf{s}_k$ concentrates at the mean of $\mathbf{r}_1$, as $k \to \infty$. Large deviation theory deals with the asymptotics of this process. If $0 < a$, how fast does $\mathsf{P}(\mathbf{s}_k \geq a)$ go to zero as $k \to \infty$? Cramér showed in 1938 that, roughly speaking, it goes as $\exp(-kI(a))$, where $I : \mathbb{R} \to \mathbb{R}$ is a certain *rate function*, [DZ93, §2.2]. I can be constructed from $\mathbf{r}_1$ by the Cramér transform: the Legendre–Fenchel transform of the logarithm of the moment generating function:

$$I(x) = \sup_{t \in \mathbb{R}} \{xt - \log(\mathsf{E}(\exp(t\mathbf{r}_1)))\} \, .$$

More generally, we have the *large deviation principle*.

Definition 6.1 ([DZ93, §1.2]) *Let $\{\mu_\epsilon \mid 0 < \epsilon\}$ be a family of probability measures on the Borel σ-algebra of a topological space X. Let $I \in (\mathbb{R}_{\min})_b(X)$*

be lower semi-continuous and satisfy $0 \leq I(x)$ for all $x \in X$. Then $\{\mu_\epsilon\}$ satisfies the large deviation principle with rate function I, if, for any closed set F and open set G of X,

$$\begin{aligned}
\limsup_{\epsilon \to 0} \epsilon \log \mu_\epsilon(F) &\leq -\inf_{x \in F} I(x) \\
\liminf_{\epsilon \to 0} \epsilon \log \mu_\epsilon(G) &\geq -\inf_{x \in G} I(x) .
\end{aligned} \qquad (6.3)$$

The papers of Akian *et al.*, [AQV], and Del Moral, [Morb, MSb, Mora], shed light on the connection between large deviations and idempotency. Theorem 5.2 of [AQV] (see also [Morb]) gives a necessary and sufficient condition for weak convergence of measures on a metric space, which is strikingly similar to (6.3). This suggests that the large deviation principle is a form of weak convergence to the rate function. Proposition 6.8 of [AQV] is asserted to be equivalent to the Gärtner–Ellis theorem, one of the main results of large deviation theory, [DZ93, Theorem 2.3.6]. Proposition 6.8 is analogous to the Lévy convergence theorem: weak convergence of distribution functions corresponds to pointwise convergence of characteristic functions, [Wil91, §18.1]. (Characteristic function is used here in the sense of probability theory, [Wil91, §16.1].) As in probability theory, the central limit theorem for cost variables, [AQV, Theorem 4.6] (see also [Morb]), falls out as a corollary of Proposition 6.8.

As this account suggests, the results of optimisation theory appear to encapsulate and reformulate the large deviation asymptotics of probability theory.

6.5   Topical functions and asymptotics

There is another way to approach the emergence of idempotency in the large deviation limit. It uses the topical functions of §4.2 and formula (6.1). We noted in §2.9 the close analogy between the spectral theory of max-plus matrices and nonnegative matrices. This seems very similar to the analogy between probability theory and optimisation theory discussed in the previous section but without the large deviation mechanism for moving from the former to the latter.

Let $A \in M_n(\mathbb{R}_{\max})$. For each $0 < h \leq 1$, define the functional $\mathcal{E}^h : M_n(\mathbb{R}_{\max}) \to \mathsf{Top}(n, n)$ by

$$\left(\mathcal{E}^h(A)(x_1, \ldots, x_n)\right)_i := h \log(\sum_{j=1}^n \exp((A_{ij} + x_j)/h)) .$$

It is easy to check, using Lemma 4.1, that $\mathcal{E}^h(A)$ is a topical function for all $h \in (0, 1]$. Furthermore, $\mathcal{E}^1(A) = \mathcal{E}(\exp(A))$, where $\exp(A) \in M_n(\mathbb{R}^{+0})$ is given by $\exp(A)_{ij} = \exp(A_{ij})$, and $\mathcal{E}$ is the functional introduced in §4.2. At the other end, for any fixed $x \in \mathbb{R}^n$, $\lim_{h \to 0^+} \mathcal{E}^h(A)(x) = Ax$, by (6.1).

We see that we have a deformation, within the space of topical functions, between $\mathcal{E}(\exp(A))$, representing the nonnegative matrix $\exp(A)$, and the max-plus matrix A. The class of topical functions provides us with a context within which both extremes, and the intervening space, can be explored.

How do the dynamics and spectral theory of $\mathcal{E}^h(A)$ vary with h? In particular, how do the dynamics and spectral theory of A emerge in the limit as $h \rightarrow 0$? Perhaps this can be viewed as a noncommutative version, appropriate for matrices, of the problems discussed in §6.4. On this enigmatic note, we bring our survey of idempotency to a close.

References

[Aki95]   M. Akian. Densities of idempotent measures and large deviations. Rapport de Recherche 2534, INRIA, April 1995.

[AQV]   M. Akian, J.-P. Quadrat, and M. Viot. Duality between probability and optimization. Appears in [Gun97].

[Aub93]   J.-P. Aubin. *Optima and Equilibria*, volume 140 of *Graduate Texts in Mathematics*. Springer-Verlag, 1993.

[AV93]   S. Abramsky and S. Vickers. Quantales, observational logic and process semantics. *Mathematical Structures in Computer Science*, 3:161–227, 1993.

[Bac94]   W. Backes. *The structure of longest paths in periodic graphs*. PhD thesis, Universität des Saarlandes, 1994.

[Bap95]   R. Bapat. Permanents, max algebra and optimal assignment. *Linear Algebra and its Applications*, 226–228:73–86, 1995.

[BC93]   F. Baccelli and M. Canales. Parallel simulation of stochastic Petri nets using recurrence equations. *ACM Transactions on Modeling and Computer Simulation*, 3:20–41, 1993.

[BCGZ84]   R. E. Burkard, R. A. Cuninghame-Green, and U. Zimmermann, editors. *Algebraic and Combinatorial Methods in Operations Research*, volume 19 of *Annals of Discrete Mathematics*. North-Holland, 1984.

[BCOQ92]   F. Baccelli, G. Cohen, G. J. Olsder, and J.-P. Quadrat. *Synchronization and Linearity*. Wiley Series in Probability and Mathematical Statistics. John Wiley, 1992.

[BFG94] F. Baccelli, S. Foss, and B. Gaujal. Structural, temporal and stochastic properties of unbounded free-choice Petri nets. Rapport de Recherche 2411, INRIA, November 1994. To appear in _IEEE Transactions on Automatic Control._

[BJ72] T. S. Blyth and M. F. Janowitz. _Residuation Theory_, volume 102 of _International Series of Monographs in Pure and Applied Mathematics_. Pergamon, 1972.

[BM] F. Baccelli and J. Mairesse. Ergodic theorems for stochastic operators and discrete event networks. Appears in [Gun97].

[BSvdD95] R. Bapat, D. P. Stanford, and P. van den Driessche. Pattern properties and spectral inequalities in max algebra. _SIAM Journal of Matrix Analysis and Applications_, 16:964–976, 1995.

[Bur90] S. M. Burns. _Performance Analysis and Optimization of Asynchronous Circuits_. PhD thesis, California Institute of Technology, 1990.

[But94] P. Butkovic. Strong regularity of matrices—a survey of results. _Discrete Applied Mathematics_, 48:45–68, 1994.

[BW92] A. Blokhuis and H. A. Wilbrink. Alternative proof of Sine's theorem on the size of a regular polygon in $\mathbb{R}^n$ with the ℓ^∞ metric. _Discrete Computational Geometry_, 7:433–434, 1992.

[Car71] B. A. Carré. An algebra for network routing problems. _Journal of the Institute of Mathematics and its Applications_, 7:273–294, 1971.

[CDQV85] G. Cohen, D. Dubois, J.-P. Quadrat, and M. Viot. A linear-system-theoretic view of discrete-event processes and its use for performance evaluation in manufacturing. _IEEE Transactions on Automatic Control_, 30(3):210–220, 1985.

[CEL84] M. G. Crandall, L. C. Evans, and P.-L. Lions. Some properties of viscosity solutions of Hamilton–Jacobi equations. _Transactions of the AMS_, 282:487–502, 1984.

[CGa] D. D. Cofer and V. K. Garg. Idempotent structures in the supervisory control of discrete event systems. Appears in [Gun97].

[CGb] R. A. Cuninghame-Green. Maxpolynomials and discrete-event dynamic systems. Appears in [Gun97].

[CG62] R. A. Cuninghame-Green. Describing industrial processes with interference and approximating their steady-state behaviour. _Operational Research Quarterly_, 13(1):95–100, 1962.

[CG79]   R. A. Cuninghame-Green. *Minimax Algebra*, volume 166 of *Lecture Notes in Economics and Mathematical Systems*. Springer-Verlag, 1979.

[CG95]   R. A. Cuninghame-Green. Minimax algebra and applications. *Advances in Imaging and Electron Physics*, 90:1–121, 1995.

[CGQ]   G. Cohen, S. Gaubert, and J.-P. Quadrat. Algebraic system analysis of timed Petri nets. Appears in [Gun97].

[CIL92]   M. G. Crandall, H. Ishii, and P.-L. Lions. User's guide to viscosity solutions of second order partial differential equations. *Bulletin of the AMS*, 27:1–67, 1992.

[CKR84]   Z.-Q. Cao, K. H. Kim, and F. W. Roush. *Incline Algebra and Applications*. Mathematics and its Applications. Ellis-Horwood, 1984.

[CMQV89]   G. Cohen, P. Moller, J.-P. Quadrat, and M. Viot. Algebraic tools for the performance evaluation of discrete event systems. *Proceedings of the IEEE*, 77(1):39–58, 1989.

[Con71]   J. H. Conway. *Regular Algebra and Finite Machines*. Chapman and Hall, 1971.

[CT80]   M. G. Crandall and L. Tartar. Some relations between nonexpansive and order preserving maps. *Proceedings of the AMS*, 78(3):385–390, 1980.

[DE95]   J. Desel and J. Esparza. *Free Choice Nets*, volume 40 of *Cambridge Tracts in Theoretical Computer Science*. Cambridge University Press, 1995.

[DJLC53]   M. L. Dubreil-Jacotin, L. Lesieur, and R. Croisot. *Leçons sur la Théorie des Treillis, des Structures Algébriques Ordonnées, et des Treillis Géométriques*. Gauthier Villars, 1953.

[dSa]   F. d'Alessandro and J. Sakarovitch. The finite power property for rational sets of a free group. Appears in [Gun97].

[DSb]   P. I. Dudnikov and S. N. Samborskiĭ. Endomorphisms of finitely generated free semimodules. Appears in [MS92].

[DZ93]   A. Dembo and O. Zeitouni. *Large Deviations Techniques and Applications*. Jones and Bartlett, 1993.

[ER95]   S. Even and S. Rajsbaum. Unison, canon and sluggish clocks in networks controlled by a synchronizer. *Mathematical Systems Theory*, 28:421–435, 1995.

[ES92]   Y. V. Egorov and M. A. Shubin, editors. *Partial Differential Equations I*, volume 30 of *Encyclopaedia of Mathematical Sciences*. Springer-Verlag, 1992.

[Eva84]   L. C. Evans. Some min-max methods for the Hamilton–Jacobi equation. *Indiana University Mathematics Journal*, 33:31–50, 1984.

[Fuc63]   L. Fuchs. *Partially Ordered Algebraic Systems*, volume 28 of *Pure and Applied Mathematics*. Pergamon, 1963.

[Gau]   S. Gaubert. On the Burnside problem for semigroups of matrices in the (max,+) algebra. To appear in Semigroup Forum.

[Gau92]   S. Gaubert. *Théorie des Systèmes Linéaires dans les dioïdes*. PhD thesis, École Nationale Supérieure des Mines de Paris, 1992.

[GJM]   B. Gaujal and A. Jean-Marie. Computational issues in recursive stochastic systems. Appears in [Gun97].

[GK90]   K. Goebel and W. A. Kirk. *Topics in Metric Fixed Point Theory*, volume 28 of *Cambridge Studies in Advanced Mathematics*. Cambridge University Press, 1990.

[GK95]   J. Gunawardena and M. Keane. On the existence of cycle times for some nonexpansive maps. Technical Report HPL-BRIMS-95-003, Hewlett-Packard Labs, 1995.

[GKS96]   J. Gunawardena, M. Keane, and C. Sparrow. Topical functions. In preparation, 1996.

[GMa]   S. Gaubert and J. Mairesse. Task resource models and (max,+) automata. Appears in [Gun97].

[GMb]   M. Gondran and M. Minoux. Linear algebra in dioids: a survey of recent results. Appears in [BCGZ84].

[GM84]   M. Gondran and M. Minoux. *Graphs and Algorithms*. Wiley-Interscience Series in Discrete Mathematics. John Wiley, 1984.

[Gol92]   J. S. Golan. *The Theory of Semirings with Applications in Mathematics and Theoretical Computer Science*, volume 54 of *Pitman Monographs and Surveys in Pure and Applied Mathematics*. Longman Scientific and Technical, 1992.

[Gona]   M. Gondran. Analyse MINMAX. To appear in *C. R. Acad. Sci. Paris*, 1996.

[Gonb]   M. Gondran. Analyse MINPLUS. To appear in *C. R. Acad. Sci. Paris*, 1996.

[GPL]   A. Gürel, O. C. Pastravanu, and F. L. Lewis. A system-theoretic approach for discrete-event control of manufacturing systems. Appears in [Gun97].

[Gro95]   Max-Plus Working Group. Max-plus algebra and applications to system theory and optimal control. In *Proceedings of the International Congress of Mathematicians, Zürich, 1994*. Birkhäuser, 1995.

[Gun93]   J. Gunawardena. Timing analysis of digital circuits and the theory of min-max functions. In *TAU'93, ACM International Workshop on Timing Issues in the Specification and Synthesis of Digital Systems*, September 1993.

[Gun94a]   J. Gunawardena. Cycle times and fixed points of min-max functions. In G. Cohen and J.-P. Quadrat, editors, *11th International Conference on Analysis and Optimization of Systems*, pages 266–272. Springer LNCIS 199, 1994.

[Gun94b]   J. Gunawardena. A dynamic approach to timed behaviour. In B. Jonsson and J. Parrow, editors, *CONCUR'94: Concurrency Theory*, pages 178–193. Springer LNCS 836, 1994.

[Gun94c]   J. Gunawardena. Min-max functions. *Discrete Event Dynamic Systems*, 4:377–406, 1994.

[Gun98]   J. Gunawardena, editor. *Idempotency*. Publications of the Isaac Newton Institute. Cambridge University Press, 1998. Proceedings of the 1994 BRIMS Workshop on Idempotency.

[GY94]   P. Glasserman and D. D. Yao. *Monotone Structure in Discrete Event Systems*. Wiley Series in Probability and Mathematical Statistics. John Wiley, 1994.

[Has79]   K. Hashiguchi. A decision procedure for the order of regular events. *Theoretical Computer Science*, 8:69–72, 1979.

[Hei95]   H. J. A. M. Heijmans. Mathematical morphology: a modern approach in image processing based on algebra and geometry. *SIAM Review*, 37(1):1–36, 1995.

[Ho89]   Y. C. Ho, editor. Special issue on Dynamics of Discrete Event Systems. *Proceedings of the IEEE*, 77(1), January 1989.

[Hop65]   E. Hopf. Generalized solutions of non-linear equations of first-order. *Journal of Mathematics and Mechanics*, 14:951–973, 1965.

[HU79]   J. E. Hopcroft and J. D. Ullman. *Introduction to Automata Theory, Languages and Computation*. Addison Wesley, 1979.

[Joh82]   P. T. Johnstone. *Stone Spaces*, volume 3 of *Studies in Advanced Mathematics*. Cambridge University Press, 1982.

[Joh83]   P. T. Johnstone. The point of pointless topology. *Bulletin of the American Mathematical Society*, 8(1):41–53, 1983.

[JT84]   A. Joyal and M. Tierney. An extension of the Galois theory of Grothendieck. *Memoirs of the AMS*, 51(309), 1984.

[Kle]   S. C. Kleene. Representation of events in nerve nets and finite automata. Appears in [SM56].

[KMa]   V. N. Kolokoltsov and V. P. Maslov. *Idempotent Analysis*. To be published by Kluwer.

[KMb]   V. N. Kolokoltsov and V. P. Maslov. A new differential equation for the dynamics of the Pareto sets. Appears in [Gun97].

[KM89]   V. N. Kolokoltsov and V. P. Maslov. Idempotent analysis as a tool of control theory and optimal synthesis I. *Functional Analysis and Applications*, 23(1):1–14, 1989.

[Kola]   V. N. Kolokoltsov. On linear, additive and homogeneous operators in idempotent analysis. Appears in [MS92].

[Kolb]   V. N. Kolokoltsov. The stochastic HJB equation and WKB method. Appears in [Gun97].

[Kro]   D. Krob. Some automata-theoretic aspects of min-max-plus semirings. Appears in [Gun97].

[KS86]   W. Kuich and A. Salomaa. *Semirings, Automata, Languages*, volume 5 of *EATCS Monographs on Theoretical Computer Science*. Springer-Verlag, 1986.

[Kui87]   W. Kuich. The Kleene and the Parikh theorem in complete semirings. In *Proceedings ICALP*, pages 212–225. Springer LNCS 267, 1987.

[Kun72]   J. Kuntzmann. *Théorie des Réseaux*. Dunod, 1972.

[Leu]   H. Leung. The topological approach to the limitedness problem on distance automata. Appears in [Gun97].

[LM]   G. L. Litvinov and V. P. Maslov. The correspondence principle for idempotent calculus and some computer applications. Appears in [Gun97].

[Mac71]   S. MacLane. *Categories for the Working Mathematician.* Graduate Texts in Mathematics. Springer-Verlag, 1971.

[Mas87a]   V. P. Maslov. *Méthodes Opératorielles.* Éditions MIR, 1987.

[Mas87b]   V. P. Maslov. On a new principle of superposition for optimization problems. *Russian Math Surveys*, 42(3):43–54, 1987.

[Min88]   H. Minc. *Nonnegative Matrices.* Wiley-Interscience Series in Discrete Mathematics and Optimization. John Wiley, 1988.

[MK94]   V. P. Maslov and V. N. Kolokoltsov. *Idempotent Analysis and its Application to Optimal Control.* Nauka, Moscow, 1994. In Russian.

[Mol88]   P. Moller. *Théorie Algébrique des Systèmes à Événements Discrets.* PhD thesis, École des Mines de Paris, 1988.

[Mora]   P. Del Moral. Maslov optimization theory: optimality versus randomness. Appears as an Appendix to [KMa].

[Morb]   P. Del Moral. Maslov optimization theory: topological aspects. Appears in [Gun97].

[MP]   G. Mascari and M. Pedicini. Types and dynamics in partially additive categories. Appears in [Gun97].

[MSa]   V. P. Maslov and S. N. Samborskiĭ. Stationary Hamilton–Jacobi and Bellman equations (existence and uniqueness of solutions). Appears in [MS92].

[MSb]   P. Del Moral and G. Salut. Maslov optimization theory. To appear in the *Russian Journal of Mathematical Physics*, 1996.

[MSc]   P. Del Moral and G. Salut. Random particle methods in (max, +) optimization problems. Appears in [Gun97].

[MS92]   V. P. Maslov and S. N. Samborskiĭ, editors. *Idempotent Analysis*, volume 13 of *Advances in Soviet Mathematics*. American Mathematical Society, 1992.

[Mul86]   C. J. Mulvey. &. *Rend. Circ. Mat. Palermo*, 12:99–104, 1986.

[Nus]   R. D. Nussbaum. Periodic points of nonexpansive maps. Appears in [Gun97].

[O'B96]   G. L. O'Brien. Sequences of capacities, with connections to large-deviation theory. *Journal of Applied Probability*, 9:19–35, 1996.

[OR88]   G. J. Olsder and C. Roos. Cramer and Cayley–Hamilton in max algebra. *Linear Algebra and its Applications*, 101:87–108, 1988.

[Pap91]   E. Pap. On non-addtive set functions. *Atti Sem. Mat. Fis. Univ. Modena*, XXXIX:345–360, 1991.

[Per90]   D. Perrin. Finite automata. In J. van Leeuwen, editor, *Handbook of Theoretical Computer Science Volume B*, pages 1–57. Elsevier, 1990.

[Pin]   J.-E. Pin. Tropical semirings. Appears in [Gun97].

[Rei68]   R. Reiter. Scheduling parallel computations. *Journal of the ACM*, 15(4):590–599, 1968.

[Rei85]   W. Reisig. *Petri Nets*, volume 4 of *EATCS Monographs on Theoretical Computer Science*. Springer-Verlag, 1985.

[RH80]   C. V. Ramamoorthy and G. S. Ho. Performance evaluation of asynchronous concurrent systems using Petri nets. *IEEE Transactions on Software Engineering*, SE-6(5):440–449, 1980.

[Rom67]   I. V. Romanovskiĭ. Optimization of stationary control of discrete deterministic process in dynamic programming. *Cybernetics*, 3, 1967.

[RS94]   S. Rajsbaum and M. Sidi. On the performance of synchronized programs in distributed networks with random processing times and transmission delays. *IEEE Transactions on Parallel and Distributed Systems*, 5:939–950, 1994.

[Sal73]   A. Salomaa. *Formal Languages*. Academic Press, 1973.

[Sam]   S. N. Samborskiĭ. The lagrange problem from the point of view of idempotent analysis. Appears in [Gun97].

[Shu]   M. A. Shubin. Algebraic remarks on idempotent semirings and the kernel theorem in spaces of bounded functions. Appears in [MS92].

[Sim78]   I. Simon. Limited subsets of a free monoid. In *Proceedings 19th IEEE Symposium on FOCS*, pages 143–150. IEEE, 1978.

[Sim88]   I. Simon. Recognizable sets with multiplicities in the tropical semiring. In *Proceedings MFCS*, pages 107–120. Springer LNCS 324, 1988.

[Sin90]   R. Sine. A nonlinear Perron–Frobenius theorem. *Proceedings of the AMS*, 109:331–336, 1990.

[SM56]   C. E. Shannon and J. McCarthy, editors. *Automata Studies*. Annals of Mathematics Studies. Princeton University Press, 1956.

[Spa96]   C. Sparrow. Existence of cycle time vectors for minmax functions of dimension 3. Technical Report HPL-BRIMS-96-008, Hewlett-Packard Labs, 1996.

[SS92]   T. Szymanski and N. Shenoy. Verifying clock schedules. In *Digest of Technical Papers of the IEEE International Conference on Computer-Aided Design of Integrated Circuits*, pages 124–131. IEEE Computer Society, 1992.

[ST]   S. N. Samborskiĭ and A. A. Tarashchan. The Fourier transform and semirings of Pareto sets. Appears in [MS92].

[Vic89]   S. Vickers. *Topology via Logic*, volume 5 of *Cambridge Tracts in Theoretical Computer Science*. Cambridge University Press, 1989.

[Vin]   J. M. Vincent. Some ergodic results on stochastic iterative DEDS. To appear in *Journal of Discrete Event Dynamics Systems*.

[Vor67]   N. N. Vorobyev. Extremal algebra of positive matrices. *Elektron Informationsverarbeitung und Kybernetik*, 3, 1967. In Russian.

[Wag]   E. Wagneur. The geometry of finite dimensional pseudomodules. Appears in [Gun97].

[Wag91]   E. Wagneur. Moduloids and pseudomodules. *Discrete Mathematics*, 98:57–73, 1991.

[WB]   E. A. Walkup and G. Borriello. A general linear max-plus solution technique. Appears in [Gun97].

[Wil91]   D. Williams. *Probability with Martingales*. Cambridge Mathematical Textbooks. Cambridge University Press, 1991.

[WXS90]   C. Wende, Q. Xiangdong, and D. Shuhui. The eigen-problem and period analysis of the discrete event system. *Systems Science and Mathematical Sciences*, 3:243–260, 1990.

[Zim81]   U. Zimmermann. *Linear and Combinatorial Optimization in Ordered Algebraic Structures*, volume 10 of *Annals of Discrete Mathematics*. North-Holland, 1981.

Tropical Semirings

Jean-Eric Pin

1 Introduction

It is a well-known fact that the boolean calculus is one of the mathematical foundations of electronic computers. This explains the important role of the boolean semiring in computer science. The aim of this paper is to present other semirings that occur in theoretical computer science. These semirings were named *tropical semirings* by Dominique Perrin in honour of the pioneering work of our Brazilian colleague and friend Imre Simon, but are also commonly known as (min, +)-semirings.

The aim of this paper is to present tropical semirings and to survey a few problems relevant to them. We shall try to give an updated status of the different questions, but detailed solutions of most problems would be too long and technical for this survey. They can be found in the other papers of this volume or in the relevant literature. We have tried to keep the paper self-contained as much as possible. Thus, in principle, there are no prerequisites for reading this survey, apart from a standard mathematical background. However, it was clearly not possible to give a full exposition of the theory of automata within 20 pages. Therefore, suitable references will be given for readers who would like to pursue the subject further and join the tropical community.

The paper is organized as follows. The main definitions are introduced in Section 2. Two apparently disconnected applications of tropical semirings are presented: the Burnside type problems in group and semigroup theory in Section 3, and decidability problems in formal language theory in Section 4. The connection between the two problems is explained in Section 5. A conclusion section ends the paper.

2 Mathematical Objects

This section is a short presentation of the basic concepts used in this paper.

2.1 Semigroups and monoids

A *semigroup* is a set equipped with an associative binary operation, usually denoted multiplicatively [11, 12, 24, 36]. Let S be a semigroup. An element

1 of S is an *identity* if, for all $s \in S$, $1s = s1 = s$. An element 0 of S is a *zero* if, for all $s \in S$, $0s = s0 = 0$. Clearly, a semigroup can have at most one identity, since, if 1 and $1'$ are two identities, then $11' = 1' = 1$. A *monoid* is a semigroup with identity. A semigroup S is commutative if, for all $s, t \in S$, $st = ts$. Given two semigroups S and T, a *semigroup morphism* $\varphi : S \to T$ is a map from S into T such that, for all $x, y \in S$, $\varphi(xy) = \varphi(x)\varphi(y)$. Monoid morphisms are defined analogously, but of course the condition $\varphi(1) = 1$ is also required.

An *alphabet* is a finite set whose elements are *letters*. A *word* (over the alphabet A) is a finite sequence $u = (a_1, a_2, \ldots, a_n)$ of letters of A. The integer n is the *length* of the word and is denoted $|u|$. In practice, the notation $(a_1, a_2, \ldots, a_n)$ is shortened to $a_1 a_2 \cdots a_n$. The empty word, which is the unique word of length 0, is denoted by 1. The (concatenation) *product* of two words $u = a_1 a_2 \cdots a_p$ and $v = b_1 b_2 \cdots b_q$ is the word $uv = a_1 a_2 \cdots a_p b_1 b_2 \cdots b_q$. The product is an associative operation on words. The set of all words on the alphabet A is denoted by A^*. Equipped with the product of words, it is a monoid, with the empty word as an identity. It is in fact the free monoid on the set A. This means that A^* satisfies the following universal property: if $\varphi : A \to M$ is a map from A into a monoid M, there exists a unique monoid morphism from A^* into M that extends φ. This morphism, also denoted φ, is simply defined by $\varphi(a_1 \cdots a_n) = \varphi(a_1) \cdots \varphi(a_n)$.

2.2 Semirings

A *semiring* is a set k equipped with two binary operations, denoted additively and multiplicatively, and containing two elements, the *zero* – denoted 0 – and the *unit* – denoted 1 – satisfying the following conditions:

1. k is a commutative monoid for addition, with the zero as identity

2. k is a monoid for multiplication, with the unit as identity

3. Multiplication is distributive over addition:
   for all $s, t_1, t_2 \in k$, $s(t_1 + t_2) = st_1 + st_2$ and $(t_1 + t_2)s = t_1 s + t_2 s$

4. The zero is a zero for the second law:
   for all $s \in k$, $0s = s0 = 0$.

A semiring is *commutative* if its multiplication is commutative. Rings are the first examples of semirings that come to mind. In particular, we denote by $\mathbb{Z}$, $\mathbb{Q}$ and $\mathbb{R}$, respectively, the rings of integers, rational and real numbers. The simplest example of a semiring which is not a ring is the boolean semiring $\mathbb{B} = \{0, 1\}$ defined by $0 + 0 = 0$, $0 + 1 = 1 + 1 = 1 + 0 = 1$, $1.1 = 1$ and $1.0 = 0.0 = 0.1 = 0$. If M is a monoid, then the set $\mathbb{P}(M)$ of subsets of M is a commutative semiring with union as addition and multiplication given by

$$XY = \{xy \mid x \in X \text{ and } y \in Y\}.$$

The empty set is the zero of this semiring and the unit is the singleton $\{1\}$. Other examples include the semiring of nonnegative integers $\mathbb{N} = (\mathbb{N}, +, \times)$ and its completion $\mathcal{N} = (\mathbb{N} \cup \{\infty\}, +, \times)$, where addition and multiplication are extended in the natural way:

$$\text{for all } x \in \mathcal{N}, \; x + \infty = \infty + x = \infty$$
$$\text{for all } x \in \mathcal{N} \setminus \{0\}, \; x \times \infty = \infty \times x = \infty$$
$$\infty \times 0 = 0 \times \infty = 0.$$

The Min-Plus semiring is $\mathcal{M} = (\mathbb{N} \cup \{\infty\}, \min, +)$. This means that in this semiring the sum is defined as the minimum and the product as the usual addition. Note that ∞ is the zero of this semiring and 0 is its unit. This semiring was introduced by Simon [43] in the context of automata theory (it is also a familiar semiring in Operations Research). Similar semirings have been considered in the literature. Mascle [32] introduced the semiring

$$\mathbb{P} = (\mathbb{N} \cup \{-\infty, \infty\}, \max, +)$$

where $-\infty + x = x + (-\infty) = -\infty$ for all x, and Leung [25, 26] the semiring

$$\overline{\mathcal{M}} = (\mathbb{N} \cup \{\omega, \infty\}, \min, +)$$

where the minimum is defined with respect to the order

$$0 < 1 < 2 < \cdots < \omega < \infty$$

and addition in the Min-Plus semiring is completed by setting $x + \omega = \omega + x = \max\{x, \omega\}$ for all x. All these semirings are called *tropical semirings*. Other extensions include the tropical integers $\mathcal{Z} = (\mathbb{Z} \cup \{\infty\}, \min, +)$, the tropical rationals $\mathcal{Q} = (\mathbb{Q} \cup \{\infty\}, \min, +)$ and the tropical reals $\mathcal{R} = (\mathbb{R} \cup \{\infty\}, \min, +)$. Mascle [31] also suggested studying the Min-Plus semiring of ordinals smaller than a given ordinal α : $\mathcal{M}_\alpha = (\{ordinals < \alpha\}, \min, +)$.

Quotients of $\mathcal{M}$ and $\mathcal{N}$ are also of interest. These quotients are

$$\mathcal{N}_r = \mathcal{N}/(r = \infty) \qquad \mathcal{N}'_{r,p} = \mathcal{N}/(r = r + p)$$
$$\mathcal{M}_r = \mathcal{M}/(r = \infty) \qquad \mathcal{M}'_r = \mathcal{M}/(r = r + 1)$$

where $(r = s)$ denotes the coarsest semiring congruence such that r and s are equivalent.

2.3 Polynomials and series

This subsection is inspired by the book of Berstel and Reutenauer [2], which is the standard reference on formal power series.

Let A be an alphabet and let k be a semiring. A *formal power series* over k with (noncommutative) variables in A is a mapping s from A^* to k. The value of s on a word w is denoted (s, w). The *range* of s is the set of words w such that $(s, w) \neq 0$. A *polynomial* is a power series of finite range. The set of power series over k with variables in A is denoted $k\langle\langle A\rangle\rangle$. It is a semiring with addition defined by

$$(s + t, w) = (s, w) + (t, w)$$

and multiplication defined by

$$(st, w) = \sum_{uv=w} (s, u)(s, v).$$

The set of polynomials, denoted $k\langle A\rangle$, forms a subsemiring of $k\langle\langle A\rangle\rangle$. If s is an element of k, one can identify s with the polynomial s defined by

$$(s, w) = \begin{cases} s & \text{if } w = 1 \\ 0 & \text{otherwise.} \end{cases}$$

The semiring k can thus be identified with a subsemiring of $k\langle A\rangle$. Similarly, one can identify A^* with a subset of $k\langle A\rangle$ by attaching to each word v the polynomial v defined by

$$(v, w) = \begin{cases} 1 & \text{if } v = w \\ 0 & \text{otherwise.} \end{cases}$$

A family of series $(s_i)_{i \in I}$ is *locally finite* if, for each $w \in A^*$, the set

$$I_w = \{i \in I \mid (s_i, w) \neq 0\}$$

is finite. In this case, the sum $s = \sum_{i \in I} s_i$ can be defined by

$$(s, w) = \sum_{i \in I_w} s_i.$$

In particular, for every series s, the family of polynomials $\big((s, w)w\big)_{w \in A^*}$ is clearly locally finite and its sum is s. For this reason, a series is usually denoted by the formal sum

$$\sum_{w \in A^*} (s, w)w.$$

If $(s, 1) = 0$, that is, if the value of s on the empty word is zero, then the family $(s^n)_{n \geq 0}$ is locally finite, since $(s^n, w) = 0$ for every $n > |w|$. Its sum is denoted s^* and is called the *star* of s. Thus

$$s^* = \sum_{n \geq 0} s^n.$$

Note that if k is a ring, $s^* = (1 - s)^{-1}$. Actually, the star often plays the role of the inverse, as in the following example. Consider the equation in X

$$X = t + sX \qquad (2.1)$$

where s and t are series and $(s, 1) = 0$. Then one can show that $X = ts^*$ is the unique solution of 2.1.

The set of *rational series* on k is the smallest subsemiring R of $k\langle\langle A\rangle\rangle$ containing $k\langle A\rangle$ and such that $s \in R$ implies $s^* \in R$. Note that if k is a ring, the rational series form the smallest subring of $k\langle\langle A\rangle\rangle$ containing $k\langle A\rangle$ and closed under inversion (whenever defined). In particular, in the one-variable case, this definition coincides with the usual definition of rational series and justifies the terminology.

2.4   Rational sets

Given a monoid M, the semiring $\mathcal{P}(M)$ can be identified with $\mathbb{B}(M)$, the boolean algebra of the monoid M. Thus union will be denoted by $+$ and the empty set by 0. It is also convenient to denote simply by m any singleton $\{m\}$. In particular, 1 will denote the singleton $\{1\}$, which is also the unit of the semiring $\mathcal{P}(M)$.

Given a subset X of M, X^* denotes the submonoid of M generated by X. Note that

$$X^* = \sum_{n \geq 0} X^n$$

where X^n is defined by $X^0 = 1$ and $X^{n+1} = X^n X$. Thus our notation is consistent with the notation s^* used for power series. It is also consistent with the notation A^* used for the free monoid over A. The *rational subsets* of M form the smallest class $\mathcal{R}at(M)$ of subsets of M such that

1. the empty set and every singleton $\{m\}$ belong to $\mathcal{R}at(M)$

2. if S and T are in $\mathcal{R}at(M)$, then so are ST and $S + T$

3. if S is in $\mathcal{R}at(M)$, then so is S^*.

In particular, every finite subset and every finitely generated submonoid of M are rational sets.

The case of free monoids is of special interest. Subsets of a free monoid A^* are often called *languages*. According to the general definition, the rational languages form the smallest class of languages containing the finite languages and closed under union, product and star. A key result of the theory, which follows from a theorem of Kleene mentioned in the next section, is that rational languages are also closed under intersection and complement. A similar result holds for the rational subsets of a free group, but not for the rational subsets of an arbitrary monoid.

Rational languages can be conveniently represented by *rational expressions*. Rational expressions on the alphabet A are defined recursively by the rules:

1. 0, 1 and a, for each $a \in A$, are rational expressions

2. if e and f are rational expressions, then so are e^*, (ef) and $(e+f)$.

For instance, if $a, b \in A$, $(a + ab)^* ab$ denotes the rational subset of A^* consisting of all elements of the form $a^{n_1}(ba)^{m_1} a^{n_2}(ba)^{m_2} \cdots a^{n_k}(ba)^{m_k} ab$, where $k \geq 0$ and $n_1, m_1, n_2, m_2, \ldots, n_k, m_k \geq 0$. It contains for instance the elements ab (take $k = 0$), $aaaab$ (take $k = 1$, $n_1 = 3$ and $m_1 = 0$) and $ababaaabaab$ (exercise!).

Two rational expressions e and f are *equivalent* ($e \equiv f$) if they denote the same rational language. For instance, if e and f are rational expressions, $e + e \equiv e$, $(e^*)^* \equiv e^*$ and $(e + f) \equiv (f + e)$, but there are much more subtle equivalences, such as $(e^* f)^* e^* \equiv (e + f)^*$. Actually, although there are known algorithms to decide whether two rational expressions are equivalent, there is no finite basis of identities of the type above that would generate all possible equivalences.

Let a^* be the free monoid on the one-letter alphabet $\{a\}$. One can show that for each rational subset R of a^*, there exist two integers i (the index) and p (the period) such that

$$R = F + G(a^p)^*$$

for some $F \subseteq \{1, a, \ldots, a^{i-1}\}$ and $G \subset \{a^i, \ldots a^{i+p-1}\}$. C. Choffrut observed that the rational sets of the form $a^n a^*$ (for $n \geq 0$) form a subsemiring of $\mathcal{R}at(a^*)$ isomorphic to $\mathcal{M}$, since $a^n a^* + a^m a^* = a^{\min\{n,m\}} a^*$ and $(a^n a^*)(a^m a^*) = a^{n+m} a^*$. Thus the tropical semiring embeds naturally into $\mathcal{R}at(a^*)$.

3 Burnside type Problems

In 1902, Burnside proposed the following problem:

Is a finitely generated group satisfying an identity of the form $x^n = 1$ necessarily finite?

The answer is yes for $n = 1, 2, 3, 4$ and 6. The case $n \leq 2$ is trivial, the case $n = 3$ was settled by Burnside [7], the case $n = 4$ by Sanov [41] and the case $n = 6$ by M. Hall [14]. Although the original problem finally received a negative answer by Novikov and Adyan in 1968 [34] (see also [3]), several related questions were proposed. At the end of this century, Burnside type problems form a very active but extremely difficult research area, recently promoted by the award of a Fields medal to the Russian mathematician E.I. Zelmanov [52, 53, 54]. Burnside type problems can also be stated for semigroups and motivate the following definitions.

A semigroup S is *periodic* (or *torsion*) if, for all $s \in S$, the subsemigroup generated by s is finite. This means that, for every $s \in S$, there exist $n, p > 0$ such that $s^n = s^{n+p}$. A semigroup is *k-generated* if it is generated by a set of k elements. It is *finitely generated* if it is k-generated for some positive integer k.

A semigroup S is *locally finite* if every finitely generated subsemigroup of S is finite. It is *strongly locally finite* if there is an *order function* f such that the order of every k-generated subsemigroup of S is less than or equal to $f(k)$.

The general Burnside problem is the following:

Is every periodic semigroup locally finite?

Morse and Hedlund [33] observed that the existence of an infinite square-free word over a three-letter alphabet [50, 51, 28] shows that the quotient of $A^* \cup \{0\}$ by the relations $x^2 = 0$ is infinite if $|A| \geq 3$. This semigroup satisfies the identity $x^2 = x^3$ and thus the answer is negative for semigroups. Actually, as shown in [6], the monoid given by $\langle A \mid x^2 = x^3 \text{ for all } x \in A^* \rangle$ is infinite even if $|A| = 2$. Note that, however, the semigroup given by $\langle A \mid x = x^2 \text{ for all } x \in A^* \rangle$ is always finite.

For groups, a negative answer was given by Golod in 1964 [13] (this follows also from the result of Novikov and Adyan mentioned above). On the positive side, Schur [42] gave a positive answer for groups of matrices over $\mathbb{C}$. Kaplansky [23] extended this result to groups of matrices over an arbitrary field and Procesi [39, 40] to groups of matrices over a commutative ring or even over a PI-ring, i.e. a ring satisfying a polynomial identity. McNaughton and Zalcstein [29] proved a similar result for semigroups of matrices over an arbitrary field. In the same paper, they announced but did not prove a similar statement for semigroups of matrices over a commutative ring or even over a PI-ring. A complete proof of these results, which do not rely on the group case, was given by Straubing [49] in 1983.

What happens for semigroups of matrices over a commutative semiring? The general question is still unsolved, but several particular instances of this problem have occurred naturally in automata theory. Mandel and Simon [30] proved that every periodic semigroup of matrices over $\mathbb{N}$ or $\mathcal{N}$ is strongly locally finite. Then Simon [43] proved that every periodic semigroup of matrices over $\mathcal{M}$ is locally finite. This result was extended by Mascle [31] to semigroups of matrices over $\mathbb{P}$ and over $\mathcal{R}at(a^*)$.

One of the key results for the study of locally finite semigroups is Brown's theorem [4, 5].

Theorem 3.1 (Brown) *Let $\varphi : S \to T$ be a semigroup morphism. If T is locally finite and, for every idempotent $e \in T$, $\varphi^{-1}(e)$ is locally finite, then S is locally finite.*

A similar result for strongly locally finite semigroups was given by Straubing [49].

Theorem 3.2 (Straubing) *Let $\varphi : S \to T$ be a semigroup morphism. If T is strongly locally finite with order function f and if, for every idempotent $e \in T$, $\varphi^{-1}(e)$ is strongly locally finite with order function g (not depending on e), then S is strongly locally finite.*

Two other problems on semigroups of matrices over a semiring can also be considered as Burnside type problems:

Finiteness problem *Given a finite set A of matrices, decide whether the semigroup S generated by A is finite or not.*

Finite section problem *Given a finite set A of square matrices of size n and $i, j \in \{1, \ldots, n\}$, decide whether the set $\{s_{i,j} \mid s \in S\}$ is finite or not, where S denotes the semigroup generated by A.*

The finiteness problem is decidable for matrices over a field (Jacob [22]), over $\mathbb{N}$ and $\mathcal{N}$ (Mandel and Simon [30]), over $\mathcal{M}$ (Simon [43]) and over $\mathbb{P}$ and $\mathcal{R}at(a^*)$ (Mascle [31, 32]). The finite section problem is decidable for matrices over a field (Jacob [22]), over $\mathbb{N}$ and $\mathcal{N}$ (Mandel and Simon [30]) and over $\mathcal{M}$ (Hashiguchi [16, 20]). It is still an open problem for matrices over $\mathcal{R}at(a^*)$.

These problems were first considered by Hashiguchi [15, 16] and Simon [43, 44] in connection with the decidability problems on rational languages presented in the next section.

4 Problems on Rational Languages

The *star height* of a rational expression, as defined by Eggan [10], counts the number of nested uses of the star operation. It is defined inductively as follows:

1. The star height of the basic languages is 0. Formally

$$h(0) = 0 \qquad h(1) = 0 \quad \text{and} \quad h(a) = 0 \text{ for every letter } a$$

2. Union and product do not affect star height. If e and f are two rational expressions, then

$$h(e + f) = h(ef) = \max\{h(e), h(f)\}$$

3. Star increases star height. For each rational expression e,

$$h(e^*) = h(e) + 1.$$

For instance

$$\left((a^* + ba^*)^* + (b^*ab^*)^*\right)^*(b^*a^* + b)^*$$

is a rational expression of star height 3. Now, the *star height* of a recognizable language L is the minimum of the star heights of the rational expressions representing L

$$h(L) = \min\{h(e) \mid e \text{ is an rational expression for } L \}.$$

The difficulty in computing the star height is that a given language can be represented in many different ways by a rational expression.

An explicit example of a language of star height n was given by Dejean and Schützenberger [9]. Given a word $u \in A^*$ and a letter $a \in A$, denote by $|u|_a$ the number of occurrences of a in u. For instance, if $u = abbabba$, $|u|_a = 3$ and $|u|_b = 4$. Let $A = \{a, b\}$ and let

$$L_n = \{u \in A^* \mid |u|_a \equiv |u|_b \bmod 2^{n-1}\}.$$

Theorem 4.1 (Dejean and Schützenberger) *For each $n \geq 1$, the language L_n is of star height n.*

It is easy to see that the languages of star height 0 are the finite languages, but the effective characterization of the other levels was left open for several years until Hashiguchi first settled the problem for star height 1 [17] and a few years later for the general case [19].

Theorem 4.2 (Hashiguchi) *There is an algorithm to determine the star height of a given rational language.*

Hashiguchi's solution for star height 1 is now well understood, and deeply relies on the solution of the finite section problem for matrices over $\mathcal{M}$. Hashiguchi's solution for arbitrary star height relies on a complicated induction, which makes the proof very difficult to follow. Let us mention another problem, the solution of which had a great influence on the theory and ultimately led to the solution of the star-height problem.

A language L has the *finite power property* (FPP for short) if there exists an integer k such that

$$X^* = 1 + X + X^2 + \cdots + X^k.$$

This means that X^* is actually a polynomial in X and in particular $h(X^*) = h(X)$. For instance, $X = a^* + (a + b)^*b$ has the FPP, since $X^* = A^* = 1 + X + X^2$, but $X = a^*(1 + b)$ does not. In 1966, Brzozowski proposed the following problem.

FPP problem *Decide whether a given rational language has the FPP.*

A solution was given independently by Simon [43] and Hashiguchi [15]. Simon's proof reduces the problem to the finiteness problem of matrices over $\mathcal{M}$. This reduction will be outlined in Section 6.

To conclude this section, let us mention yet another problem on rational languages. Let $\mathcal{R}$ be a set of languages. A language L belongs to the *polynomial closure of* $\mathcal{R}$ if it is a finite union of products of languages of $\mathcal{R}$. For instance, if $\mathcal{R} = \{R_1, R_2\}$ then $R_1 + R_2R_1R_2 + R_2R_2$ belongs to the polynomial closure of $\mathcal{R}$. The following problem was proposed by Hashiguchi [18].

Polynomial closure problem *Given a finite set $\mathcal{R}$ of rational languages and a rational language R, decide whether R belongs to the polynomial closure of $\mathcal{R}$.*

Note that the FPP problem is a particular instance of this problem. Indeed, the FPP problem amounts to deciding whether, given a rational language L, L^* belongs to the polynomial closure of the set $\{1, L\}$. It was shown by Hashiguchi [18] that the polynomial closure problem reduces to the finite section problem for matrices over $\mathcal{M}$ and is therefore decidable. See also [37] for a survey.

A brief introduction to finite automata and formal languages is in order to explain the connection between the FPP problem and the Burnside type problems of Section 3.

5 Finite Automata and Recognizable Sets

This section is a brief introduction to the theory of finite automata. A more extensive presentation can be found in [11, 35, 36, 38].

5.1 Finite automata

A *finite (nondeterministic) automaton* is a quintuple $\mathcal{A} = (Q, A, E, I, F)$ where Q is a finite set (the set of *states*), A is an alphabet, E is a subset of $Q \times A \times Q$, called the set of *edges* (also called *transitions*) and I and F

are subsets of Q, called the set of *initial* and *final* states, respectively. Two edges (p, a, q) and (p', a', q') are *consecutive* if $q = p'$. A *path* in $\mathcal{A}$ is a finite sequence of consecutive edges

$$e_0 = (q_0, a_0, q_1), \quad e_1 = (q_1, a_1, q_2), \quad \ldots, \quad e_{n-1} = (q_{n-1}, a_{n-1}, q_n)$$

also denoted

$$q_0 \xrightarrow{a_0} q_1 \xrightarrow{a_1} q_2 \cdots q_{n-1} \xrightarrow{a_{n-1}} q_n.$$

The state q_0 is the *origin* of the path, the state q_n is its *end*, and the word $x = a_0 a_1 \cdots a_{n-1}$ is its *label*. It is convenient to have also, for each state q, an empty path of label 1 from q to q. A path in $\mathcal{A}$ is *successful* if its origin is in I and its end is in F.

The language *recognized* by $\mathcal{A}$ is the set, denoted $|\mathcal{A}|$, of the labels of all successful paths of $\mathcal{A}$. A language X is *recognizable* if there exists a finite automaton $\mathcal{A}$ such that $X = |\mathcal{A}|$. Two automata are said to be *equivalent* if they recognize the same language. Automata are conveniently represented by labelled graphs, as in the example below. Incoming arrows indicate initial states and outgoing arrows indicate final states.

Example. Let $\mathcal{A} = (\{1, 2\}, \{a, b\}, E, \{1\}, \{2\})$ be an automaton, with $E = \{(1, a, 1), (1, b, 1), (1, a, 2)\}$. The path $(1, a, 1)(1, b, 1)(1, a, 2)$ is a successful path of label aba. The path $(1, a, 1)(1, b, 1)(1, a, 1)$ has the same label but is unsuccessful since its end is 1.

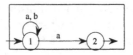

An automaton.

The set of words accepted by $\mathcal{A}$ is $|\mathcal{A}| = A^* a$, the set of all words ending with an a.

Kleene's theorem states the equivalence between automata and rational expressions. Its proof can be found in most books on automata theory [11, 21].

Theorem 5.1 (Kleene) *A language is rational if and only if it is recognizable.*

An automaton is *deterministic* if it has exactly one initial state, usually denoted q_0, and if E contains no pair of edges of the form $(q, a, q_1), (q, a, q_2)$ with $q_1 \neq q_2$.

The forbidden pattern in a deterministic automaton.

In this case, each letter a defines a partial function from Q to Q, which associates with every state q the unique state qa, if it exists, such that $(q, a, qa) \in E$. This can be extended into a right action of A^* on Q by setting, for every $q \in Q$, $a \in A$ and $u \in A^*$:

$$q1 = q$$
$$q(ua) = \begin{cases} (qu)a & \text{if } qu \text{ and } (qu)a \text{ are defined} \\ \text{undefined} & \text{otherwise.} \end{cases}$$

Then the language accepted by $\mathcal{A}$ is

$$|\mathcal{A}| = \{u \in A^* \mid q_0 u \in F\}.$$

It can be shown that every finite automaton is equivalent to a deterministic one. This result has an important consequence.

Corollary 5.1 *Recognizable languages are closed under union, intersection and complementation.*

States which cannot be reached from the initial state or from which one cannot access any final state are clearly useless. This leads to the following definition. A deterministic automaton $\mathcal{A} = (Q, A, E, q_0, F)$ is *trim* if for every state $q \in Q$ there exist two words u and v such that $q_0 u = q$ and $qv \in F$. It is not difficult to see that every deterministic automaton is equivalent to a trim one.

Let $\mathcal{A} = (Q, A, E, q_0, F)$ and $\mathcal{A}' = (Q', A, E', q'_0, F')$ be two deterministic automata. A *covering* from $\mathcal{A}$ onto $\mathcal{A}'$ is a surjective function $\varphi : Q \to Q'$ such that $\varphi(q_0) = q'_0$, $\varphi^{-1}(F') = F$ and, for every $u \in A^*$ and $q \in Q$, $\varphi(qu) = \varphi(q)u$. We denote $\mathcal{A}' \leq \mathcal{A}$ if there exists a covering from $\mathcal{A}$ onto $\mathcal{A}'$. This defines a partial order on deterministic automata. One can show that, amongst the trim deterministic automata recognizing a given recognizable language L, there is a minimal one for this partial order. This automaton is called the *minimal automaton* of L. Again, there are standard algorithms for minimizing a given finite automaton [21].

5.2 Transducers

The modelling power of finite automata can be enriched by adding an output function [1, 11]. Let k be a semiring. The definition of a *k-transducer* (or *automaton with output in k*) is quite similar to that of a finite automaton. It is also a quintuple $\mathcal{A} = (Q, A, E, I, F)$, where Q (resp. I, F) is the set of states (resp. initial and final states) and A is the alphabet. But the set of edges E, instead of being a subset of $Q \times A \times Q$, is a finite subset of

$Q \times A \times k \times Q$. An edge $(q, a, x, q') \in Q \times A \times k \times Q$ is graphically represented by

$$q \xrightarrow{a \,|\, x} q'.$$

The *output* of a path

$$q_0 \xrightarrow{a_1 \,|\, x_1} q_1 \xrightarrow{a_2 \,|\, x_2} q_2 \cdots \xrightarrow{a_k \,|\, x_k} q_k$$

is the product $x_1 x_2 \cdots x_k$. The output $\|\mathcal{A}\| w$ of a word w is the sum of the outputs of all successful paths of label w. If there is no successful path of label w the output is 0^1. This defines a function $\|\mathcal{A}\|$ from A^* into k, called the *output function* of $\mathcal{A}$.

Example. Let $k = \mathcal{M}$ and let $\mathcal{A} = (\{1, 2, 3\}, \{a, b\}, E, \{1\}, \{2, 3\})$, with $E = \{(1, a, 0, 1), (1, a, 2, 2), (2, b, 5, 2), (1, a, 1, 3), (3, b, 0, 1), (3, a, 3, 2)\}$.

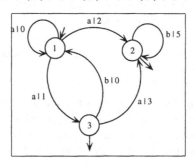

An automaton with output.

The label of the path $1 \xrightarrow{a \,|\, 1} 3 \xrightarrow{b \,|\, 0} 1 \xrightarrow{a \,|\, 0} 1 \xrightarrow{a \,|\, 2} 2 \xrightarrow{b \,|\, 5} 2$ is 8. There are three successful paths of label aaa:

$$1 \xrightarrow{a \,|\, 0} 1 \xrightarrow{a \,|\, 0} 1 \xrightarrow{a \,|\, 2} 2, \; 1 \xrightarrow{a \,|\, 0} 1 \xrightarrow{a \,|\, 0} 1 \xrightarrow{a \,|\, 1} 3 \text{ and } 1 \xrightarrow{a \,|\, 0} 1 \xrightarrow{a \,|\, 1} 3 \xrightarrow{a \,|\, 3} 2.$$

Therefore the output of aaa is $\|\mathcal{A}\|(aaa) = \min \{2, 1, 4\} = 1$.

5.3   Matrix representation

It is convenient to compute the output function by using matrices. Let $\mathcal{A} = (Q, A, E, I, F)$ be a k-transducer. The set $M_Q k$ of $Q \times Q$ matrices over the semiring k forms a semiring under the usual addition and multiplication of matrices, defined by

$$\begin{aligned} (m + m')_{p,q} &= m_{p,q} + m'_{p,q} \\ (m m')_{p,q} &= \textstyle\sum_{r \in Q} m_{p,r} m'_{r,q}. \end{aligned}$$

---

[1]This is consistent with the standard convention $\sum_{i \in \emptyset} x_i = 0$

Define a monoid morphism $\mu : A^* \to M_Q k$ by setting, for each $a \in A$,

$$\mu(a)_{p,q} = \sum_{(p,a,x,q)\in E} x$$

where, according to a standard convention, $\sum_{x\in\emptyset} x = 0$. Finally, let λ be the row matrix defined by

$$\lambda_q = \begin{cases} 1 & \text{if } q \in I \\ 0 & \text{otherwise} \end{cases}$$

and let ν be the column matrix defined by

$$\nu_q = \begin{cases} 1 & \text{if } q \in F \\ 0 & \text{otherwise.} \end{cases}$$

Then the output function is computed by the following fundamental formula

$$||A||w = \lambda\mu(w)\nu.$$

Example. The matrix representation of the transducer of Example 5.2 is given by[2]

$$\mu(a) = \begin{pmatrix} 0 & 2 & 1 \\ \infty & \infty & \infty \\ \infty & 3 & \infty \end{pmatrix} \qquad \mu(b) = \begin{pmatrix} \infty & \infty & \infty \\ \infty & 5 & \infty \\ 0 & \infty & \infty \end{pmatrix}.$$

Therefore

$$\mu(aaa) = \begin{pmatrix} 0 & 2 & 1 \\ \infty & \infty & \infty \\ \infty & \infty & \infty \end{pmatrix}.$$

The vectors λ and ν are given by

$$\lambda = \begin{pmatrix} 0 & \infty & \infty \end{pmatrix} \qquad \nu = \begin{pmatrix} \infty \\ 0 \\ 0 \end{pmatrix}.$$

Thus the output of aaa, given by $\lambda\mu(aaa)\nu$, is equal to

$$\begin{pmatrix} 0 & \infty & \infty \end{pmatrix} \begin{pmatrix} 0 & 2 & 1 \\ \infty & \infty & \infty \\ \infty & \infty & \infty \end{pmatrix} \begin{pmatrix} \infty \\ 0 \\ 0 \end{pmatrix} = \begin{pmatrix} 0 & 2 & 1 \end{pmatrix} \begin{pmatrix} \infty \\ 0 \\ 0 \end{pmatrix} = 1.$$

---

[2]The slight ambiguity in the role of the symbol 0 may confuse the reader. Here the semiring is the tropical semiring, its zero is ∞ and its unit is 0.

6   Reduction of the FPP Problem

In this section, we briefly outline the reduction of the FPP problem to the finiteness problem for semigroups of matrices over $\mathcal{M}$. Since a language L has the FPP if and only if $(L\backslash\{1\})^*$ has the FPP, one may assume that L does not contain the empty word. Next, by a simple construction, left to the reader, one may assume that L is recognized by an automaton $\mathcal{A} = (Q, A, E, \{1\}, F)$ with a unique initial state 1 and no edge arriving in this initial state, as in the example below:

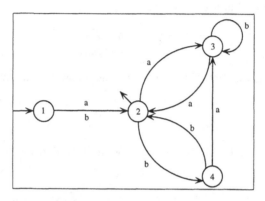

The automaton $\mathcal{A}$.

We claim that an automaton $\mathcal{A}'$ recognizing L^* is obtained by taking 1 as the unique final state and by adding an edge $(s, a, 1)$ for each edge $(s, a, q) \in E$ such that $a \in A$ and $q \in F$. In our example, one would add the edges $(1, a, 1)$, $(1, b, 1)$, $(3, a, 1)$ and $(4, a, 1)$. Let us first verify that every word of L^* is accepted by $\mathcal{A}'$. A word of L^* is a product $u = u_1 \cdots u_k$ of words of L. Since $\mathcal{A}$ does not accept the empty word, each u_is is the label of some nonempty successful path p_i, whose last edge reaches a final state. Replace this last edge (s, a, q), with $q \in F$, by $(s, a, 1)$. One gets a path p'_i from 1 to 1 and the product $p'_1 \cdots p'_k$ is a successful path of label u. Therefore u is accepted by $\mathcal{A}'$.

Conversely, every successful path can be factorized as a product of elementary paths around 1. Necessarily, the last edge of such an elementary path is one of the new edges $(s, a, 1)$ of $\mathcal{A}'$. Thus there is an edge of the form (s, a, q) such that $q \in F$. Therefore the label of the elementary path belongs to L and the label of the full path to L^*. Thus $\mathcal{A}'$ recognizes exactly L^*.

Actually, the previous argument shows that a word belongs to L^k if and only if it is the label of a path of $\mathcal{A}'$ containing exactly k new edges. Therefore, one can convert $\mathcal{A}'$ into a $\mathcal{M}$-transducer whose output on a word $u \in L^*$ is the smallest k such that $u \in L^k$. It suffices to have an output 0 on the edges of $\mathcal{A}$ and output 1 on the new edges. This can be interpreted as a cost to pay to go back to the initial state. In our example, one obtains the following transducer

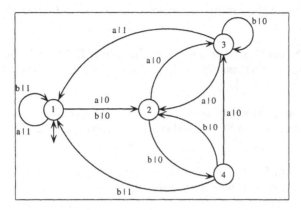

The automaton $\mathcal{B}$.

Now, if $w \in L^*$ then $||\mathcal{B}||w$ is exactly the least k such that $w \in L^k$; otherwise $||\mathcal{B}||w = \infty$. Thus L has the FPP if and only if the image of the function $||\mathcal{B}||$ is finite. Since $||\mathcal{B}||w = \lambda\mu(w)\nu = \mu_{1,1}(w)$, the equivalence of the first two conditions of the following statement has been established.

Theorem 6.1 *Let $\mathcal{A}$ be a finite automaton and L be the language recognized by $\mathcal{A}$. The following conditions are equivalent:*

1. L has the FPP

2. the associated semigroup of matrices has a finite section in $(1,1)$

3. the associated semigroup of matrices is finite.

The equivalence with the third condition is left as an exercise for the reader. It follows from the fact that all edges with output 1 arrive in state 1.

7 Conclusion

The examples presented in this paper do not exhaust the problems on semigroups or languages connected with tropical semirings, and the reader is invited to read the literature on this domain, in particular the recent article of Simon [48]. Roughly speaking, tropical semirings provide an algebraic setting to decide whether a collection of objects is finite or infinite. But, as illustrated by the FPP problem, it is usually a nontrivial task to reduce a given problem to a proper algebraic formulation.

References

[1] J. Berstel, 1979, *Transductions and Context-Free Languages*, Teubner, Stuttgart.

[2] J. Berstel and C. Reutenauer, 1984, *Les Séries Rationnelles et leurs Langages*, Masson. English edition, 1988 *Rational Series and Their Languages*, Springer-Verlag, Berlin.

[3] J.L. Britton, the existence of infinite Burnside groups, in W. W. Boone, F. B. Cannonito and R. C. Lyndon (eds.), *Word problems*, North Holland, 67–348.

[4] T. C. Brown (1969), On Van der Waerden's theorem on arithmetic progressions, *Notices Amer. Math. Soc.* **16**, 245.

[5] T. C. Brown (1971), An interesting combinatorial method in the theory of locally finite semigroups, *Pacific J. Math.* **36**, 285–289.

[6] J. A. Brzozowski, K. Čulik II and A. Gabrielian (1971), Classification of noncounting events, *J. Comput. Syst. Sci.* **5**, 41–53.

[7] W. Burnside (1902), On an unsettled question in the theory of discontinuous groups, *Q. J. Pure Appl. Math.* **33**, 230–238.

[8] Chan and Ibarra (1983), On the finite-valuedness problem for sequential machines, *Theoretical Comput. Sci.* **23**, 95–101.

[9] F. Dejean and M. P. Schützenberger (1966), On a question of Eggan, *Information and Control* **9**, 23–25.

[10] L. C. Eggan (1963), Transition graphs and the star height of regular events, *Michigan Math. J.* **10**, 385–397.

[11] S. Eilenberg (1974), *Automata, Languages and Machines*, Vol. A, Academic Press, New York.

[12] S. Eilenberg (1976), *Automata, Languages and Machines*, Vol. B, Academic Press, New York.

[13] E. S. Golod (1964), On nil algebras and finitely approximable groups, *Izv. Acad. Nauk SSSR Ser. Matem.* **28**, 273–276.

[14] M. Hall Jr. (1957), Solution of the Burnside problem for exponent 6, *Proceedings Nat. Acad. Sci. USA*, **43**, 751–753.

[15] K. Hashiguchi (1979), A decision procedure for the order of regular events, *Theoretical Comput. Sci.* **8**, 69–72.

[16] K. Hashiguchi (1982), Limitedness theorem on finite automata with distance functions, *J. Comput. System Sci.* **24**, 233–244.

[17] K. Hashiguchi (1982), Regular languages of star height one, *Information and Control* **53**, 199–210.

[18] K. Hashiguchi (1983), Representation theorems on regular languages, *J. Comput. System Sci.* **27**, 101–115.

[19] K. Hashiguchi (1988), Algorithms for determining relative star geight and star height, *Information and Computation* **78**, 124–169.

[20] K. Hashiguchi (1990), Improved limitedness theorems on finite automata with distance functions *Theoretical Comput. Sci.* **72**.

[21] J. E. Hopcroft and J. D. Ullman (1979), *Introduction to Automata Theory, Languages and Computation*, Addison Wesley.

[22] G. Jacob (1978), La finitude des représentations linéaires de semi-groupes est décidable, *Journal of Algebra* **52**, 437–459.

[23] I. Kaplansky (1965), *Fields and Rings*, University of Chicago.

[24] G. Lallement (1979), *Semigroups and Combinatorial Applications*, Wiley, New York.

[25] H. Leung (1987), *An algebraic method for solving decision problems in finite automata theory*, PhD thesis, Department of Computer Science, The Pennsylvania State University.

[26] H. Leung (1988), On the topological structure of a finitely generated semigroup of matrices, *Semigroup Forum* **37**, 273–287.

[27] M. Linna (1973), Finite power property of regular languages, in *Automata, Languages and Programming*, M. Nivat (ed.), North-Holland, Amsterdam, 87–98.

[28] M. Lothaire (1983), *Combinatorics on Words*, Encyclopedia of Mathematics and its Applications **17**, Addison Wesley.

[29] R. McNaughton and Y. Zalcstein (1975), The Burnside problem for semigroups, *Journal of Algebra* **34**, 292–299.

[30] A. Mandel and I. Simon (1977), On finite semigroups of matrices, *Theoretical Comput. Sci.* **5**, 101–112.

[31] J.P. Mascle (1985), Quelques résultats de décidabilité sur la finitude des semigroupes de matrices, LITP Report 85-50.

[32] J.P. Mascle (1986), Torsion matrix semigroups and recognizable transductions, in L. Kott ed., *Automata, Languages and Programming, Lecture Notes in Computer Science* **226**, 244–253.

[33] M. Morse and G. Hedlund (1944), Unending chess, symbolic dynamics and a problem in semigroups, *Duke Math. J.* **11**, 1–7.

[34] P.S. Novikov and S.I. Adyan (1968), On infinite periodic groups, I, II, III, *Izv. Acad. Nauk SSSR Ser. Matem.* **32**, 212–244, 251–524, 709–731.

[35] D. Perrin (1990), *Automata*, Chapter 1 in Van Leeuwen, J. ed., *Handbook of Theoretical Computer Science, Vol B: Formal Models and Semantics*, Elsevier.

[36] J.-E. Pin (1984), *Variétés de langages formels*, Masson, Paris; English translation: (1986), *Varieties of formal languages*, Plenum, New York.

[37] J.-E. Pin (1990), Rational and recognizable langages, in Ricciardi (ed.), *Lectures in Applied Mathematics and Informatics*, Manchester University Press, 62–106.

[38] J.-E. Pin (1993), Finite semigroups and recognizable languages: an introduction, in J. Fountain and V. Gould (eds.), *NATO Advanced Study Institute Semigroups, Formal Languages and Groups*, Kluwer academic publishers, to appear.

[39] C. Procesi (1966), The Burnside problem, *J. of Algebra* **4**, 421–425.

[40] C. Procesi (1973), *Rings with Polynomial Identities*, Marcel Dekker.

[41] I.N. Sanov (1940), Solution of the Burnside problem for exponent 4, *Uch. Zapiski Leningrad State University, Ser. Matem.* **10**, 166–170.

[42] I. Schur (1911), Über Gruppen periodischer Substitutionen, *Sitzungsber. Preuss. Akad. Wiss.*, 619–627.

[43] I. Simon (1978), Limited subsets of a free monoid, in *Proc. 19th Annual Symposium on Foundations of Computer Science*, Piscataway, N.J., Institute of Electrical and Electronics Engineers, 143–150.

[44] I. Simon (1988) Recognizable sets with multiplicities in the tropical semiring, in Chytil, Janiga and Koubek (eds.), *Mathematical Foundations of Computer Science, Lecture Notes in Computer Science* **324**, Springer Verlag, Berlin, 107–120.

[45] I. Simon (1990), Factorization forests of finite height, *Theoretical Comput. Sci.* **72**, 65–94.

[46] I. Simon (1990), The nondeterministic complexity of a finite automaton, in M. Lothaire (ed.), *Mots – mélanges offerts à M.P. Schützenberger*, Hermes, Paris, 384–400.

[47] I. Simon (1993), The product of rational languages, *Proceedings of ICALP 1993, Lecture Notes in Computer Science* **700**, 430–444.

[48] I. Simon (1994), On semigroups of matrices over the tropical semiring, *Informatique Théorique et Applications* **28**, 277–294.

[49] H. Straubing (1983), The Burnside problem for semigroups of matrices, in L.J. Cummings (ed.), *Combinatorics on Words, Progress and Perspectives*, Acad. Press, 279–295.

[50] A. Thue (1906), Über unendliche Zeichenreihen, *Norske Vid. Selsk. Skr. I. Math. Nat. Kl.*, Christiania **7**, 1–22.

[51] A. Thue (1912), Über die gegenseitige Lage gleicher Teile gewisser Zeichenreihen, *Norske Vid. Selsk. Skr. I. Math. Nat. Kl.*, Christiania **1**, 1–67.

[52] E.I. Zelmanov (1990), The solution of the restricted Burnside problem for groups of odd exponent, *Izvestia Akad. Nauk SSSR* **54**.

[53] E.I. Zelmanov (1991), The solution of the restricted Burnside problem for 2-groups, *Mat. Sbornik* **182**.

[54] E.I. Zelmanov (1993), On additional laws in the Burnside problem on periodic groups, *International Journal of Algebra and Computation* **3**, 583–600.

Some Automata-Theoretic Aspects of Min–Max–Plus Semirings

Daniel Krob

Abstract

This paper is devoted to the survey of some automata-theoretic aspects of different exotic semirings, i.e. semirings whose underlying set is some subset of $\mathbb{R}$ equiped with min, max or + as sum and/or product. We here address three types of properties related to rational series with multiplicities in such semirings: structure of supports, decidability of equality and inequality problems, and Fatou properties.

1   Introduction

Min–max–plus computations are used in several areas. These techniques appeared initially in the seventies in the context of Operations Research for analyzing discrete event systems (cf. Chapter 3 of [5]; see also [1] for a survey of these aspects of the theory). In another direction, the (min / max, +) semirings were also used in mathematical physics in the study of several partial differential equations which, like the Hamilton–Jacobi equation, appeared to be (min, +)-linear (see for instance the last chapter of Maslov's book [9]). It is also interesting to observe that similar objects were studied for Artificial Intelligence purposes: the fuzzy calculus involves indeed essentially (min, max) semirings (see [3] for more details and extensive references on this area).

More recently, the min–max–plus techniques have also appeared in formal language theory: the so-called tropical semiring, i.e. $\mathcal{M} = (\mathbb{N} \cup \{+\infty\}, \min, +)$, played indeed a central role in the study and solution of the finite power problem for rational languages (which is the problem of deciding whether the star of a given rational language L is equal to some finite union of iterated concatenations of L; cf. [6]). Moreover, the same semiring can be used for studying other fundamental notions of automata theory such as nondeterminism (cf [11]) or infinite behaviour of finite automata (see [10] for a survey of the use of $\mathcal{M}$ in the context of automata theory).

All these results motivated several studies of rational series with multiplicities in min–max–plus semirings in order to more deeply understand how these algebraic structures interact with automata. The purpose of this short paper is to give a survey of some aspects of these studies. We first present some structural results on the supports of the rational series that appear in these contexts. We then deal with equality problems for automata with costs in these semirings. Finally we adress some Fatou questions for these algebraic structures.

2 Preliminaries

2.1 Exotic semirings

There are several semirings whose underlying set is some subset of the real numbers equipped with min, max or + as sum and/or product. Among them, we shall especially consider in this paper the following:[1]

- the *tropical semiring* $\mathcal{M} = (\mathbb{N} \cup \{+\infty\}, \min, +)$,

- the *polar semiring* $\mathcal{N} = (\mathbb{N} \cup \{-\infty\}, \max, +)$,

- the *K-tropical semiring* $\mathcal{M}_K = (K \cup \{+\infty\}, \min, +)$,

- the *fuzzy semiring* $\mathcal{F} = (\mathbb{N} \cup \{+\infty\}, \min, \max)$,

- the *K-fuzzy semiring* $\mathcal{F}_K = (K \cup \{+\infty\}, \min, \max)$,

where K denotes some subset of $\mathbb{R}$ stable under min, + or max according to the case which is considered (typically $K = \mathbb{Z}$, $\mathbb{Q}$ or $\mathbb{R}$).

We can of course define a K-polar semiring $\mathcal{N}_K = (K \cup \{-\infty\}, \max, +)$, but the reader will easily check that $\mathcal{N}_K$ is isomorphic to $\mathcal{M}_{-K}$ where $-K$ denotes the set $\{-k, \ k \in K\}$, an effective isomorphism between these two semirings being realized by the mapping $x \longrightarrow -x$. Note that $\mathcal{N}$ can therefore be identified with the subsemiring of $\mathcal{M}_{\mathbb{Z}}$ (also called sometimes the *equatorial semiring*) based on $\mathbb{N}^- \cup \{+\infty\}$. The following picture summarizes the inclusion relations between all these structures (an arrow denotes here an inclusion).

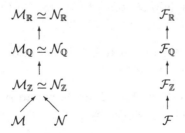

---

[1] In all these definitions, the first (resp. the second) operation is always to be understood as the sum (resp. the product) of the considered semiring.

2.2   Automata with multiplicities in a semiring

Let $\mathcal{K}$ be a semiring and let A be an alphabet. We then denote by $\mathcal{K}\langle\langle A \rangle\rangle$ the $\mathcal{K}$-algebra of formal power series over A with multiplicities in $\mathcal{A}$. An element S of $\mathcal{K}\langle\langle A \rangle\rangle$ is usually represented by a formal sum of the form

$$S = \sum_{w \in A^*} (S|w)\, w \;,$$

where $(S|w) \in \mathcal{K}$ denotes the coefficient (or the multiplicity) of the series S on the word $w \in A^*$. The sum and product by an element of $\mathcal{K}$ are defined componentwise on $\mathcal{K}\langle\langle A \rangle\rangle$, which is also equipped by a product defined with the usual Cauchy rule. One can also define the star of a proper series S (i.e. a series with $(S|1) = 0$) by the relation

$$S^* = \sum_{n=0}^{+\infty} S^n \;.$$

The $\mathcal{K}$-algebra of $\mathcal{K}$-*rational series* is then the smallest sub-$\mathcal{K}$-algebra, denoted $\mathcal{K}Rat(A)$, of $\mathcal{K}\langle\langle A \rangle\rangle$ that contains all letters $a \in A$ and that is stable under star (of proper series).

Let us now recall that a $\mathcal{K}$-representation of order n of the free monoid A^* is just a monoid morphism of A^* into the monoid of square matrices of order n with entries in $\mathcal{K}$. A K-representation μ of A^* is therefore completely defined by the images $(\mu(a))_{a \in A}$ of the letters $a \in A$.

A $\mathcal{K}$-*automaton* of order n is then a triple (I, μ, T) where μ is a $\mathcal{K}$-representation of order n of A^* and where I and T are respectively a row and a column vector of order n with entries in $\mathcal{K}$. One can represent graphically any $\mathcal{K}$-automaton $\mathcal{A} = (I, \mu, T)$ of order n by a graph $\mathcal{G}(\mathcal{A})$ defined as follows (see also Figure 1):

- the set of vertices of $\mathcal{G}(\mathcal{A})$ is $[1, n]$;

- for every letter $a \in A$ and every pair (i, j) of vertices in $[1, n]$, there is an oriented edge in $\mathcal{G}(\mathcal{A})$ going from i to j and labelled by the pair $\mu(a)_{i,j}\, a \in \mathcal{K} \times A$;

- for every vertex $i \in [1, n]$, there is an input-arrow labelled by $I_i \in \mathcal{K}$ that points onto the vertex i;

- for every vertex $i \in [1, n]$, there is an output-arrow labelled by $T_i \in \mathcal{K}$ that issues from the vertex i.

By convention, one does not represent an edge or arrow when the scalar $\mu(a)_{i,j}$, I_i or T_j of $\mathcal{K}$ that labels it is equal to 0.

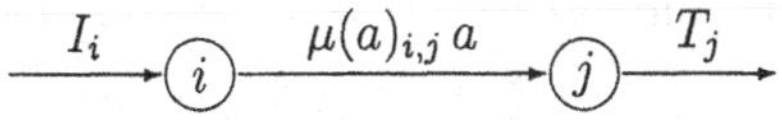

Figure 1

A series $S \in \mathcal{K}\langle\langle A\rangle\rangle$ is then said to be $\mathcal{K}$-*recognizable* if and only if there exists a $\mathcal{K}$-automaton $\mathcal{A} = (I, \mu, T)$ such that $(S|w) = I\,\mu(w)\,T$ for every word $w \in A^*$ (one says that the series S is recognized by $\mathcal{A}$). Let us also recall the Kleene–Schützenberger theorem, which states that a series of $\mathcal{K}\langle\langle A\rangle\rangle$ is $\mathcal{K}$-recognizable iff it is $\mathcal{K}$-rational. The interested reader may refer to [2] or [4] for more details.

Let us end this subsection by a last definition. If L is any language of A^*, its characteristic series in $\mathcal{K}\langle\langle A\rangle\rangle$ is the series $\underline{L}$ defined by

$$(\underline{L}|w) = \begin{cases} 1_{\mathcal{K}} & \text{if } w \in L, \\ 0_{\mathcal{K}} & \text{if } w \notin L, \end{cases}$$

for every word w of A^*.

3 Structure of Rational Series

3.1 k-supports of a series

Let $\mathcal{K}$ be a semiring and let k be an element of $\mathcal{K}$. If S is a series of $\mathcal{K}\langle\langle A\rangle\rangle$, the k-*support* of S is the language of A^* defined by

$$\text{supp}(S, k) = \{\, w \in A^*,\ (S|w) = k \,\}.$$

Observe that one can always write

$$S = \bigoplus_{k \in \mathcal{K}} k \otimes \underline{\text{supp}(S, k)}$$

when $\mathcal{K}$ denotes one of the exotic semirings considered in Section 2.1.

3.2 The K-tropical semiring case

The following proposition is folklore (cf. [7] for instance). It shows that the structures of k-supports of rational series with multiplicities in (min/max, +) semirings based on $\mathbb{N}$ are always rational.

Proposition 3.1 *Let S be a rational series with multiplicities in $\mathcal{M}$ or $\mathcal{N}$. Then all k-supports of S are effective rational languages of A^*.*

On the other hand, the situation is completely different when one allows negative integers in the support of a K-tropical semiring.

Proposition 3.2 (Krob; [8]) *Let K be a subset of $\mathbb{R}$ containing $\{-1,0,1\}$ and stable under* min *and* $+$. *Then one has*

1. *If $|A| = 1$, every k-support of an $\mathcal{M}_K$-rational series is an effective rational one-letter language;*

2. *If $|A| \geq 2$, there exists an $\mathcal{M}_K$-rational series whose 0-support is non-recursive.*

3.3 The K-fuzzy semiring case

In the K-fuzzy semiring case, one can give a very explicit description of $\mathcal{F}_K$-rational series.

Proposition 3.3 *Let K be a subset of $\mathbb{R}$ stable under* min *and* max. *Then the following two assertions are equivalent:*

1. *S is an $\mathcal{F}_K$-rational series.*

2. *There exists a finite family $(k_i)_{i=1,N}$ of elements of K and a finite family $(L_i)_{i=1,N}$ of disjoint effective rational languages such that*

$$(S|w) = \begin{cases} k_i & \text{if } w \in L_i \text{ with } i = 1, N , \\ +\infty & \text{if } w \notin \cup_{i=1,N} L_i . \end{cases} \tag{3.1}$$

Proof Let S be an $\mathcal{F}_K$ rational series. Then one has

$$(S|w) = \min_{k_1,\ldots,k_{n+1}} \max (I_{k_1}, (\mu(a_i)_{k_i,k_{i+1}})_{i=1,n}, T_{k_{n+1}})$$

for every word $w = a_1 \cdots a_n$ of A^*. It follows clearly from this last relation that $(S|w)$ can only take a finite number of values which are necessarily entries of the matrices $\mu(a)$ or of the vectors I and T. Let $F = \{k_1,\ldots,k_N\}$ be the set of these values. To arrive at our result, it clearly suffices to prove that $\text{supp}(S,k)$ is an effective rational language for every $k \in F$.

Let $k \in F$. Let us consider the semiring $\mathcal{G} = (\{-\infty,0,+\infty\},\min,\max)$. We can define a projection π_k of $\mathcal{F}_K$ onto $\mathcal{G}$ by the following rules

$$\pi_k(x) = \begin{cases} -\infty & \text{if } x <_\mathbb{R} k , \\ 0 & \text{if } x = k , \\ +\infty & \text{if } x >_\mathbb{R} k . \end{cases}$$

The reader will then easily check that the k-support of the series S is exactly equal to the 0-support of the series $\pi_k(S)$ (which is recognized by any automaton $\pi_k(\mathcal{A})$ obtained from an automaton $\mathcal{A}$ that recognizes S by taking the projections of all entries in the matrices $\mu(a)$ and in the vectors I and T).

We can now consider the two projections π^+ and π^- of $\mathcal{G}$ into the boolean semiring $\mathcal{B} = (\{0,1\}, +, \cdot)$ (where $1 + 1 = 1$) defined by

$$\left\{ \begin{array}{l} \pi^+(-\infty) = \pi^+(0) = 0, \ \pi^+(+\infty) = 1 \ , \\ \pi^-(-\infty) = 0, \ \pi^-(0) = \pi^-(+\infty) = 1 \ . \end{array} \right.$$

It is not difficult to see that one has

$$\left\{ \begin{array}{l} \mathrm{supp}(\pi_k(S), -\infty) = \mathrm{supp}(\pi^-(\pi_k(S)), 0) \ , \\ \mathrm{supp}(\pi_k(S), +\infty) = \mathrm{supp}(\pi^+(\pi_k(S)), 1) \ . \end{array} \right.$$

These two last identities clearly show that the $+\infty$ and the $-\infty$ supports of $\pi_k(S)$ are effective rational languages. It follows that

$$\mathrm{supp}(\pi_k(S), 0) = A^* - (\mathrm{supp}(\pi_k(S), -\infty) \cup \mathrm{supp}(\pi_k(S), +\infty))$$

is also an effective rational language. This ends our proof. $\qquad\qquad\square$

Note 3.4 The last proposition says equivalently that every $\mathcal{F}_K$ rational series S can be written as

$$S = \min_{i=1}^{N} \ \max(k_i, \underline{L_i}) \ ,$$

where $(L_i)_{i=1,N}$ is a disjoint family of rational languages and where the elements k_i are constants of K.

4 Decidability Questions

4.1 Equality and inequality problems

Let $\mathcal{K}$ be a semiring totally ordered by the order $<_\mathcal{K}$. Several classical decision problems concerning $\mathcal{K}$-rational series can then be stated.

1. *Equality problem:*

    Instance $S, T \in \mathcal{K}Rat(A)$

    Problem Does the following property hold?

    $$\forall \ w \in A^*, \ (S|w) = (T|w) \qquad\qquad (Eq)$$

2. *Inequality problem*:

   Instance $S, T \in \mathcal{K}Rat(A)$

   Problem Does the following property hold?

   $$\forall\ w \in A^*,\ (S|w) \leq_{\mathcal{K}} (T|w) \qquad\qquad (Ineq)$$

3. *Local equality problem*:

   Instance $S, T \in \mathcal{K}Rat(A)$

   Problem Does the following property hold?

   $$\exists\ w \in A^*,\ (S|w) = (T|w) \qquad\qquad (LocalEq)$$

4. *Local inequality problem*:

   Instance $S, T \in \mathcal{K}Rat(A)$

   Problem Does the following property hold?

   $$\exists\ w \in A^*,\ (S|w) \leq_{\mathcal{K}} (T|w) \qquad\qquad (LocalIneq)$$

When $\mathcal{K}$ is an arbitrary semiring, these decidability problems are in general not connected. For instance, when $\mathcal{K}$ is the semiring $\mathbb{N}$ of integers (equipped with the usual sum and product), the equality problem is decidable when the inequality and the local inequality problems are undecidable (cf. [4] for instance). In the K-tropical case, all these problems are in fact related with respect to decidability in the following way

$$(Eq) \Longleftrightarrow (Ineq) \Longleftrightarrow (LocalIneq) \Longleftarrow (LocalEq)\ ,$$

an arrow $P \Longrightarrow Q$ meaning here that the decidability of problem P implies the decidability of problem Q.

Moreover it can be shown that the decidability status of the equality problems for $\mathcal{M}$, $\mathcal{N}$ and $\mathcal{M}_{\mathbb{Z}}$ is the same. In other words, these three equality problems are necessarily all decidable or all undecidable.

4.2   The K-tropical semiring case

We can now state the following main undecidability result for rational series with multiplicities in the equatorial semiring.

Theorem 4.1 (Krob; [8]) *Let A be an alphabet with $|A| \geq 2$. The following problem is then undecidable:*

   Instance $S \in \mathcal{M}_{\mathbb{Z}}Rat(A)$

   Problem *Does the following property hold?*

   $$\exists\ w \in A^*,\ (S|w) \geq 0$$

Sketch of the proof It is first easy to prove that the decidability status of the considered problem does not depend on the cardinality of A (when $|A| \geq 2$). Let then $A = \{a, b, c, d\}$ be a four letter alphabet. One can construct an encoding which associates in an injective way a word $w(\underline{n})$ of A^* with every vector $\underline{n} = (n_1, \ldots, n_N)$ of $\mathbb{N}^N$. Using this encoding, one can associate with every polynomial $P \in \mathbb{Z}[x_1, \ldots, x_N]$ an $\mathcal{M}_\mathbb{Z}$-rational series S_P over A which has the following properties :

$$(S_P|w) \begin{cases} \leq -1 & \text{if } w \neq w(\underline{n}) \text{ ,} \\ = -|P(n_1, \ldots, n_N)| & \text{if } w = w(\underline{n}) \text{ .} \end{cases}$$

These properties show that one has

$$\exists (n_1, \ldots, n_N) \in \mathbb{N}^N, \ P(n_1, \ldots, n_N) = 0 \iff \exists w \in A^*, (S|w) \geq 0 \text{ .}$$

These relations show that Hilbert's 10th problem can be reduced to the restricted local inequality problem for $\mathcal{M}_\mathbb{Z}$-rational series considered here. This implies immediately that this last problem is undecidable. □

Corollary 4.2 *The equality, inequality, local equality and local inequality problems for rational series with multiplicities in $\mathcal{M}$ and $\mathcal{N}$ are*

1. *decidable when $|A| = 1$;*

2. *undecidable when $|A| \geq 2$.*

Note 4.3 (1) All the above decidability problems are still undecidable even if the entries of the considered automata only belong to $\{0, 1, +\infty\}$.

(2) All the above decidability problems become decidable if one only considers $\mathbb{N}$-MinMax series, i.e. series of $\mathbb{N}\langle\!\langle A \rangle\!\rangle$ which are both $\mathcal{M}$- and $\mathcal{N}$-rational. It is an open problem to characterize such series.

4.3 The K-fuzzy semiring case

On the other hand, the fuzzy case does not contain any technical difficulty.

Proposition 4.4 *Let K be any subset of $\mathbb{R}$ stable under min and max. Then the equality, inequality, local equality and local inequality problems are decidable for rational series with multiplicities in $\mathcal{F}_K$.*

Proof This is an obvious consequence of Proposition 3.3. □

5   Fatou Results

5.1   Fatou extensions

Let $\mathcal{K} \subset \mathcal{L}$ be two semirings. The semiring $\mathcal{L}$ is then said to be a *Fatou extension* of the semiring $\mathcal{K}$ if and only if every $\mathcal{L}$-rational series of $\mathcal{K}\langle\!\langle A \rangle\!\rangle$ is a $\mathcal{K}$-rational series, i.e. if and only if one has

$$\mathcal{L}Rat(A) \cap \mathcal{K}\langle\!\langle A \rangle\!\rangle \;=\; \mathcal{K}Rat(A) \;.$$

The reader will find more details on Fatou extensions in [2].

5.2   The K-tropical semiring case

One can give the following Fatou properties of $(\min/\max, +)$ semirings which can be proved by a simple modification of the argument used in [7].

Proposition 5.1 *Let K be a subset of $\mathbb{R}$ stable under* min *and* $+$. *Let $K^+ = K \cap \mathbb{R}^+$ and $K^- = K \cap \mathbb{R}^-$. We suppose moreover that $K^+ = -K^-$. Then the two following properties hold:*

1. *$\mathcal{M}_K$ is an effective Fatou extension of $\mathcal{M}_{K^+}$;*

2. *$\mathcal{M}_K$ is not a Fatou extension of $\mathcal{M}_{K^-}$.*

5.3   The K-fuzzy semiring case

As for our previous results, the fuzzy case is immediate.

Proposition 5.2 *Let $K \subset L$ be two subsets of $\mathbb{R}$ stable under* min *and* max. *Then $\mathcal{F}_L$ is a Fatou extension of $\mathcal{F}_K$.*

Proof This property is a simple consequence of Proposition 3.3.   □

References

[1] F. Baccelli, G. Cohen, G.J. Olsder, J.P. Quadrat (1992), *Synchronization and linearity – An algebra for discrete event systems*, Wiley.

[2] J. Berstel, C. Reutenauer (1986), *Rational series and their languages*, Springer.

[3] D. Dubois, H. Prade (1980), *Fuzzy sets and systems*, Academic Press.

[4] S. Eilenberg (1974), *Automata, languages and machines*, Vol. A, Academic Press.

[5] M. Gondran, M. Minoux (1979), *Graphes et algorithmes*, Eyrolles.

[6] K. Hashigushi (1982), 'Limitedness theorem on finite automata with distance functions', J. of Comput. and Syst. Sci., **24**, (2), 233–244.

[7] D. Krob (1994), 'Some consequences of a Fatou property of the tropical semiring', *J. of pure and appl. alg.*, **93**, 231–249.

[8] D. Krob (1994), 'The equality problem for rational series with multiplicities in the tropical semiring is undecidable', *J. of Alg. and Comput.*, **4**, (3), 405–425.

[9] V. Maslov (1973), *Méthodes opérationnelles*, MIR.

[10] I. Simon (1988), 'Recognizable sets with multiplicities in the tropical semiring', Proceedings of MFCS'88 (M.P. Chytil et al., Eds.), Lect. Notes in Comput. Sci., **324**, 107–120, Springer.

[11] I. Simon (1990), 'The nondeterministic complexity of finite automata', Mots (M. Lothaire, Ed.), 384–400, Hermès.

The Finite Power Property for Rational Sets of a Free Group

Flavio d'Alessandro and
Jacques Sakarovitch

1 Introduction

As already explained in this volume, the original problem that eventually led people working in automata theory to use and study idempotent semirings was posed by J. Brzozowki in 1966 at the 7th SWAT Conference: he raised the question whether the finite power property was decidable for a *rational* set of a free monoid A^*. Recall that a subset L of A^* has the *finite power property* if there exists an integer n such that

$$L^* = L^{\leq n}$$

where $L^{\leq n}$ naturally denotes the set $1 \cup L \cup L^2 \cup \cdots \cup L^n$.

The problem was shown to be decidable in 1978 independently by K. Hashiguchi [6] and I. Simon [12], the solution of the latter author being based on the decidability of the finiteness of a monoid of matrices with coefficients in the *tropical semiring*. Clearly the finite power property may be stated for any family of (effectively defined) subsets of a free monoid or even more generally of any monoid (in which multiplication and star of subsets are effective operations). In [9] it has been shown that the problem is undecidable for the class of context-free languages, the class next to the one of rational languages in the classical "Chomsky hierarchy" of language families. In this paper, we address, and solve, the problem for rational sets of a free group. We prove indeed the following.

Theorem 1 *It is decidable whether a rational set of a free group has the finite power property or not.*

It should be noted that the family of rational subsets of the free group defines a family of deterministic context-free languages for which the finite power property is thus decidable. On the other hand the techniques involved allow us to extend the result from the free group to a whole class of monoids obtained as the quotient of a free monoid by congruences generated by a Thue system of certain type.

2   Distance Automata and the Finite Power Property

2.1   Representation of finite automata

We basically follow the definitions and notations of [4] and [11] for automata and for rational and recognizable sets, as they have been recalled in the contribution of J.-E. Pin in this volume [10]. The *label* of a computation c of an automaton $\mathcal{A} = (Q, A, E, I, T)$ is denoted by $|c|$. The *behaviour* of $\mathcal{A}$ is the subset $|\mathcal{A}|$ of A^* consisting of the labels of the successful computations of $\mathcal{A}$. An automaton $\mathcal{A} = (Q, A, E, I, T)$ can be equivalently described by a *representation* (λ, μ, ν) where $\mu : A^* \to \mathbb{B}^{Q \times Q}$ is a morphism from A^* into the monoid of Boolean matrices of dimension Q, and where λ and ν are two Boolean row and column vectors of dimension Q. We then have

$$|\mathcal{A}| = \{f \in A^* \mid \lambda \cdot \mu(f) \cdot \nu = 1\}.$$

2.2   Distance automata

In the sequel, and as in [10], we denote by $\mathcal{M} = (\mathbb{N} \cup \infty; \oplus, \otimes)$ the idempotent "min, plus" semiring. A *distance automaton* over a free monoid A^* is an automaton over the alphabet A with multiplicities in the semiring $\mathcal{M}$; that is to say, an automaton $\mathcal{A} = (Q, A, E, I, T)$ equipped with a (distance) mapping $\sigma : E \to \mathcal{M}$: every edge (p, a, q) in E is given a multiplicity or a *coefficient* $(p, a, q)\sigma$ in $\mathcal{M}$.

We follow the notations of [13]: for a computation c in $\mathcal{A}$, its *label*, denoted by $|c|$, is the product of letters of its edges; and the *multiplicity* of c, denoted by $\|c\|$, is the product of multiplicities of its edges; that is, since we are in $\mathcal{M}$, the sum of integers that are the coefficients of the edges. As for (classical) automata we denote by $|\mathcal{A}|$ the set of labels of successful computations. The *behaviour* of $\mathcal{A}$, denoted by $\|\mathcal{A}\|$, is a mapping $\|\mathcal{A}\| : A^* \to \mathcal{M}$, *i.e.* a series of $\mathcal{M}\langle\langle A \rangle\rangle$. For every word f of A^*, its multiplicity $f\|\mathcal{A}\|$ is the sum, that is — since we are in $\mathcal{M}$ — the *minimum*, of the multiplicities $\|c\|$ for all successful computations c, the label $|c|$ of which is equal to f.

A distance automaton $\mathcal{A}$ can be equivalently described by a *representation* (η, κ, ζ) where $\kappa : A^* \to \mathcal{M}^{Q \times Q}$ is a morphism from A^* into the monoid of matrices of dimension Q with entries in $\mathcal{M}$, and where η and ζ are two row and column vectors of dimension Q with entries in $\mathcal{M}$. For every f in A^* we then have $f\|\mathcal{A}\| = \eta \cdot \kappa(f) \cdot \zeta$.

Note that every classical automaton $\mathcal{A}$ with representation (λ, μ, ν) is easily and canonically turned into a distance automaton, denoted again by $\mathcal{A}$, with representation (η, κ, ζ): $0_{\mathbb{B}}$ is replaced by $0_{\mathcal{M}} = \infty$ and $1_{\mathbb{B}}$ is replaced by

$1_{\mathcal{M}} = 0$ in the matrices and vectors. It then holds that for every f in $|\mathcal{A}|$, $f\|\mathcal{A}\| = 0$ (and for every f not in $|\mathcal{A}|$, $f\|\mathcal{A}\| = \infty$).

A distance automaton $\mathcal{A}$ is said to be *bounded* if there exists an integer M such that, for every $f \in |\mathcal{A}|$, $f\|\mathcal{A}\| < M$. And the following result has already been recalled several times.

Theorem 2 (Hashiguchi [7], [8]) *It is decidable whether a distance automaton is bounded or not.*

2.3 Intersection

The intersection of two recognizable subsets of A^* is a recognizable set. This result can be generalized via the *Hadamard product* to behaviour of distance automata, that is to recognizable series over A^* with coefficients in $\mathcal{M}$. The Hadamard product $s \odot t$ of two series in $\mathcal{M}\langle\!\langle A^* \rangle\!\rangle$ is the series defined by

$$\forall f \in A^*, \ <s \odot t, f> = <s, f> \otimes <t, f> = <s, f> + <t, f>$$

and the following result holds.

Theorem 3 (Schützenberger [14]) *Let $\mathbb{K}$ be a commutative semiring. Then the Hadamard product of two recognizable series s and t with coefficients in $\mathbb{K}$ is a recognizable series, the representation of which is effectively constructible from the representation of s and t.*

As an immediate application of this result we have the following.

Proposition 1 *Let $\mathcal{A}$ be a distance automaton over A^* and K be a recognizable set of A^*. Then a distance automaton $\mathcal{B}$ is effectively constructible such that $|\mathcal{B}| = |\mathcal{A}| \cap K$ and for every $f \in |\mathcal{B}|$ $f\|\mathcal{B}\| = f\|\mathcal{A}\|$.*

Proof   Let $\mathcal{K}$ be the classical automaton which recognizes K and turn it into a distance automaton as indicated in the previous section. It is then immediate to check that $\|\mathcal{B}\| = \|\mathcal{A}\| \odot \|\mathcal{K}\|$. The construction of $\mathcal{B}$ is effective from the fact that the previous theorem is effective.   □

2.4 Finite power property

Let L be a language of A^* and let f be a word in L^*. We define the *order* of f with respect to L, $o(f)$, to be the least integer n such that f is in L^n:

$$o(f) = \min\{n \in \mathbb{N} \mid f \in L^n\}.$$

In other words, L has the finite power property if and only if there exists a bound for the order of every word of L^*. Let us recall the construction, due to I. Simon [12], that, given an automaton $\mathcal{A}$ which recognizes L, builds the distance automaton $\mathcal{S}_L$ that computes the order of every word of L^*.

Proposition 2 [12] *Let L be a recognizable set of A^* accepted by an automaton $\mathcal{A}$. Then a distance automaton $\mathcal{S}_L$ is effectively constructible from $\mathcal{A}$ with the property that $|\mathcal{S}_L| = L^*$ and $f\|\mathcal{S}_L\| = o(f)$.*

Proof Starting from $\mathcal{A} = (Q, A, E, I, T)$ the classical construction first gives $\mathcal{S}_L = (Q', A, E', I', T')$ that recognizes L^*: let q be a state not in Q, $Q' = Q \cup q$ and $I' = T' = \{q\}$. The edge set E' of $\mathcal{S}_L$ is defined by

$$\begin{aligned} E' \;=\; & E \;\cup\; \{(q, a, q) \mid (i, a, t) \in E, i \in I, t \in T\} \\ & \cup \{(q, a, p) \mid (i, a, p) \in E, i \in I\} \;\cup\; \{(p, a, q) \mid (p, a, t) \in E, t \in T\}. \end{aligned}$$

We then turn $\mathcal{S}_L$ into a distance automaton with the distance $\sigma : E' \to \mathcal{M}$ defined by the following:

$$\sigma(i, a, j) = \begin{cases} \infty & \text{if} \quad (i, a, j) \notin E' \\ 0 & \text{if} \quad (i, a, j) \in E' \quad \text{and} \quad j \neq q \\ 1 & \text{if} \quad (i, a, j) \in E' \quad \text{and} \quad j = q. \end{cases}$$

It then follows, by induction on the length of w, that $w\|\mathcal{S}_L\| = o(w)$. $\quad\square$

From Theorem 2 or using a direct proof relying on the special form of the distance automaton constructed above, the following holds.

Theorem 4 (Hashiguchi [6], Simon [12]) *It is decidable whether a given rational subset of a free monoid has the finite power property or not.*

3 Rational Sets in a Free Group

3.1 Reduction in the free group $F(A)$

We fix here the notations to deal with the elements of the free group and we recall a combinatorial property. Let A be a finite alphabet, A^{-1} a disjoint copy of A and let $B = A \cup A^{-1}$. It is known that $F(A)$, the free group with base A, is the quotient of B^* by the congruence generated by the relations $\{aa^{-1} = 1_{B^*} \mid a \in B\}$. Let us denote by $\alpha : B^* \to F(A)$ the canonical morphism defined by this congruence. A word of B^* is called *reduced* if it does not contain any factor of the form aa^{-1} with $a \in B$. Every element w of B^* is congruent to a unique reduced word, denoted by $\rho(w)$, and this defines a mapping $\rho : B^* \to B^*$ called *Dyck's reduction*. Since, moreover, $\rho(u) = \rho(v)$ implies $\alpha(u) = \alpha(v)$, there is a (unique) injective function $\iota : F(A) \to B^*$ such that $\iota \circ \alpha = \rho$. Let δ' be the relation over B^* defined by

$$(u, v) \in \delta' \text{ if and only if } u = u_1 aa^{-1} u_2, v = u_1 u_2, \quad u_1, u_2 \in B^*, \ a \in B$$

and let δ be the reflexive and transitive closure of δ'. We denote by K the set $\rho(B^*)$ of *reduced words* of B^*. The subset $\alpha^{-1}(1_{F(A)})$ is known as the *Dyck language* and is usually denoted by D^*. The relation δ is characterized by the following.

Lemma 3.1 *Let* $g = a_1a_2 \cdots a_n \in \delta(f)$, *with* $f \in B^*$. *Then there exist* $w_0, w_1, \ldots, w_n \in D^*$ *for which* $f = w_0a_1w_1a_2 \cdots a_nw_n$.

The following combinatorial property of D^* states roughly that *"sufficiently long"* words of D^* contain a factor that is either a product of an arbitrary number of (nonempty) elements of D^* or a arbitrarily long series of "embedded" words of D^*. More precisely, a word of D^* is said to contain a *k-factorization*, with $k \geq 0$, if $w = \alpha d_1 d_2 \cdots d_k \beta$, with $d_i \in D^*$, and $\alpha\beta \in D^*$. A word f is said to have *height at least k*, with $k \geq 0$, if there exists a sequence $d_0 = f, d_1, d_2, \ldots, d_k$ of nonempty words of D^* for which $d_{(i-1)} = \alpha_i d_i \beta_i$, with $|\alpha_i\beta_i| > 0$, for $i = 1, \ldots, k$.

Proposition 3 (Autebert & Beauquier [1]) *Let* h, p *be two integers* ≥ 0. *Then if* $N(h,p) = 2p(p^{(h+1)} - 1)/(p - 1)$, *for every* $f \in D^*$, $|f| \geq N(h,p)$ *implies that* f *admits a p-factorization or has height greater than* h.

3.2   Rational sets in $F(A)$

Rational sets are defined in the free group as in any other monoid (cf. [4, 10]). They are characterized by the following.

Theorem 5 (Benois [2]) *Let* X *be a subset of* $F(A)$, R *a subset of* B^*. *Then:*
(i) $X \in Rat(F(A))$ *if and only if* $\iota(X) \in Rat(B^*)$;
(ii) If $R \in Rat(B^*)$, *then* $\rho(R) \in Rat(B^*)$.

Conditions (i) and (ii) in the above statement imply easily each other. The key step in the proof of (ii) is the following.

Proposition 4 (Fliess [5]) *Let* L *be a rational subset of* B^* *accepted by an automaton* $\mathcal{A}$ *with representation* (λ, μ, ν). *Let* H *be the image of the Dyck language by* μ, *i.e.*

$$H = \sum_{f \in D^*} \mu(f).$$

Then $\delta(L)$ *is recognized by the automaton* $\mathcal{B}$ *with representation* (λ, π, ζ) *where* $\pi(a) = H \cdot \mu(a)$ *for every letter* a *of* B *and* $\zeta = H \cdot \nu$.

The matrix H is well-defined since the Boolean semiring is complete. Moreover there exists a finite subset $U \subseteq D^*$ such that $H = \sum_{f \in U} \mu(f)$ and the computation of H is effective as stated in the following.

Proposition 5 (Benois & Sakarovitch [3]) *There exists an algorithm which computes the matrix* H, *and thus the representation* (λ, π, ζ), *in at most* $O(m^3)$ *steps, where* m *is the dimension of* π, *i.e. the cardinality of the state set of* $\mathcal{A}$.

4   The Finite Power Property in $F(A)$

Before proving our main result, we give some preliminary propositions.

Proposition 6 *Let X be a subset of $F(A)$. For every integer $m \geq 0$, $X^* = X^{\leq m}$ if and only if $\rho(\iota(X)^*) = \rho(\iota(X))^{\leq m})$.*

Proof For every positive integer n we have

$$X^n = (\alpha(\iota(X)))^n = \alpha((\iota(X))^n) = \alpha(\rho(\iota(X))^n)) \qquad (4.1)$$

since α is a morphism and $\alpha = \alpha \circ \rho$. Therefore $\rho(\iota(X)^*) = \rho(\iota(X)^{\leq m})$ implies $X^* = X^{\leq m}$. Conversely, taking the image of equation 4.1 by ι implies $\iota(X^n) = \rho((\iota(X))^n)$ since $\iota \circ \alpha$ is the identity. Thus $X^* = X^{\leq m}$ implies $\rho(\iota(X)^*) = \rho(\iota(X)^{\leq m})$.   □

Proposition 7 *Let L be a rational subset of B^* and $\mathcal{S}_L = (\eta, \kappa, \zeta)$ be the distance automaton that computes the order with respect to L. Let $C = \bigoplus_{w \in D^*} \kappa(w)$. Then the automaton $\mathcal{R}_L = (\eta, \sigma, \chi)$, defined by $\sigma(a) = C \cdot \kappa(a)$ for every letter a in B and $\chi = C \cdot \zeta$, is such that $|\mathcal{R}_L| = \delta(L^*)$ and, for every $w \in \delta(L^*)$, $w\|\mathcal{R}_L\| = \min\{n \mid w \in \delta(L^n)\}$.*

Proof From Proposition 2, it holds that, for every word $w \in L^*$, $w\|\mathcal{S}_L\| = \min\{n \geq 0 \mid w \in L^n\}$. From Proposition 4, there exists an automaton $\mathcal{R}_L$ such that $|\mathcal{R}_L| = \delta(L^*)$. Further, $\mathcal{R}_L$ is effectively constructible from Proposition 5. Now consider the matrix $C = \bigoplus_{w \in D^*} \kappa(w)$ and set $\chi = C \cdot \zeta$, $\sigma(a) = C \cdot \kappa(a)$, for $a \in B$. Then, from Lemma 3.1 it follows that $w\|\mathcal{R}_L\| = \eta \cdot \sigma(w) \cdot \chi = \bigoplus_{\substack{f \in B^* \\ w \in \delta(f)}} f\|\mathcal{S}_L\| = \bigoplus_{\substack{f \in B^* \\ w \in \delta(f)}} o(f) = \min\{n \in \mathbb{N} \mid \exists f, f \in L^n, w \in \delta(f)\}$.   □

Proposition 8 *The matrix C is effectively constructible.*

Proof Let $N = N(h, p)$, be the number defined in Proposition 3 with $p = r$ and $h = r^2$ where r is the cardinality of the state set Q of $\mathcal{R}_L$. We prove that, for $T = D^* \cap B^N$, then $C = \bigoplus_{w \in T} \kappa(w)$. For every (i, j) such that $C_{ij} < \infty$, let w be a word of D^* of minimal length for which $C_{ij} = \kappa(w)_{ij}$. Suppose for a contradiction that $|w| > N$. Then by Proposition 3, w admits a k-factorization with $k \geq r$ or it has height greater than r^2.

In the first case, we have $w = \alpha d_1 \cdots d_k \beta$, with $d_i \in D^* \backslash 1_{B^*}$, $\alpha\beta \in D^*$. Let c be a computation from q_i to q_j with label w:

$$c = q_i \xrightarrow{\alpha} q_1 \xrightarrow{d_1} q_2 \xrightarrow{d_2} \cdots \xrightarrow{d_k} q_{k+1} \xrightarrow{\beta} q_j.$$

Since $k > r$ there exist two distinct integers n and m, $1 \leq n < m \leq k$ such that $q_m = q_n$, so that the computation $e = q_n \xrightarrow{d_n} q_{n+1} \xrightarrow{d_{n+1}} \cdots \xrightarrow{d_{m-1}} q_{m-1} \xrightarrow{d_m} q_m$

is a cycle. Therefore $w' = \alpha d_1 d_2 \cdots d_{n-1} d_{m+1} \cdots d_k$ is in D^* and is the label of the computation c' from q_i to q_j:

$$c' = q_i \xrightarrow{\alpha} q_1 \xrightarrow{d_1} q_2 \to \cdots \to q_{n-1} \xrightarrow{d_{n-1}} q_n \xrightarrow{d_{m+1}} q_{m+1} \xrightarrow{d_{m+2}} q_{m+2} \to \cdots \to q_{k+1} \xrightarrow{\beta} q_j.$$

We have $\|c'\| \le \|c\|$ and then $\kappa(w')_{ij} = \kappa(w)_{ij}$. Since $|w'| < |w|$, a contradiction is achieved.

In the other case, by definition of a word of height k, there exists a sequence of words of D^*: $w = d_0, d_1, \ldots, d_k$ such that $d_{i-1} = \alpha_i d_i \beta_i$, with $|\alpha_i \beta_i| > 0$. Then $w = \alpha_0 \alpha_1 \cdots \alpha_k d_k \beta_k \beta_{k-1} \cdots \beta_1 \beta_0$. Let

$$c = q_i \xrightarrow{\alpha_0} p_0 \xrightarrow{\alpha_1} p_1 \to \cdots \to p_{k-1} \xrightarrow{\alpha_k} p_k \xrightarrow{d_k} q_k \xrightarrow{\beta_k} q_{k-1} \to \cdots \to q_1 \xrightarrow{\beta_1} q_0 \xrightarrow{\beta_0} q_j$$

be a computation from q_i to q_j with label w. Since $k > r^2$ there exist two distinct integers n and m, $0 \le n < m \le k$ such that $p_m = p_n$ and $q_n = q_m$. We have $d_n = (\alpha_{n+1} \cdots \alpha_m) d_m (\beta_m \cdots \beta_{n+1})$ and the computations

$$e = p_n \xrightarrow{\alpha_{n+1}} p_{n+1} \to \cdots \to p_{m-1} \xrightarrow{\alpha_m} p_m \quad \text{and} \quad f = q_m \xrightarrow{\beta_m} q_{m-1} \to \cdots \to q_{n+1} \xrightarrow{\beta_n} q_n$$

are two cycles. Therefore $w' = (\alpha_0 \alpha_1 \cdots \alpha_n) d_m (\beta_n \cdots \beta_1 \beta_0)$ is the label of a computation c' from q_i to q_j. As above, $\|c'\| \le \|c\|$, $\kappa(w')_{ij} = \kappa(w)_{ij}$ and $|w'| < |w|$, a contradiction.

Thus, if $w \in D^*$ is of minimal length for which $C_{ij} = \kappa(w)_{ij} < \infty$, then $|w| \le N$ so that $w \in T$. Therefore $C = \bigoplus_{w \in T} \kappa(w)$.   □

Proof of Theorem 1 The proof – every step of which is effective – is a walk through all the results gathered so far. Let X be a rational subset of $F(A)$. From Proposition 6 it suffices to decide whether there exists a positive integer m such that $\rho(\iota(X)^*) = \rho(\iota(X)^{\le m})$. Since $X \in Rat(F(A))$ it follows from Theorem 5 that $\iota(X) \in Rat(B^*)$. Let $\mathcal{S}_{\iota(X)}$ be the distance automaton defined in Proposition 2 that recognizes $\iota(X)^*$ and computes the order of any word w with respect to $\iota(X)^*$. From Proposition 7, there exists a distance automaton $\mathcal{T}_{\iota(X)}$ that recognizes $\delta(\iota(X)^*)$, and that computes, for every word w in $\delta(\iota(X)^*)$, its *reduced order*, i.e. the smallest integer n such that w belongs to $\delta(\iota(X)^n)$. By Proposition 8, $\mathcal{T}_{\iota(X)}$ is effectively computable. As in Section 3.1, let $K = \rho(B^*)$ be the set of reduced words of B^*. From Proposition 1, a distance automaton $\mathcal{U}$ is effectively constructible that computes on its domain $\delta(\iota(X)^*) \cap K = \rho(\iota(X)^*)$ the same function of $\mathcal{T}_{\iota(X)}$, i.e. for every w in $\rho(\iota(X)^*)$, $w\|\mathcal{U}\|$ is the least integer n such that w is in $\rho(\iota(X)^n)$. Thus, as announced, X has the finite power property if and only if $\mathcal{U}$ is (distance) bounded – which is decidable by Hashiguchi's result (Theorem 2).   □

References

[1] J.M. Autebert and J. Beauquier (1974), 'Une caractérisation des générateurs standard', *RAIRO, R-1*, 63–83.

[2] M. Benois (1969), 'Parties Rationnelles du Groupe Libre', *CR Acad. Sci. Paris Ser. A*, 1188–1190.

[3] M. Benois and J. Sakarovitch (1986), 'On the complexity of some extended word problems defined by cancellation rules', *Information Proc. Letters*, **23**, 281–287.

[4] S. Eilenberg (1974), *Automata, Languages, and Machines, Vol. A*. Academic Press.

[5] M. Fliess (1971), 'Deux applications de la représentation matricielle d'une série non commutative', *J. of Algebra*, **19**, 344–353.

[6] K. Hashiguchi (1979), 'A decision procedure for the order of regular events', *Theoret. Comput. Sci.*, **72**, 27–38.

[7] K. Hashiguchi (1982), 'Limitedness theorem on finite automata with distance functions', *J. Comput. Syst. Sci.*, **24**, 233–244.

[8] K. Hashiguchi (1990), 'Improved limitedness theorems on finite automata with distance functions', *Theoret. Comput. Sci.*, **72**, 27–38.

[9] C.E. Hughes and S.M. Selkow (1981), 'The finite power property for context-free languages', *Theoret. Comput. Sci.*, **15**, 111–114.

[10] J.-E. Pin (1997), 'Tropical Semirings', *this volume*.

[11] J. Berstel and Ch. Reutenauer (1988), *Rational Series and their Languages*. Springer.

[12] I. Simon (1978), 'Limited Subsets of a Free Monoid', *Proc. 19th Annual Symposium on Foundations of Computer Science*, 143-150.

[13] I. Simon (1988), 'Recognizable sets with multiplicities in the tropical semiring', *LNCS*, **324**, 107-120.

[14] M.P. Schützenberger (1962), 'On a theorem of R. Jungen', *Proc. Amer. Math. Soc.*, **13**, 885–889.

The Topological Approach to the Limitedness Problem on Distance Automata

Hing Leung

1 Introduction

The finite automaton [11] is a mathematical model of a computing device. It is very simple and captures the essential features of a sequential circuit. The only difference is that a finite automaton does not produce outputs. Its main objective is to decide whether to accept a given input. The study of finite automata is closely related to the study of regular languages in formal language theory.

Hashiguchi [5] and Simon [18] studied the finite power property problem of regular languages. Hashiguchi's solution is a direct application of the pigeonhole principle. Simon solved the problem by reducing it to the problem of deciding whether a given finitely generated semigroup of matrices over the tropical semiring is finite.

Hashiguchi studied the star height problem and the representation problems for regular languages in ([6], [8], [9]). His solutions relied heavily on the decidability result of the limitedness problem ([7], [10]) for distance automata. Conceptually, the distance automaton is an extension of the finite automaton such that the automaton not only decides whether an input is accepted, but also assigns a measure of cost for the effort to accept the input. In fact, the finite power property problem can be considered as a special case of the limitedness problem.

The nondeterministic behavior of finite automata in connection with their inner structure was considered in ([12], [3], [4]). One important decision problem that arose from this study is a slightly weaker version of the limitedness problem for distance automata.

Motivated by Simon's solution for the finite power property problem, Leung ([14], [15]) also solved the limitedness problem for distance automata by introducing a topological approach and by extending Simon's technique. Further applications of Leung's solution can be found in [16].

Simon proposed a very powerful combinatorial framework using factorization forests for solving the limitedness problem in ([21], [22]). A very good survey of these topics is given by Simon in [19].

Let us briefly highlight the differences among the three solutions to the limitedness problem. Let n denote the number of states in a distance automaton. Hashiguchi's solution is based on the pigeonhole principle. He showed that there exists a function $f(n)$ which is double exponential in n such that the distance automaton is limited in distance if and only if its distance is at most $f(n)$. Implementing this idea, the decision algorithm will run in triple exponential time, which is not efficient. Leung's solution relied on the use of a powerful combinatorial result of Brown's [1]. Its solution runs in time $2^{O(n^2)}$ which is almost the best we can hope for since it is shown [15] that the limitedness problem is PSPACE-hard. But Leung's solution does not offer any upper bound like the function $f(n)$ given by Hashiguchi's solution. Simon obtained the same algorithm as Leung's. In addition, his method returns a double exponential upper bound in n^2 which is only slightly weaker than the bound offered by Hashiguchi.

In this paper, we present the topological approach to the limitedness problem on distance automata. Our techniques have been improved so that Brown's result is no longer needed. The main mathematical tools used are the local structure theory for finite semigroups [13] and some basic topological ideas. Most of the technical results in this paper except Lemma 3.9 and Lemma 3.10 were obtained in ([14], [15]). We include all the proofs, both for completeness and because many of them have been re-worked for better presentations.

In Section 2, we introduce the concepts of distance automata and the limitedness problem with the use of examples. We show how the limitedness problem is reduced to the problem of computing a certain homomorphic image of the topological closure of a finitely generated semigroup of matrices over the tropical semiring. The concept of sublinear automata ([20], [4]) is also introduced. In Section 3, we first give some basic results on finite semigroup theory. Then step by step we derive a solution to our problem. In Section 4, two open problems are presented.

2   Limitedness Problem on Distance Automata

Let $\mathbb{N}$ denote the set of nonnegative integer and $\mathbb{N}^+$ denote the set of positive integer. A distance automaton is a 5-tuple (Q, Σ, d, Q_I, Q_F) where Q is the set of states, Σ is the alphabet set, $d : Q \times \Sigma \times Q \longrightarrow \mathbb{N} \cup \{\infty\}$ is the distance function, $Q_I \subseteq Q$ is the set of starting states and $Q_F \subseteq Q$ is the set of final states.

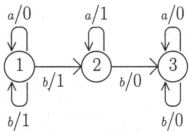

Figure 1: Example of a distance automaton A.

Figure 1 presents the structure of a distance automaton A in the form of a state transition diagram where Q consists of three states numbered 1, 2 and 3, Σ consists of two letters a and b, and d is expressed by the labelled edges (transitions) of the diagram. For example, an edge going from state 1 to state 2 with label $b/1$ indicates that $d(1, b, 2) = 1$; intuitively, we say that the transition $(1, b, 2)$ requires a cost of 1. More exactly, from the transition diagram, the distance function $d\ :\ \{1, 2, 3\} \times \{a, b\} \times \{1, 2, 3\} \longrightarrow \mathbb{N} \cup \{\infty\}$ is defined such that $d(1, a, 1) = 0$, $d(1, b, 1) = 1$, $d(1, b, 2) = 1$, $d(2, a, 2) = 1$, $d(2, b, 3) = 0$, $d(3, a, 3) = 0$ and $d(3, b, 3) = 0$, whereas $d(i, x, j) = \infty$ when there is no edge from state i to state j with the label x/k for $x \in \Sigma$ and $k \in \mathbb{N}$. However, the diagram has not yet completely specified a distance automaton. Depending on how we define the set of starting states and the set of final states, we obtain different finite automata.

Let Σ^* denote the set of all finite sequences (strings) over of the set of symbols (letters) in Σ. Consider the example when $\Sigma = \{a, b\}$. In $\{a, b\}^*$, there are two different strings of length one, namely a and b, and four different strings of length two, which are aa, ab, ba and bb. In general, we have 2^k different strings of length k in $\{a, b\}^*$. In addition, there is a string of length zero in Σ^*. We call it the empty string and denote it by ϵ. Two strings are considered equal if and only if they have the same length and exactly the same sequence of letters. Concatenation is the binary operation $\cdot\ :\ \Sigma^* \times \Sigma^* \longrightarrow \Sigma^*$ such that $x \cdot y$ is the string that we obtain if we write down x first and then follow it with y where $x, y \in \Sigma^*$. Note that we define $\epsilon \cdot x = x$ and $x \cdot \epsilon = x$ for all $x \in \Sigma^*$. If x and y have lengths i and j respectively, then $x \cdot y$ has length $i + j$. Example: $aba \cdot bb = ababb = ab \cdot abb$. It is immediate to see that $\cdot$ is associative and ϵ is an identity. Thus, $(\Sigma^*, \cdot, \epsilon)$ forms a monoid.

The mission of a distance automaton is to process strings over the alphabet Σ. An automaton processes a string symbol by symbol. Consider A again. Given a string $x = abbb$, A processes the first symbol a of x first, followed by the second symbol b, then the third symbol b and the last symbol b. At any moment during the processing of a string, one of the states of the automaton is supposed to be active (current). If A begins with the state 2 being active, it could process the first symbol a of x by following the transition $(2, a, 2)$ with the distance 1 and arrive back at state 2. The new current state after the processing of the first symbol is again state 2. Next, traversing the transition $(2, b, 3)$ with distance 0, the automaton reaches a new current state 3. It continues to remain in state 3 by twice using the transition $(3, b, 3)$ with distance 0 to process the third and the last symbol. That is, A processes the string $abbb$ by traversing the sequence of transition moves $(2, a, 2)$ $(2, b, 3)$ $(3, b, 3)$ $(3, b, 3)$ with a total distance of 1. There are five other possible sequences of transitions for processing $abbb$ with finite distances:

1. sequence $(1, a, 1)$ $(1, b, 1)$ $(1, b, 1)$ $(1, b, 1)$ with a total distance 3,

2. sequence $(1, a, 1)$ $(1, b, 1)$ $(1, b, 1)$ $(1, b, 2)$ with a total distance 3,

3. sequence $(1, a, 1)$ $(1, b, 1)$ $(1, b, 2)$ $(2, b, 3)$ with a total distance 2,

4. sequence $(1, a, 1)$ $(1, b, 2)$ $(2, b, 3)$ $(3, b, 3)$ with a total distance 1

5. sequence $(3, a, 3)$ $(3, b, 3)$ $(3, b, 3)$ $(3, b, 3)$ with a total distance 0.

All other sequences of moves for processing $abbb$ require infinite distances. Even though sequences 3 and 4 both begin with state 1 and end at state 3, they require different distances. In that case, the automaton is assumed to favor the sequence requiring smaller distance. That is, we say that the distance from state 1 to state 3 consuming $abbb$ is 1, which is the minimum of the two distances 1 and 2. The distance behavior of A for processing $abbb$ is summarized in the following table:

Distance required		to state 1	2	3
from state	1	3	3	1
	2	∞	∞	1
	3	∞	∞	0

Table 1: The distance behavior of A for processing $abbb$.

Let $\mathcal{M}$ denote the *tropical semiring* with support $\mathbb{N} \cup \{\infty\}$ and operations $a \oplus b = \min\{a, b\}$ and $a \otimes b = a + b$ where ∞ and 0 are the identities of $\oplus$ and $\otimes$ respectively. Note that the linear ordering $0 < 1 < 2 < ... < \infty$ is assumed for the minimum operation. We denote by $M_n\mathcal{M}$ the multiplicative monoid of $n \times n$ matrices with coefficients in $\mathcal{M}$.

To capture the distance behavior of a distance automaton, we introduce a function $D : \Sigma \longrightarrow M_n\mathcal{M}$ where n is the number of states in the automaton, and for $1 \le i, j \le n$ and $a \in \Sigma$, $D(a)(i, j)$ is defined to be $d(i, a, j)$. D can be extended to a homomorphism $D : \Sigma^* \longrightarrow M_n\mathcal{M}$ with $D(\epsilon)$ defined to be the identity of $M_n\mathcal{M}$, which has the value 0 in the diagonal and the value ∞

in all other entries, and $D(x \cdot y) = D(x) \cdot D(y)$ for all $x, y \in \Sigma^*$. With respect to A, $D(a)$ and $D(b)$ are given as follows:

$D(a)$	1	2	3
1	0	∞	∞
2	∞	1	∞
3	∞	∞	0

$D(b)$	1	2	3
1	1	1	∞
2	∞	∞	0
3	∞	∞	0

Table 2: Distance matrices for $D(a)$ and $D(b)$.

Thus, the distance behavior of A for the processing of the string *abbb* is captured by the matrix $D(abbb)$ which can be calculated by computing $D(ab) = D(a) \cdot D(b)$, $D(bb) = D(b) \cdot D(b)$ and $D(abbb) = D(ab) \cdot D(bb)$. The results of the computation are as follows:

$D(ab)$	1	2	3
1	1	1	∞
2	∞	∞	1
3	∞	∞	0

$D(bb)$	1	2	3
1	2	2	1
2	∞	∞	0
3	∞	∞	0

$D(abbb)$	1	2	3
1	3	3	1
2	∞	∞	1
3	∞	∞	0

Table 3: Distance matrices for $D(ab)$, $D(bb)$ and $D(abbb)$.

Observe that the matrix for $D(abbb)$ is the same as the one we obtained in Table 1. In general, we see that the semigroup of matrices generated by $D(a)$ and $D(b)$, denoted $< D(a), D(b) >$, captures the distance behavior of A on all the different finite strings.

So far we have not considered the significance of the set of starting states Q_I and the set of final states Q_F on the processing of a string by a distance automaton. Given a string $x \in \Sigma^*$ and $k \in \mathbb{N}$, we say that the string x is accepted with distance at most k, written as $d(x) \leq k$, if there exists a sequence of transitions with total distance less than or equal to k that begins with a state in Q_I and ends at a state in Q_F; we say that the string x is accepted with distance k, written as $d(x) = k$, if x is accepted with distance at most k but it is not accepted with distance at most $k - 1$; we say that the string x is accepted if x is accepted with some distance $k \in \mathbb{N}$; we say that the string x is not accepted, written as $d(x) = \infty$, if x is not accepted with distance k for all $k \in \mathbb{N}$. Remark: the same notation d is chosen to denote both the distance of a string accepted by a finite automaton and the distance function of a distance automaton. The interpretation can be decided by the context in which d is being used.

It is clear that $d(x)$ can be expressed as $\min_{i \in Q_I;\ j \in Q_F} D(x)(i,j) = u_I D(x) v_F$ where $u_I \in \mathcal{M}^{1 \times n}$ is a $1 \times n$ row vector over $\mathcal{M}$ and $v_F \in \mathcal{M}^{n \times 1}$

is an $n \times 1$ column vector over $\mathcal{M}$ such that $u_I(i) = 0$ if $i \in Q_I$, $u_I(i) = \infty$ if $i \notin Q_I$, $v_F(i) = 0$ if $i \in Q_F$ and $v_F(i) = \infty$ if $i \notin Q_F$.

With respect to A, if we let $Q_I = \{1, 2\}$ and $Q_F = \{2, 3\}$, then $d(abbb) = \min_{i=1,2; \ j=2,3}\{D(abbb)(i,j)\} = [0, 0, \infty] \cdot D(abbb) \cdot [\infty, 0, 0]^T = [3, 3, 1][\infty, 0, 0]^T = 1$; if we let $Q_I = \{2, 3\}$ and $Q_F = \{2\}$, then $d(abbb) = \min_{i=2,3; \ j=2}\{D(abbb)(i,j)\} = [\infty, 0, 0] \cdot D(abbb) \cdot [\infty, 0, \infty]^T = [\infty, \infty, 0][\infty, 0, \infty]^T = \infty$.

Given a distance automaton, the limitedness problem asks if there is a $k \in \mathbb{N}$ such that for all x accepted by the automaton, x is accepted with distance at most k. If such a k exists, then we say that the distance automaton is limited in distance.

Consider again our example automaton A. If we let $Q_I = \{3\}$ and $Q_F = \{2\}$, A is limited in distance since no strings are accepted. If we choose $Q_I = \{2, 3\}$ and $Q_F = \{3\}$, A is again limited in distance since all strings can be accepted with distance 0 with a sequence of transitions that begins with state 3 and ends with state 3.

If $Q_I = \{1\}$ and $Q_F = \{1, 2, 3\}$, A becomes not limited in distance. We want to show by induction that for each $k \in \mathbb{N}^+$, there exists a string $s(k)$ accepted by A with distance k. When $k = 1$, we choose $s(1)$ to be b. Assume that the statement is true for $k = m$. We claim that $ba^m s(m)$ is accepted with distance $m + 1$, hence we let $s(m+1) = ba^m s(m)$. To process ba^m beginning with state 1, A may go into two possible paths; one path ends at state 1 with distance 1 and the other path ends at state 2 with distance $m+1$. If A chooses the first path for the processing of the first $m + 1$ symbols of $ba^m s(m)$, A will continue the processing of $s(m)$ from state 1, which by induction requires a distance m; the total distance required is therefore $m+1$. On the other hand, if A chooses the second path for the processing of the first $m + 1$ symbols of $ba^m s(m)$, we know that A already requires a distance of $m + 1$, not counting the distances that may be needed to continue the processing of $s(m)$ from state 2. Hence we are done, since the distance of a string is defined to be the minimum distance over all the distances of accepting sequences of transition moves.

We can easily strengthen the previous result and show that for each $k \geq 1$, $s(k)$ defined above is in fact the (unique) shortest string that requires distance k. The proof [4] is based on the observation that if $s(k)$ does not begin with ba^{k-1}, then either $s(k)$ is not the shortest string or $s(k)$ does not require distance k to be processed. The details are left to the reader as an exercise.

By the definition that $s(1) = b$ and $s(k) = ba^{k-1}s(k-1)$ for $k > 1$, we deduce that the length (number of symbols) of $s(k)$ is $\Theta(k^2)$. Thus for any accepted string x, $d(x) = O(\sqrt{|x|})$. Such an automaton is called a sublinear distance automaton ([20], [4]) when the automaton is not limited in distance and the growth of distances is sublinear with respect to the growth in the lengths of strings. It has been shown ([20], [4]) that for each $p \in \mathbb{N}^+$, there exists a distance automaton A_p not limited in distance such that the shortest string $s_p(k)$ that requires distance k has length $\Theta(k^p)$.

In general, how are we going to decide if a given distance automaton is not limited in distance? In order to arrive at the conclusion that a certain distance automaton is indeed not limited in distance, we must furnish some proofs that for each $k \geq 1$, there exists a string $s(k)$ such that the distance to accept $s(k)$ is at least k. However, the existence of sublinear distance automata immediately tells us that our algorithm has to be quite clever in order to discover the generic form of $s(k)$, whose length may grow polynomially as k increases.

We need to extend the semiring $\mathcal{M}$ to a semiring $\mathcal{T}$ with support $\mathbb{N} \cup \{\omega, \infty\}$ and operations $a \oplus b = \min\{a, b\}$ and $a \otimes b = a + b$ where the linear ordering $0 < 1 < 2 < \cdots < \omega < \infty$ is assumed and for any $x \in \mathcal{T}$, $\omega + x$ and $x + \omega$ are defined to be $\max(\omega, x)$. We consider a one-point compactification of the discrete topology over $\mathbb{N} \cup \{\infty\}$ on $\mathcal{T}$ where ω is the point at infinity[1]. Formally, this topology consists of open sets that are either subsets of $\mathbb{N} \cup \{\infty\}$ or co-finite subsets that contain ω. Intuitively, ω represents an unbounded finite value whereas ∞ represents an infinite value. In this topology, a sequence a_n has a limit ω if for all $k \in \mathbb{N}$ there exists $i_k \in \mathbb{N}$ such that $a_n \in \mathbb{N} \cup \{\omega\} - \{0, 1, \ldots, k\}$ for all $n > i_k$; a sequence a_n has a limit $a \in \mathbb{N} \cup \{\infty\}$ if $a_n = a$ for all large enough n. It is easy to see that $\mathcal{T}$ is metrizable and compact. We denote by $M_n\mathcal{T}$ the multiplicative monoid of $n \times n$ matrices with coefficients in $\mathcal{T}$. Then $M_n\mathcal{T}$ with the product topology is metrizable and compact. Also, the matrix multiplication is continuous. A sequence $A_k \in M_n\mathcal{T}$ has a limit iff for all $1 \leq i, j \leq n$ the sequence $A_k(i, j) \in \mathcal{T}$ has a limit.

We want to characterize distance automata that are limited in distance in terms of the topological closure of the corresponding set of distance matrices $D(x)$ where $x \in \Sigma^*$.

Assuming that a distance automaton is not limited in distance, there must exist a sequence of matrices $D(x_m)$ such that $d(x_m) = u_I D(x_m) v_F$ grows unboundedly. Since $M_n\mathcal{T}$ is metrizable and compact, there exists a subsequence $D(x_{\alpha_m})$ such that the limit exists. Let us denote this limit by M. Hence, $u_I M v_F = \omega$.

On the other hand, suppose there is a matrix M in the topological closure of $\{D(x) \mid x \in \Sigma^*\}$ such that $u_I M v_F = \omega$. It follows that there exists a sequence of matrices $D(x_m)$ such that the limit is M. Hence the distance automaton is not limited in distance since $\lim_{m \to \infty} d(x_m) = \lim_{m \to \infty} u_I D(x_m) v_F = u_I \left(\lim_{m \to \infty} D(x_m)\right) v_F = u_I M v_F = \omega$.

---

[1]We refer the reader to [23] for all the topological concepts used in this paper.

Therefore, a distance automaton is not limited in distance iff there is a matrix M in the topological closure of the distance matrices $D(x)$ where $x \in \Sigma^*$ such that $u_I M v_F = \omega$.

Let $\mathcal{R}$ be the semiring with support $\{0, 1, \omega, \infty\}$ and operations $a \oplus b = \min\{a, b\}$ and $a \otimes b = \max\{a, b\}$ where the linear ordering $0 < 1 < \omega < \infty$ is assumed. We define a projection function $\Psi : \mathcal{T} \longrightarrow \mathcal{R}$ such that $\Psi(a) = 1$ if a is a positive integer, otherwise $\Psi(a) = a$. We extend Ψ to be a function $M_n \mathcal{T} \longrightarrow M_n \mathcal{R}$ in the natural way, where $M_n \mathcal{R}$ is the multiplicative monoid of $n \times n$ matrices with coefficients in $\mathcal{R}$. Moreover, $\Psi : M_n \mathcal{T} \longrightarrow M_n \mathcal{R}$ is a homomorphism. Caution: Ψ is not continuous under the assumption that a discrete topology is placed on $\mathcal{R}$. We say that two matrices $s, t \in M_n \mathcal{T}$ have the same structure if $\Psi(s) = \Psi(t)$. We say that $t \in M_n \mathcal{T}$ has an idempotent structure if $\Psi(t) = \Psi(t)\Psi(t)$ is an idempotent in $M_n \mathcal{R}$.

Let $< D(a) \mid a \in \Sigma >^c$ denote the topological closure of the semigroup generated by the matrices $D(a)$ for all $a \in \Sigma$. Since $\{D(x) \mid x \in \Sigma^*\}^c = < D(a) \mid a \in \Sigma >^c \cup \{D(\epsilon)\}$, a distance automaton is not limited in distance iff there is a matrix M in $\Psi(< D(a) \mid a \in \Sigma >^c)$ such that $u_I M v_F = \omega$. Note that $\Psi(< D(a) \mid a \in \Sigma >^c) \subseteq M_n \mathcal{R}$ is finite. That is, we can decide if a given finite automaton is limited in distance if we know how to compute algorithmically $\Psi(< D(a) \mid a \in \Sigma >^c)$.

3 Computing the Topological Closure

3.1 The algorithm

In the previous section, we have reduced the limitedness problem of a distance automaton to the problem of computing $\Psi(< D(a) \mid a \in \Sigma >^c) \subseteq M_n \mathcal{R}$. Note that $< D(a) \mid a \in \Sigma >^c \subseteq M_n \mathcal{T}$ whereas $D(a) \in M_n \mathcal{M}$ for all $a \in \Sigma$.

Throughout this section, we assume a generalized version of the problem, namely that we are given a finite subset T of $M_n \mathcal{T}$ and our aim is to develop an algorithm for computing $\Psi(<T>^c)$.

Let $t \in M_n \mathcal{T}$. We write $\|t\|$ to denote $\max_{1 \leq i, j \leq n}\{t(i, j) \mid t(i, j) \in \mathbb{N}^+\}$ if the maximum exists, otherwise $\|t\| = 1$. We write Δ to denote $\max_{t \in T} \|t\|$.

We define a partial ordering $\leq$ on $M_n \mathcal{T}$ (respectively, $M_n \mathcal{R}$) such that for $t_1, t_2 \in M_n \mathcal{T}$ (respectively, $M_n \mathcal{R}$), $t_1 \leq t_2$ iff for all $1 \leq i, j \leq n$, if $t_1(i, j) \in \mathbb{N}^+$ then $t_1(i, j) \leq t_2(i, j) \in \mathbb{N}^+ \cup \{\omega\}$ otherwise $t_1(i, j) - t_2(i, j)$. It can be easily shown that for both $M_n \mathcal{T}$ and $M_n \mathcal{R}$, the matrix multiplication is monotonic with respect to the partial ordering $\leq$, that is, $t_1 \leq t_2$ and $t_3 \leq t_4$ implies that $t_1 t_3 \leq t_2 t_4$.

The first thing that we want to study is how the entry coefficients, which belong to $\mathcal{T}$, in each matrix in T may affect the result of $\Psi(<T>^c)$. We want to argue that if in each matrix we change every entry that is of positive integer value to the value 1, then we still end up with the same answer for $\Psi(<T>^c)$. More exactly, let $T_1 = \{t_1 \in M_nT \mid t_1 = \Psi(t), t \in T\}$ and let $T_\Delta = \{t_\Delta \in M_nT \mid t_\Delta(i,j) = \Delta$ if $t(i,j) \in \mathbb{N}^+$, otherwise $t_\Delta(i,j) = t(i,j),\ t \in T\}$. It is easy to observe that for every $s_1 \in <T_1>^c$ there is a corresponding $s_\Delta \in <T_\Delta>^c$ and vice versa such that for $1 \leq i, j \leq n$, either $s_\Delta(i,j) = s_1(i,j) \in \{0, \omega, \infty\}$ or $s_1(i,j) \cdot \Delta = s_\Delta(i,j) \in \mathbb{N}^+$. Thus, $\Psi(<T_1>^c) = \Psi(<T_\Delta>^c)$. Since for every $t \in T$ there exist $t_1 \in T_1$ and $t_\Delta \in T_\Delta$ such that $t_1 \leq t \leq t_\Delta$, we deduce that for every $s \in <T>^c$, there must exist $s_1 \in <T_1>^c$ and $s_\Delta \in <T_\Delta>^c$ such that $s_1 \leq s \leq s_\Delta$. Together with the fact that $\Psi(<T_1>^c) = \Psi(<T_\Delta>^c)$, this implies that $\Psi(<T_1>^c) = \Psi(<T>^c) = \Psi(<T_\Delta>^c)$.

Let us consider a special case when T is a singleton set containing only one element $t \in M_nT$. For example, let t be a 3×3 matrix such that $t(1,3) = t(2,2) = t(3,1) = 1$ and $t(i,j) = \infty$ otherwise. We see that for $k \geq 1$, $t^{2k}(i,j) = 2k$ if $i = j$ and $t^{2k}(i,j) = \infty$ otherwise. On the other hand, for $k \geq 0$, $t^{2k+1}(i,j) = 2k+1$ if $i+j = 4$ and $t^{2k+1}(i,j) = \infty$ otherwise. Thus, the sequence t^k for $k \geq 1$ does not have a limit since t^k alternates between two different structures as k increases. However, t^{2k} always has an idempotent structure; moreover $\lim_{k \to \infty} t^{2k}$ exists and has ω values in the diagonal and ∞ values elsewhere. Since matrix multiplication is continuous in M_nT and $\lim_{k \to \infty} t^{2k}$ exists, we have $\lim_{k \to \infty} t^{2k+1} = t(\lim_{k \to \infty} t^{2k})$. Since $\Psi(<t>) = <\Psi(t)>$, and taking account of the above analyis, $\Psi(<t>^c) = <\Psi(t), \Psi(s)>$ where $s = \lim_{k \to \infty} t^{2k}$.

The example points to us the following two questions.

For an arbitrary t, does there exist $m \in \mathbb{N}^+$ such that t^m has an idempotent structure? Since Ψ is a homomorphism, it is equivalent to ask if there exists $m \in \mathbb{N}^+$ such that $\Psi(t)^m$ is an idempotent in $M_n\mathcal{R}$. Since $<\Psi(t)> \subseteq M_n\mathcal{R}$ is finite, and by the pigeonhole principle, there exist α, β such that $\Psi(t)^x = \Psi(t)^{x+\beta}$ for all $x \geq \alpha$. By picking m to be a multiple of β and $m \geq \alpha$, then $\Psi(t)^m = \Psi(t)^{2m}$. Thus the answer to the first question is positive.

The second question is: Does the limit of t^k always exist if t has an idempotent structure? The following lemma answers this question positively.

Lemma 3.1 *If $t \in M_nT$ has an idempotent structure, then $\lim_{k \to \infty} t^k$ exists.*

Proof Let $\delta = 2\|t\|$. Let $1 \leq i, j \leq n$.

Case 1. Suppose that $t(i,j) \in \{0, \omega, \infty\}$. Then for every $k \geq 1$, since t^k always preserves the same idempotent structure as that of t, $t^k(i,j)$ must equal $t(i,j)$. Then $\lim_{k \to \infty} t^k(i,j) = t(i,j)$ exists.

Case 2. Suppose that $t(i,j) \in \mathbb{N}^+$ and there exist p, q such that $t(i,p) \in \mathbb{N}$, $t(p,q) = 0$ and $t(q,j) \in \mathbb{N}$. Then $\lim_{k \to \infty} t^k(i,j)$ exists if we can show that for all $m \geq 3$, $t^{\delta+m}(i,j) \leq \delta$ and $t^{\delta+m}(i,j) \geq t^{\delta+m+1}(i,j)$.

Firstly, $t^{\delta+m}(i,j) \leq t(i,p) + t^{\delta+m-2}(p,q) + t(q,j) = t(i,p) + t(q,j) \leq \delta$ since $0 = t(p,q) = t^2(p,q) = \cdots = t^{\delta+m-2}(p,q)$ by the fact that t has an idempotent structure.

Secondly, since $t^{\delta+m}(i,j) \leq \delta$ and $m \geq 3$, there exist $1 \leq g, h \leq n$ and $\alpha, \beta \geq 1$ such that $t^{\delta+m}(i,j) = t^{\alpha}(i,g) + t(g,h) + t^{\beta}(h,j)$ where $t(g,h) = 0$ and $\alpha + \beta + 1 = \delta + m$. Since t has an idempotent structure, $t(g,h) = 0$ implies that $t^2(g,h) = 0$. Thus, $t^{\delta+m}(i,j) = t^{\alpha}(i,g) + t^2(g,h) + t^{\beta}(h,j) \geq t^{\alpha+2+\beta}(i,j) = t^{\delta+m+1}(i,j)$.

Case 3. Suppose that $t(i,j) \in \mathbb{N}^+$ and there do not exist p,q such that $t(i,p) \in \mathbb{N}$, $t(p,q) = 0$ and $t(q,j) \in \mathbb{N}$. We want to show that $t^k(i,j) \geq k$ for $k \geq 1$. Thus, $\lim_{k\to\infty} t^k(i,j) = \omega$.

Assume on the contrary that $t^k(i,j) \leq k - 1$ for some $k \geq 2$. There exist g, h such that $t^k(i,j) = t^{\alpha}(i,g) + t(g,h) + t^{\beta}(h,j)$ where $t(g,h) = 0$, $\alpha + \beta + 1 = k$ and $\alpha, \beta \geq 0$. If $\alpha = 0$ then technically we can assume that $g = i$ and t^0 is the identity matrix such that $t^0(i,i) = 0$. Similarly, if $\beta = 0$ then assume that $h = j$ and $t^0(j,j) = 0$. Since t has an idempotent structure and $t(g,h) = 0$, we have $t^3(g,h) = 0$. It follows that there exist p,q such that $t(g,p) = t(p,q) = t(q,h) = 0$. Again since t has an idempotent structure, $t^{\alpha}(i,g) \in \mathbb{N}$ and $t(g,h) = 0$ implies that $t(i,h) \in \mathbb{N}$. Similarly, $t(q,h) = 0$ and $t^{\beta}(h,j) \in \mathbb{N}$ implies that $t(q,j) \in \mathbb{N}$. Thus $t(i,p) \in \mathbb{N}$, $t(p,q) = 0$ and $t(q,j) \in \mathbb{N}$. But this contradicts our assumption of case 3. □

From the proof of Lemma 3.1, we see that $\Psi(\lim_{k\to\infty} t^k)$ depends only on the structure of t. That is, for any $t_1, t_2 \in M_n\mathcal{T}$, if both of them have the same idempotent structure then $\Psi(\lim_{k\to\infty} t_1^k) = \Psi(\lim_{k\to\infty} t_2^k)$.

Therefore, the following operation #, called stabilization, on the idempotents $e = e^2$ in $M_n\mathcal{R}$ is well-defined:

$$e^{\#} \overset{\text{def}}{=} \Psi\left(\lim_{k\to\infty} t^k\right) \text{ where } t \in \Psi^{-1}(e).$$

$e^{\#}$ can be computed mechanically as follows: $e^{\#}(i,j) = \omega$ if $e(i,j) = 1$ and there do not exist $1 \leq p, q \leq n$ such that $e(i,p) \in \{0,1\}$, $e(p,q) = 0$ and $e(q,j) \in \{0,1\}$; otherwise $e^{\#}(i,j) = e(i,j)$.

Observe that $e \leq e^{\#}$. We say that an idempotent $e \in M_n\mathcal{R}$ is stable if $e = e^{\#}$, and unstable if $e < e^{\#}$.

Lemma 3.2 *For $t \in M_n\mathcal{T}$, $\Psi(<t>^c) = <Psi(t), \Psi(t^m)^{\#}>$ where m is some positive integer such that t^m has an idempotent structure.*

Proof It has been argued that there exists $m \in \mathbb{N}^+$ such that t^m has an idempotent structure. Given that $\lim_{k\to\infty}(t^m)^k$ exists by Lemma 3.1, we have for all $1 \leq p \leq m-1$, $\lim_{k\to\infty} t^{mk+p} = t^p \lim_{k\to\infty}(t^m)^k$. Thus for all $1 \leq p \leq m-1$, $\Psi(\lim_{k\to\infty} t^{mk+p}) = \Psi(t^p \lim_{k\to\infty}(t^m)^k) = \Psi(t)^p \Psi(\lim_{k\to\infty}(t^m)^k) = \Psi(t)^p \Psi(t^m)^{\#} \in <\Psi(t), \Psi(t^m)^{\#}>$. Since $<\Psi(t), \Psi(t^m)^{\#}>$ includes the Ψ

images of all the limit points, $\Psi(<t>^c) \subseteq \; <\Psi(t), \Psi(t^m)^\# >$. On the other hand, $\Psi(<t>^c) \supseteq \Psi(<t, \lim_{k \to \infty}(t^m)^k >) = \; <\Psi(t), \Psi(\lim_{k \to \infty}(t^m)^k)> = \; < \Psi(t), \Psi(t^m)^\# >$. $\square$

From the discussion of the case when T is a singleton set, we learn that the matrices in $<T>$ with idempotent structures play an important role in determining the topological structure of the semigroup generated by T. Can this observation be generalized to the case when T may not be a singleton set? We answer this question positively in Theorem 1, whose proof will be established in the rest of this section.

Theorem 1 *For a finite subset T of $M_n\mathcal{T}$, $\Psi(<T>^c) = \Psi(T)^{+,\#}$ where $\Psi(T)^{+,\#}$ denotes the algebraic closure of $\Psi(T)$ under both multiplication and stabilization operations.*

An easy part of the proof is to show that $\Psi(<T>^c) \supseteq \Psi(T)^{+,\#}$, whereas the part $\Psi(<T>^c) \subseteq \Psi(T)^{+,\#}$ is more difficult.

Lemma 3.3 *For a finite subset T of $M_n\mathcal{T}$, $\Psi(<T>^c) \supseteq \Psi(T)^{+,\#}$.*

Proof We want to show that for all $z \in \Psi(T)^{+,\#}$ there exists $t \in <T>^c$ such that $\Psi(t) = z$. Since $\Psi(T)^{+,\#}$ is finite, let us number all the elements in $\Psi(T)^{+,\#}$ and call them $z_1, z_2, ..., z_p$ such that for each $1 \le k \le p$, either

(a) $z_k \in \Psi(T)$, or

(b) there exist $1 \le i, j \le k - 1$ such that $z_k = z_i z_j$, or

(c) there exists $1 \le i \le k - 1$ such that z_i is an idempotent and $z_k = z_i^\#$.

Our proof is by induction on i where $1 \le i \le p$. The base case is when $i = 1$. z_1 must be in $\Psi(T)$, which implies that there exists $t \in T$ such that $\Psi(t) = z_1$. Thus, there exists $t \in <T>^c$ such that $\Psi(t) = z_1$. Assume that the statement is true for all $1 \le i \le k - 1$ where $k \ge 2$. Consider z_k. If case (a) applies to z_k, then we are done by the same argument as in the base case. If case (b) applies to z_k, then by the induction hypothesis there exist $t_i, t_j \in <T>^c$ such that $\Psi(t_i) = z_i$ and $\Psi(t_j) = z_j$. Let $t = t_i t_j \in <T>^c$. Since Ψ is a homomorphism, we have $\Psi(t) = \Psi(t_i t_j) = \Psi(t_i)\Psi(t_j) = z_i z_j = z_k$. If case (c) applies to z_k, then by the induction hypothesis there exists $t_i \in <T>^c$ with the idempotent structure z_i. Let $t = \lim_{h \to \infty}(t_i)^h \in <T>^c$. We have $\Psi(t) = \Psi(\lim_{h \to \infty}(t_i)^h) = \Psi(t_i)^\# = z_i^\# = z_k$. $\square$

Consider the special case when T is a singleton set $\{t\}$. By the definition of $\Psi(T)^{+,\#}$, $<\Psi(t), \Psi(t^m)^\# > \subseteq \Psi(t)^{+,\#}$. Next by Lemma 3.2 and Lemma 3.3, $\Psi(t)^{+,\#} \subseteq \; <\Psi(t), \Psi(t^m)^\# >$. That is, $\Psi(<t>^c) = \; <\Psi(t), \Psi(t^m)^\# > = \Psi(t)^{+,\#}$ and we have proved Theorem 1 when T is a singleton set.

Let us demonstrate how to compute $\Psi(T)^{+,\#}$, using as an example $T = \{D(a), D(b)\}$ where $D(a)$ and $D(b)$ are defined to be the distance matrices for letters a and b of the distance automaton A given in Figure 1.

We can argue that for every $t \in \Psi(T)^{+,\#}$, the matrix t must be upper triangular in the sense that $t(2,1) = t(3,1) = t(3,2) = \infty$. In order to save space, instead of presenting t in its usual two-dimensional form we are going to write the relevant matrix entries in one sequence as follows: $[t(1,1), t(1,2), t(1,3); t(2,2), t(2,3); t(3,3)]$.

Initially, T has two matrices: $[1]\ [0, \infty, \infty; 1, \infty; 0]$ and $[2]\ [1, 1, \infty; \infty, 0; 0]$. Note that matrix $[1]$ is $D(a)$ and matrix $[2]$ is $D(b)$. Next, we compute the algebraic closure of T using matrix multiplication and we collect three more matrices:

$[3][1, 1, \infty; \infty, 1; 0]$, $[4]\ [1, 1, 1; \infty, 0; 0]$, $[5]\ [1, 1, 1; \infty, 1; 0]$.

Matrices $[1]$, $[4]$ and $[5]$ are idempotents. By applying stabilizations to the three idempotents, we obtain three new matrices:

$[6][0, \infty, \infty; \omega, \infty; 0]$, $[7]\ [\omega, \omega, 1; \infty, 0; 0]$, $[8]\ [\omega, \omega, 1; \infty, 1; 0]$.

Now taking the algebraic closure again using multiplications, we obtain 15 more matrices:

$[9]\ [1, \omega, \infty; \infty, 0; 0]$, $[10]\ [1, 1, \infty; \infty, \omega; 0]$, $[11]\ [1, \omega, \infty; \infty, 1; 0]$,

$[12]\ [1, \omega, \infty; \infty, \omega; 0]$, $[13]\ [1, 1, \omega; \infty, 0; 0]$, $[14]\ [1, \omega, \omega; \infty, 0; 0]$,

$[15]\ [1, 1, \omega; \infty, \omega; 0]$, $[16]\ [1, \omega, \omega; \infty, \omega; 0]$, $[17]\ [1, 1, \omega; \infty, 1; 0]$,

$[18]\ [1, \omega, \omega; \infty, 1; 0]$, $[19]\ [1, \omega, 1; \infty, 0; 0]$, $[20]\ [1, \omega, 1; \infty, 1; 0]$,

$[21]\ [1, 1, 1; \infty, \omega; 0]$, $[22]\ [1, \omega, 1; \infty, \omega; 0]$, $[23]\ [\omega, \omega, 1; \infty, \omega; 0]$.

Among the new matrices generated, matrices $[14]$, $[15]$, $[16]$, $[18]$, $[19]$, $[20]$, $[21]$, $[22]$ and $[23]$ are idempotents. We apply stabilizations to these 9 idempotents. Not all of them give us new matrices. The stabilizations on matrices $[14]$, $[15]$, $[16]$ and $[18]$ give us three new matrices:

$[24][\omega, \omega, \omega; \infty, 0; 0]$, $[25]\ [\omega, \omega, \omega; \infty, \omega; 0]$, $[26]\ [\omega, \omega, \omega; \infty, 1; 0]$.

Finally, we do not gain any new matrices by taking another algebraic closure using multiplication.

To compute $\Psi(T)^{+,\#}$ for $T = \{D(a), D(b)\}$, we found that we need to begin with T and repeatedly compute the closure under multiplication followed by stabilization on idempotents, for two times before we stop.

We consider again the distance automaton A with $Q_I = \{1\}$ and $Q_F = \{1, 2, 3\}$. Let M be the matrix $[24]$ in $\Psi(T)^{+,\#}$. Then $u_I M v_F = [0, \infty, \infty] M [0, 0, 0]^T = \omega$. Hence, we have verified that the distance automaton is not limited in distance. Observe that if we do not compute the second round of stabilizations then for all the matrices $[k]$ where $1 \le k \le 23$, $u_I [k] v_F \ne \omega$; that is, we would not be able to verify that the distance automaton is not limited in distance.

Furthermore, we can express matrix $[24]$ as $[14]^\#$, matrix $[14]$ as $[9][9]$, matrix $[9]$ as $[2][6]$, and matrix $[6]$ as $[1]^\#$. Together, $[24] = (([2][1]^\#)([2][1]^\#))^\#$. Corresponding to each of the matrices $[1]$, $[6]$, $[9]$, $[14]$ and $[24]$, we define a set of strings indexed by $k \in \mathbb{N}^+$ as follows: $s_1(k) = a$, $s_6(k) = a^k$, $s_9(k) = ba^k$, $s_{14}(k) = ba^k ba^k$ and $s_{24}(k) = (ba^k ba^k)^k$. It is established in the proof of Lemma 3.1 that $e^\#(i,j) = \omega$ iff $t^k(i,j) \geq k$ for all $k \geq 1$ where e is an idempotent and $t \in \Psi^{-1}(e)$. Using this result and the definition of matrix multiplication we can argue that for $h = 1, 6, 9, 14, 24$, $[h](i,j) = \omega$ iff $D(s_h(k))(i,j) \geq k$ for every k. Since $u_I[24]v_F = \omega$, we have $d(s_{24}(k)) = u_I D(s_{24}(k))v_F \geq k$ for every k. Thus, $s_{24}(k)$ is a set of strings that demonstrates that A is not limited in distance.

In fact, this technique can always be applied to a distance automaton not limited in distance and returns to us a set of strings (witnesses) such that the distances required to accept them grow unboundedly. Such a set of strings of the form $((ba^k)(ba^k))^k$ resembles the multiplicative rational expression $((ba^*)(ba^*))^*$ except that we allow only the substitution of $*$ symbols by the same constant k. The same characterization of distance automata that are not limited in distance is obtained in ([10], [15], [22]).

Observe that in the expression $(((2][1]^\#)([2][1]^\#))^\#$ for matrix $[24]$, the symbol $\#$ is nested for two levels which corresponds to the fact that we need to wait for the second round of applying stabilization before we can obtain $[24]$. This number of nested levels of $\#$'s is also directly related to the fact the the length of the string $s_{24}(k)$ grows quadratically as k increases. Since it is known ([20], [4]) that for each $p \in \mathbb{N}^+$, there exists A_p not limited in distance such that the shortest string $s_p(k)$ that requires distance k has length $\Theta(k^p)$, we conclude that for any $p \in \mathbb{N}^+$, there exists a finite set $T \in M_n \mathcal{T}$ such that the number of rounds needed to calculate $\Psi(T)^{+,\#}$ is at least p.

Next we will show that for a finite subset T of $M_n \mathcal{T}$, $\Psi(<T>^c) \subseteq \Psi(T)^{+,\#}$.

3.2   Semigroup theory

We are going to introduce a *minimal* amount of the local structure theory of finite semigroups, which will be sufficient for our work in this paper. All the topics discussed in this subsection can be found in Chapter 2 (Green's relations) of Lallement [13].

Let S be a finite semigroup. We define S^1 as follows: If S has an identity, then $S^1 = S$; otherwise we obtain S^1 by adding a new element 1 to S and the underlying binary operation is extended such that 1 behaves as an identity for S^1. We define[2] three equivalence relations $\mathcal{R}$, $\mathcal{L}$ and $\mathcal{D}$ on S as follows:

$$a\mathcal{R}b \Longleftrightarrow aS^1 = bS^1$$
$$a\mathcal{L}b \Longleftrightarrow S^1a = S^1b$$
$$a\mathcal{D}b \Longleftrightarrow S^1aS^1 = S^1bS^1.$$

$a\mathcal{R}b$ can be expressed equivalently as: either $a = b$ or there exist $x, y \in S$

---

[2]More accurately, the definition given here for $\mathcal{D}$ is called $\mathcal{J}$ in the literature. But there is no danger involved here since $\mathcal{J}$ and $\mathcal{D}$ are known to be the same for finite semigroups.

such that $ax = b$ and $by = a$. Similarly $a\mathcal{L}b$ can be expressed as: either $a = b$ or there exist $x, y \in S$ such that $xa = b$ and $yb = a$. $a\mathcal{D}b$ is equivalent to $b \in S^1aS^1$ and $a \in S^1bS^1$, that is, a can be transformed to b by multiplying by some elements (possibly empty) from S on both its left and right sides, and b can also be transformed to a. It is trivial to see that $\mathcal{R} \subseteq \mathcal{D}$ and $\mathcal{L} \subseteq \mathcal{D}$.

We say that a $\mathcal{D}$-class D is regular if there is an idempotent in D, otherwise it is nonregular. Let a be an element of S. We write R_a (respectively L_a and D_a) to denote the $\mathcal{R}$-class (respectively $\mathcal{L}$-class and $\mathcal{D}$-class) represented by a. We define a partial ordering on the $\mathcal{D}$-classes such that $D_a \leq_{\mathcal{D}} D_b$ iff $S^1aS^1 \subseteq S^1bS^1$. Equivalently, $\mathcal{D}_a <_{\mathcal{D}} \mathcal{D}_b$ if a can be reached from b by multiplying with elements on the left and right sides, but not the other way around. Warning: the reader should distinguish $\leq_{\mathcal{D}}$ from the partial ordering $\leq$ on the matrices that we defined in the previous section.

We illustrate the concepts by giving the local structure for the finite semigroup $(\Psi(D(a)), \Psi(D(b)))^{+,\#}$ in Figure 2. The 26 matrices in the semigroup are grouped under 11 $\mathcal{D}$-classes, of which three contain six matrices each and the other eight are singleton sets. The partial ordering of the $\mathcal{D}$-classes is also presented, the "larger" $\mathcal{D}$-classes being positioned higher in the diagram. Within a single $\mathcal{D}$-class, matrices that are $\mathcal{R}$-equivalent are placed in the same row and matrices that are $\mathcal{L}$-equivalent are placed in the same column. For example, the bottom $\mathcal{D}$-class represented by matrix 7 is regular and has six matrices that are all $\mathcal{L}$-equivalent whereas for the $\mathcal{D}$-class represented by matrix 13, matrices 13 and 14 (respectively 15 and 16, 17 and 18) are $\mathcal{R}$-equivalent and matrices 13, 15 and 17 (respectively 14, 16 and 18) are $\mathcal{L}$-equivalent.

We need the following technical lemma for this section. However, we are not going to prove the result starting from first principles.

Lemma 3.4 *Let S be a finite semigroup. Let D be a $\mathcal{D}$-class of S and let $a, b \in D$. Then $ab \in D$ implies that $ab \in R_a \cap L_b$ and $R_b \cap L_a$ contains an idempotent, hence D is regular.*

Proof The lemma is an immediate corollary of the result of Miller and Clifford (see Proposition 2.5 of [13]) which states that $ab \in R_a \cap L_b$ iff $R_b \cap L_a$ contains an idempotent, and of Lemma 6.3 of [21] which states that for any a, b in a finite semigroup, $a\mathcal{D}ab$ iff $a\mathcal{R}ab$, and $b\mathcal{D}ab$ iff $b\mathcal{L}ab$. □

We need the following simple result frequently in our discussions. Let e be an idempotent. Let $a\mathcal{L}e$ and $b\mathcal{R}e$. Then there exist $x, y \in S^1$ such that $a = xe$ and $b = ey$. Hence, $ae = xee = xe = a$ and $eb = eey = ey = b$.

Observe that we cannot have two different $\mathcal{D}$-classes D_1 and D_2 both at the bottom. Suppose the contrary and let $a \in D_1$ and $b \in D_2$. If $ab \notin D_1$ then D_1 is not at the bottom; otherwise D_2 is not at the bottom. In each case we have a contradiction. Thus, the bottom $\mathcal{D}$-class is unique. If we take two elements a, b from the bottom $\mathcal{D}$-class then ab is also in the same $\mathcal{D}$-class since it is at the bottom. Then by Lemma 3.4, the bottom $\mathcal{D}$-class is regular.

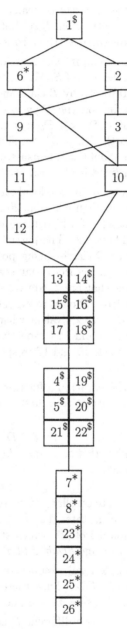

Figure 2: The local structure of the semigroup $(\Psi(D(a)), \Psi(D(b)))^{+,\#}$. Symbols $ and $*$ denote unstable and stable idempotents respectively.

3.3 More about stabilization

Lemma 3.5 *Let e be an idempotent in $M_n\mathcal{R}$. Then $e^\# = e^\# e = ee^\# = e^\# e^\# = (e^\#)^\#$.*

Proof Use the definition of $e^\#$ and the facts that Ψ is a homomorphism and the matrix multiplication is continuous. □

Lemma 3.6 *Let e and f be idempotents in the same $\mathcal{D}$-class of a subsemigroup S of $M_n\mathcal{R}$. Then e and f are either both stable or both unstable.*

Proof Let[3] $r_1 \in R_e \cap L_f$. Since $r_1 \mathcal{R} e$, there exist $x, y \in S^1$ such that $r_1 x = e$ and $ey = r_1$. Let $r_2 = fx$. We claim that $r_1 r_2 = e$ and $r_2 r_1 = f$. Since $r_1 \mathcal{L} f$, there exists $s, t \in S^1$ such that $sr_1 = f$ and $tf = r_1$. Thus, $r_1 r_2 = r_1 fx = tffx = tfx = r_1 x = e$ and $r_2 r_1 = fxr_1 = sr_1 xr_1 = ser_1 = seey = sey = sr_1 = f$.

Let $a \in \Psi^{-1}(r_1)$ and $b \in \Psi^{-1}(r_2)$. Then $ab \in \Psi^{-1}(r_1 r_2) = \Psi^{-1}(e)$ and $ba \in \Psi^{-1}(r_2 r_1) = \Psi^{-1}(f)$. Next $e^\# = \Psi(\lim_{k\to\infty}(ab)^k) = \Psi(a(\lim_{k\to\infty}(ba)^k)b)$ $= \Psi(a)\Psi(\lim_{k\to\infty}(ba)^k)\Psi(b) = r_1 f^\# r_2$. Similarly, $f^\# = \Psi(\lim_{k\to\infty}(ba)^k) = r_2 e^\# r_1$. Thus, if e is stable then $f^\# = r_2 e^\# r_1 = r_2 e r_1 = r_2 r_1 = f$, where $er_1 = r_1$ since $r_1 \in R_e$, and hence f is also stable. Similarly, f is stable implies that $e^\# = r_1 f^\# r_2 = r_1 f r_2 = rffx = r_1 fx$ which has previously been shown to be e. □

Lemma 3.7 *Let e be an idempotent in a subsemigroup of $M_n\mathcal{R}$. Then $e^\# \leq_\mathcal{D} e$. Furthermore, $e^\# <_\mathcal{D} e$ if e is unstable.*

Proof By Lemma 3.5 $e^\# = ee^\#$. Thus $e^\# \leq_\mathcal{D} e$. Again by Lemma 3.5, $e^\#$ is an idempotent which is stable. If e is unstable, then e and $e^\#$, which are not both stable or both unstable, cannot be $\mathcal{D}$-equivalent by Lemma 3.6. □

As a corollary, the bottom $\mathcal{D}$-class must be both regular and stable.

We want to extend the definition of stabilization to all elements in a regular $\mathcal{D}$-class D of a semigroup S. Let $a, e \in D$ and let e be an idempotent. Let $u, v \in D$ such that $a = uev$. We define $a^\# = ue^\# v$. By the following lemma, $a^\#$ is well defined[4].

Lemma 3.8 *Let e and f be idempotents in the same $\mathcal{D}$-class D of a subsemigroup S of $M_n\mathcal{R}$. Let $a \in D$. Let $u, v, x, y \in D$ such that $a = uev = xfy$. Then $ue^\# v = xf^\# y$.*

---

[3]Let a, b be elements in the same $\mathcal{D}$-class. Then (Chapter 2 of [13]) $R_a \cap L_b \neq \emptyset$.

[4]We can also define $a^\# = ue^\# v$ where $u, v \in S^1$ and $a = uev$. This more general version can be shown [14] to be well defined.

Proof Suppose that we can prove the special case of the lemma when $u = ue$, $v = ev$, $x = xf$ and $y = fy$. Then the lemma can be proved as follows. Since $(ue) = (ue)e$, $(ev) = e(ev)$, $(xf) = (xf)f$, $(fy) = f(fy)$ and $(ue)e(ev) = uev = a = xfy = (xf)f(fy)$, by the special case of the lemma $(ue)e^{\#}(ev) = (xf)f^{\#}(fy)$. Thus by Lemma 3.5, $ue^{\#}v = u(ee^{\#}e)v = (ue)e^{\#}(ev) = (xf)f^{\#}(fy) = x(ff^{\#}f)y = xf^{\#}y$.

Therefore, our aim is to prove the special case of the lemma when $u = ue$, $v = ev$, $x = xf$ and $y = fy$. By Lemma 3.4, $u\mathcal{L}e$, $v\mathcal{R}e$, $x\mathcal{L}f$ and $y\mathcal{R}f$.

By Lemma 3.4 and the given fact that $a = uev = xfy$, we have $a\mathcal{R}u\mathcal{R}x$ and $a\mathcal{L}v\mathcal{L}y$. $u\mathcal{R}x$ implies that there exists t such that $ut = x$. Define $r_1 = et$. Then $ur_1 = uet = ut = x$. From $r_1 = et$, $e \in D$ and $ur_1 = x \in D$, it follows that $r_1 \in D$. By Lemma 3.4, $r_1 = et = e(et) = er_1$ implies that $r_1\mathcal{R}e$. Also by Lemma 3.4, $ur_1 = x$ implies that $x\mathcal{L}r_1$. $v\mathcal{L}y$ implies that there exists s such that $sv = y$. Define $r_2 = se$. Then $r_2v = sev = sv = y$. From $r_2 = se$, $e \in D$ and $r_2v = y \in D$, it follows that $r_2 \in D$. By Lemma 3.4, $r_2e = (se)e = se = r_2 \in D$ implies that $r_2\mathcal{L}e$. Again by Lemma 3.4, $r_2v = y$ implies that $y\mathcal{R}r_2$. To summarize, $r_1\mathcal{L}x\mathcal{L}f$, $r_2\mathcal{L}e\mathcal{L}u$, $r_2\mathcal{R}y\mathcal{R}f$ and $r_1\mathcal{R}e\mathcal{R}v$.

We want to show that $r_1y = v$. Since $u\mathcal{L}e$, there exists p such that $e = pu$. Then $r_1y = er_1y = pur_1y = pxy = pxfy = pa = puev = puv = ev = v$. Next we want to show that $xr_2 = u$. Since $v\mathcal{R}e$, there exists q such that $e = vq$. Then $xr_2 = xr_2e = xr_2vq = xyq = xfyq = aq = uevq = uvq = ue = u$.

Since $f\mathcal{R}y$, $f\mathcal{L}x$ and $r_2\mathcal{L}e$, there exist b, c, d such that $yb = f$, $cx = f$ and $dr_2 = e$. Then $r_2r_1 = fr_2r_1f = cxr_2r_1yb = cxr_2er_1yb = cuevb = cxfyb = fff = f$ and $r_1r_2 = er_1r_2 = dr_2r_1r_2 = dfr_2 = dr_2 = e$. Knowing that $r_2r_1 = f$ and $r_1r_2 = e$, we have $f^{\#} = r_2e^{\#}r_1$ by using the same proof as given in the proof of Lemma 3.6. Therefore, $xf^{\#}y = xr_2e^{\#}r_1y = ue^{\#}v$. $\square$

3.4   Strategy for computing the topological closure

In subsection 3.1, we computed $\Psi(T)^{+,\#}$ for $T = \{D(a), D(b)\}$. We generated the resulting matrices in the order [1], [2], ..., [26]. However, when we consider the positions of these matrices in Figure 2, we see that the ordering of the matrices [1] through [26] does not correspond to the partial ordering defined by the $\mathcal{D}$-classes. We know that matrices in M_nT with structures belonging to the higher $\mathcal{D}$-classes could generate by multiplication and stabilization new matrices in M_nT with structures belonging to the lower $\mathcal{D}$-classes, but not the other way around. Therefore, in order to show that $\Psi(<T>^c) \subseteq \Psi(T)^{+,\#}$, we follow the partial ordering defined by the $\mathcal{D}$-classes to inspect the elements in $<T>^c$ for their Ψ images. That is, the elements in $<T>^c$ with structures belonging to the higher $\mathcal{D}$-classes are considered earlier for their Ψ images than the elements in $<T>^c$ with structures belonging to the lower $\mathcal{D}$-classes.

Let us look at our usual example with $T = \{D(a), D(b)\}$. To begin with, we have two matrices $D(a)$ and $D(b)$ in M_nT. $D(a)$ has the structure of matrix [1] whereas $D(b)$ has the structure of matrix [2]. Since the structure of $D(a)$ is located higher in the partial ordering, we begin our computation for $<T>^c$ there. Matrix [1] is an idempotent. Thus $D(a)$ has an idempotent structure. The computation of $<D(a)>$ generates an infinite set of matrices. They are $[0, \infty, \infty; k, \infty; 0]$ for all $k \in \mathbb{N}^+$. This gives rise to one new limit point $[0, \infty, \infty; \omega, \infty; 0]$; this has the structure of matrix [6], which is below matrix [1] in the partial ordering of the $\mathcal{D}$-classes. We verify that matrix [6] belongs to $\Psi(T)^{+,\#}$. Before we proceed to compute more elements in $<T>^c$, we find that we have already generated an infinite number of elements.

We are not exactly interested in computing $<T>^c$. It is the image of $<T>^c$ under Ψ that we want to analyse. Since this image is finite, there exists $K \in \mathbb{N}^+$ such that $\Psi(<T>^c) = \Psi(\{t \in <T>^c \mid \|t\| \leq K\})$. Note that $\{t \in <T>^c \mid \|t\| \leq K\}$ is finite. In the following discussions, we assume that the value of K is known. Remark: we do not assume that the value of K can be determined algorithmically. With this assumption, our aim is compute $\{t \in <T>^c \mid \|t\| \leq K\}$ and argue that $\Psi(\{t \in <T>^c \mid \|t\| \leq K\}) \subseteq \Psi(T)^{+,\#}$. Thus $\Psi(<T>^c) \subseteq \Psi(T)^{+,\#}$.

We return to our example. The current collection of matrices, denoted C, consists of $[0, \infty, \infty; k, \infty; 0]$ for all $k \in \mathbb{N}^+$ and $[0, \infty, \infty; \omega, \infty; 0]$. Before we continue to generate new elements based on C by multiplications and taking limits, we can first throw away from C the elements $[0, \infty, \infty; k, \infty; 0]$ for all $k > K$. This is because our aim is to compute $\{t \in <T>^c \mid \|t\| \leq K\}$. The elements removed are playing a very similar role to $[0, \infty, \infty; \omega, \infty; 0]$. If we need to "use" (as part of a multiplication or as an element in a limit sequence) a certain matrix $[0, \infty, \infty; k', \infty; 0]$ where $k' > K$ to generate a matrix t in $\{t \in <T>^c \mid \|t\| \leq K\}$, then replacing the use of $[0, \infty, \infty; k', \infty; 0]$ by $[0, \infty, \infty; \omega, \infty; 0]$ would not change the result. Therefore it does not hurt to remove from C the matrices $[0, \infty, \infty; k, \infty; 0]$ for all $k > K$. The collection C after trimming is finite.

Our strategy is keep the current collection C finite. As new matrices are generated by taking limits, we verify that the new matrices have structures in $\Psi(T)^{+,\#}$. We follow the partial ordering of the $\mathcal{D}$-classes in our analysis.

Suppose that so far we are "successful" in the analysis (all matrices generated so far have structures in $\Psi(T)^{+,\#}$ and the current collection C is finite) and we are down to a certain $\mathcal{D}$-class D. That is, all the $\mathcal{D}$-classes that are above D (and possibly some other $\mathcal{D}$-classes not related to D) have already been considered for generating new matrices. Many of the matrices in C have structures that belong to $\mathcal{D}$-classes above D whereas the rest of the elements have structures that belong to D or to some $\mathcal{D}$-classes further below D or unrelated to D. The reason that we may have collected elements with structures that belong to $\mathcal{D}$-classes below D is because we could have generated some new limit points previously that are of structures far below its current $\mathcal{D}$-class position; for example, matrix [14] is unstable and, since $[14]^\# = [24]$, it gives rise to new limit points with the structure of matrix [24], whose position in the partial ordering is two $\mathcal{D}$-classes below.

We partition our current collection C into three finite subcollections: $C_{>D}$ consists of elements with structures that are above D, C_D consists of elements with structures that are in D, and C_{rest} consists of elements with structures that are below D or unrelated to D. Since $C_{>D}$ has been trimmed, $C_{>D}$ is a finite subset of $\{t \in <T>^c \mid D <_D D_{\Psi(t)}\}$ such that for $t_1, t_2 \in C_{>D}$, $D <_D D_{\Psi(t_1 t_2)}$ implies that either $t_1 t_2 \in C_{>D}$ or $t_1 t_2$ is not "needed"[5] in computing $\{t \in <T> \mid \|t\| \leq K\}$. We want to to compute $\{t \in <C_{>D}, C_D> \mid \Psi(t) \in D\}$ before we determine the new limit points, verify that their Ψ images are in $\Psi(T)^{+,\#}$ and do the trimming.

Let $t = t_1 t_2 \ldots t_k$ and $\Psi(t) \in D$ where $t_i \in C_{>D} \cup C_D$ for $1 \leq i \leq k$. We are interested in the shortest sequence $t_1 t_2 \ldots t_k$ that gives the result t. We claim that for $1 \leq i < k$, $\Psi(t_i t_{i+1}) \in D$. First, $\Psi(t_i t_{i+1})$ cannot belong to a $\mathcal{D}$-class that is below D or unrelated to D because otherwise $\Psi(t) \notin D$. Next we suppose that $D <_D D_{\Psi(t_i t_{i+1})}$. If $t_i t_{i+1} \in C_{>D}$ then we could shorten the sequence $t_1 t_2 \ldots t_k$ by replacing the subsequence $t_i t_{i+1}$ with an element having the same value as $t_i t_{i+1}$ directly from $C_{>D}$. If $t_i t_{i+1} \notin C_{>D}$ then, since we have assumed that $D <_D D_{\Psi(t_i t_{i+1})}$, it must be that $t_i t_{i+1}$ has been trimmed as not needed for computing $\{t \in <T>^c \mid \|t\| \leq K\}$. In this case, either t can be computed in another way or t is also not need for computing $\{t \in <T>^c \mid \|t\| \leq K\}$.

Let C_2 denote $\{t \in C_{>D} C_{>D} \mid \Psi(t) \in D\}$. If k is even, then t is composed of $t_1 t_2, t_3 t_4, \ldots, t_{k-1} t_k$ where $t_i t_{i+1} \in C_D C_D \cup C_D C_{>D} \cup C_{>D} C_D \cup C_2$. If k is odd, we can still group the t_i's as before except that the last three symbols are grouped together such that $t_{k-2} t_{k-1} t_k \in C_D C_D C_D \cup C_D C_D C_{>D} \cup C_D C_{>D} C_D \cup C_{>D} C_D C_D \cup C_{>D} C_D C_{>D} \cup C_2 C_D \cup C_D C_2 \cup C_2 C_{>D}$. Let

$$C_{\text{new}} = C_D \cup C_D C_{>D} \cup C_{>D} C_D \bigcup C_{>D} C_D C_{>D} \cup C_2 \cup C_2 C_{>D}$$

Then C_{new} is finite and $\Psi(C_{\text{new}}) \subseteq D$. By computing $\{t \in <C_{\text{new}}> \mid \Psi(t) \in D\}$, we "essentially" compute $\{t \in <C_{>D}, C_D> \mid \Psi(t) \in D\}$ in the sense that the differences of the two sets are matrices that are not needed for computing $\{t \in <T>^c \mid \|t\| \leq K\}$.

In the next subsection, C_{new} is renamed as C_D. We will describe how many new limit points could arise from the computation of $< C_D >$, what the Ψ images of the limit points are, and how we could trim our collection correctly. Warning: to perform trimming, one has to be more careful than naively throwing out all matrices t with $\|t\| > K$.

---

[5] $t_i t_{i+1}$ was trimmed from the collection C previously because its "use" can be replaced by another element in C.

3.5   Computing the topological closure for one $\mathcal{D}$-class

From the discussion in the previous subsection, we can assume that we are given a finite set of elements C_D that all have structures belonging to the same $\mathcal{D}$-class D. Applying multiplication, we may end up with an infinite set of elements. Then we need to determine the new limit points; we are going to argue that there are only finitely many of them. The Ψ images can be seen to be in $\Psi(T)^{+,\#}$. Finally, we need to trim our set of elements if it is infinite.

Recall that by Lemma 3.6 the idempotents for a regular $\mathcal{D}$-class are either all stable or all unstable. We say that a regular $\mathcal{D}$-class is stable if one of its elements is stable, otherwise it is unstable. Therefore, there are three types of $\mathcal{D}$-classes: D is nonregular, D is regular and stable, and D is regular but unstable.

Consider the case when D is nonregular. Let $t_1, t_2 \in C_D$. That is, $\Psi(t_1), \Psi(t_2) \in D$. Since D is nonregular and by Lemma 3.4, $\Psi(t_1 t_2) = \Psi(t_1)\Psi(t_2) \notin D$. Therefore, we do not get any new elements with structures in D by multiplying elements in C_D. Since C_D is finite, we do not get any new limit points and there is no need for trimming.

Consider the case when D is regular. We want to compute the set $C_D^* = \{t \in <C_D> \mid \Psi(t) \in D\}$ and analyse its new limit points.

We need a technical lemma (Lemma 3.9) before we continue.

Lemma 3.9 *Let e be an idempotent in a regular $\mathcal{D}$-class D. Let $e = \lambda\beta\mu$ where $\lambda, \beta, \mu \in D$. Then $e^{\#}(r, s) \in \{0, 1\}$ implies that there exist p, q such that $\lambda(r, p) \in \{0, 1\}$, $\beta(p, q) = 0$ and $\mu(q, s) \in \{0, 1\}$.*

Proof Suppose $e^{\#}(r, s) = 0$. Then $e(r, s) = 0$. Since $e = \lambda\beta\mu$, it follows that there exist p, q such that $\lambda(r, p) = 0$, $\beta(p, q) = 0$ and $\mu(q, s) = 0$.

Assume that $e^{\#}(r, s) = 1$. Since $e^{\#}(r, s) = 1$ and by the computational definition of $e^{\#}$, there exist g, h such that $e(r, g) \in \{0, 1\}$, $e(g, h) = 0$ and $e(h, s) \in \{0, 1\}$. By the facts that $e(g, h) = 0$ and $e = \lambda\beta\mu$, there exist p, q such that $\lambda(g, p) = \beta(p, q) = \mu(q, h) = 0$. Again using the facts that $e = \lambda\beta\mu$ and $\lambda, \beta, \mu, e \in D$ and by Lemma 3.4, we have $e\mathcal{R}\lambda$ and $e\mathcal{L}\mu$ which imply that $e\lambda = \lambda$ and $\mu e = \mu$ since e is an idempotent. Putting together $e(r, g) \in \{0, 1\}$, $\lambda(g, p) = 0$ and $e\lambda = \lambda$, we obtain $\lambda(r, p) \in \{0, 1\}$. Similarly by putting together $\mu(q, h) = 0$, $e(h, s) \in \{0, 1\}$ and $\mu e = \mu$, we obtain $\mu(q, s) \in \{0, 1\}$. □

Lemma 3.10 *Let $\Delta_{C_D} = \max_{t \in C_D} \|t\|$. Let $t = t_1 t_2 \dots t_k \in C_D^*$ where $t_i \in C_D$ for $1 \leq i \leq k$. Let $1 \leq i, j \leq n$ and $\Psi(t)(i, j) = 1$. Then $t(i, j) \leq 2\Delta_{C_D}$ if $\Psi(t)^{\#}(i, j) = 1$, and[6]*

$$t(i, j) \geq k - 2 \quad \text{if } \Psi(t)^{\#}(i, j) = \omega.$$

---

[6]More accurately, one could prove $t(i, j) \geq k$. But this requires more works in the proof. Moreover, the stronger statement is not really needed for the rest of the paper.

Proof Let us first consider the case when $\Psi(t)^{\#}(i,j) = 1$. When $k \leq 2$, it is trivial to see that $t(i,j) \leq 2\Delta_{C_D}$. Let $k \geq 3$. We want to show that there exist $1 \leq p, q \leq n$ such that $t_1(i,p) \in \mathbb{N}$, $(t_2 t_3 \ldots t_{k-1})(p,q) = 0$ and $t_k(q,j) \in \mathbb{N}$. Equivalently, we want to show that there exist $1 \leq p, q \leq n$ such that $\alpha(i,p) \in \{0,1\}$, $\beta(p,q) = 0$ and $\gamma(q,j) \in \{0,1\}$ where $\alpha = \Psi(t_1)$, $\beta = \Psi(t_2 t_3 \ldots t_{k-1})$ and $\gamma = \Psi(t_k)$. As a consequence, $t(i,j) \leq t_1(i,p) + t_k(q,j) \leq 2\Delta_{C_D}$.

Since $\alpha, \beta, \gamma, \alpha\beta\gamma \in D$, we have $\alpha\beta, \beta\gamma \in D$. By Lemma 3.4 there exist idempotents $e \in L_\alpha \cap R_\beta$ and $f \in L_\beta \cap R_\gamma$. $e\mathcal{R}\beta$ implies that there exists δ such that $\beta\delta = e$. Let $\mu = f\delta$. Since $\mu = f\delta$ where $f \in D$ and $\beta\mu = \beta f\delta = \beta\delta = e \in D$, we deduce $\mu \in D$. By the facts that $\mu = f\delta$ (which implies $f\mathcal{R}\mu$ by Lemma 3.4) and $f\mathcal{R}\gamma$, we have $\mu\mathcal{R}\gamma$. Thus, there exists $\xi \in D$ such that[7]. $\mu\xi = \gamma$ From the definitions and $f\mathcal{L}\beta$, $\Psi(t) = \alpha\beta\gamma = \alpha\beta\mu\xi = \alpha\beta f\delta\xi = \alpha\beta\delta\xi = \alpha e\xi$. Since $\Psi(t)^{\#} = \alpha e^{\#}\xi$, by the fact that $\Psi(t)^{\#}(i,j) = 1$ it follows that there exist r, s such that $\alpha(i,r) \in \{0,1\}$, $e^{\#}(r,s) \in \{0,1\}$ and $\xi(s,j) \in \{0,1\}$. Since $e = \beta\delta = \beta f\delta = \beta\mu = e\beta\mu$ and $e^{\#}(r,s) \in \{0,1\}$, by Lemma 3.9 there exist p, q such that $e(r,p) \in \{0,1\}$, $\beta(p,q) = 0$ and $\mu(q,s) \in \{0,1\}$. Putting together $\alpha\mathcal{L}e$, $\alpha(i,r) \in \{0,1\}$ and $e(r,p) \in \{0,1\}$, we get $\alpha(i,p) = (\alpha e)(i,p) \in \{0,1\}$. Putting together $\gamma = \mu\xi$, $\mu(q,s) \in \{0,1\}$ and $\xi(s,j) \in \{0,1\}$, we get $\gamma(q,j) = (\mu\xi)(q,j) \in \{0,1\}$. Therefore, we are done and $\alpha(i,p) \in \{0,1\}$, $\beta(p,q) = 0$ and $\gamma(q,j) \in \{0,1\}$.

Next we want to show that if $\Psi(t)^{\#}(i,j) = \omega$ then $t(i,j) \geq k - 2$. If $k \leq 2$ then $t(i,j) \geq 1$ since $\Psi(t)(i,j) = 1$. Thus $t(i,j) \geq k - 2$ is verified for $k \leq 2$. Let $k \geq 3$. Consider any sequence $i_1, i_2, \ldots, i_{k-1} \in \{1, \ldots, n\}$ such that $t_1(i,i_1) + t_2(i_1,i_2) + \cdots + t_k(i_{k-1},j) = t(i,j) \in \mathbb{N}^+$. It is sufficient to show that $t_p(i_{p-1}, i_p) \neq 0$ for all $2 \leq p \leq k - 1$.

Suppose on the contrary that there exists $2 \leq p \leq k - 1$ such that $t_p(i_{p-1}, i_p) = 0$. Let $\alpha = \Psi(t_1 \ldots t_{p-1})$, $\beta = \Psi(t_p)$ and $\gamma = \Psi(t_{p+1} \ldots t_k)$. From the given condition, we have $\alpha\beta\gamma = \Psi(t)$ and $\alpha(i,i_{p-1}) \in \{0,1\}$, $\beta(i_{p-1}, i_p) = 0$ and $\gamma(i_p, j) \in \{0,1\}$. Since $\alpha\beta \in D$, by Lemma 3.4 there exists an idempotent $e \in L_\alpha \cap R_\beta$. Next $e\mathcal{R}\beta$ implies that $e\beta = \beta$. Thus, $(e\beta)(i_{p-1}, i_p) = 0$. That is, there exists $1 \leq r \leq n$ such that $e(i_{p-1}, r) = \beta(r, i_p) = 0$. Since $\beta(r, i_p) = 0$ and $\gamma(i_p, j) \in \{0,1\}$, we get $(\beta\gamma)(r,j) \in \{0,1\}$. Since $\Psi(t) = \alpha e\beta\gamma$, by the definition of stabilization we have $\Psi(t)^{\#} = \alpha e^{\#}\beta\gamma$ where e is an idempotent and $\alpha, \beta\gamma \in D$. Note that $e(i_{p-1}, r) = 0$ implies that $e^{\#}(i_{p-1}, r) = 0$. Putting together $\alpha(i, i_{p-1}) \in \{0,1\}$, $e^{\#}(i_{p-1}, r) = 0$ and $(\beta\gamma)(r,j) \in \{0,1\}$, then $\Psi(t)^{\#}(i,j) = (\alpha e^{\#}\beta\gamma)(i,j) \neq \omega$ which contradicts our assumption. $\square$

Consider the case when D is stable. Let $t \in C_D^*$. It follows that $\Psi(t)(i,j) = 1$ implies $\Psi(t)^{\#}(i,j) = 1$. By Lemma 3.10, $t(i,j) \in \mathbb{N}$ means that $t(i,j) \leq$

---

[7]By definition, $\mu\mathcal{R}\gamma$ implies that there exists ξ' such that $\mu\xi' = \gamma$. Let $\xi = e\xi'$. Since $\beta\mu = e$ and e is an idempotent, $\mu\mathcal{L}e$ by Lemma 3.4 and $\mu e = \mu$. Then $\mu\xi = \mu e\xi' = \mu\xi' = \gamma$. From $\xi = e\xi'$ and $\mu\xi = \gamma \in D$, we have $D = D_\gamma \leq_D D_\xi \leq_D D_e = D$; that is, $\xi \in D$.

$2\Delta_{C_D}$. Thus, C_D^* is finite. Taking the topological closure of C_D^* would not give any new limit point.

We are now left with the case when D is unstable. By Lemma 3.10, the new limit points obtained when taking the topological closure of C_D^* would have structures that are in $\{a^\# \mid a \in D\}$. Let $a \in D$. Let t be a new limit point with $\Psi(t) = a^\#$. Then again by Lemma 3.10, for any i, j such that $a(i,j) \neq \omega$ we have $t(i,j) \in \{\infty, 0, 1, 2, \ldots, 2\Delta_{C_D}\}$. Therefore, the number of new limit points with the structure $a^\#$ is finite. Since D is a finite set, the overall number of new limit points is also finite.

After we have determined the finite set of new limit points $(C_D^*)^c - C_D^*$, we consider how to trim the set C_D^*.

Let $t \in C_D^*$. We define t^ω such that $t^\omega(i,j) = \omega$ if $\Psi(t)^\#(i,j) = \omega$ otherwise $t^\omega(i,j) = t(i,j)$. We are going to partition the elements in C_D^* into groups such that t_1, t_2 belong to the same group iff $t_1^\omega = t_2^\omega$. By Lemma 3.10, we have only a finite number of partitions. We consider the partitions one by one. If a certain partition has only a finite number of elements, then we do not apply any trimming to this partition. Consider a partition that has an infinite number of elements. Let t be an element in this partition. By Lemma 3.10, it is immediate to see that t^ω exists in $(C_D^*)^c - C_D^*$. We can now remove any element t in this partition that has a representation $t_1 t_2 \ldots t_{k+2}$ where $k > K$ and $t_i \in C_D$ for $1 \le i \le k+2$. This is because by Lemma 3.10 $t(i,j) \ge k > K$ if $\Psi(t)^\#(i,j) = \omega$ and therefore any use (as part of a multiplication or as an element in a limit sequence) of t in computing some element in $\{x \in <T>^c \mid \|x\| \le K\}$ can be replaced by $t^\omega \in (C_D^*)^c - C_D^*$ without changing the result obtained. Thus, the partition will be left with elements that have representations of lengths shorter than or equal to $K + 2$. Hence the partition after trimming has only a finite number of elements. Overall, the number of elements left in C_D^* after trimming is also finite.

We repeat the process from one D-class to the next, observing the partial ordering. This process stops when we reach the bottom D-class which must be regular and stable. Hence for the last D-class that we considered, the number of elements in C_D^* is finite.

From the above discussions, we have argued that the Ψ images of $\{x \in <T>^c \mid \|x\| \le K\}$ all belong to $\Psi(T)^{+,\#}$. Since by assumption $\Psi(<T>^c) = \Psi(\{x \in <T>^c \mid \|x\| \le K\})$, therefore $\Psi(<T>^c) \subseteq \Psi(T)^{+,\#}$. Together with Lemma 3.3, Theorem 1 is proved.

In our previous analysis for the single D-class case, the idempotents do not really behave differently from other elements. On the other hand, it is much easier to reason with them. We get a lot more insight into the structure of the problem by focusing on the idempotents.

4   Open Problems

There are two main open problems, which are related. It has been proven [15] that the limitedness problem for distance automata is PSPACE-hard. The first open problem is whether the limitedness problem can be solved in polynomial space. The second open problem is about the upper bound on the distance of a distance automaton given that it is limited in distance. The best upper bound; it is double exponential in n, was obtained by Hashiguchi [10] where n is the number of states in a distance automaton. However, we do not find any examples that match Hashiguchi's upper bound. Weber [24] found a family of distance automata that are limited in distance such that the distances are $2^n - 2$. Leung [17] also found another family of distance automata with single exponential finite distances in n. It is open whether in the general case the upper bound should be single exponential in n. If this can be shown then one important consequence, whose proof we omit, is that we can derive a nondeterministic algorithm running in polynomial space to decide the limitedness problem for distance automata. Hence, by Savitch's theorem (Theorem 7.12 of [2]) we can solve the limitedness problem in deterministic polynomial space.

References

[1] T.C. Brown (1971), 'An interesting combinatorial method in the theory of locally finite semigroups', *Pacific J. Math.*, **36** 285–289.

[2] M.R. Garey and D.S. Johnson (1979), *Computers and intractability*, Freeman, San Francisco, CA.

[3] J. Goldstine, C.M.R. Kintala and D. Wotschke (1990), 'On measuring nondeterminism in regular languages', *Inform. and Comput.*, **86** 179–194.

[4] J. Goldstine, H. Leung and D. Wotschke (1992), 'On the relation between ambiguity and nondeterminism in finite automata', *Inform. and Comput.*, **100** 261–270.

[5] K. Hashiguchi (1979), 'A decision procedure for the order of regular events', *Theor. Comput. Sci.*, **8** 69–72.

[6] K. Hashiguchi (1982), 'Regular languages of star height one', *Inform. and Control*, **53** 199–210.

[7] K. Hashiguchi (1982), 'Limitedness theorem on finite automata with distance functions', *J. Comput. Syst. Sci.*, **24** 233–244.

[8] K. Hashiguchi (1983), 'Representation theorems on regular languages', *J. Comput. System Sci.*, **27** 101–115.

[9] K. Hashiguchi (1988), 'Algorithms for determining relative star height and star height', *Inform. and Comput.*, **78** 124–169.

[10] K. Hashiguchi (1990), 'Improved limitedness theorem on finite automata with distance functions', *Theor. Comput. Sci.*, **72** 27–38.

[11] J. Hopcroft and J. Ullman (1979), *Introduction to automata theory, languages, and computation*, Addison-Wesley, Reading, MA.

[12] C.M.R. Kintala and D. Wotschke (1980), 'Amount of nondeterminism in finite automata', *Acta Inf.*, **13** 199–204.

[13] G. Lallement (1979), *Semigroups and combinatorial applications*, John Wiley and Sons, New York.

[14] H. Leung (1988), 'On the topological structure of a finitely generated semigroup of matrices', *Semigroup Forum*, **37** 273–287.

[15] H. Leung (1991), 'Limitedness theorem on finite automata with distance functions: an algebraic proof', *Theor. Comput. Sci.*, **81** 137–145.

[16] H. Leung (1991), 'On some decision problems in finite automata', in J. Rhodes (ed.), *Monoids and semigroups with applications*, 509–526, World Scientific, Singapore.

[17] H. Leung (1992), 'On finite automata with limited nondeterminism', *Lect. Notes Comput. Sci.*, **629** 355–363.

[18] I. Simon (1978), 'Limited Subsets of a Free Monoid', *Proc. Nineteenth Annual IEEE Symposium on Foundations of Computer Science*, 143–150.

[19] I. Simon (1988), 'Recognizable sets with multiplicities in the tropical semiring', *Lect. Notes Comput. Sci.*, **324** 107–120.

[20] I. Simon (1990), 'The nondeterministic complexity of a finite automaton', in M. Lothaire (ed.), *Mots – mélanges offerts à M. P. Schützenberger*, 384–400, Hermes, Paris.

[21] I. Simon (1990), 'Factorization forests of finite height', *Theor. Comput. Sci.*, **72** 65–94.

[22] I. Simon (1994), 'On semigroups of matrices over the tropical semiring', *RAIRO ITA*, **28** 277–294.

[23] B.T. Sims (1976), *Fundamentals of topology*, Macmillan, New York.

[24] A. Weber (1993), 'Distance automata having large finite distance or finite ambiguity', *Math. Syst. Theory*, **26** 169–185.

Types and Dynamics
in Partially Additive Categories

Gianfranco Mascari and Marco Pedicini

1 Introduction

A challenging problem in studying complex systems is to model the internal structure of the system and its dynamical evolution in an integrated view. An algorithm computes a function with domain consisting of all its possible inputs and as codomain all possible outputs. In order to "compare" algorithms computing the same function (e.g. with respect to time or space complexity) it is necessary to model the "dynamic behavior" of the algorithm. The main approach for modeling the dynamic behavior $Bh(\mathbf{P})$ of an algorithm is "the dynamic system paradigm": $Bh(\mathbf{P})$ is obtained as the action of a (control) monoid on a (state) space. Various mathematical structures have been considered for attacking such a problem; among others, deterministic automata and stochastic automata. In order to capture the fine structure of computation, an attempt to describe the "macroscopic" global dynamic behavior of an algorithm in terms of the "microscopic" local dynamic behavior of its components has recently been proposed within a logical approach to computing called "Geometry of Interaction" developed within C^*-algebras. We propose an algebraic approach to "structured dynamics" inspired by the Geometry of Interaction and based on a "many objects" generalization of semirings: partially additive categories.

The paper is organized as follows: in the next section we present a general view of structural dynamics inspired by proof theory. In Section 3 we present the execution of pseudoalgorithms in the framework of partially additive categories as a kind of iteration of endomorphisms. In Section 4 we consider deadlock-free pseudoalgorithms as characterized by a summability condition, and the theorem stating modularity with respect to dynamics is proved. Finally, in Section 5, concepts about structural aspects are organized via type discipline and type compatible dynamics.

2 Structural Dynamics

The methodological approach emerging from recent research in proof theory suggests dealing with structural aspects of systems by means of logical connectives, and dynamical properties by means of transforming logical proofs into simpler ones with respect to logical connectives.

In this section we illustrate the idea of structural dynamics by means of a fragment of linear logic where logical rules capture the structural aspects, and the identity group the dynamical ones.

The logical approach to computing, based on the so called Curry–Howard isomorphism, can be summarized as follows:

Type T_1	Type $T_1 \to T_2$	Formula A	Formula $A \to B$
Data $v : T_1$	Program $u : T_1 \to T_2$	Proof v of A	Proof u of $A \to B$
	Execution of $\mathrm{APP}(u,v)$		Normalization of $\mathrm{MP}(u,v)$

In the first case $\mathrm{APP}(u,v)$ is the ensemble consisting of the input passed to the program with the program itself. In the second case, $\mathrm{MP}(u,v)$ is the proof obtained from u and v by applying the basic logical rule *modus ponens* "From A and $A \to B$, deduce B", and the normalization of a proof obtained by modus ponens consists in transforming it in a proof obtained by modus ponens of "lower complexity". Such a paradigm is illustrated by considering the Multiplicative fragment of Linear Logic (MLL). This study is to be considered a preliminary step, because of the very weak expressive power of the multiplicative fragment; more interesting fragments and derivations of linear logic to be considered are: the MELL (Multiplicative with Exponential Linear Logic), ELL (Elementary Linear Logic), LLL (Light Linear Logic) and obviously LL (full Linear Logic with all connectives).

- **Logical Rules**

Times
$$\frac{\vdash A, \Gamma, [\Delta] \quad \vdash B, \Gamma', [\Delta']}{\vdash A \otimes B, \Gamma, \Gamma', [\Delta, \Delta']} \otimes$$

Par
$$\frac{\vdash A, B, \Gamma, [\Delta]}{\vdash A \wp B, \Gamma, [\Delta]} \wp$$

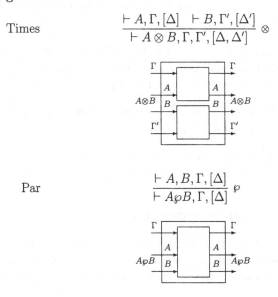

- **Identity Group**

Axiom $\vdash A^{\perp}, A$

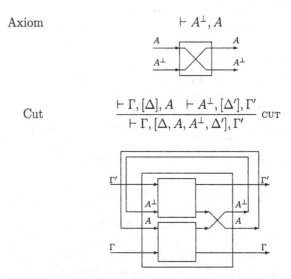

Cut $$\dfrac{\vdash \Gamma, [\Delta], A \quad \vdash A^{\perp}, [\Delta'], \Gamma'}{\vdash \Gamma, [\Delta, A, A^{\perp}, \Delta'], \Gamma'} \ \text{CUT}$$

The crucial point for obtaining the structural dynamics of proof systems is to give an interpretation of the cut rule (i.e. modus ponens) by means of a feedback loop, as will be shown in Theorem 5.1.

3   Execution of Pseudoalgorithms

Dynamical and structural aspects can be expressed by means the following concepts:

- a notion of convergence capturing the evolutionary law of dynamics represented by the monoid action on a state space; this is a motivation to consider partially additive categories [34].

- a binary construction that allows the combination of systems, this suggests the use of biproducts.

Several mathematical structures (e.g. inverse semigroups [16], C^*-algebras [21] and complete partial orders [2]) have been used inside such paradigms.

3.1   Partially additive categories

Let us fix a graphical language for categories; in order to make more explicit the link between programs, flow diagrams and morphisms, statements in category theory will be also presented in a flow diagrams metaphor.

- A morphism $f : a \to b$ will be

- Identity will be the void flow diagram

- Composition of $f : a \to b$ with $g : b \to c$ is the compositional statement:

$$\xrightarrow{a} \boxed{f} \xrightarrow{b} \boxed{g} \xrightarrow{c}$$

Moreover, in a category with coproduct, morphisms such as $f : a_1 \amalg \ldots \amalg a_n \to b_1 \amalg \ldots \amalg b_m$ will be

$$\begin{array}{c} a_n \\ \vdots \quad \boxed{f} \quad \vdots \\ a_1 \end{array} \begin{array}{c} b_m \\ \\ b_1 \end{array}$$

Definition 3.1 *Let I a countable set, a **partially additive monoid** over a set M is (M, Σ) where*

$$\Sigma : M^* = \{(x_i)_{i \in I} \mid x_i \in M\} \to M$$

is a partial function defined over countable families. A family $(x_i)_{i \in I}$ is said **summable** *if $\sum_{i \in I} x_i$ is defined. The following axioms must be satisfied by Σ:*

- Partition-associative *Given $(x_i)_{i \in I} \in M^*$ and a partition $(I_j)_{j \in J}$ of I then $(x_i)_{i \in I}$ is summable if and only if every family $(I_j)_{j \in J}$ is summable and the family $(\sum_{i \in I_j} x_i)_{j \in J}$ too. In this case $\sum_{i \in I} x_i = \sum_{j \in J}(\sum_{i \in I_j} x_i)$.*

- Singleton summability *If $I = \{i_0\}$, $\sum_{i \in I} x_i$ is defined to be equal to x_{i_0}.*

- Limit sums *$(x_i)_{i \in I} \in M^*$, if for every $(x_i)_{i \in F}$ with $F \subset I$ and $\#(F) < \infty$, $\sum_{i \in F} x_i$ is defined then $\sum_{i \in I} x_i$ is defined too.*

Remark Equalities between morphisms are to be considered in the strong sense: the left hand side and right hand side expressions are both defined and equal, or both undefined.

Definition 3.2 *A category $\mathcal{M}$ is **partially preadditive** if for every hom-set there is a partially additive monoid such that composition is distributive over sums: if $(f_i : a \to b)_{i \in I}$ is a summable family then for every $g : c \to a$ and $h : b \to d$ the families $(f_i g)_{i \in I}$ and $(h f_i)_{i \in I}$ are summable and*

$$\left(\sum_{i \in I} f_i\right) g = \sum_{i \in I} (f_i g)$$

$$h\left(\sum_{i \in I} f_i\right) = \sum_{i \in I} (h f_i).$$

For a given family $(f_i)_{i \in I}$ such that $f_i : a \to b$ the morphism $\sum_{i \in I} f_i$ will be represented in flow diagrams as

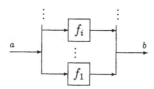

Definition 3.3 *In a category $\mathcal{C}$ a **family of zero morphisms** is $0_{a,b}$ where $a, b \in Obj(\mathcal{C})$ such that for every $f : a \to b$ and $g : c \to d$*

$$0_{a,c} g = f 0_{b,d}.$$

Proposition 3.1 *A partially preadditive category $\mathcal{M}$ has zero morphisms.*

Definition 3.4 *Given a category $\mathcal{C}$ with countable coproducts and zero morphisms, for every countable family $(a_i)_{i \in I}$, $a_i \in Obj(\mathcal{C})$ and for every $J \subset I$:*

1. *The **quasi-projection** $\pi_J : \coprod_{i \in I} a_i \to \coprod_{i \in J} a_i$ is defined by*

$$\pi_J in_i = \begin{cases} in_i & i \in J \\ 0 & i \notin J. \end{cases}$$

   When $J = \{i_0\}$, by a slight abuse of notation π_J will be written π_{i_0}.

2. *The **diagonal injection** $\Delta : \coprod_{i \in I} a_i \to \coprod_{i \in I} (\coprod_{i \in I} a_i)$ is defined by*

$$\Delta in_j = in'_j in_j$$

   where $in_j : a_j \to \coprod_{i \in I} a_i$, and $in'_j : \coprod_{i \in I} a_i \to \coprod_{i \in I} (\coprod_{i \in I} a_i)$.

3. *When $a_i = a$, the* **codiagonal morphism** $\nabla \coprod_{i \in I} a \to a$ *is defined by:*

$$\nabla in_i = id_a$$

Definition 3.5 *A* **partially additive category** *is a partially preadditive category C which has countable coproducts and satisfies the following two axioms:*

- Compatible-sum axiom *If $(f_i)_{i \in I}$ is a countable family such that $f_i \in Hom(a, b)$, if there exists $f : a \to \coprod_{i \in I} b$ such that $\pi_i f = f_i$ then $(f_i)_{i \in I}$ is summable*

- Untying axiom *If $f, g : a \to b$ are summable then also $in_1 f, in_2 g : a \to b \amalg b$ are summable.*

Let us recall some uniqueness results useful for working in partially additive categories.

Theorem 3.1 *If C is a partially additive category then a family $(f_i)_{i \in I}$ of arrows $f_i : a \to b$ is summable if and only if it is compatible. So, if this is the case there exists a unique $f : a \to \coprod_{i \in I} b$ where $\pi_i f = f_i$ and*

$$\sum_{i \in I} f_i = f \nabla^b.$$

Proposition 3.2 *Given a family $(f_i)_{i \in I}$ and $(g_i)_{i \in I}$ where $f_i : a \to b$ and $g_i : b \to c$, if $\sum_{i \in I} f_i$ is defined then $\sum_{i \in I} g_i f_i$ is defined.*

Proposition 3.3 *Given $f : a \to \coprod_{i \in I} b_i$ in a partially additive category C,*

there exists a unique family $(f_i)_{i \in I}$ with $f_i : a \to b_i$ such that

$$f = \sum_{i \in I} in_i f_i$$

A structure of direct sum having at the same time the property of product and coproduct can be added to a partially additive category.

Definition 3.6 *Given a partially additive category C a **biproduct diagram** for objects a_1, a_2 is a tuple (c, i_1, i_2, p_1, p_2) such that*

$$a_1 \underset{i_1}{\overset{p_1}{\rightleftarrows}} c \underset{i_2}{\overset{p_2}{\rightleftarrows}} a_2$$

with arrows which satisfy the following identities:

$$\begin{cases} p_k i_k = 1_{a_k}, \ k \in \{1,2\} \\ p_k i_j = 0_{a_k a_j}, \ k \neq j \\ i_1 p_1 + i_2 p_2 = 1_c. \end{cases}$$

Remark Given a partially additive category C then for every $a, b \in C$ the tuple $(a \amalg b, in_1, in_2, \pi_1, \pi_2)$ is a biproduct diagram. Conversely for every biproduct diagram, $(a \oplus b, i_1, i_2)$ is a coproduct.

Proposition 3.4 *The biproduct object for $a_1, a_2 \in Obj(C)$ is determined uniquely up to an isomorphism.*

Definition 3.7 *Given a partially additive category C, if for all $a_1, a_2 \in Obj(C)$ there exists a biproduct diagram, the choice of a $c = a_1 \oplus a_2$ determines a bifunctor $\oplus : C \times C \to C$, with $f_1 \oplus f_2$ defined by the equations*

$$\pi'_j(f_1 \oplus f_2) = f_j \pi_j \qquad j \in \{1, 2\}$$

where $\pi'_j : b_1 \oplus b_2 \to b_j$ is the projection on the codomain biproduct.

The sums in hom-sets and biproducts can *simulate* each other.

Proposition 3.5 *Given two arrows $f_1, f_2 : a \to b$ in a partially additive category $(C, \oplus)$, then*

$$f_1 + f_2 = \nabla^b(f_1 \oplus f_2)\Delta_a$$

where $\nabla^b : b \amalg b \to b$ is the codiagonal map and $\Delta_a : a \to a \times a$ is the diagonal.

$$a \overset{a}{\longrightarrow} \boxed{\Delta_a} \overset{a}{\underset{a}{\longrightarrow}} \boxed{f_1 \oplus f_2} \overset{b}{\underset{b}{\longrightarrow}} \boxed{\nabla^b} \overset{b}{\longrightarrow}$$

The usual matrix calculus in additive categories can be extended to partially additive categories.

Proposition 3.6 *Given a morphism in a partially additive category* $f : a_1 \oplus a_2 \to b_1 \oplus b_2$, *let* $f_{ji} = \pi_i f i n_j$, *then* $\sum i n_i f_{ji} \pi_i$ *is equal to* f. *If further* $g : b_1 \oplus b_2 \to c_1 \oplus c_2$ *then the composition morphism* $gf : a_1 \oplus a_2 \to c_1 \oplus c_2$ *is equal to the matrix product of morphisms.*

Proof $f = 1_{b_1 \oplus b_2} f 1_{a_1 \oplus a_2}$. By definition of biproduct $a_1 \oplus a_2$, $f = 1_{b_1 \oplus b_2} f(in_1 \pi_1 + in_2 \pi_2)$. By distributivity, $f = 1_{b_1 \oplus b_2} f in_1 \pi_1 + 1_{b_1 \oplus b_2} f in_2 \pi_2$. By definition of biproduct $b_1 \oplus b_2$, $f = (in_1 \pi_1 + in_2 \pi_2) f in_1 \pi_1 + (in_1 \pi_1 + in_2 \pi_2) f in_1 \pi_1$ and so $f = \sum_{i,j} in_i f_{ji} \pi_i$ as claimed.

The matrix product is the sum: $in_1(f_{11}g_{11} + f_{12}g_{21})\pi_1 + in_1(f_{11}g_{12} + f_{12}g_{22})\pi_2 + in_2(f_{21}g_{11} + f_{22}g_{21})\pi_1 + in_2(f_{21}g_{12} + f_{22}g_{22})\pi_2$. By zero morphism properties, when $h_{11}^1 = f_{11}$, $h_{12}^2 = f_{12}$, $h_{21}^3 = f_{21}$, $h_{22}^4 = f_{22}$ and otherwise $h_{ij}^k = 0$. Hence $f = \sum_{i,j} \sum_{k=1}^{4} in_i h_{ij}^k \pi_j$. By subfamilies summability for every choice of index $h_{ij} = \sum_{k=1}^{4} in_i h_{ij}^k \pi_j$ are summable morphisms. For the analogously defined $\tilde{h}_{ij}$ w.r.t. g_{ij} the following is summable

$$\tilde{h}_{11} + \tilde{h}_{12} + \tilde{h}_{21} + \tilde{h}_{22}.$$

Hence the matrix product is equal to

$$h_{11}\tilde{h}_{11} + (h_{12} + h_{22})\tilde{h}_{12} + (h_{21} + h_{11} + h_{21})\tilde{h}_{21} + (h_{12} + h_{22})\tilde{h}_{22}.$$

By proposition 3.2 we deduce its summability. □

In a similar way, the compatibility of matrix summability with componentwise summability can be tested.

Proposition 3.7 *Let* $(f_n)_{n \in \omega}$ *be a family of matrix morphisms,* $f_n : a_1 \oplus \cdots \oplus a_n \to b_1 \oplus \cdots \oplus b_m$; *then* f_n *is summable if and only if the families* $((f_n)_{i,j})_{n \in \omega}$ $i \in \{1, \ldots, n\}$ *and* $j \in \{1, \ldots m, \}$ *are summable.*

Proof By the distributive laws, given a family of matrix morphisms $\sum f_n$ we can prove the summability of the component i, j by $\pi_j(\sum f_n)in_i = \sum(\pi_j f_n in_i) = \sum(f_n)_{i,j}$. Conversely, the summability of the family $(in_i \pi_j)_{i,j \in \{1,2\}}$ and of the component families $((f_n)_{i,j})_{n \in \omega}$ allows us to apply proposition 3.2 and deduce the summability of $(\sum_{i,j} in_i(f_n)_{i,j} \pi_i in_i \pi_j)_{n \in \omega}$, i.e. $(\sum_{i,j} in_i(f_n)_{i,j}\pi_j)_{n \in \omega}$. From the partition axiom we deduce the summability of the family $(in_i(f_n)_{i,j}\pi_j)_{(i,j,n)}$, i.e. the resulting matrix morphism. □

3.2 Dynamics of pseudoalgorithms

In this subsection the mathematical machinery introduced above will be applied in order to define an abstract notion of algorithm taking into account that the dynamics associated with the execution of an algorithm can follow three possible behaviors: termination, failure over data or nontermination. Here, on the main road pointed out by J.Y. Girard in linear logic, we show how the algebraic setting of partially additive categories can be used in order to model these behaviors. The machine to execute is the iteration of a certain dynamics, so we are involved, as shown in Theorem 3.1, with solutions of fixed-point equations. The theorem allows us to consider every *iterable* morphism (endomorphisms) as an algorithm, i.e. something with an execution attached to it. What we are introducing are the basic tools:

- for distinguishing terminating execution algorithms from the rest;

- to manage safely (compose) algorithms with good behaviors (i.e. a notion of type).

Notation The following notations will be used : $\vec{a} = \bigoplus_{i=1}^{n} a_i$, $\vec{b} = \bigoplus_{i=1}^{m} b_i$, where $a_i, b_i \in Obj(\mathcal{C})$.

Definition 3.8 *An $(\vec{a}, \vec{b})$-**pseudoalgorithm** is $f \in End(\vec{a} \oplus \vec{b})$.*

Notation: An $(\vec{a}, \emptyset)$-pseudoalgorithm will be called an $\vec{a}$-pseudoalgorithm.

Lemma 3.1 *Given $f : a \to a \amalg b$ in a partially additive category, let $f = in_1 f_1 + in_2 f_2$ where f_1 and f_2 are the unique functions stated in Proposition 3.3. Then there exists the **iteration** function $f^\dagger : a \to b$*

$$f^\dagger = \sum_{i \in \omega} f_2 f_1^i : a \to b.$$

This morphism $f^\dagger$ is to be considered as the solution of the equation $x = x f_1 + f_2$.

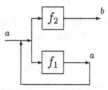

Definition 3.9 *The **execution** of an $(\vec{a}, \vec{b})$-pseudoalgorithm f with respect to $\vec{b}$ is an $\vec{a}$-pseudoalgorithm defined in the following way:*

$$EX_{\vec{b}}(f) = f_{\vec{a},\vec{a}} + f^\dagger_{\vec{b},\vec{a}\oplus\vec{b}} f_{\vec{a},\vec{b}}.$$

Theorem 3.2 *For every $(\vec{a}, \vec{b})$-pseudoalgorithm f, the $\vec{a}$-pseudoalgorithm $EX_{\vec{b}}(f)$ is defined.*

Proof From the fact that f is a matrix morphism the family $(in_i f_{ij} \pi_j)_{i,j \in \{\vec{a}, \vec{b}\}}$ is summable, so by distributivity $f in_{\vec{a}} = in_{\vec{a}} f_{\vec{a}, \vec{a}} + in_{\vec{b}} f_{\vec{a}, \vec{b}}$ is summable, moreover for the lemma 3.1 $\sum f_{\vec{b}, \vec{a}} f_{\vec{b}, \vec{b}}^n$ is defined, and by proposition 3.2 also $\mathrm{EX}_{\vec{b}}(f)$ is defined. □

Remark A summable family $(g f_i)_{i \in I}$ has two different ways to be summable according to whether or not the family $(f_i)_{i \in I}$ is summable. The second one will be the central condition to be requested to morphisms to be algorithms as will be seen in section 4. The crucial point is to formulate the notion of (weak) nilpotency introduced by J.Y.Girard in our setting, in order to select a meaningful subclass of pseudoalgorithms.

4 Execution of Deadlock-Free Pseudoalgorithms

From theorem 3.2 we know that for every $(\vec{a}, \vec{b})$-pseudoalgorithm f is possible to define its execution: $\mathrm{EX}_{\vec{b}}(f)$, among these the next definition distinguishes those such that summability of EX can be deduced from a suitable *local* condition on $f_{\vec{b}, \vec{b}}$.

Definition 4.1 *An $(\vec{a}, \vec{b})$-pseudoalgorithm f is **deadlock-free** with respect to $\vec{b}$ if and only if $((f_{\vec{b}, \vec{b}})^n)_{n \in \omega}$ is summable.*

The basic property of the execution consists in a sort of "associative law" which states that there is no need of a global clock in order to perform a calculus and has as consequence its local and asynchronous execution.

Theorem 4.1 *Let f be a $(\vec{a}, \vec{b} \oplus \vec{b}')$-pseudoalgorithm. Then*

   i) f is deadlock-free w.r.t. $\vec{b} \oplus \vec{b}'$ if and only if
      f is deadlock-free w.r.t. $\vec{b}$ and the $(\vec{a}, \vec{b}')$-pseudoalgorithm $\mathrm{EX}_{\vec{b}}(f)$ is
      deadlock-free w.r.t. $\vec{b}'$.

   ii) $\mathrm{EX}_{\vec{b} \oplus \vec{b}'}(f) = \mathrm{EX}_{\vec{b}'}(\mathrm{EX}_{\vec{b}}(f))$

Proof Let

$$f = \begin{pmatrix} f_{\vec{a}, \vec{a}} & f_{\vec{b}, \vec{a}} & f_{\vec{b}', \vec{a}} \\ f_{\vec{a}, \vec{b}} & f_{\vec{b}, \vec{b}} & f_{\vec{b}', \vec{b}} \\ f_{\vec{a}, \vec{b}'} & f_{\vec{b}, \vec{b}'} & f_{\vec{b}', \vec{b}'} \end{pmatrix}$$

In order to show i) and ii) let'us compare morphisms obtained from $\mathrm{EX}_{\vec{b} \oplus \vec{b}'}(f)$ and $\mathrm{EX}_b'(\mathrm{EX}_b(f))$ forgetting factors from/to a (i.e. $f_{\vec{a}, \vec{a}}, f_{\vec{b}, \vec{a}}, f_{\vec{b}', \vec{a}}, f_{\vec{a}, \vec{b}}$ and $f_{\vec{a}, \vec{b}'}$); will be showed that terms in these two sums are exactly the same.

Let us define $\alpha_1 = \begin{pmatrix} f_{\vec{b},\vec{b}} & 0 \\ 0 & 0 \end{pmatrix}$, $\alpha_2 = \begin{pmatrix} 0 & f_{\vec{b'},\vec{b}} \\ 0 & 0 \end{pmatrix}$, $\alpha_3 = \begin{pmatrix} 0 & 0 \\ f_{\vec{b},\vec{b'}} & 0 \end{pmatrix}$ and $\alpha_4 = \begin{pmatrix} 0 & 0 \\ 0 & f_{\vec{b'},\vec{b'}} \end{pmatrix}$ so

$$f^n_{\vec{b}\oplus\vec{b'},\vec{b}\oplus\vec{b'}} = \sum_{\hat{q}_1,\hat{q}_2,\hat{q}_3} \mu_{n,\hat{q}_1,\hat{q}_2,\hat{q}_3}$$

with the sum over all the possible tuple $(\hat{q}_1,\hat{q}_2,\hat{q}_3)$ being $\hat{q}_i \subset \{1,\ldots,n\}$ and $\hat{q}_i \cap \hat{q}_j = \emptyset$ and where $\mu_{n,\hat{q}_1,\hat{q}_2,\hat{q}_3} = a_1 \ldots a_m$, with $a_i = \begin{cases} \alpha_1 & \text{for } i \in \hat{q}_1 \\ \alpha_2 & \text{for } i \in \hat{q}_2 \\ \alpha_3 & \text{for } i \in \hat{q}_3 \\ \alpha_4 & \text{for } i \notin \hat{q}_1 \cup \hat{q}_3 \cup \hat{q}_3 \end{cases}$

On the other hand, let us calculate $\text{EX}_{b'}(\text{EX}_b(f))$, since we have that

$$\text{EX}_{\vec{b}}(f) = \begin{pmatrix} f_{\vec{a},\vec{a}} & f_{\vec{b'},\vec{a}} \\ f_{\vec{a},\vec{b'}} & f_{\vec{b'},\vec{b'}} \end{pmatrix} + \sum_{n\in\omega} \left(\begin{pmatrix} f_{\vec{b},\vec{a}} \\ f_{\vec{b},\vec{b'}} \end{pmatrix} f^n_{\vec{b},\vec{b}} \right) \begin{pmatrix} f_{\vec{a},\vec{b}} & f_{\vec{b'},\vec{b}} \end{pmatrix}$$

then

$$\text{EX}_{\vec{b'}}(\text{EX}_{\vec{b}}(f)) = f_{\vec{a},\vec{a}} + \sum_n f_{\vec{b},\vec{a}} f^n_{\vec{b},\vec{b}} f_{\vec{a},\vec{b}} +$$
$$+ \sum_k (f_{\vec{b'},\vec{a}} + \sum_n f_{\vec{b},\vec{a}} f^n_{\vec{b},\vec{b}} f_{\vec{b'},\vec{b}})(f_{\vec{b'},\vec{b'}} + \sum_n f_{\vec{b},\vec{b'}} f^n_{\vec{b},\vec{b}} f_{\vec{b'},\vec{b}})^k (f_{\vec{a},\vec{b'}} + \sum_n f_{\vec{b},\vec{b'}} f^n_{\vec{b},\vec{b}} f_{\vec{a},\vec{b}})$$

excluding from every term the part form and to a, we obtain the following expression:

$$\sum_n f^n_{\vec{b},\vec{b}} + \sum_k (\sum_n f^n_{\vec{b},\vec{b}} f_{\vec{b'},\vec{b}})(f_{\vec{b'},\vec{b'}} + \sum_n f_{\vec{b},\vec{b'}} f^n_{\vec{b},\vec{b}} f_{\vec{b'},\vec{b}})^k (\sum_n f_{\vec{b},\vec{b'}} f^n_{\vec{b},\vec{b}}) + \quad (4.1)$$

$$+ \sum_k (\sum_n f^n_{\vec{b},\vec{b}} f_{\vec{b'},\vec{b}})(f_{\vec{b},\vec{b}} + \sum_n f_{\vec{b},\vec{b'}} f^n_{\vec{b},\vec{b}} f_{\vec{b'},\vec{b}})^k + \quad (4.2)$$

$$+ \sum_k (f_{\vec{b'},\vec{b'}} + \sum_n f_{\vec{b},\vec{b'}} f^n_{\vec{b},\vec{b}} f_{\vec{b'},\vec{b}})^k (\sum_n f_{\vec{b},\vec{b'}} f^n_{\vec{b},\vec{b}}) + \quad (4.3)$$

$$+ \sum_k (f_{\vec{b'},\vec{b'}} + \sum_n f_{\vec{b},\vec{b'}} f^n_{\vec{b},\vec{b}} f_{\vec{b'},\vec{b}})^k \quad (4.4)$$

which is actually a matrix-morphism over $\vec{b} \oplus \vec{b'}$.

Let be $\xi_1 \ldots \xi_k = \beta(k,i_1,\ldots,i_k)$ where $i_j = \begin{cases} -1 & \text{for } \xi_j = f_{\vec{b'},\vec{b'}} \\ n & \text{for } \xi_j = f_{\vec{b},\vec{b'}} f^n_{\vec{b},\vec{b}} f_{\vec{b'},\vec{b}} \end{cases}$, supposing defined $\sum_k (f_{\vec{b'},\vec{b'}} + \sum_n f_{\vec{b},\vec{b'}} f^n_{\vec{b'},\vec{b'}} f_{\vec{b'},\vec{b}})^k$ we can write general terms of the series (1), (2), (3) and (4): , that will match the following typologies:

1. a) $f^n_{\vec{b},\vec{b}}$

    b) $f^{n_1}_{\vec{b},\vec{b}} f_{\vec{b'},\vec{b}} \beta(k,i_1,\ldots,i_k) f_{\vec{b},\vec{b'}} f^{n_2}_{\vec{b},\vec{b}}$

2. $f^{n_1}_{\vec{b},\vec{b}} f_{\vec{b'},\vec{b}} \beta(k,i_1,\ldots,i_k)$

3. $\beta(k, i_1, \ldots, i_k) f_{\vec{b}, \vec{b}'} f_{\vec{b}, \vec{b}}^{n_2}$

4. $\beta(k, i_1, \ldots, i_k)$

Finally, we show by induction over m that for every element in the family $(\mu_{m, \hat{q}_1, \hat{q}_2, \hat{q}_3})_{m \in \omega}$ there exists a corresponding element in (1), (2), (3) or (4) equal to the only component of $\mu_{m, \hat{q}_1, \hat{q}_2, \hat{q}_3} \in End(\vec{b} \oplus \vec{b}')$ which is different from zero.

Base of Induction:
if $m = 1$ then there are only four possibilities:

- if $\mu_{1, \{1\}, \emptyset, \emptyset} = \alpha_1$ then case (1.a) with $n = 1$,

- if $\mu_{1, \emptyset, \{1\}, \emptyset} = \alpha_2$ then case (2) with $n_1 = 1$ and $k = 0$,

- if $\mu_{1, \emptyset, \emptyset, \{1\}} = \alpha_3$ then case (3) with $k = 0$ and $n_2 = 0$,

- if $\mu_{1, \emptyset, \emptyset, \emptyset} = \alpha_4$ then case (4) with $k = 1$, $\beta(1, -1)$.

Inductive Step:
Suppose $a_1 \ldots a_m \neq 0$ and such that there exists an element in (1), (2), (3) or (4) equal to it, let us show that for every choice of α_i holds $a_1 \ldots a_m \alpha_i = 0$ or there exists a term equal to it.

First of all, you know that if $a_1 \ldots a_m \neq 0$ there are only two choises of α_i in order to obtain somenthing of different of zero:

- if $a_1 \ldots a_m$ is of type (1) then can be multiplied by α_1 or α_2 ,

- if $a_1 \ldots a_m$ is of type (2) then can be multiplied by α_3 or α_4,

- if $a_1 \ldots a_m$ is of type (3) then can be multiplied by α_1 or α_2,

- if $a_1 \ldots a_m$ is of type (4) then can be multiplied by α_3 or α_4 .

1. – a) means that $a_1 \ldots a_m = \begin{pmatrix} f_{\vec{b}, \vec{b}}^{n'} & 0 \\ 0 & 0 \end{pmatrix}$:

   – since $a_1 \ldots a_m \alpha_1 = \begin{pmatrix} f_{\vec{b}, \vec{b}}^{n'+1} & 0 \\ 0 & 0 \end{pmatrix}$, it is again of type (1.a) when n=n'+1,

   – since $a_1 \ldots a_m \alpha_2 = \begin{pmatrix} 0 & f_{\vec{b}, \vec{b}}^{n'} f_{\vec{b}', \vec{b}} \\ 0 & 0 \end{pmatrix}$ it is of type (2) when $n_1 = n'$ and $k = 0$.

   – b) means that $a_1 \ldots a_m = \begin{pmatrix} f_{\vec{b}, \vec{b}}^{n'_1} f_{\vec{b}', \vec{b}} \beta(k', i'_1, \ldots, i'_{k'}) f_{\vec{b}, \vec{b}'} f_{\vec{b}, \vec{b}}^{n'_2} & 0 \\ 0 & 0 \end{pmatrix}$:

   – since $a_1 \ldots a_m \alpha_1 = \begin{pmatrix} f_{\vec{b}, \vec{b}}^{n'_1} f_{\vec{b}', \vec{b}} \beta(k', i'_1, \ldots, i'_{k'}) f_{\vec{b}, \vec{b}'} f_{\vec{b}, \vec{b}}^{n'_2+1} & 0 \\ 0 & 0 \end{pmatrix}$, it is again of type (1.b) when $n_1 = n'_1$, $k = k'$, $i_j = i'_j$ and $n_2 = n'_2 + 1$,

– since $a_1 \ldots a_m \alpha_2 = \begin{pmatrix} 0 & f_{\bar{b},\bar{b}}^{n'_1} f_{\bar{b}',\bar{b}} \beta(k', i'_1, \ldots, i'_{k'}) f_{\bar{b},\bar{b}'} f_{\bar{b},\bar{b}}^{n'_2} f_{\bar{b}',\bar{b}} \\ 0 & 0 \end{pmatrix}$, it is of

type (2) when $n_1 = n'_1$, $k = k'$, $i_j = i'_j$ when $j \in \{1, \ldots k'\}$, and
$i_{k'+1} = n'_2$.

2. means that $a_1 \ldots a_m = \begin{pmatrix} 0 & f_{\bar{b},\bar{b}}^{n'_1} f_{\bar{b}',\bar{b}} \beta(k', i'_1, \ldots, i'_{k'}) \\ 0 & 0 \end{pmatrix}$:

   – since $a_1 \ldots a_m \alpha_3 = \begin{pmatrix} f_{\bar{b},\bar{b}}^{n'_1} f_{\bar{b}',\bar{b}} \beta(k', i'_1, \ldots, i'_{k'}) f_{\bar{b},\bar{b}'} & 0 \\ 0 & 0 \end{pmatrix}$, it is of type

   (1.b) when $n_1 = n'_1$, $k = k'$, $i_j = i'_j$ and $n_2 = 0$,

   – since $a_1 \ldots a_m \alpha_4 = \begin{pmatrix} 0 & f_{\bar{b},\bar{b}}^{n'_1} f_{\bar{b}',\bar{b}} \beta(k', i'_1, \ldots, i'_{k'}) f_{\bar{b}',\bar{b}'} \\ 0 & 0 \end{pmatrix}$, it of type is (2)

   when $n_1 = n'_1$, $k = k' + 1$, $i_j = i'_j$ for $j \in \{1, \ldots, k'\}$, $i_k = -1$, and
   $n = 0$.

3. means that $a_1 \ldots a_m = \begin{pmatrix} 0 & 0 \\ \beta(k', i'_1, \ldots, i'_{k'}) f_{\bar{b},\bar{b}'} f_{\bar{b},\bar{b}}^{n'_2} & 0 \end{pmatrix}$:

   – since $a_1 \ldots a_m \alpha_1 = \begin{pmatrix} 0 & 0 \\ \beta(k', i_1, \ldots, i_{k'}) f_{\bar{b},\bar{b}'} f_{\bar{b},\bar{b}}^{n'_2} f_{\bar{b},\bar{b}} & 0 \end{pmatrix}$, it is of type (3)

   with $n_2 = n'_2 + 1$, $k = k'$, and $i_j = i'_j$,

   – since $a_1 \ldots a_m \alpha_2 = \begin{pmatrix} 0 & 0 \\ 0 & \beta(k', i'_1, \ldots, i'_{k'}) f_{\bar{b},\bar{b}'} f_{\bar{b},\bar{b}}^{n'_2} f_{\bar{b}',\bar{b}} \end{pmatrix}$, it is of type (4)

   with $k = k' + 1$, $i_j = i'_j$ for $j \in \{1, \ldots, k'\}$ and $i_k = n'_2$.

4. means that $a_1 \ldots a_m = \begin{pmatrix} 0 & 0 \\ 0 & \beta(k', i'_1, \ldots, i'_{k'}) \end{pmatrix}$:

   – since $a_1 \ldots a_m \alpha_3 = \begin{pmatrix} 0 & 0 \\ \beta(k', i'_1, \ldots, i'_{k'}) f_{\bar{b},\bar{b}'} & 0 \end{pmatrix}$, it is of type (3) with

   $k = k'$, $i_j = i'_j$ and $n_2 = 0$,

   – since $a_1 \ldots a_m \alpha_4 = \begin{pmatrix} 0 & 0 \\ 0 & \beta(k', i'_1, \ldots, i'_{k'}) f_{\bar{b}',\bar{b}'} \end{pmatrix}$ it is of type (4) with

   $k = k' + 1$, $i_j = i'_j$ for $j \in \{1, \ldots, k'\}$, and $i_k = -1$).

Finally, if f if deadlock-free w.r.t. $b' \oplus b$ then it is also with respect to b because $\sum_n f_{\bar{b},\bar{b}}^n$ is summable as subsum of (1), moreover $\mathrm{EX}_{\bar{b}}(f)$ is dead-locks free w.r.t. $\bar{b}'$ because $\sum_k [\mathrm{EX}_{\bar{b}}(f)]_{\bar{b}',\bar{b}'}^k$ is equal to (4). Conversely, if $\mathrm{EX}_{\bar{b}}$ is deadlocks free w.r.t $\bar{b}'$ then the family $((f_{\bar{b}',\bar{b}'} + \sum_n f_{\bar{b},\bar{b}'} f_{\bar{b},\bar{b}}^n f_{\bar{b}',\bar{b}})^k)_{k \in \omega}$ is summable: by distributivity, such are the families in (1),(2),(3) and (4) and such is $(f_{\bar{b} \oplus b',\bar{b} \oplus b'}^n)_{n \in \omega}$. $\qquad\square$

5 Types and Dynamics of Algorithms

In this paragraph we reach the final step of our construction: an algebraic version of "structural dynamics".

Let us first consider:

- the notion of *orthogonality* between endomorphisms capturing the "on going" behaviour expressed by convergence of infinite sums,

- the notion of *type* as expressed by a biclosure condition respect to orthogonality,

- the notion of *algorithm* having two aspects: internal behaviour expressed by deadlock-freeness and compositional capabilities of the execution according to types discipline.

5.1 Linear Types

Let $\mathcal{C}$ a partially additive category. In order to provide the set of endomorphisms of the logical structure the same standard construction, used to build the semantics of formulas in linear logic, will be applied. The algebraic view of structural dynamics will be modelled by a closure operator on the quantale constructed on the power set of endomorphisms. Results on quantales that will be used, are proved in [38] and [36].

Let us define the following operations on $\mathcal{P}(End(\mathcal{C}))$,

Definition 5.1 *Given* $X, Y \in \mathcal{P}(End(\mathcal{C}))$, *let be*

- $X \oplus Y = \{f \oplus g \mid f \in X, g \in Y\}$

- *and* $X \multimap Y = \{f \mid \forall g \in X, f \oplus g \in Y\}$

Proposition 5.1 $\{\mathcal{P}(End(\mathcal{C})), \oplus, \multimap, \subset\}$ *is a closed poset.*

Let us define the dualizing element $\perp \in \mathcal{P}(End(\mathcal{C}))$ starting from the dynamical notion of summability, which will provide the searched connection between dynamics (the notion of convergence) and structure (the algebraic notion of type):

Definition 5.2

$$\perp = \{f_1 \oplus f_2 \mid ([T_{a \oplus a}(f_1 \oplus f_2)]^n)_{n \in \omega} \text{ is summable and } f_1, f_2 \in End(b)$$
$$\text{for some } b\}$$

where

$$T_{b\oplus b} = \begin{pmatrix} 0 & 1_b \\ 1_b & 0 \end{pmatrix}$$

and let be

$$X^\perp = X \multimap \perp$$

Remark The last definition can be particularized to morphisms: let f and g be b-pseudoalgorithms, where $b \in Obj(\mathcal{C})$ then

$$f \perp g \text{ iff the family } ((T_{b\oplus b}(f \oplus g))^i)_{i\in\omega} \text{ is deadlock-free w.r.t } b \oplus b.$$

and this condition is equivalent to the following one:

$$\text{the family } ((fg)^i)_{i\in\omega} \text{ is summable.}$$

Moreover, given $X \subset End(a)$ where $a \in Obj(\mathcal{C})$ the **orthogonal** can be defined in the following way:

$$X^\perp = \{f \in End(a) \mid \forall g(g \in X \Rightarrow f \perp g)\}.$$

is equivalent to that one in definition 5.2

From these definitions follows that:

Proposition 5.2 *Given $A_i \in \mathcal{P}(End(\mathcal{C}))$ and σ a permutation of $(1, \ldots, n)$*

1. *the element $\perp$ is cyclic i.e.*
   if $A_1 \oplus \ldots \oplus A_n \subset \perp$ then $A_{\sigma(1)} \oplus \ldots \oplus A_{\sigma(n)} \subset \perp$,

2. *the operation $(\quad)^{\perp\perp}$ is a quantic nucleous i.e. an operator preserving the increasing order on $\mathcal{P}(End(\mathcal{C}))$ and such that*

$$X^{\perp\perp} \oplus Y^{\perp\perp} \subset (X \oplus Y)^{\perp\perp}$$

for every $X, Y \in \mathcal{P}(End(\mathcal{C}))$.

Proof 1 is obtained by applying the permutation to the operator T and 2 follows from 1 by proposition 3.3.4 in [38].

Proposition 5.3 $Q' = (\{X \mid X = X^{\perp\perp}\}, \otimes_{\perp\perp}, \mathsf{C})$ *is a quantale where* $X \otimes_{\perp\perp} Y = (X \oplus Y)^{\perp\perp}$.

Proof Again see [38] theorem 3.1.2.

Finally $\mathcal{P}(End(\mathcal{C}))$ has been structured via orthogonality as a quantale in a such way that it is now possible to define:

- operations (*linear connectives*) on sets of endomorphisms which are compatible with the notion of types in such a way as to give to the set of algorithms the same logical structure as the multiplicative fragment of linear logic,

- a binary operation (called *application*) among pairs of "compatible" endomorphisms whose execution gives rise to the structural dynamics of algorithms corresponding to the logical procedure of cut-elimination.

The set of types will be Q' itself, i.e. by notation:

$$\textbf{TYPES} = Q'$$

and consequentily,

Definition 5.3 *Given an object $a \in C$ an a-**type** will be an element $X \in Q'$ such that $X \subset End(a)$ and so*

$$\textbf{TYPES}(a) = \{X \mid X \text{ is an } a\text{-type}\}$$

and connectives on the set of types will be the two operations on **TYPES** $\otimes_{\perp\perp}$ and $\multimap$ above defined.

5.2 Dynamics of algorithms

The basic idea of modeling algorithms in a way such that type structure and dynamics are intrinsically connected, is summarized in the following steps.

First of all let us give the definition that states the link between execution of pseudoalgorithm and types as defined in the previous section:

Definition 5.4 *A $\vec{b}$-**algorithm** f of type $Y \in \textbf{TYPES}(\vec{a})$ is a $(\vec{a}, \vec{b})$-pseudoalgorithm such that*

- *it is deadlock-free w.r.t. $\vec{b}$*

- $EX_{\vec{b}}(f) \in Y$.

Now, the dynamical meaning of modus ponens can be defined and its compatibility with the two notions of type and execution stated.

Definition 5.5 *Let $X \in \mathcal{P}(End(b))$ and $Y \in \mathcal{P}(End(a))$, $u \in X \multimap Y \in \mathcal{P}(End(a \oplus b))$ and $v \in X$ then the **application** of u to v is*

$$APP(u, v) = u_{b \oplus a, a} + (T_{2b} \oplus 0_a)(v \oplus u)$$

Theorem 5.1 *Let* $X \in$ **TYPES**(b) *and* $Y \in$ **TYPES**(a). *Then* $u \in X \multimap Y$ *if and only if*

- $u_{b,b} \in X^{\perp}$

- *if* $v \in X$ *then* $\mathrm{APP}(u,v)$ *is a 2b-algorithm of type* Y.

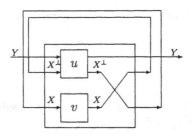

Sketch of proof By definition, $u \in Y \multimap Z = \{f \oplus g \mid f \in Y, g \in Z^{\perp}\}^{\perp}$ is equivalent to summability of

$$
\begin{pmatrix} u_{\vec{b},\vec{b}}f & u_{\vec{a},\vec{b}}g \\ u_{\vec{b},\vec{a}}f & u_{\vec{a},\vec{a}}g \end{pmatrix}^{m}.
$$

This is equivalent to the summability of H^m where

$$
H = \begin{pmatrix} 0 & u_{\vec{a},\vec{a}} & u_{\vec{b},\vec{a}} & 0 \\ g & 0 & 0 & 0 \\ 0 & 0 & 0 & f \\ 0 & u_{\vec{a},\vec{b}}f & u_{\vec{b},\vec{b}} & 0 \end{pmatrix}
$$

By theorem 4.1 H is deadlock-free w.r.t. $\vec{b} \oplus \vec{a}$ if and only if

$$
[H]_{\vec{b},\vec{b}} = \begin{pmatrix} 0 & f \\ u_{\vec{b},\vec{b}} & 0 \end{pmatrix}
$$

is deadlock-free w.r.t. $\vec{b}$, if and only if $(u_{\vec{b},\vec{b}}f)^{n}$ is summable and $\mathrm{EX}_{\vec{b}}(H)$ is deadlock-free w.r.t. $\vec{a}$.   □

The approach followed in this paper can be directly applied in order to give a semantics of reduction for the multiplicative fragment of linear logic: in order to capture larger fragments it should be extended dealing with infinitary objects and more generally, this approach allows to investigate an algebraic theory of computation based on the cut-elimination of proofs.

6   Conclusions and Related Works

The problem of bridging the gap between mathematics and computer science can be attacked by a "discrepancy" between the mathematical structures used in computer science to model the dynamical behavior of algorithms and those considered in mathematics within which specific algorithms are designed. In this paper we have showed an approach to such a key problem by using a "many point" objects generalization of semirings.

Promising works in the same direction come from the Idempotent Calculus [32] , [37] and Category Theory [14] , [18], [31] , [33], [39].

From a methodological point of view semirings have already been used in Computer Science and System Theory

- semirings and dynamical systems [5] [24]

- semirings and automata theory [26] [27] [28] [40]

- semirings and semantics of programs [7] [10] [11] [17] [29]

Moreover the issue of modeling the dynamics behavior of computational systems is central for various classes of such systems.

- transition systems [6] [9] [42] [44] [25] [8] [1]

- proof systems [2] [16] [21] [20] [19] [22] [43]

- constraints/hybrid systems [4], [12], [23], [41]

A long way remains to be explored in Computational Mathematics following several directions:

- modularity issue in time [3], [13]

- modularity issue in space [30], [15]

- expressive power of algorithmic systems [35]

indicated in this paper.

References

[1] K. Aberer (1993), 'Combinatory Models and Symbolic Computations ', *Lecture Notes in Computer Science*, Springer Verlag **721**.

[2] S. Abramsky and Radha Jagadeesan (1994), 'New Foundations for the Geometry of Interaction', *Information and Computation*, **111,1** 178–193 Academic Press.

[3] M. Adamo and G.F. Mascari, 'Algebraic Dynamics of Proofs', *Studia Logica* invited paper in Special Issue in honour of Prof. H. Rasiowa.

[4] E. Asarin and O. Maler (1994), 'On some Relations between Dynamical Systems and Transition Systems', *Lecture Notes in Computer Science*, **820** 59–72.

[5] F. Baccelli, G. Cohen, G.J. Olsder and J.P. Quadrat (1992), 'Synchronization and Linearity', Wiley.

[6] M. Bartha (1992), 'Foundations of a theory of synchronous systems', *Theoretical Computer Science*, **100** 325–346.

[7] D.B. Benson (1989), 'Bialgebras: Some Foundations for Distributed and Concurrent Computation', *Fundamenta Informaticae*, **XII** 427–486.

[8] E. Börger, Y. Gurevich and D. Rosenzweig (1995), 'The Bakery Algorithm: Yet Another Specification and Verification', *E. Borger (ed.) Specification and Validation Methods*, Oxford University Press.

[9] C. Böhm and G. Jacopini (1966). 'Flow diagrams, Turing machines, and languages with only two formation rules'. *Communications of the A.C.M.* **9** 366–371.

[10] S.L. Bloom, Z. Esik (1993), 'Matrix and Matricial Iteration Theories, Part I', *Journal of Computer and System Sciences*, **46** 381–408.

[11] L. Blum, M. Shub and S. Smale (1989), 'On a theory of Computation and Complexity over the Real Numbers', *Bulletin of the American Mathematical Society*, **21,1** 1–46.

[12] A. Borning (Ed.) (1994), 'Principles and Practice of Constraint Programming', *Lecture Notes in Computer Science* **874**, Springer Verlag.

[13] N. Buratti and G.F. Mascari, 'Ordinals and Dynamics in Categories with Direct Limits', *submitted for publication.*

[14] A. Carboni and R.F.C. Walters (1987), 'Cartesian Bicategories I', *Journal of Pure and Applied Algebra* , **45** 127–141.

[15] F. Costantini and G.F.Mascari, 'A Category of Algorithms', *in preparation.*

[16] V. Danos (1989), 'Logique linéaire: une représentation algébrique du calcul', *Gaz. Math. (Soc. Math. de France)* **41** 55–64.

[17] C.C. Elgot (1975), 'Monadic Computations and iterative algebraic theories', *Logic Colloquium 73* 175–230.

[18] P.J. Freyd and A. Scedrov (1990), 'Categories, Allegories', North-Holland

[19] J.-Y. Girard (1987), 'Linear logic', *Theoretical Computer Science* **50** 1–102.

[20] J.-Y. Girard (1987). 'Multiplicatives' *Rendiconti del Seminario Matematico dell' Università e Politecnico Torino, Special Issue on Logic and Computer Science*, pages 11–33.

[21] J.-Y. Girard (1989), 'Geometry of interaction I Interpretation of system F', in *Logic Colloquium '88*, Amsterdam. North-Holland.

[22] J.-Y. Girard (1990), 'Geometry of interaction II Deadlock-free algorithms', in *Lecture Notes in Computer Science* **417**, Springer Verlag.

[23] R.L. Grossman, A. Nerode, A.P. Ravn and H. Rischel (Eds.) (1994), 'Hybrid Systems ', *Lecture Notes in Computer Science* **736**, Springer Verlag.

[24] J.Gunawardena (1994), 'A Dynamic Approach to Timed Behaviour', *Lecture Notes in Computer Science* **836** 178–193, Springer Verlag.

[25] N.A. Harman and J.V. Tucker (1993), 'Algebraic Models and the Correctness of Microprocessors', *Lecture Notes in Computer Science* **683** 92–108, Springer Verlag.

[26] W. Kuich (1991), 'Automata and Languages generalized to ω-continuous semirings', *Theoretical Computer Science* **79** 137–150.

[27] W. Kuich and A. Salomaa (1986), 'Semirings, Automata, Languages', *Springer Verlag.*

[28] D. Krob (1987), 'Monoïdes et semi-annaux complets', *Semigroup Forum* **36** 323–339.

[29] Y.E. Ionnidis and E. Wong (1991), 'Towards an Algebraic Theory of Recursion', *Journal of the A.C.M.* **38** 329–381.

[30] A. Labella and G.F. Mascari, 'Generalized Calculi and Enriched Categories', *in preparation.*

[31] F.W. Lawere (1990), 'Some Thoughts on the Future of Category Theory', *Lecture Notes in Mathematics* **1488** 1–13.

[32] G.I. Litvinov and V.P. Maslov (1993), 'Correspondence Principle for Idempotent Calculus and some Computer Applications', *Proceedings Newton Institute Idempotency Workshop.*

[33] S. MacLane (1988), 'Categories for the Working Mathematician', Springer Verlag.

[34] E.G. Manes, M.A. Arbib (1986), *Algebraic Approaches to Program Semantics*, Springer Verlag, Berlin.

[35] G.F. Mascari, M. Pedicini (1994), 'Head Linear Reduction and Pure Proof Net Extraction', *Theoretical Computer Science*, **135** 111–137.

[36] G.F. Mascari, F.Pucci (1993), 'Autonomous Poset and Quantales', *Theoretical Informatics and Applications* vol.27,n.6, 483–501.

[37] V.P. Maslov,S.N.Samborskii (Editors) (1992), 'Idempotent Analysis', *Advances in Soviet Mathematics* Volume **13**, American Mathematical Society.

[38] K.I. Rosenthal (1990) , 'Quantales and their applications', *Pitman Research Notes in Mathematics Series*, Longman Scientific and Technical.

[39] S.H. Schanuel (1990), 'Negative Sets have Euler Characteristic and Dimension', *Lecture Notes in Mathematics*, **1488** 380–385.

[40] I. Simon (1988), 'Recognizable Sets with Multiplicities in the Tropical Semiring ', *Lecture Notes in Computer Science*, **324** 107–120.

[41] M.Sintzoff, F. Geurts (1995), 'Analysis of Dynamical Systems using Predicate Transformers: Attraction and Composition', *Lecture Notes in Computer Science*, **888** 227–260.

[42] E.W. Stark (1995), 'An Algebra of Data Flow Networks', *Fundamenta Informaticae*, **22** 167–185.

[43] A.S. Troelstra (1992), 'Lectures on Linear Logic', *CSLI Lecture Notes* No. **29**. Center for the Study of Language and Information, Stanford University.

[44] E.R. Wagner, W. Khalil and R.F.C. Walters (1995), 'Fix-point Semantics for Programs in Distributive Categories', *Fundamenta Informaticae*, **22** 187–202.

Task Resource Models and (max,+) Automata

Stéphane Gaubert and Jean Mairesse

Abstract

We show that a typical class of timed concurrent systems can be modeled as automata with multiplicities in the (max,+) semiring. This representation can be seen as a timed extension of the logical modeling in terms of trace monoids. We briefly discuss the applications of this algebraic modeling to performance evaluation.

1 Introduction

Different variations of (stochastic) queuing networks with precedence-based relations between customers have been studied for quite a long time in the performance evaluation community, see [3, 5, 20]. In the combinatorics community on the other hand, concurrent systems are usually modeled in terms of traces – elements of free partially commutative monoids –, see [8, 11]. An equivalent formalism is that of *heaps of pieces* [19].

One of the purposes of this note is to bridge the gap between the two approaches. In the first part of the paper, we establish the relations between the models. An important feature is that execution times of these models can be represented as finite dimensional (max,+) linear dynamical systems. In an essentially equivalent way, they are recognized by automata with multiplicities in the (max,+) semiring. The existence of similar (max,+) models has already been noticed in the context of queuing theory [20, 7]. Their analogue for trace monoids seems to be new.

In the second part of the paper, we apply this algebraic modeling to performance evaluation problems. We present asymptotic results on the existence of mean execution time for random schedules, and for optimal and worst schedules. They are obtained by appealing to subadditive arguments borrowed from the theory of random (max,+) matrices [1].

Finally, we apply the machinery of (max,+) rational series to the exact computation of the asymptotic worst case mean execution time, when the set of admissible schedules is given by a rational language.

Some generalizations of Task Resource models will be considered in a forthcoming paper [16] (heaps of pieces with arbitrary shapes, for which all the results can be extended). These models provide an algebraic framework to handle scheduling problems.

2 Basic Task Resource Model

2.1 General presentation

Definition 2.1 *A (timed)* Task Resource *system is a 4-uple* $\mathcal{T} = (\mathcal{A}, \mathcal{R}, R, h)$ *where:*

- $\mathcal{A}$ *is a finite set whose elements are called* tasks.
- $\mathcal{R}$ *is a finite set whose elements are called* resources.
- $R : \mathcal{A} \rightarrow \mathcal{P}(\mathcal{R})$ *gives the subset of resources required by a task. We assume that each task requires at least one resource:* $\forall a \in \mathcal{A}, R(a) \neq \emptyset$.
- $h : \mathcal{A} \rightarrow \mathbb{R}^+$ *gives the execution time of a task.*

A length n *schedule* is a sequence of n tasks $a_1, \dots, a_n$, that we will write as a word[1] $w = a_1 \dots a_n$. The functioning of the system under the schedule w is as follows.

1. All the resources become initially available at time zero.

2. Task a_i begins as soon as all the required resources $r \in R(a_i)$ used by the earlier tasks $a_j, j < i$, become free, say at time t_i.

3. Task a_i uses each resource $r \in R(a_i)$ during $h(a_i)$ time units. Thus, resource r is released at time $t_i + h(a_i)$.

The *execution time* or *makespan* of the schedule $w = a_1 \dots a_n$ is the completion time of the latest task of the schedule (which is not necessarily a_n):

$$y(w) \stackrel{\text{def}}{=} \max_{1 \leqslant i \leqslant n} (t_i + h(a_i)). \tag{2.1}$$

Task Resource systems are intimately related with the classical *trace monoids* that we next define.

Definition 2.2 *A dependence alphabet is an alphabet $\mathcal{A}$ equipped with a reflexive symmetric relation called* dependence relation, *denoted D, and written graphically* —. *We denote by I the complement of D (called* independence *relation).*

Definition 2.3 *The* trace *monoid* $\mathbb{M}(\mathcal{A}, D)$ *is the quotient of the free monoid $\mathcal{A}^*$ by the congruence $\sim$ generated by the relations* $ab = ba, \forall a \, I \, b$. *The elements of* $\mathbb{M}(\mathcal{A}, D)$ *will be called* traces.

---

[1]We recall the following usual notation. Given a finite set (alphabet) $\mathcal{A}$, we denote by $\mathcal{A}^n$ the set of words of length n on $\mathcal{A}$. We denote by $\mathcal{A}^*$ the free monoid on $\mathcal{A}$, that is, the set of finite words equipped with concatenation. The unit (empty word) will be denoted e. We denote by $\mathcal{A}^+ = \mathcal{A}^* \setminus \{e\}$ the free semigroup on $\mathcal{A}$. The length of the word w will be denoted $|w|$. We shall write $|w|_a$ for the number of occurrences of a given letter a in w.

Let alph(w) denote the set of letters appearing in the word w. The word $\overline{w} \sim w$ is a *Cartier–Foata normal form* of w [8, 11] if we have a factorization $\overline{w} = u_1 \ldots u_p$, $u_i \in \mathcal{A}^+$, such that:

$$a, b \in \text{alph}(u_i) \Rightarrow a \, I \, b, \quad a \in \text{alph}(u_i) \Rightarrow \exists b \in \text{alph}(u_{i-1}), a \, D \, b. \tag{2.2}$$

Such a normal form is unique up to a reordering of the letters inside factors. We shall denote by $\ell(w) = p$ the length (number of factors) of the normal form of w.

With each Task Resource system is associated a dependence relation over the alphabet $\mathcal{A}$; tasks are dependent when they share some resource:

$$a \, D \, b \Leftrightarrow R(a) \cap R(b) \neq \emptyset. \tag{2.3}$$

Conversely, starting from an arbitrary trace monoid $\mathbb{M}(\mathcal{A}, D)$, one can build an associated Task Resource system. For example, one can consider $\mathcal{T} = (\mathcal{A}, \mathcal{R}, R, h \equiv 1)$ with $\mathcal{R} = \{\{a, b\} \mid a \, D \, b\}$ and $R(a) = \{r \in \mathcal{R} \mid a \in r\}$.

Proposition 2.4 (i) *When $h \equiv 1$, $y(w) = \ell(w)$: the makespan is equal to the length of the Cartier–Foata normal form of w.* (ii) *For general execution times h,*

$$y(w) = \max \sum_{j=1}^{p} h(a_{i_j}), \tag{2.4}$$

where the max is taken over the subwords $a_{i_1} \ldots a_{i_p}$ of $w = a_1 \ldots a_n$, composed of consecutive dependent letters (i.e. $a_{i_j} \, D \, a_{i_{j+1}}$).

The first assertion is classical [9]. It implies in particular that the makespan of Task–Resource systems with $h \equiv 1$ can be represented in a more intrinsic way in terms of trace monoids. The second assertion can easily be proved by elementary means, or deduced from the (max,+)-linear representation given below. It provides an alternative formula for (2.1).

Example 2.5 For the sequential dependence alphabet $a \, D \, b$, we have $y(w) = h(a)|w|_a + h(b)|w|_b$. For the purely parallel dependence alphabet $a \, I \, b$, we have $y(w) = \max(h(a)|w|_a, h(b)|w|_b)$.

Example 2.6 (Ring network) Consider a ring shaped communication network with k stations $\mathcal{R} = \{r_1, \ldots, r_k\}$. Messages can be sent between neighbor stations. The possible messages are $\mathcal{A} = \{a_1, \ldots, a_k\}$ where a_i corresponds to a communication between r_i and r_{i+1} (with the convention $k + 1 = 1$). Therefore, we have $R(a_i) = \{r_i, r_{i+1}\}$. This system can also be viewed as a variant of the classical dining philosophers model [12] (replace stations by chopsticks, messages by philosophers). E.g., for $k = 5$, $y(a_1 a_2 a_4 a_1 a_5) = \max(2h(a_1) + h(a_2) + h(a_5), h(a_4) + h(a_5))$ (direct application of 2.4(ii) since the maximal dependent subwords taken from $a_1 a_2 a_4 a_1 a_5$ are $a_1 a_2 a_1 a_5$ and $a_4 a_5$).

2.2   Linear representation over the (max,+) semiring

Definition 2.7 *The* (max,+) *semiring* $\mathbb{R}_{\max}$ *is the set* $\mathbb{R} \cup \{-\infty\}$*, equipped with* max*, written additively (i.e.* $a \oplus b = \max(a,b)$*) and the usual sum, written multiplicatively (i.e.* $a \otimes b = a + b$*). We write* $\varepsilon = -\infty$ *for the zero element, and* $e = 0$ *for the unit element.*

We shall use throughout the paper the matrix and vector operations induced by the semiring structure[2]. The identity matrix ($\mathrm{I}_{ii} = e; \mathrm{I}_{ij} = \varepsilon, i \neq j$) with entries indexed by X will be denoted by I_X. The row vector with entries indexed by X and all equal to e will be denoted by e_X. We denote by $\|M\| = \bigoplus_{ij} M_{ij}$ (resp. $\|v\| = \bigoplus_i v_i$) the (max,+) *norm* of a matrix M (vector v).

A (max,+) *automaton*[3] of dimension k over the alphabet $\mathcal{A}$ is a triple $(\alpha, \mathcal{M}, \beta)$, where $\alpha \in \mathbb{R}_{\max}^{1 \times k}$, $\beta \in \mathbb{R}_{\max}^{k \times 1}$, and $\mathcal{M}$ is a morphism from $\mathcal{A}^*$ to the multiplicative monoid of matrices $\mathbb{R}_{\max}^{k \times k}$. A map $y : \mathcal{A}^* \to \mathbb{R}_{\max}$ is *recognizable* if there is an automaton such that $y(w) = \alpha \mathcal{M}(w) \beta$.

In a spirit closer to discrete event systems theory, automata may be seen as (max,+) linear systems whose dynamics are indexed by letters. Indeed, introducing the "state vector" $x(w) \overset{\text{def}}{=} \alpha \mathcal{M}(w) \in \mathbb{R}_{\max}^{1 \times k}$, we get

$$x(e) = \alpha, \quad x(wa) = x(w)\mathcal{M}(a), \quad y(w) = x(w)\beta, \quad \text{or} \qquad (2.5)$$

$$y(a_1 \ldots a_n) = \alpha \mathcal{M}(a_1) \ldots \mathcal{M}(a_n)\beta. \qquad (2.6)$$

Definition 2.8 (Task & resource daters) *A dater over the alphabet* $\mathcal{A}$ *is a scalar map* $\mathcal{A}^* \to \mathbb{R} \cup \{-\infty\}$*. With each task* $a \in \mathcal{A}$ *is associated a task dater* x_a*:* $x_a(w)$ *gives the time of completion of the last task of type* a *in the schedule* w*. With each resource* $r \in \mathcal{R}$ *is associated a resource dater* x_r*:* $x_r(w)$ *gives the last instant of release of the resource* r *under the schedule* w*. We shall denote by* $x_{\mathcal{A}}$ *and* $x_{\mathcal{R}}$ *the vectors of task and resource daters.*

Note the important duality relations

$$x_a(w) = \bigoplus_{r \in R(a)} x_r(w), \quad x_r(w) = \bigoplus_{a \in R^{-1}(r)} x_a(w). \qquad (2.7)$$

We identify each subset $R(a)$ with a boolean matrix of size $|\mathcal{R}| \times |\mathcal{A}|$ denoted $\mathcal{I}(a)$.

$$\forall a \in \mathcal{A}, \mathcal{I}(a)_{rb} = \begin{cases} e & \text{if } r \in R(a) \text{ and } b = a \\ \varepsilon & \text{otherwise.} \end{cases}$$

---

[2]I.e. for matrices A, B of appropriate sizes, $(A \oplus B)_{ij} = A_{ij} \oplus B_{ij} = \max(A_{ij}, B_{ij})$, $(A \otimes B)_{ij} = \bigoplus_k A_{ik} \otimes B_{kj} = \max_k(A_{ik} + B_{kj})$, and for a scalar a, $(a \otimes A)_{ij} = a \otimes A_{ij} = a + A_{ij}$. We will abbreviate $A \otimes B$ to AB as usual.

[3]This is a specialization to the $\mathbb{R}_{\max}$ case of the notion of automaton with multiplicities over a semiring (or equivalently, of recognizable series over a semiring). See [13, 6].

We define the following matrices:

$$\forall a \in \mathcal{A}, \quad \mathcal{M}_\mathcal{R}(a) = \mathbf{I}_\mathcal{R} \oplus h(a)\mathcal{I}(a)\mathcal{I}(a)^T, \tag{2.8}$$

$$\mathcal{M}_\mathcal{A}(a) = \mathbf{I}_\mathcal{A} \oplus h(a)\left(\bigoplus_b \mathcal{I}(b)^T\right)\mathcal{I}(a), \tag{2.9}$$

or more explicitly

$$\mathcal{M}_\mathcal{R}(a)_{rs} = \begin{cases} e & \text{if } r = s, s \notin R(a), \\ h(a) & \text{if } r \in R(a), s \in R(a), \\ \varepsilon & \text{otherwise.} \end{cases} \tag{2.10}$$

$$\mathcal{M}_\mathcal{A}(a)_{bc} = \begin{cases} e & \text{if } a \neq (b = c), \\ h(a) & \text{if } a = c, bDc, \\ \varepsilon & \text{otherwise.} \end{cases} \tag{2.11}$$

We extend $\mathcal{M}_\mathcal{A}$ (resp. $\mathcal{M}_\mathcal{R}$) to a morphism $\mathcal{A}^* \to \mathbb{R}_{\max}^{\mathcal{A} \times \mathcal{A}}$ (resp. $\mathcal{A}^* \to \mathbb{R}_{\max}^{\mathcal{R} \times \mathcal{R}}$).

Theorem 2.9 *The dater functions of task resource systems admit the following linear representations over the (max,+) semiring:*

$$x_\mathcal{R}(wa) = x_\mathcal{R}(w)\mathcal{M}_\mathcal{R}(a), \quad x_\mathcal{R}(e) = e_\mathcal{R}, \tag{2.12}$$

$$x_\mathcal{A}(wa) = x_\mathcal{A}(w)\mathcal{M}_\mathcal{A}(a), \quad x_\mathcal{A}(e) = e_\mathcal{A}, \tag{2.13}$$

$$y(w) = \|x_\mathcal{A}(w)\| = \|x_\mathcal{R}(w)\| = \|\mathcal{M}_\mathcal{A}(w)\| = \|\mathcal{M}_\mathcal{R}(w)\|. \tag{2.14}$$

In other words, y is recognized both by the *resource automaton* $(e_\mathcal{R}, \mathcal{M}_\mathcal{R}, e_\mathcal{R}^T)$ and by the *task automaton* $(e_\mathcal{A}, \mathcal{M}_\mathcal{A}, e_\mathcal{A}^T)$.

Proof We have

$$x_a(wb) = \begin{cases} x_a(w) & \text{if } a \neq b, \\ \max_{r \in R(a)} x_r(w) + h(a) & \text{if } a = b, \end{cases} \tag{2.15}$$

$$x_r(e) = x_a(e) = e. \tag{2.16}$$

These relations are a simple translation of the functioning of the system, as described after Definition 2.1 (items 1,2,3). Eliminating x_r in (2.15) using (2.7), we get the *task equation*

$$x_a(wb) = \begin{cases} x_a(w) & \text{if } a \neq b \\ \max_{cDa} x_c(w) + h(a) & \text{if } a = b. \end{cases} \tag{2.17}$$

Dually, it is not difficult to obtain the *resource equation*

$$x_r(wa) = \begin{cases} x_r(w) & \text{if } R(a) \not\ni r \\ \max_{s \in R(a)} x_s(w) + h(a) & \text{if } R(a) \ni r. \end{cases} \tag{2.18}$$

Rewriting (2.17) and (2.18) with the semiring notations, we get (2.12),(2.13). $\square$

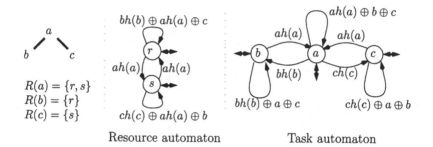

Resource automaton   Task automaton

Figure 1: Task and resource automata for b—a—c.

Example 2.10 We consider a Task Resource model with dependence alphabet b—a—c. In Fig. 1, we have represented[4] the resource automaton $(e_{\mathcal{R}}, \mathcal{M}_{\mathcal{R}}, e_{\mathcal{R}}^T)$ and the task automaton $(e_{\mathcal{A}}, \mathcal{M}_{\mathcal{A}}, e_{\mathcal{A}}^T)$ associated with the dependence alphabet b—a—c. The matrices associated with the resource automaton are:

$$\mathcal{M}_{\mathcal{R}}(a) = \begin{bmatrix} h(a) & h(a) \\ h(a) & h(a) \end{bmatrix}, \mathcal{M}_{\mathcal{R}}(b) = \begin{bmatrix} h(b) & \varepsilon \\ \varepsilon & e \end{bmatrix}, \mathcal{M}_{\mathcal{R}}(c) = \begin{bmatrix} e & \varepsilon \\ \varepsilon & h(c) \end{bmatrix}.$$

The makespan $y(w)$ is equal to the maximal weight of a path labeled w between two arbitrary nodes of the graph. E.g. $y(cba) = \max(h(c)+h(a), h(b)+h(a))$.

2.3   Interpretation in terms of heaps of pieces

There is a useful geometrical interpretation of Task Resource models in terms of *heaps of pieces*. This interpretation was first noticed by Viennot for trace monoids. The reader is referred to [19] for a more formal presentation. Imagine an horizontal axis with as many slots as resources. With each letter a is associated a *piece*, i.e. a solid "rectangle" occupying the slots $r \in R(a)$, with height $h(a)$. The heap associated with the word $w = a_1 \ldots a_n$ is built by piling up the pieces $a_1, \ldots, a_n$, in this order. The makespan $y(w)$ coincides with the height of the heap. The vector $x_{\mathcal{R}}(w) = e_{\mathcal{R}}\mathcal{M}_{\mathcal{R}}(w)$ can be interpreted as the upper contour of the heap. Adding one piece above the heap amounts to right multiplication by the corresponding matrix.

---

[4]An automaton $(\alpha, \mathcal{M}, \beta)$ of dimension k over an alphabet $\mathcal{A}$ is usually represented as a graph with nodes $1, \ldots, k$, and three kinds of labeled and weighted arcs. There is an *internal arc* $i \to j$ with label $a \in \mathcal{A}$ and weight $\mathcal{M}(a)_{ij}$ whenever $\mathcal{M}(a)_{ij} = t \neq \varepsilon$. We will write $x \overset{at}{\to} y$ but we omit the unit valuations (when $t = e$). When there are two arcs $x \to y$ with respective labels a, b and weights t, t', we shall write $x \overset{at \oplus bt'}{\to} y$ as a shorthand for the two arcs $x \overset{at}{\to} y$, $x \overset{bt'}{\to} y$. There is an *input arc* at node i with weight α_i, whenever $\alpha_i \neq \varepsilon$. *Output arcs* are obtained in a dual way from β.

Example 2.11 Consider the ring model of Example 2.6 with $k = 4$ and $h \equiv 1$. We have represented in Fig. 2(I) the heap associated associated with the word $a_1a_2a_3a_4a_4a_3a_2a_1$.

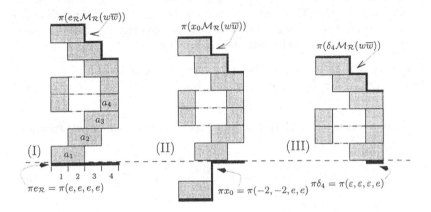

Figure 2: Heaps of pieces for a ring model.

3 Performance Evaluation

3.1 Stochastic case

The simplest[5] stochastic extension of task resource systems arises when the sequence of tasks is given by a sequence of random variables $a(n) \in \mathcal{A}$: we get the random schedule $w_n = a(1) \ldots a(n)$, and consider the asymptotics of $y(w_n), x(w_n)$, which we shall shorten to $y(n), x(n)$. For stochastic Task Resource models, we propose two types of asymptotic results.

1. First order limits or mean execution times $x(n)_i/n$.

2. Second order limits or asymptotics of relative delays $x(n)_i - x(n)_j$ (e.g. differences of last occupation times of the different resources).

Second order quantities are best defined in terms of (max,+) projective space. The (max,+) projective space $\mathbb{PR}_{\max}^k$ is the quotient of $\mathbb{R}^k$ by the parallelism relation $x \simeq y \Leftrightarrow \exists \lambda \in \mathbb{R}, \ x = \lambda y$. We write $\pi : \mathbb{R}_{\max}^k \to \mathbb{PR}_{\max}^k$ for the canonical projection. The relative delays $x(n)_i - x(n)_j$ can be computed from $\pi x(n)$. Geometrically, $\pi x(n)$ corresponds to *the upper shape of the heap (the*

---

[5]In order to simplify the presentation, we shall not consider more general cases with random initial conditions, random execution times and random arrival times, which can be dealt with along the same lines.

quotient by $\simeq$ identifies two heaps with the same upper contour but different heights, see Fig. 2(I)).

We assume that the random variables $a(n)$ are defined on a common probability space $(\Omega, \mathcal{F}, P)$, equipped with a stationary and ergodic shift θ. We consider a *connected* Task Resource system, i.e. such that the graph of the dependence relation is connected (if this is not the case, the theorem has to be applied to each connected sub-system).

Theorem 3.1 *Let $\{a(n), n \in \mathbb{N}\}$ be a stationary and ergodic sequence (i.e. $a(n + 1, \omega) = a(n, \theta(\omega)))$ of integrable random variables, such that $\forall b \in \mathcal{A}, P(a(1) = b) > 0$.*

1. There exists a constant $\lambda_E \in \mathbb{R}$ (stochastic Lyapunov exponent) such that, $\forall i \in \mathcal{A} \cup \mathcal{R}$,

$$\lim_n \frac{x(n)_i}{n} = \lim_n E\left(\frac{x(n)_i}{n}\right) = \lambda_E \quad P\text{-a.s.} \tag{3.1}$$

2. Moreover, if the sequence $\{a(n), n \in \mathbb{N}\}$ is i.i.d. then the random variable $\pi x(n)$ converges in total variation to a unique stationary distribution.

Proof In order to prove point 1, the main tool is the subadditivity of the sequence $\{y(w) = \|x(w)\|\}$, more precisely:

$$\forall w_1, w_2 \in \mathcal{A}^*, \quad y(w_1 w_2) \leqslant y(w_1) + y(w_2). \tag{3.2}$$

This property enables us to apply Kingman's subadditive ergodic theorem, see [1]. More generally, this result is just a special case of a general theorem on homogeneous and monotone operators, see [20] or [4].

We show point 2 for the resource dater $x_{\mathcal{R}}(w) = e_{\mathcal{R}}\mathcal{M}_{\mathcal{R}}(w)$ (the behavior of $x_{\mathcal{A}}$ can be deduced easily from that of $x_{\mathcal{R}}$ by appealing to (2.7)). The following necessary and sufficient condition for the existence and uniqueness of a stationary distribution for $\pi x_{\mathcal{R}}(w(n))$ is stated in [17]:

There is a word w such that the matrix $\mathcal{M}_{\mathcal{R}}(w)$ is of rank one, with non-ε entries.

The matrix $\mathcal{M}_{\mathcal{R}}(w)$ constitutes a regeneration pattern for the model. Indeed, the rank one condition is equivalent to a forgetting of the initial condition:

$$\forall x_0, x_0', \quad \pi(x_0 \mathcal{M}_{\mathcal{R}}(w)) = \pi(x_0' \mathcal{M}_{\mathcal{R}}(w)). \tag{3.3}$$

This pattern enables us to use regeneration theory to obtain stability of the model. The existence of the pattern is guaranteed by the following lemma.

Lemma 3.2 *Let $w = a_1 \ldots a_n$ be a path in the graph of the dependence relation (i.e $a_i D a_{i+1}$), visiting all the nodes. Let $\tilde{w} = a_n \ldots a_1$ denote the mirror image of w. The matrix $\mathcal{M}_{\mathcal{R}}(w\tilde{w})$ is of rank one with non-ε entries.*

Rather than proving the result formally (which can be done using representation (2.8), (2.12) and the fact that $\mathcal{I}(a)$ has rank one), we provide a geometrical justification using heaps of pieces. Condition (3.3) is equivalent to the following: *the upper shape of the heap is independent of the shape of the ground* (which corresponds to the initial condition). The property $a_i \, D \, a_{i+1}$ of the word $w\tilde{w}$ means that the heap is staircase shaped. It implies condition (3.3) as illustrated by the different heaps (I),(II),(III) shown in Fig. 2 (corresponding to the respective initial conditions $e_{\mathcal{R}}$, $(-2, -2, e, e)$, $(\varepsilon, \varepsilon, \varepsilon, e)$). $\square$

Remark 3.3 A result analogous to Theorem 3.1, point 2 was proved by Saheb [18] for trace monoids, using a Markovian argument. The advantage of the method presented here is that it can be applied to the various extensions mentioned in footnote 5.

3.2 Optimal case and worst case

Given a language $L \subset \mathcal{A}^*$ describing the set of admissible schedules, a natural problem consists in finding an admissible schedule of length n with minimal or maximal makespan. The following theorem shows the existence of an asymptotic mean execution time, under optimal or worst case schedules. It can be seen as a (weak) analogue for optimization problems of the first order ergodic theorem 3.1.1.

Theorem 3.4

1. *For a language L such that $L^2 \subset L$, the following limit (optimal Lyapunov exponent) exists*

$$\lambda_{\min}(L) \overset{\text{def}}{=} \lim_{n \to \infty, \, \mathcal{A}^n \cap L \neq \emptyset} \, \min_{w \in \mathcal{A}^n \cap L} \frac{y(w)}{n} = \inf_{w \in L} \frac{y(w)}{|w|} \, . \tag{3.4}$$

2. *For a bifix language L (such that $uv \in L \Rightarrow u, v \in L$), the following limit (worst Lyapunov exponent) exists*

$$\lambda_{\max}(L) \overset{\text{def}}{=} \lim_{n \to \infty} \, \max_{w \in \mathcal{A}^n \cap L} \frac{y(w)}{n} = \inf_{n \geqslant 1} \, \max_{w \in \mathcal{A}^n \cap L} \frac{y(w)}{n} \, . \tag{3.5}$$

Proof Let $m_n = \inf_{w \in \mathcal{A}^n \cap L} y(w)$. Since $L^2 \subset L$, $w \in L \cap \mathcal{A}^n, z \in L \cap \mathcal{A}^p \Rightarrow wz \in L \cap \mathcal{A}^{n+p}$. Using the subadditivity property (3.2), we get $m_{n+p} \leqslant m_n + m_p$, from which (3.4) readily follows. The argument for $\lambda_{\max}$ is similar. $\square$

The assumption that $L^2 \subset L$ for the optimal case is practically reasonable. For instance, for usual scheduling problems, it is natural to impose a fixed proportion of the different tasks, i.e. $L = \{w \mid |w|_a = r_a |w|\}$, for some fixed $r_a \in \mathbb{R}^+, \sum_a r_a = 1$. Such a language satisfies $L^2 \subset L$. The restriction to bifix languages for the worst case behavior is an artefact due to the subadditive argument.

The following theorem shows that the worst case performance can be exactly computed for the subclass of rational schedule languages. The reader is referred to [6] for the notation concerning series.

Theorem 3.5 *Consider the* generating series of the worst case behavior, $z = \bigoplus_{n \in \mathbb{N}} z_n x^n \in \mathbb{R}_{\max}[[x]]$, *where* $z_n = \sup_{w \in \mathcal{A}^n \cap L} y(w)$. *If the admissible language L is rational, the series z is rational.*

Proof Let $\mathrm{char} L \in \mathbb{R}_{\max}\langle\langle \mathcal{A} \rangle\rangle$ denote the characteristic series[6] of the language L. Then $\mathrm{char} L$ is rational. Introduce the morphism $\varphi : \mathbb{R}_{\max}\langle\langle \mathcal{A} \rangle\rangle \to \mathbb{R}_{\max}[[x]]$ such that $\forall a, \varphi(a) = x$. Recall that the *Hadamard product* of series is defined by $(s \odot t)(w) = s(w)t(w)$. Since rational series are closed under alphabetical morphisms and Hadamard product, $z = \varphi(\mathrm{char} L \odot y) \in \mathbb{R}_{\max}[[x]]$ is rational. □

Corollary 3.6 *Let α, μ, β denote a trim linear representation of $\mathrm{char} L$. Then*

$$\limsup_n \frac{z_n}{n} = \rho(A), \quad A = \bigoplus_{a \in \mathcal{A}} \mu(a) \otimes^t \mathcal{M}_{\mathcal{R}}(a), \qquad (3.6)$$

where ρ denotes the (max,+) maximal eigenvalue and $\otimes^t$ the tensor product of matrices.

This is an immediate consequence of the (max,+) spectral theorem, together with the fact [13, 6] that $\mathrm{char} L \odot y$ is recognized by the tensor product of the representations (α, μ, β), $(e_{\mathcal{R}}, \mathcal{M}_{\mathcal{R}}, e_{\mathcal{R}}^T)$ (see [15, §3.2] for details).

Remark 3.7 More generally, Theorem 3.5 holds for an algebraic (=context-free) language L and not only for a rational one. Indeed, it is an easy extension[7] of Parikh's theorem [10] that *algebraic series in several commuting indeterminates, with coefficients in $\mathbb{R}_{\max}$, are rational*. Since algebraic series are closed under Hadamard product with recognizable series and alphabetical morphism, the above proof shows that, when L is algebraic, the series $z = \varphi(\mathrm{char} L \odot y)$ is algebraic, hence rational. This shows that *the generating series z of the worst case behavior of an algebraic language L is rational*. In this case, the effective computation of z along the lines of [10, Chapter XI] is less immediate, since it requires solving (max,+) commutative rational equations.

---

[6]The coefficient of $\mathrm{char} L$ at w is equal to e if $w \in L$, ε otherwise.

[7]By algebraic series, we mean *constructive* algebraic series as defined in [14]. The argument given in [10, Chapter XI] can be adapted to algebraic series in commuting indeterminates with coefficients in commutative idempotent semirings.

Example 3.8 Consider the dependence alphabet b—a—c, together with the set of admissible schedules $L = (a \oplus bc^*b)^*$. Its characteristic series is recognized by

$$\alpha = [e, \varepsilon], \beta = [e, \varepsilon]^T, \mu(a) = \begin{bmatrix} e & \varepsilon \\ \varepsilon & \varepsilon \end{bmatrix}, \ \mu(b) = \begin{bmatrix} \varepsilon & e \\ e & \varepsilon \end{bmatrix}, \ \mu(c) = \begin{bmatrix} \varepsilon & \varepsilon \\ \varepsilon & e \end{bmatrix}.$$

We get from Example 2.10 and (3.6),

$$A = \begin{bmatrix} h(a) & h(a) & h(b) & \varepsilon \\ h(a) & h(a) & \varepsilon & e \\ h(b) & \varepsilon & e & \varepsilon \\ \varepsilon & e & \varepsilon & h(c) \end{bmatrix}, \quad \rho(A) = h(a) \oplus h(c) \oplus h(b),$$

where $\rho(A)$ is obtained from its characterization as maximal mean weight of the circuits of A [2]. Note that the different terms in $\rho(A)$ are attained asymptotically for the sequences of schedules $a^n, n \in \mathbb{N}$; $bc^n b, n \in \mathbb{N}$; $b^{2n}, n \in \mathbb{N}$ (whose periodic parts correspond to circuits of A).

Remark 3.9 Cérin and Petit [9] study the absolute worst case behavior $\overline{\lambda}_{\max} \stackrel{\text{def}}{=} \sup_{w \in L} |w|^{-1} \times y(w)$. This can be obtained along the same lines:

$$\overline{\lambda}_{\max} = \rho(A) \oplus \bigoplus_{1 \leq i \leq \dim A} cA^i b, \tag{3.7}$$

where $c = \alpha \otimes^t e_{\mathcal{R}}, b = \beta \otimes^t e_{\mathcal{R}}^T$. These quantities can be computed in $O((\dim A)^3)$ steps (using Karp's algorithm [2] for $\rho(A)$). Observe that the dual quantity $\inf_{w \in L} y(w)/|w|$ treated in [9] cannot be obtained by such simple arguments due to its "min-max" structure.

References

[1] F. Baccelli. Ergodic theory of stochastic Petri networks. *Annals of Probability*, 20(1):375–396, 1992.

[2] F. Baccelli, G. Cohen, G.J. Olsder and J.P. Quadrat. *Synchronization and Linearity*. John Wiley & Sons, New York, 1992.

[3] F. Baccelli and Z. Liu. On a class of stochastic recursive equations arising in queuing theory. *Annals of Probability*, 21(1):350–374, 1992.

[4] F. Baccelli and J. Mairesse. Ergodic theorems for stochastic operators and discrete event systems. This volume.

[5] N. Bambos and J. Walrand. Scheduling and stability aspects of a general class of parallel processing systems. *Adv. Appl. Prob.*, 25:176–202, 1993.

[6] J. Berstel and C. Reutenauer. *Rational Series and their Languages.* Springer, 1988.

[7] M. Brilman and J.M. Vincent. Synchronization by resources sharing: a performance analysis. 1995.

[8] P. Cartier and D. Foata. *Problèmes combinatoires de commutation et réarrangements.* Number 85 in Lecture Notes in Mathematics. Springer Verlag, 1969.

[9] C. Cérin and A. Petit. Speedup of recognizable trace languages. In *Proc. MFCS 93*, number 711 in Lecture Notes in Computing Science. Springer, 1993.

[10] J.H. Conway. *Regular algebra and finite machines.* Chapman and Hall, 1971.

[11] V. Diekert. *Combinatorics on traces.* Number 454 in Lecture Notes in Computing Science. Springer, 1990.

[12] E. Dijkstra. *Cooperating sequential processes.* In *Programming languages.* Academic Press, 1968.

[13] S. Eilenberg. *Automata, Languages and Machines*, volume A. Acad. Press, 1974.

[14] M. Fliess. *Sur certaines familles de séries formelles.* Thèse de doctorat d'etat, Université Paris VII, 1972.

[15] S. Gaubert. Performance evaluation of (max,+) automata. *IEEE Transactions on Automatic Control*, to appear. (Preliminary version: Rapport de Recherche 1922, INRIA, May 1993).

[16] S. Gaubert and J. Mairesse. Scheduling and simulation of timed Petri nets using (max,+) automata. In preparation, 1995.

[17] J. Mairesse. Products of irreducible random matrices in the (max,+) algebra. Rapport de Recherche 1939, INRIA, June 1993. To appear in Advances in Applied Prob.

[18] N. Saheb. Concurrency measure in commutation monoids. *Discrete Applied Mathematics*, 24:223–236, 1989.

[19] X. Viennot. Heaps of pieces, I: Basic definitions and combinatorial lemmas. In Labelle and Leroux, editors, *Combinatoire Énumérative*, number 1234 in Lect. Notes in Math., pages 321–350. Springer, 1986.

[20] J.M. Vincent. Some ergodic results on stochastic iterative DEDS. Rapport de Recherche 4, Apache, IMAG, 1993.

Algebraic System Analysis of Timed Petri Nets

Guy Cohen, Stéphane Gaubert and Jean-Pierre Quadrat

Abstract

We show that Continuous Timed Petri Nets (CTPN) can be modeled by generalized polynomial recurrent equations in the (min,+) semiring. We establish a correspondence between CTPN and Markov decision processes. We survey the basic system theoretical results available: behavioral (input–output) properties, algebraic representations, asymptotic regime. Particular attention is paid to the subclass of stable systems (with asymptotic linear growth).

1 Introduction

The fact that a subclass of Discrete Event Systems equations can be written linearly in the (min,+) or in the (max,+) semiring is now almost classical [9, 2]. The (min,+) linearity allows the presence of synchronization and saturation features but unfortunately prohibits the modeling of many interesting phenomena such as "birth" and "death" processes (multiplication of tokens) and concurrency. The purpose of this paper is to show that after some simplifications, these additional features can be represented by polynomial recurrences in the (min,+) semiring.

We introduce a fluid analogue of general Timed Petri Nets (in which the quantities of tokens are real numbers), called Continuous Timed Petri Nets (CTPN). We show that, assuming a stationary routing policy, the counter variables of a CTPN satisfy recurrence equations involving the operators min, +, ×. We interpret CTPN equations as dynamic programming equations of classical Markov Decision Problems: CTPN can be seen as the dedicated hardware executing the value iteration.

We set up a hierarchy of CTPN which mirrors the natural hierarchy of optimization problems (deterministic vs. stochastic, discounted vs. ergodic). For each level and sublevel of this hierarchy, we recall or introduce the required algebraic and analytic tools, provide input–output characterizations and give asymptotic results.

The paper is organized as follows. In §2, we give the dynamic equations satisfied by general Petri Nets under the earliest firing rule. The counter equations given here are much more tractable than the dater equations obtained previously [1]. Similar equations have been introduced by Baccelli *et al.* [3] in a stochastic context.

In §3, we introduce the continuous analogue of Timed Petri Nets. We discuss various natural routing policies, and show that they lead to simple recurrence equations.

In §4, we present the first level of the hierarchy: Continuous Timed Event Graphs with Multipliers (CTEGM), characterized by the absence of routing decisions. We single out several interesting subclasses. 1. Ordinary Timed Event Graphs (TEG) are probably the simplest and best understood class of Timed Discrete Event Systems. TEG are exactly causal finite dimensional recurrent linear systems over the (min,+) semiring. They correspond to deterministic decision problems with finite state and additive undiscounted cost. Their asymptotic theory is mere translation of the (min,+) spectral theory. Their input–output relations are inf-convolutions with (min,+) rational sequences. 2. We introduce the subclass of CTEGM *with potential*, which reduce to TEG after a change of units (they are linearized by a nonlinear change of variable in the (min,+) semiring). The importance and tractability of the (noncontinuous) version of these systems, called *expansible* [23], were first recognized by Munier. 3. α-discounted TEG are the TEG-analogue of uniformly discounted deterministic optimization problems. They represent systems with constant birth (or death) rate α. 4. We consider general CTEGM. Their input–output relations are affine convolutions (minima of affine functions of the delayed input). The transfer operators are rational series with coefficients in the semiring of piecewise affine concave monotone maps. To CTEGM correspond deterministic decision problems where the actualization rate (and not only the transition cost) is controlled. Lastly, certain routing policies, called *injective*, reduce CTPN to CTEGM. Related resource optimization problems (optimizing the allocation of the initial marking) are discussed in §4.7.

In §5, we examine the second level of the hierarchy: general CTPN, which correspond to *stochastic* decision problems. Algebraically, CTPN are (min,+) polynomial systems whose outputs admit Volterra series expansions. They are characterized by simple behavioral properties (essentially monotonicity and concavity). We focus on the following tractable subclasses. 1. *Undiscounted TPN* are the Petri Net analogue of stochastic control problems with undiscounted (ergodic) cost. They are characterized by a structural condition (as many input as output arcs at each place) plus a compatibility condition on routings. Undiscounted TPN admit an asymptotically linear growth. The asymptotic behavior can be obtained by transferring the results known for the value iteration: we give a "critical circuit" formula similar to the TEG case (the circuits have to be replaced by recurrent classes of stationary policies). 2. Similar results exist for TPN *with potential* (obtained from undiscounted

TPN by a diagonal change of variable). 3. CTPN with a fixed birth/death rate α correspond to the well studied class of discounted Dynamic Programming recurrences.

2 Recurrence Equations of Timed Petri Nets

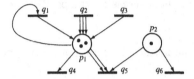

Figure 1: Notation for Petri Nets. $\mathcal{P} = \{p_1, p_2\}$, $\mathcal{Q} = \{q_1, \ldots, q_6\}$, $p_1^{\text{out}} = \{q_1, q_4, q_5\}$, $p_1^{\text{in}} = \{q_1, q_2, q_3\}$, $p_2^{\text{out}} = \{q_5, q_6\}$, $M_{q_5 p_1} = 2$, $M_{p_1 q_2} = 3$, $m_{p_1} = 3$, $m_{p_2} = 1$.

Definition 2.1 (TPNM) A Timed Petri Net with Multipliers (TPNM) is a valued bipartite graph given by a 5-tuple $\mathcal{N} = (\mathcal{P}, \mathcal{Q}, M, m, \tau)$.

1. The finite set $\mathcal{P}$ is called the set of places. *A place may contain tokens which travel from place to place according to a firing process described later on.*

2. The finite set $\mathcal{Q}$ is called the set of transitions. *A transition may fire. When it fires, it consumes and produces tokens.*

3. $M \in \mathbb{N}^{\mathcal{P} \times \mathcal{Q} \cup \mathcal{Q} \times \mathcal{P}}$. M_{pq} *(resp. M_{qp}) gives the number of edges from transition q to place p (resp. from place p to transition q). In particular, the zero value for M corresponds to the absence of edge.*

4. $m \in \mathbb{N}^{\mathcal{P}}$: m_p *denotes the number of tokens that are initially in place p (initial marking).*

5. $\tau \in \mathbb{N}^{\mathcal{P}}$: τ_p *gives the minimal time a token must spend in place p before becoming available for consumption by downstream transitions[1]. It will be called the* holding time *of the place throughout this paper.*

We denote by r^{out} the set of vertices (places or transitions) downstream of a vertex r, and by r^{in} the set of vertices upstream of r. Formally,

$$r^{\text{out}} = \{s \mid M_{sr} \neq 0\}, \quad r^{\text{in}} = \{s \mid M_{rs} \neq 0\}.$$

In order to specify a unique behavior of the system, we equip TPN with *routing policies.*

---

[1]Without loss of modeling power, the firing of transitions is supposed to be instantaneous (i.e. it involves no delay in consuming and producing tokens).

Definition 2.2 (Routing policy) A routing policy at place p is a family $\{m_{qp}, \Pi^p_{qq'}\}_{q \in p^{\text{out}}, q' \in p^{\text{in}}}$, where

1. $m_p = \sum_{q \in p^{\text{out}}} m_{qp}$ is an integer partition of the initial marking. m_{qp} *tells the number of tokens of the initial marking reserved for transition q.*

2. $\{\Pi^p_{qq'}\}_{q \in p^{\text{out}}}$ is a partition of the flow from q'. That is, $\Pi^p_{qq'}(n)$ *tells the number of tokens routed from q' to q via p among the first n ones.* More formally, $\Pi^p_{qq'}$ are nondecreasing maps $\mathbb{N} \to \mathbb{N}$ such that $\forall n$, $\sum_{q \in p^{\text{out}}} \Pi^p_{qq'}(n) = n$.

A routing policy for the net is a collection of routing policies for places.

Then the *earliest behavior* of the system is defined as follows. As soon as a token enters a place, it is *reserved* for the firing of a given downstream transition according to the routing policy. A transition q must *fire* as soon as all the places p upstream of q contain enough tokens (M_{qp}) that are reserved for transition q and have spent at least τ_p units of time in place p (by convention, the tokens of the initial marking are present from time $-\infty$, so that they are immediately available at time 0). When the transition fires, it consumes the corresponding upstream tokens and immediately produces an amount of tokens equal to M_{pq} in each place p downstream of q.

We next give the dynamic equations satisfied by the Timed Petri Net. We associate *counter functions* to nodes and arcs of the graph: $Z_p(t)$ denotes the cumulated number of tokens that have entered place p up to time t, *including the initial marking*; $Z_q(t)$ denotes the number of firings of transition q that have occurred up to time t; $W_{pq}(t)$ denotes the cumulated number of tokens that have arrived at place p from transition q up to time t; $W_{qp}(t)$ denotes the cumulated number of tokens that have arrived at place p up to time t (including the initial marking) reserved for the firing of transition q. We introduce the notation

$$\mu_{pq} \stackrel{\text{def}}{=} M_{pq}, \quad \mu_{qp} \stackrel{\text{def}}{=} M_{qp}^{-1},$$

and we set $\lfloor x \rfloor = \sup\{n \in \mathbb{Z} \mid n \le x\}$.

Assertion 2.3 *The counter variables of a Timed Petri Net under the earliest firing rule satisfy the following equations*[2]

$$Z_q(t) = \min_{p \in q^{\text{in}}} \lfloor \mu_{qp} W_{qp}(t - \tau_p) \rfloor, \tag{2.1a}$$

$$W_{pq}(t) = \mu_{pq} Z_q(t), \tag{2.1b}$$

---

[2]We adopt the convention $\sum_{q \in \emptyset}() = 0$, so that (2.1c) becomes $Z_p(t) = m_p$ when $p^{\text{in}} = \emptyset$. The transitions q such that $q^{\text{in}} = \emptyset$ will be considered as *input transitions* whose behavior is given externally. Thus, (2.1a) should be ignored whenever q has no predecessors.

$$Z_p(t) = m_p + \sum_{q \in p^{\text{in}}} W_{pq}(t), \tag{2.1c}$$

$$W_{qp} = m_{qp} + \sum_{q' \in p^{\text{in}}} \Pi^p_{qq'}(W_{pq'}). \tag{2.1d}$$

We deduce from (2.1) the *transition-to-transition* equation

$$Z_q(t) = \min_{p \in q^{\text{in}}} \left\lfloor \mu_{qp} \left(m_{qp} + \sum_{q' \in p^{\text{in}}} \Pi^p_{qq'} \left(\mu_{pq'} Z_{q'}(t - \tau_p) \right) \right) \right\rfloor. \tag{2.2}$$

If $\tau_p = 0$ for some places, this equation becomes implicit and we may have difficulties in proving the existence of a finite solution. We say that the TPN is *explicit* if there are no circuits containing only places with zero holding times. This ensures the uniqueness of the solution of (2.1) and (2.2) under any routing policy Π.

Input–output Partition We partition the set of transitions $\mathcal{Q} = \mathcal{U} \cup \mathcal{X} \cup \mathcal{Y}$ where $\mathcal{U}$ is the set of transitions with no predecessors (input transitions), $\mathcal{Y}$ is the set of transitions with no successors (output transitions) and $\mathcal{X} = \mathcal{Q} \setminus (\mathcal{U} \cup \mathcal{Y})$. We denote by u (resp. x, y) the vector of input (resp. state, output) counters $Z_q, q \in \mathcal{U}$ (resp. $\mathcal{X}$, $\mathcal{Y}$). Throughout the paper, we will study the *input–output* behavior of the system. That is, we look for the minimal trajectory (x, y) generated by the input history $u(t), t \in \mathbb{Z}$. This encompasses the *autonomous regime* traditionally considered in the Petri Net literature, when the system is frozen at an initial condition $Z_q(t) = v_q \in \mathbb{R}$ for negative t, and evolves freely according to the dynamics (2.1) for $t \geq 0$. This can be obtained as a specialization of the input–output case by adjoining an input transition q' upstream of each original transition q and setting $u_{q'}(t) = v_q$ for $t < 0$, $u_{q'}(t) = +\infty$ otherwise.

3 Modeling of Continuous Timed Petri Nets

We shall address the *continuous* version of TPN (in which the numbers of tokens are real numbers instead of integers). Such continuous models occur naturally when fluids rather than tokens flow in networks (see [2, §1.2.7], [24] for an elementary example). They also arise as approximations of (discrete) Petri Nets since they provide an upper bound for the real behavior.

A continuous TPN (CTPN) is defined as a TPN, but the marking m, the multipliers M and the counter functions are real-valued (the multipliers must be nonnegative: $M_{rs} \in \mathbb{R}^+$). This allows one to define some simple stationary routing policies. We shall single out three classes of policies.

General stationary routing A *stationary* routing policy is of the form $\Pi^p_{qq'}(n) = \rho^p_{qq'} \times n$ for some constants $\rho^p_{qq'} \geq 0$ such that for all $q' \in p^{in}$, $\sum_{q \in p^{out}} \rho^p_{qq'} = 1$. That is, the flow from q' at place p goes to q with proportion $\rho^p_{qq'}$. The counter functions of a CTPN satisfy the equations

$$Z_q(t) = \min_{p \in q^{in}} \mu_{qp} W_{qp}(t - \tau_p), \tag{3.1a}$$

$$W_{qp}(t) = m_{qp} + \sum_{q' \in p^{in}} \rho^p_{qq'} W_{pq'}(t), \tag{3.1b}$$

together with (2.1c), (2.1b). Eliminating W, we get a transition-to-transition equation

$$Z_q(t) = \min_{p \in q^{in}} \left(\mu_{qp} m_{qp} + \sum_{q' \in p^{in}} \mu_{qp} \rho^p_{qq'} \mu_{pq'} Z_{q'}(t - \tau_p) \right). \tag{3.2}$$

Dually, an equation involving only the variables W_{qp} can be obtained:

$$W_{qp} = m_{qp} + \sum_{q' \in p^{in}} \min_{p' \in (q')^{in}} \left(\rho^p_{qq'} \mu_{pq'} \mu_{q'p'} W_{q'p'}(t - \tau_{p'}) \right). \tag{3.3}$$

The following special cases of stationary routing are worth mentioning.

Origin independent routing When the routing at place p does not take into account the origin of the token but only its numbering, we get the condition

$$\forall p, q, \quad \forall q', q'' \in p^{in}, \quad \rho^p_{qq'} = \rho^p_{qq''}, \quad \rho^p_{qq'} m_p = m_{qp}. \tag{3.4}$$

We shorten $\rho^p_{qq'}$ to ρ^p_q. The dynamics of the system (3.1) can be rewritten with the aggregated variables Z_p (instead of W_{qp}):

$$Z_q(t) = \min_{p \in q^{in}} \mu_{qp} \rho^p_q Z_p(t - \tau_p), \tag{3.5a}$$

$$Z_p(t) = m_p + \sum_{q \in p^{in}} \mu_{pq} Z_q. \tag{3.5b}$$

Such routing policies depending only on the numbering of tokens (and leading to similar equations) have been studied by Baccelli *et al.* in a stochastic context [3]. We note that when $\tau_p \equiv 1$, (3.5) reads as the coupling of a conventional linear system with a (min, ×) linear system, namely[3]

$$Z_Q(t) = \mu'_{QP} \otimes Z_P(t - 1), \tag{3.6}$$

$$Z_P(t) = m + \mu_{PQ} Z_Q(t), \tag{3.7}$$

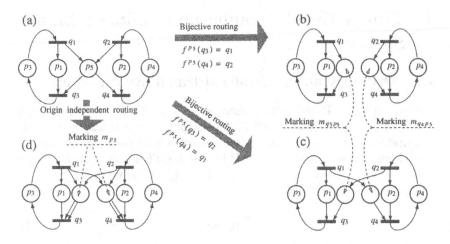

Figure 2: A balanced Petri net under various routing policies.

where $(A \otimes x)_i = \bigoplus_j A_{ij} \otimes x_j = \min_j A_{ij}x_j$ is the matrix product of the dioid[4] $\mathbb{R}_{\min,\times} \overset{\text{def}}{=} (\mathbb{R}^{+*} \cup \{+\infty\}, \min, \times)$.

Example 3.1 The origin independent routing $\rho_{q_3}^{p_5} = \rho_{q_4}^{p_5} = 1/2$ reduces the CTPN in Fig. 2a to that of Fig. 2d.

Injective routing We say that the routing function ρ^p at place p is *injective* if there is a map $f^p : p^{\text{out}} \to p^{\text{in}}$ such that

$$\forall q, \ \rho_{qq'}^p \neq 0 \Rightarrow q' = f^p(q). \tag{3.8}$$

That is, all the tokens routed to q at place p come for a single transition $f(q)$. Such routings occur frequently when tokens correspond to *resources* (e.g. pallets) which follow some well defined physical routes. An injective routing exists iff[5] $|p^{\text{out}}| \geq |p^{\text{in}}|$. Indeed, the following stronger condition is often satisfied in practice (e.g. in Fig. 2a).

Definition 3.2 (Balanced TPN) A TPN is *balanced* if $\forall p, |p|^{\text{out}} = |p|^{\text{in}}$.

In this particular case, we shall speak of *bijective* routing policies (since f^p becomes a bijection $p^{\text{out}} \to p^{\text{in}}$). We shall see later on that injective and bijective routing policies lead to tractable classes of systems.

---

[3]We denote by Z_Q (resp. Z_P) the restriction of Z to transitions (resp. to places). The convention for μ_{pq} is similar. We have set $(\mu'_{QP})_{qp} = \rho_q^p \mu_{qp}$.

[4]A *dioid* [9, 2] is a semiring whose addition is idempotent: $a \oplus a = a$.

[5]We denote by $|X|$ the cardinal of a set X.

4   Timed Event Graphs and (min,+) Linear Systems

4.1   Ordinary and generalized timed event graphs

Definition 4.1 (Timed event graphs) A *Continuous Timed Event Graph with Multipliers* (CTEGM) is a CTPN such that there is exactly one transition upstream and one transition downstream of each place. An (ordinary) *Continuous Timed Event Graph* (CTEG) is a CTEGM such that all arcs have multiplier one: $M_{pq}, M_{qp} \in \{0, 1\}$. More generally, we define the *place multipliers*[6]

$$\alpha_p \stackrel{\text{def}}{=} \mu_{p^{\text{out}}p}\mu_{pp^{\text{in}}}. \tag{4.1}$$

A *(rate α) CTEG* is a CTEGM with unit holding times and constant place multipliers. A CTEGM admits a *potential* if there exists a vector $v \in (\mathbb{R}^{+*})^{\mathcal{Q} \cup \mathcal{P}}$ (potential) such that

$$\forall r, s \in \mathcal{Q} \cup \mathcal{P}, \quad r \in s^{\text{out}} \Rightarrow v_r = \mu_{rs} v_s. \tag{4.2}$$

We set

$$\nu_p \stackrel{\text{def}}{=} \mu_{p^{\text{out}}p} m_p. \tag{4.3}$$

Assertion 4.2 *The dynamics of a CTEGM can be written as*

$$Z_q(t) = \min_{p \in q^{\text{in}}} \left(\nu_p + \alpha_p Z_{p^{\text{in}}}(t - \tau_p) \right). \tag{4.4a}$$

We have the following specializations:

$$Z_q(t) = \min_{p \in q^{\text{in}}} \left(\nu_p + Z_{p^{\text{in}}}(t - \tau_p) \right) \qquad \text{(TEG case),} \tag{4.4b}$$

$$Z_q(t) = \min_{p \in q^{\text{in}}} \left(\nu_p + \alpha Z_{p^{\text{in}}}(t - 1) \right) \qquad \text{(rate α case),} \tag{4.4c}$$

$$Z_q(t) = v_q \min_{p \in q^{\text{in}}} \left(v_p^{-1} m_p + v_{p^{\text{in}}}^{-1} Z_{p^{\text{in}}}(t - \tau_p) \right) \qquad \text{(potential case).} \tag{4.4d}$$

The last equation shows that a CTEGM with potential reduces to ordinary CTEG after the diagonal change of variable $Z_q = v_q Z_q'$. This change of variables should be interpreted as a *change of units* (v_q firings of transition q being counted as a single one).

---

[6]Since p^{out} and p^{in} are singletons, the notation will be used to designate their single members.

Example 4.3 If one mixes white and red paints in equal proportions to produce pink paint, the main concern is to say that with 3 liters of red for a single liter of white, there is 2 liters of red which is useless (that is, the min is the appropriate operator) but then 2 liters of pink can be produced, hence the right thing to do is to count pink paint by *pairs* of liters.

Theorem 4.4 *CTPN under injective routing policies reduce to CTEGM. Balanced CTPN with unit multipliers reduce to (ordinary) TEG.*

Proof Define the new set of places $\mathcal{P}' = \mathcal{Q} \times \mathcal{P}$, with the incidence relation $q^{\text{in}} = \{(qp) \mid p \in p^{\text{in}}\}$, $(qp)^{\text{in}} = f^p(q)$. Then the dynamics (3.2) reduces to (4.4a), with $\alpha_{qp} = \mu_{qp}\mu_{pf^p(q)}$. The specialization to the TEG case is immediate. $\qquad\square$

Example 4.5 The Petri Net of Fig. 2a admits two possible bijective routing policies at place p_5 which lead to the two Timed Event Graphs of Fig. 2b and 2c respectively.

4.2 Dynamic programming interpretation of CTEGM

We exhibit a correspondence between the above classes of Event Graphs and classical deterministic decision problems.

Given a CTEGM, we consider the discrete time controlled process q_n over a horizon t with

1. finite state space $\mathcal{Q}$;

2. set of admissible control histories $\mathcal{P}_{\text{ad}} = \{p_1, \dots, p_t \mid \forall n, p_n \in q_n^{\text{in}}\}$;

3. backward dynamics $q_{n-1} = p_n^{\text{in}}$ where $p_n \in q_n^{\text{in}}$.

In other words, the controlled process follows the edges of the net with the reverse orientation, backward in time. The control at state (transition) q consists in choosing a place p upstream of q, which leads to the (unique) transition q' upstream of p.

We shall consider the following three deterministic cost structures.

Additive

$$J^{\text{add}}(p,t) = Z(0)_{q_0} + \sum_{n=1}^{t} \nu_{p_n}. \tag{4.5}$$

Note that the *initial cost* $Z(0)$ coincides with the initial value of the counter function of the CTEGM.

Additive with constant discount rate

$$J^{\text{disc}}(p,t) = \alpha^t Z(0)_{q_0} + \sum_{n=1}^{t} \alpha^{t-n} \nu_{p_n}. \tag{4.6}$$

Additive with controlled discount rate

$$J^{\text{c-disc}}(p,t) = \left(\prod_{j=1}^{t} \alpha_{p_j} \right) Z(0)_{q_0} + \sum_{n=1}^{t} \left(\prod_{j=n+1}^{t} \alpha_{p_j} \right) \nu_{p_n}. \tag{4.7}$$

The *value function* associated with any of the above cost functions J is the map

$$Z_q(t) = \min_{p \in \mathcal{P}_{\text{ad}}, \; q_t = q} J(p,t).$$

Theorem 4.6 *When $\tau_p \equiv 1$,*

1. *The counter of a CTEG coincides with the value function for the additive cost J^{add}.*

2. *The counter of a (rate α) CTEG coincides with the value function for the discounted cost J^{disc}.*

3. *The counter of a CTEGM coincides with the value function for the cost with controlled discount rate $J^{\text{c-disc}}$.*

Remark 4.7 Minimizing $J^{\text{c-disc}}$ is known as a problem of shortest path with gains. See [17, Chapter 3, §7] and the references therein.

4.3   Operatorial representation of CTEGM

We introduce the set of signals $\mathcal{S} \stackrel{\text{def}}{=} (\mathbb{R} \cup \{+\infty\})^{\mathbb{Z}}$ to represent counter functions (we do not require the signals to be either positive-valued or nondecreasing, although this will be the case in most applications).

Definition 4.8 An operator $f : \mathcal{S} \to \mathcal{S}$ is

1. *additive* if it satisfies the min-superposition property

$$f(\min(x, x')) = \min(f(x), f(x')); \tag{4.8}$$

2. *linear* if it is additive and satisfies the homogeneity property

$$f(\lambda + x) = \lambda + f(x).$$

Of course, "linear" refers to the (min,+) dioid $\mathbb{R}_{\min} \stackrel{\text{def}}{=} (\mathbb{R} \cup \{+\infty\}, \min, +)$. Throughout the paper, we shall freely use the dioid notation $a \oplus b$ for $\min(a, b)$, $a \otimes b$ for $a + b$, $\varepsilon = +\infty$ for the zero element, $e = 0$ for the unit.

The following three families of operators play a central role in CTEGM:

$$
\begin{aligned}
\gamma^{\nu} &: \quad \gamma^{\nu} x(t) \stackrel{\text{def}}{=} x(t) + \nu \quad \text{(shift in counting)} \\
\delta^{\tau} &: \quad \delta^{\tau} x(t) \stackrel{\text{def}}{=} x(t - \tau) \quad \text{(shift in dating)} \\
\mu &: \quad \mu x(t) \stackrel{\text{def}}{=} \mu \times x(t) \quad \text{(scaling)},
\end{aligned}
\tag{4.9}
$$

where $\nu \in \mathbb{R}, \tau \in \mathbb{N}, \mu \in \mathbb{R}^{+*}$. We note that γ and δ are linear while μ is only additive. We have the commutation rules:

$$
\gamma^{\nu} \delta^{\tau} = \delta^{\tau} \gamma^{\nu}, \tag{4.10a}
$$

$$
\mu \delta^{\tau} = \delta^{\tau} \mu, \tag{4.10b}
$$

$$
\mu \gamma^{\nu} = \gamma^{\mu\nu} \mu. \tag{4.10c}
$$

Additive operators equipped with pointwise min and composition form an idempotent semiring, which we denote by $\mathcal{O}$. The following subsemirings of $\mathcal{O}$ are central.

1. The semiring generated by $\gamma^{\nu}; \nu \in \mathbb{R}$ is isomorphic to $\mathbb{R}_{\min}$ via the identification of ν with γ^{ν}.

2. The semiring generated by $\gamma^{\nu}, \delta^{\tau}; \nu \in \mathbb{R}, \tau \in \mathbb{R}$ is isomorphic to the semiring of polynomials in the indeterminate δ, $\mathbb{R}_{\min}[\delta]$ (via the same identification).

3. The semiring generated by $\gamma^{\nu}; \nu \in \mathbb{R}^{+}$ and by the powers of $\alpha\delta$, where α is a given and fixed value of μ, will be denoted by $\mathbb{R}_{\min}[\alpha\delta]$. It is a particular instance of a classical structure in difference algebra: *Ore polynomials*[7] [26, 19, 13].

4. The semiring generated by $\gamma^{\nu}, \mu; \nu \in \mathbb{R}, \mu \in \mathbb{R}^{+*}$ is isomorphic to the semiring of *nondecreasing concave piecewise affine* maps $\mathbb{R} \cup \{+\infty\} \to \mathbb{R} \cup \{+\infty\}$, which we denote by $\mathcal{A}_{\min}$. A generic element in $\mathcal{A}_{\min}$ is a map $p = \bigoplus_{i=1}^{k} \mu_i \gamma^{\nu_i}$,

$$
p(x) = \min_{1 \leq i \leq k} (\nu_i + \mu_i x).
$$

5. Finally, the semiring generated by $\gamma^{\nu}, \delta^{\tau}, \mu; \nu \in \mathbb{R}, \tau \in \mathbb{N}, \mu \in \mathbb{R}^{+*}$ is isomorphic to the semiring of polynomials $\mathcal{A}_{\min}[\delta]$.

---

[7]We recall that given a semiring $\mathcal{S}$ equipped with an automorphism $\alpha : \mathcal{S} \to \mathcal{S}$, the semiring of *Ore polynomials* in the indeterminate X, denoted by $\mathcal{S}[X; \alpha]$, is the set of finite formal sums $\sum_n s_n X^n$ (all but a finite number of s_n are zero), equipped with the usual componentwise sum $(s \oplus s')_n \stackrel{\text{def}}{=} s_n \oplus s'_n$ and the skew Cauchy product $(s \otimes s')_n \stackrel{\text{def}}{=} \bigoplus_{p+q=n} s_p \otimes \alpha^p(s_q)$. This product is determined by the rule $Xa = \alpha(a)X$ for all $a \in \mathcal{S}$. Identifying X with $\alpha\delta$ and setting $\alpha(\nu) \stackrel{\text{def}}{=} \alpha \times \nu$ for $\nu \in \mathbb{R}_{\min}$, we see that $X\nu = \alpha(\nu)X$ is just the rule $\alpha\delta\gamma^{\nu} = \gamma^{\alpha\nu}\alpha\delta$ which follows from (4.10).

We extend the operatorial notation to matrices by setting for $A \in \mathcal{O}^{n \times p}$ and $x \in \mathcal{S}^p$,

$$(Ax)_i \overset{\text{def}}{=} \min_j A_{ij}(x_j). \tag{4.11}$$

Note that for operator matrices A, A', B and vectors of counters x, x' of appropriate sizes

$$(AB)x = A(Bx), \quad (A \oplus A')x = Ax \oplus A'x, \quad A(x \oplus x') = Ax \oplus Ax'.$$

More formally, vectors of counter functions are a *left semimodule* under the action of additive matrix operators.

Theorem 4.9 *The counter equations of a CTEGM can be written as*

$$x = Ax \oplus Bu, \quad y = Cx \oplus Du \tag{4.12}$$

where A, B, C, D are matrices with entries in $\mathcal{O}$. More precisely,

1. *the entries of A, B, C, D belong to $\mathbb{R}_{\min}[\delta]$ for an ordinary CTEG;*

2. *the entries belong to $\mathbb{R}_{\min}[\alpha\delta]$ for a (rate α)-CTEG;*

3. *the entries belong to $\mathcal{A}_{\min}[\delta]$ for a general CTEGM.*

Theorem 4.10 (Convolution representation) *An explicit SISO[8] CTEGM admits an input output relation of the form*

$$y(t) = \inf_{\tau \in \mathbb{N}}[h(\tau) + u(t - \tau)] \qquad \text{(ordinary CTEG)} \tag{4.13}$$

$$y(t) = v_y \inf_{\tau \in \mathbb{N}}[h(\tau) + v_u^{-1}u(t - \tau)] \qquad \text{(CTEG with potential)} \tag{4.14}$$

$$y(t) = \inf_{\tau \in \mathbb{N}}[h(\tau) + \alpha^\tau u(t - \tau)] \qquad \text{(CTEG with rate α)} \tag{4.15}$$

$$y(t) = \inf_{i \in I}[\nu_i + \mu_i u(t - \tau_i)] \qquad \text{(general case)} \tag{4.16}$$

where h is a map $\mathbb{N} \to \mathbb{R} \cup \{+\infty\}$, $v_u, v_y \in \mathbb{R}^{+*}$, and where the family $\{\nu_i \in \mathbb{R}, \mu_i \in \mathbb{R}^{+*}, \tau_i \in \mathbb{N}\}$ is such that there are only finitely many i such that $\tau_i = \tau$ for any $\tau \in \mathbb{N}$.

We postpone the proof: these representation results will appear as consequences of the more general behavioral properties of CTEGM operators given in §4.4.

Theorem 4.9 established a connection between various algebras of polynomial type and various classes of Event Graphs. Theorem 4.10 now establishes a similar connection between input–output representations and certain formal

---

[8]Single Input Single Output. The extension to the Multiple Inputs Multiple Outputs (MIMO) case is immediate.

series algebras. Let us recall that given a semiring $\mathcal{K}$ and an indeterminate δ, we denote by $\mathcal{K}[[\delta]]$ the semiring of series with coefficients in $\mathcal{K}$ (set of formal sums $\bigoplus_{t\in\mathbb{N}} h_t \delta^t$ with $h_t \in \mathcal{K}$, equipped with pointwise sum and Cauchy product). The generic series of $\mathcal{A}_{\min}[[\delta]]$ can be written as

$$h = \bigoplus_{\tau\in\mathbb{N}} h_\tau \delta^\tau = \bigoplus_{\tau} \left(\bigoplus_{i\in I_\tau} \mu_{i\tau} \gamma^{\nu_{i\tau}} \right) \delta^\tau$$

where for all τ, I_τ is finite. Such series act naturally on $\mathcal{S}$ by interpreting the indeterminate δ as the shift operator:

$$hu(t) = \bigoplus_{\tau\in\mathbb{N}} h_\tau(u(t-\tau)) = \inf_{\tau\in\mathbb{N}} \min_{i\in I_\tau} (\nu_{i\tau} + \mu_{i\tau} u(t-\tau)).$$

Theorem 4.10 asserts that (i) CTEGM operators correspond to the action of $\mathcal{A}_{\min}[[\delta]]$ on counter functions, (ii) CTEG operators correspond to the action of $\mathbb{R}_{\min}[[\delta]]$, (iii) α-CTEG operators correspond to the action of the dioid of *Ore series* $\mathbb{R}_{\min}[[\alpha\delta]]$ (defined as Ore polynomials, without the finiteness condition).

4.4 Behavioral characterizations of CTEGM

Theorem 4.11 *The input–output map* $\mathcal{H} : u \rightarrow y$ *of a SISO explicit CTEGM satisfies the following properties.*

1. Stationarity. $\mathcal{H}\delta^\tau = \delta^\tau\mathcal{H}$.

2. Causality. $\forall t \leq \tau, u(t) = v(t) \Rightarrow \forall t \leq \tau,\ \mathcal{H}u(t) = \mathcal{H}v(t)$.

3. Additivity. $\mathcal{H}(\min(u,v)) = \min(\mathcal{H}u, \mathcal{H}v)$.

4. Scott continuity. For any filtered[9] family $\{u_i\}_{i\in I}$, $\mathcal{H}(\inf_{i\in I} u_i) = \inf_{i\in I} \mathcal{H}u_i$.

5. Concavity. $\forall \lambda_i \geq 0, \sum_i \lambda_i = 1, \mathcal{H}(\sum_{i=1}^n \lambda_i u_i) \geq \sum_i \lambda_i \mathcal{H}u_i$.

A CTEG with rate α *satisfies the additional property*

6. α-homogeneity. For all constant λ, $\mathcal{H}(\lambda\alpha^t + u) = \lambda\alpha^t + \mathcal{H}u$, *with an obvious convention[10].*

A CTEGM with potential satisfies the alternative additional property[11]

7. (v_u, v_y)-homogeneity. For all $\lambda \in \mathbb{R}$, $\mathcal{H}(\lambda v_u + u) = \lambda v_y + \mathcal{H}(u)$.

---

[9]A family is filtered if any finite subfamily admits a lower bound in the family. Note that the Scott continuity together with additivity is equivalent to the preservation of arbitrary inf: $\mathcal{H}(\inf_i u_i) = \inf_i \mathcal{H}u_i$ for an arbitrary family. The Scott topology is presented in detail in [16]. What we call here Scott continuity is in fact Scott continuity with respect to the algebraic order $\preceq$ of the (min,+) semiring, defined by $a \preceq b \Leftrightarrow a \oplus b = b$ (which is reversed with respect to natural order).

[10]α^t denotes the map $t \mapsto \alpha^t$.

[11]$\lambda + u$ denotes the signal $t \mapsto \lambda + u(t)$.

Note that the specialization of the α-homogeneity to $\alpha = 1$ gives the standard homogeneity property $\lambda + u \to \lambda + y$. So does the specialization of the (v_u, v_y)-homogeneity to the case of constant potential v.

Proof The additivity of $\mathcal{H}$ is an immediate consequence of the additivity of A, B, C, D and the uniqueness of the solution of $x = Ax \oplus Bu, y = Cx \oplus Du$. The other properties can be proved along the same lines by transferring to $\mathcal{H}$ the properties valid for A, B, C, D.   □

The following converse theorem shows that the properties listed are accurate.

Theorem 4.12 *A map $\mathcal{H}$ which satisfies properties 1–5 in Theorem 4.11 is a nonincreasing limit of CTEGM operators[12]. An operator which satisfies 1–6 (resp. 1–5,7) is a nonincreasing limit of rate α CTEG operators (resp. with potential v).*

The main point of the proof consists in the following general "convolution" representation lemma for additive continuous stationary operators.

Lemma 4.13 *Let $\mathcal{D}$ denote a complete[13] dioid, $\mathcal{H} : \mathcal{D}^{\mathbb{Z}} \to \mathcal{D}^{\mathbb{Z}}$. The following assertions are equivalent.   1. $\mathcal{H}$ is stationary, causal, additive, and Scott continuous; 2. there exists a family of additive Scott continuous maps $h_\tau, \mathcal{D} \to \mathcal{D}, \tau \in \mathbb{N}$ such that*

$$\mathcal{H}u(t) = \bigoplus_{\tau \in \mathbb{N}} h_\tau(u(t - \tau)). \tag{4.17}$$

Proof Clearly, 2⇒1. For the converse we introduce the Dirac function

$$\mathbf{e} : \mathbb{Z} \to \mathcal{D}, \quad \mathbf{e}(t) = \begin{cases} e & \text{if } t = 0 \\ \varepsilon & \text{otherwise.} \end{cases}$$

We have the decomposition of an arbitrary signal $u \in \mathcal{D}^{\mathbb{Z}}$ on the basis of shifted Dirac functions:

$$u = \bigoplus_{\tau \in \mathbb{Z}} u(\tau)\delta^\tau \mathbf{e}.$$

The additivity, stationarity and Scott continuity assumptions yield

$$\mathcal{H}u = \bigoplus_{\tau \in \mathbb{Z}} \delta^\tau \mathcal{H}(u(\tau)\mathbf{e}). \tag{4.18}$$

---

[12]I.e. there exists a nonincreasing sequence $\mathcal{H}_i \geq \mathcal{H}_{i+1}, i \in \mathbb{N}$ of input–output operators of CTEGM such that $\mathcal{H} = \inf_{i \in I} \mathcal{H}_i$.

[13]A dioid $\mathcal{D}$ is complete if an arbitrary subset admits a least upper bound (for the order $a \preceq b \Leftrightarrow a \oplus b = b$) and if the product is Scott continuous.

Now, let us decompose the output corresponding to $u = x\mathbf{e}$ (with $x \in \mathcal{D}$) on the basis $\{\delta^\tau \mathbf{e}\}_{\tau \in \mathbb{Z}}$:

$$\mathcal{H}(x\mathbf{e}) = \bigoplus_{\tau \in \mathbb{Z}} h_\tau(x)\delta^\tau \mathbf{e}.$$

This together with (4.18) gives

$$\mathcal{H}u = \bigoplus_{\tau, \tau' \in \mathbb{Z}} h_\tau(u(\tau))\delta^{\tau + \tau'} \mathbf{e}$$

i.e.

$$\mathcal{H}u(t) = \bigoplus_{\tau \in \mathbb{Z}} h_\tau(u(t - \tau)).$$

The sum can obviously be restricted to $\tau \in \mathbb{N}$ due to causality. The additivity and continuity of h_τ are immediate. $\qquad\qquad\qquad\qquad\qquad$ □

To complete the proof of Theorem 4.12, it suffices to observe that the additivity, concavity, and potential properties, valid for $\mathcal{H}$, transfer to each h_τ. Then, the concave monotone real-valued map h_τ admits a representation as a denumerable infimum of increasing affine functions:

$$h_\tau(x) = \inf_{n \in \mathbb{N}} (\nu_{n\tau} + \mu_{n\tau} x), \quad \text{where } \nu_n \in \mathbb{R} \cup \{+\infty\}, \mu_{n\tau} > 0.$$
$$(4.19)$$

The operator $\mathcal{H}^n = \bigoplus_{\tau \le n, k \le n} \gamma^{\nu_{k\tau}} \mu_{k\tau} \delta^\tau$ arises from a CTEGM operator (since it is obtained by a finite number of parallel/series compositions of elementary γ, μ, δ operators). It follows from (4.17)–(4.19) that $\lim_n \downarrow \mathcal{H}_n u = \mathcal{H}u$. This proves the first assertion of Theorem 4.12. The α-rate and potential special cases are immediate. $\qquad\qquad\qquad\qquad\qquad$ □

Finally, we note that the construction of the above proof explicitly yields the convolution representations stated in Theorem 4.10, with the exception of the additional finiteness condition that h_τ is a finite sum of $\gamma^{\nu_i} \mu_i$. This last result stems from the *rationality* features that we next introduce.

4.5 Rational operators

A natural problem is to characterize the subclass of series of $\mathcal{A}_{\min}[[\delta]]$ which arise as transfer operators of CTEGM (called *transfer series*). We recall that given a semiring of formal series $\mathcal{K}[[\delta]]$, the semiring of *rational* series [4] denoted by $\mathcal{K}^{\mathrm{rat}}[[\delta]]$ is the least subsemiring containing polynomials and stable under the operations $\oplus, \otimes, *$, where $a^* \overset{\text{def}}{=} \bigoplus_{n \in \mathbb{N}} a^n$ is defined only on series with zero constant coefficient. An immediate fixed-point argument[14] shows

---

[14]The unique solution of $x = Ax \oplus Bu$ is $x = A^*Bu$. The existence of A^* and the uniqueness of the solution follow from the assumption that there are no circuits with zero holding times.

160   *Guy Cohen, Stéphane Gaubert and Jean-Pierre Quadrat*

that the input and output counters given by (4.12) satisfy $y = hu$, where $h = CA^*B \oplus D$ is the transfer series of the system. Therefore, rephrasing the Kleene–Schützenberger theorem [4], we claim that transfer series and rational series coincide.

Assertion 4.14 *The transfer series of explicit SISO CTEGM (resp. α-CTEG, CTEG) are precisely the elements of $\mathcal{A}_{\min}^{\mathrm{rat}}[[\delta]]$ (resp. $\mathbb{R}_{\min}^{\mathrm{rat}}[[\alpha\delta]]$, $\mathbb{R}_{\min}^{\mathrm{rat}}[[\delta]]$).*

One important problem is to characterize these particular classes of rational series. The answer is known in the case of $\mathbb{R}_{\min}[[\delta]]$ and $\mathbb{R}_{\min}[[\alpha\delta]]$. We say that a series is *ultimately periodic with rate α* if there exists a constant λ and a positive integer c (cyclicity) such that for t large enough

$$h_{t+c} = \lambda \frac{1 - \alpha^c}{1 - \alpha} + \alpha^c h_t. \tag{4.20}$$

When $\alpha < 1$, this periodicity property means that h_t converges towards $\lambda/(1-\alpha)$ with rate α and that the rate is attained exactly after a finite time. The specialization to $\alpha = 1$ (in fact, $\alpha = 1^-$) yields $h_{t+c} = \lambda c + h_t$. The *merge* of k series $h^{(0)}, \dots, h^{(k-1)}$ is the series with coefficients $h_{i+nk} = h_n^{(i)}$ for $0 \le i \le k-1, n \in \mathbb{N}$.

Theorem 4.15 *A series in $\mathbb{R}_{\min}[[\alpha\delta]]$ is rational iff it is a merge of ultimately α-periodic series.*

The CTEG case (i.e. $\alpha = 1$) is proved in [9, 2] for the subclass of monotone[15] series $h_{n+1} \ge h_n$. It was already noticed by Moller [22] in the non-monotone case. It is essentially known to the tropical community [20]. The α-generalization was announced in [13]. The proof will appear in a paper in preparation [15].

No such simple characterization seems to exist for $\mathcal{A}_{\min}^{\mathrm{rat}}[[\delta]]$: the coefficient h_τ of δ^τ in h is an element of $\mathcal{A}_{\min}$, but its complexity[16] grows in general as $\tau \to \infty$.

4.6   Asymptotic behavior of CTEGM

We consider the autonomous case $Z = AZ$ with boundary condition $\forall t \le 0, Z(t) = v \in \mathbb{R}^Q$, where A belongs to one of the above matrix operator algebras. We associate several additive *weights* with the circuit $C =$

---

[15]The results are stated in the so called $\mathcal{M}_{\mathrm{in}}^{\mathrm{ax}}[[\gamma, \delta]]$ dioid which is isomorphic to the dioid of series in one indeterminate δ with coefficients in $\overline{\mathbb{R}}_{\min} \stackrel{\mathrm{def}}{=} (\mathbb{R} \cup \{\pm\infty\}, \min, +)$ such that $h_{n+1} \ge h_n$.

[16]The minimal number of monomials in a sum $h_\tau = \bigoplus_i \gamma^{\nu_i} \mu_i$.

$(q_1, p_1, q_2, \ldots, q_k, p_k),$

$$\begin{array}{rcll} |\mathcal{C}|_\nu & = & \sum_i \nu_{q_i p_i} & \text{Total normalized marking} \\ |\mathcal{C}|_\tau & = & \sum_i \tau_{p_i} & \text{Total holding time} \\ |\mathcal{C}|_l & = & \sum_i 1 = k & \text{Length} \\ |\mathcal{C}|_{m,v} & = & \sum_i m_{p_i} v_{p_i}^{-1} & \text{Total weighted marking} \end{array}$$

where the last quantity will be used only when the graph admits a potential v. The following periodicity theorem is central. The CTEG case is a consequence of the (max,+)-Perron–Frobenius theorem [25, 8, 2, 10]. Another proof has been given by Chretienne [7]. The inequality variant below (4.24) can be found in [12, Chapter IV, Lemma 1.3.8], [14]. The α-discounted case is due to Braker and Resing [5, 6].

Theorem 4.16 *Consider a strongly connected CTEG. There exists $N \geq 0$ and $c \geq 1$ (cyclicity) such that, for all initial conditions v,*

$$t \geq N \Rightarrow Z(t + c) = \lambda c + Z(t), \tag{4.21}$$

where

$$\lambda = \min_{\mathcal{C}} \frac{|\mathcal{C}|_\nu}{|\mathcal{C}|_\tau} \tag{4.22}$$

(the minimum is taken over the elementary circuits of the graph). Alternatively, λ is the unique scalar for which there exists a finite vector solution v of the spectral problem[17]

$$v_q = \min_{p \in q^{\text{in}}} \left(\nu_{qp} - \lambda \tau_p + v_{p^{\text{in}}} \right), \tag{4.23}$$

or it is the solution of the LP problem

$$\lambda \to \max, \quad \forall p \in q^{\text{in}}, \ v_q \leq \nu_{qp} - \lambda \tau_p + v_{p^{\text{in}}}. \tag{4.24}$$

For a strongly connected CTEG with potential, the periodicity property (4.21) becomes $Z_r(t + c) = \lambda_r c + Z_r(t)$, where

$$\lambda_r v_r^{-1} = \min_{\mathcal{C}} \frac{|\mathcal{C}|_{m,v}}{|\mathcal{C}|_\tau}. \tag{4.25}$$

For a strongly connected CTEG with rate α, the periodicity property (4.21) becomes

$$t \geq N \Rightarrow Z_q(t + c) = \lambda_q \frac{1 - \alpha^c}{1 - \alpha} + \alpha^c Z_q(t) \tag{4.26}$$

where $\lambda_q \in \mathbb{R}^+$ (the dependence in q is essential).

---

[17] With the (min,+) notation, when $\tau_p \equiv 1$, (4.23) becomes $Av = \lambda \otimes v$ where $A_{qq'} = \bigoplus_{p \in q^{\text{in}} \cap (q')^{\text{out}}} \nu_{qp}$.

The asymptotic behavior of general CTEGM is more subtle. We shall not attempt to treat it here.

Remark 4.17 When $\alpha < 1$, from (4.26) we get $\lim_{t\to\infty} Z_q(t) = \lambda_q/(1-\alpha)$. It is well known that one obtains the average cost value as the limit of the discounted case, i.e. $\forall q$, $\lim_{\alpha\to 1^-} \lambda_q = \lambda$.

Remark 4.18 When the graph has a potential v, for all circuits $\mathcal{C}$, the quantity $|\mathcal{C}|_{m,v}$ used in the periodic throughput formula is an *invariant* of the net (the firing of one transition leads to a new marking m' with the same weight).

4.7   Resource optimization problems

As a by-product of the above characterizations of the throughput λ, it is possible to address *resource optimization* problems. The most classical problem [8, 18, 21, 12] relating to TEG consists in optimizing a linear cost function $J(m)$ associated with the initial marking, under the constraint $\lambda \geq \lambda_0$. Physically, the initial marking represents resources (number of machines, pallets, processors, storage capacities), and the problem consists in minimizing the cost of the resources in order to guarantee a given throughput λ_0. By appealing to (4.24), this class of problems reduces to linear programming, with integer and real variables.

We will discuss here new resource optimization problems which arise for more general TPN due to the presence of routing decisions. We restrict to *balanced* TPN with unit multipliers. When a bijective routing f is fixed, the only remaining decision consists in the assignment of the initial marking m_p to the downstream transitions: $m_p = \sum_{q\in p^{\text{out}}} m_{qp}$. We thus consider the problem of finding the allocation of the initial marking which maximizes the performance of the system. We only consider internally *stable* systems in the sense of [2] (such that tokens do not accumulate indefinitely in places). Then, there is a single periodic throughput λ_r associated with every simply connected component r of the graph (characterized by (4.22)). We denote by $\mathcal{R}$ the set of simply connected components. The most natural performance measure to be optimized will be a linear combination of these throughputs, $c\lambda \stackrel{\text{def}}{=} \sum_{r\in\mathcal{R}} c_r \lambda_r$ where $c_r \geq 0$ are given weights.

Theorem 4.19 *The resource assignment problem for a balanced CTPN with unit multipliers under the bijective policy f reduces to the following linear programming problem. Given $\{m_p, \tau_p\}_{p\in\mathcal{P}}$, c and f, and denoting by $r(q)$ the simply connected component of transition q under policy f, solve*

$$\max_{v_q, \lambda_r, m_{qp}} c\lambda,$$

$$\begin{cases} m_p = \sum_{q\in p^{\text{out}}} m_{qp}, & \forall p, \\ v_q \leq m_{qp} - \lambda_{r(q)}\tau_p + v_{fp(q)}, & \forall q, \forall p \in q^{\text{in}}, \end{cases}$$

where $\{v_q\}_{q \in \mathcal{Q}}$, $\{m_{qp}\}_{q \in p^{\mathrm{out}}, p \in \mathcal{P}}$, and $\{\lambda_r\}_{r \in \mathcal{R}}$ are real (finitely) valued variables.

Proof Easy consequence of the characterization (4.24). $\qquad\qquad\square$

The same resource assignment problem for discrete (noncontinuous) TEG leads to a similar LP problem with mixed integer and real variables.

Example 4.20 For the routing policy of Fig. 2b, we obtain two strongly connected components with rates

$$\lambda_1 = \min\left(\frac{m_{q_3 p_5} + m_{p_3}}{\tau_{p_5} + \tau_{p_3}}, \kappa_1\right), \qquad \text{where } \kappa_1 = \frac{m_{p_1} + m_{p_3}}{\tau_{p_1} + \tau_{p_3}} \quad (4.27)$$

$$\lambda_2 = \min\left(\frac{m_{q_4 p_5} + m_{p_4}}{\tau_{p_5} + \tau_{p_4}}, \kappa_2\right), \qquad \text{where } \kappa_2 = \frac{m_{p_2} + m_{p_4}}{\tau_{p_2} + \tau_{p_4}}. \quad (4.28)$$

Maximizing the throughput in place p_5 reduces to

$$\max_{m_{q_3 p_5} + m_{q_4 p_5} = m_{p_5}} (\lambda_1 + \lambda_2). \quad (4.29)$$

The bijective policy shown of Fig. 2c gives a unique strongly connected component and a throughput

$$\lambda = \min\left(\kappa_1, \kappa_2, \frac{m_{p_3} + m_{p_4} + m_{p_5}}{\tau_{p_3} + \tau_{p_4} + 2\tau_{p_5}}\right) \quad (4.30)$$

independent of the allocation of m_{p_5}.

5 Time Behavior of Continuous Timed Petri Nets

5.1 Stochastic control interpretation

We interpret the evolution equations of a CTPN as the dynamic programming equation of the following stochastic extension of the deterministic decision process described in §4.2. The control at state (transition) q selects an upstream place $p \in q^{\mathrm{in}}$. Then, q moves randomly (in backward time) to one of the upstream transitions $q' \in p^{\mathrm{in}}$. More precisely:

1. The dynamics is given by a controlled Markov chain in backward time: the probability $P_{qq'}^p$ of the transition $q \to q'$ from time n to time $n-1$ under the decision p is given by

$$P_{qq'}^p = \alpha_{qp}^{-1} \mu_{qp} \rho_{qq'}^p \mu_{pq'}$$

where $\alpha_{qp} > 0$ is a normalization factor[18] (chosen such that $\sum_{q' \in p^{\mathrm{in}}} P_{qq'}^p = 1$).

---

[18]Note that in the CTEGM case, for $q = p^{\mathrm{out}}$, we have $\alpha_{qp} = \mu_{p^{\mathrm{out}}p} \mu_{pp^{\mathrm{in}}}$ so that α_{qp} coincides with α_p as defined in (4.1).

2. The set $\mathcal{P}_{ad}$ of admissible control histories is the set of sequences $p_1, \ldots, p_t$ such that $p_n \in q_n^{in}$ and the decision p_n is a feedback of q_n.

3. We consider a mean cost at state q of the form

$$J(p, t, q) = \mathbb{E}\left(\Big(\prod_{j=1}^{t} \alpha_{q_j p_j}\Big) Z(0)_{q_0} + \sum_{n=1}^{t} \Big(\prod_{j=n+1}^{t} \alpha_{q_j p_j} \Big) \nu_{q_n p_n} \Big| q_t = q \right).$$

Assertion 5.1 *For a CTPN such that $\tau_p \equiv 1$, the counter function coincides with the value function:*

$$Z_q(t) = \inf_{p \in \mathcal{P}_{ad}} J(p, t, q). \tag{5.1}$$

As in the case of Event Graphs, we shall pay a particular attention to simple cost functions.

Definition 5.2 A CTPN is *undiscounted* if $\alpha_{qp} \equiv 1$. It is *α-discounted* if $\tau_p \equiv 1$ and $\alpha_{qp} \equiv \alpha$. It admits a *potential* if there exists a vector $v \in (\mathbb{R}^{+*})^{\mathcal{Q}}$ such that the change of variable $Z_q = v_q Z_q'$ makes the CTPN undiscounted.

Clearly, the cost function of an undiscounted (resp. α-discounted) CTPN is given by

$$J(p, t, q) = \mathbb{E}\left(Z(0)_{q_0} + \sum_{n=1}^{t} \nu_{q_n p_n} \Big| q_t = q \right), \tag{5.2}$$

$$\text{resp.}\quad J(p, t, q) = \mathbb{E}\left(\alpha^t Z(0)_{q_0} + \sum_{n=1}^{t} \alpha^{t-n} \nu_{q_n p_n} \Big| q_t = q \right). \tag{5.3}$$

Theorem 5.3 *1. A CTPN becomes undiscounted under a stationary routing iff it satisfies the following equilibrium condition:*

$$\forall p, \quad \sum_{q \in p^{out}} M_{qp} = \sum_{q \in p^{in}} M_{pq}. \tag{5.4}$$

In this case, the only origin independent routing policy which makes the net undiscounted is given by[19]:

$$\forall q' \in p^{in}, \quad \rho_{qq'}^{p} = \frac{M_{qp}}{\sum_{q'' \in p^{out}} M_{q''p}}. \tag{5.5}$$

2. A CTPN with $\tau_p \equiv 1$ becomes α-discounted under a stationary routing iff

$$\forall p, \quad \sum_{q \in p^{in}} M_{qp} = \alpha \Big(\sum_{q \in p^{out}} M_{pq} \Big). \tag{5.6}$$

---

[19]This is a fairness condition which states that tokens are routed equally to the downstream arcs, counted with their multiplicities.

3. *There exists a stationary routing under potential v iff*

$$\forall p, \quad \sum_{q \in p^{\text{out}}} v_q M_{qp} = \sum_{q \in p^{\text{in}}} M_{pq} v_q. \tag{5.7}$$

4. *A CTEGM with routing ρ admits a potential v iff for all $q \in \mathcal{Q}, p \in q^{\text{in}}$,*

$$v_q = \sum_{q' \in p^{\text{in}}} \mu_{qp} \rho^p_{qq'} \mu_{pq'} v_{q'}. \tag{5.8}$$

Proof We prove item 3 (which contains item 1 as a special case). The CTPN has potential v iff for all p the matrix

$$P^p_{qq'} = v_q^{-1} M_{qp}^{-1} \rho^p_{qq'} M_{pq'} v_{q'}$$

is stochastic. Summing over $q' \in p^{\text{in}}$, we get $v_q M_{qp} = \sum_{q' \in p^{\text{in}}} \rho^p_{qq'} M_{pq'} v_{q'}$. Summing over $q \in p^{\text{out}}$ and using the fact that the transpose of $\rho^p_{\cdot \cdot}$ is stochastic, we get the necessary condition (5.7). Then the origin independent routing policy

$$\rho^p_{qq'} = \frac{v_q M_{qp}}{\sum_{q'' \in p^{\text{out}}} v_{q''} M_{q''p}} \quad \forall q' \in p^{\text{in}} \tag{5.9}$$

turns out to be admissible, which shows that the condition is also sufficient. The other points are left to the reader. $\qquad \Box$

5.2 Input–output representation of CTPN

Pursuing the program previously illustrated with additive systems (CTEGM), we provide an algebraic input–output representation for CTPN. In view of the dynamics of CTPN (see (3.2)), we introduce (min,+) polynomials and formal series in *several* commutative indeterminates. Given a family of indeterminates $\{z_i\}_{i \in \mathcal{I}}$ (not necessarily finite), we denote by $(\mathbb{R}^+)^{(\mathcal{I})}$ the set of *almost zero* sequences $\alpha_i \in \mathbb{R}^+, i \in \mathcal{I}$ (such that $I(\alpha) \overset{\text{def}}{=} \{i \in \mathcal{I} \mid \alpha_i \neq 0\}$ is finite). A generalized[20] formal series in the commutative indeterminates z_i with coefficients in $\mathbb{R}_{\min}$ is a sum

$$s = \bigoplus_{\alpha \in (\mathbb{R}^+)^{(\mathcal{I})}} s_\alpha \bigotimes_{i \in I(\alpha)} z_i^{\alpha_i}, \quad s_\alpha \in \mathbb{R}_{\min}. \tag{5.10}$$

This is a polynomial whenever $s_\alpha = \varepsilon$ for all but a finite number of α. The numerical function associated with a series s is the map $S : \mathbb{R}^{\mathcal{I}} \to \mathbb{R} \cup \{\pm\infty\}$,

$$S(z) = \inf_\alpha \left(s_\alpha + \sum_{i \in I(\alpha)} \alpha_i z_i \right). \tag{5.11}$$

---

[20] We allow nonnegative real valued exponents α_i, not only integer ones.

When s is a nonzero polynomial, the infimum in (5.11) is finite. This defines a proper notion of finitely valued (min,+) *polynomial function*. Polynomial functions are stable under pointwise min, pointwise sum and composition. It is clear that (3.2) is just a polynomial induction of the form

$$x(t) \; = \; A(x(t), \dots , x(t - \overline{\tau}), u(t), \dots , u(t - \overline{\tau})), \qquad (5.12)$$
$$y(t) \; = \; C(x(t), \dots , x(t - \overline{\tau}), u(t), \dots , u(t - \overline{\tau})), \qquad (5.13)$$

where A, C are polynomial functions and $\overline{\tau} \overset{\text{def}}{=} \max_p \tau_p$. Thus, *CTPN and* (min, +) *recurrent stationary polynomial systems coincide*. For simplicity, we shall limit ourselves to SISO systems (the MIMO case is not more difficult, although the notation is more intricate). We introduce the family of indeterminates $u_\tau, \tau \in \mathbb{N}$. The series s given by (5.10) is a *Volterra series* [11] if for all τ the series is a polynomial in the indeterminate u_τ (equivalently, if the indeterminate u_τ appears in (5.10) with a finite number of exponents). The *evaluation su* of the Volterra series s at the input u is obtained by substituting $u(t - \tau)$ for the indeterminate u_τ.

Theorem 5.4 (Volterra expansion) *The output of an explicit SISO CTPN is obtained as the evaluation of a Volterra series:*

$$y(t) = su(t) = \inf_\alpha \Big(a_\alpha + \sum_{\tau \in I(\alpha)} \alpha_\tau u(t - \tau) \Big). \qquad (5.14)$$

A case of particular interest arises for inputs with finite past: $u(\tau) = \varepsilon$ for $\tau \leq T_0$. Then, for all t, the Volterra expansion of $y(t)$ is obviously finite.

5.3 Behavioral properties of CTPN

Theorem 5.5 *The input–output map $\mathcal{H}$ of a MIMO CTPN is*

1. *stationary,*

2. *causal,*

3. *monotone:* $u \leq v \Rightarrow \mathcal{H}u \leq \mathcal{H}v$,

4. *Scott continuous,*

5. *concave (see Theorem 4.11 for the definitions).*

Undiscounted CTPN satisfy the following property.

6. *Homogeneity:* $\mathcal{H}(\lambda + u) = \lambda + \mathcal{H}(u)$.

CTPN with potential v satisfy the following.

7. (v_u, v_y)-*homogeneity:* $\mathcal{H}(\lambda v_u + u) = \lambda v_y + \mathcal{H}u$.

All these properties are immediate consequences of the (MIMO extension) of the Volterra expansion (5.14). Again, these properties are accurate: it could be shown that an map satisfying the above properties is a limit of CTPN operators, but we shall not attempt to detail this statement here.

5.4 Asymptotic properties of undiscounted Petri nets

Theorem 5.6 *For a strongly connected undiscounted CTPN, we have*

$$\lim_{t \to \infty} \frac{1}{t} Z_q(t) = \lambda, \quad \forall q,$$

where λ is a constant. The periodic throughput λ *is characterized as the unique value for which a finite vector v is solution of*

$$v = \min_p \left(\nu_{\cdot p} - \lambda \tau_p + P^p v \right). \tag{5.15}$$

Indeed, the asymptotic behavior of $Z(t)$ is known in much more detail [27]. Note that the effective computation of λ from (5.15) proceeds from standard algorithms (Policy Improvement [28], Linear Programming).

Proof This is an adaptation of standard stochastic control results [28, Chapter 33, Theorem 4.1]. The growth rate λ is independent of the initial point q for the subclass of communicating systems[21]. This assumption is equivalent to the strong connectivity of the net. □

There is an equivalent characterization of λ which exhibits the analogy with the CTEG case in a better way. A *feedback policy* (or policy[22], for short) is a map $u : \mathcal{Q} \to \mathcal{P}$. The policy is *admissible* if $u(q) \in q^{\text{in}}$, that is, if setting $p_n = u(q_n)$ yields an admissible policy for the stochastic control problem presented in §5.1. With a policy u are associated the following vectors and matrices

$$\nu_q^u \overset{\text{def}}{=} \nu_{qu(q)}, \qquad \tau_q^u \overset{\text{def}}{=} \tau_{u(q)}, \qquad P_{qq'}^u \overset{\text{def}}{=} P_{qq'}^{u(q)}.$$

We denote by $\mathcal{R}(u)$ the set of final classes[23] of the matrix P^u. For each class $r \in \mathcal{R}(u)$, we have a unique invariant measure π^{ru} with support r (i.e. $\pi^{ru} P^u = \pi^{ru}$, and $\pi_q^{ru} = 0$ if $q \notin r$.)

Theorem 5.7 *For a strongly connected undiscounted CTPN, we have*

$$\lambda = \min_u \min_{r \in \mathcal{R}(u)} \frac{\pi^{ru} \nu^u}{\pi^{ru} \tau^u}. \tag{5.16}$$

Thus, λ is the minimal ratio of the mean marking over the mean holding time in the places visited following a stationary policy. In the CTEG case, the final classes are precisely circuits and the invariant measures are uniform on the final classes, so that (5.16) reduces to the well known (4.22).

The proof of Theorem 5.7 uses the fact that the rate λ is obtained asymptotically for stationary policies, together with the following lemma.

---

[21]The system is communicating if for all q, q', there is a policy u and an integer k such that $(P_{qq'}^u)^k > 0$ —i.e. q has access to q'.

[22]This feedback policy has nothing to do with the *routing* policy introduced in §3.

[23]The *classes* of a matrix A are by definition the strongly connected components of the graph of A. A class is *final* if there is no other class downstream.

Lemma 5.8 *Let u denote a policy such that P^u admits a positive invariant measure π. The unique λ such that there exists a finite vector v:*

$$v = \nu^u - \lambda \tau^u + P^u v \tag{5.17}$$

is given by

$$\lambda = \frac{\pi \nu^u}{\pi \tau^u}. \tag{5.18}$$

Proof Left multiplying (5.17) by the row vector π, we get that λ is necessarily equal to (5.18). Conversely, we need only prove the existence of a solution (λ, v) when P^u is irreducible. Then 1 is a simple eigenvalue of P^u, hence, $\mathrm{Im}(P^u - I)$ is $|\mathcal{Q}| - 1$ dimensional. Moreover, $\tau^u \notin \mathrm{Im}(P^u - I)$ (since $\tau^u = P^u v - v \Rightarrow \pi \tau^u = \pi(P^u v - v) = 0$, a contradiction). Hence, $\mathbb{R}\tau^u + \mathrm{Im}(P^u - I) = \mathbb{R}^{\mathcal{Q}}$. □

It is not surprising that the terms on the right-hand side of (5.16) are indeed *invariants* of the net.

Theorem 5.9 (Invariants) *Given an undiscounted CTPN, for all policies u and for all final classes r associated with u,*

$$I^{ur} \stackrel{\text{def}}{=} \pi^{ur} \nu^u = \sum_{q \in r} \pi^{ur} \nu_q^u \tag{5.19}$$

is invariant under firing of transitions.

Proof After one firing of the transition $q \in r$ (the case when $q \notin r$ is trivial), I^{ur} increases by

$$-\pi_q^{ur} + \sum_{q' \in (q^{\mathrm{out}})^{\mathrm{out}} \cap r} \pi_{q'}^{ur} P_{q'q}^u$$

which is zero because π^{ur} is an invariant measure of P^u with support r. □

Example 5.10 The CTPN shown in Fig. 2a is equivalent to that of Fig. 2d under a fair routing policy independent of the origin of the tokens. In this particular case, we obtain the same periodic throughput λ as in the case of the bijective routing shown in Fig. 2c (see (4.30)). This can be seen from the following table and formula (5.16).

Policy	Final classes	Invariant measures	Invariants
$u_1(q_3) = p_1$	$r_1 = \{q_1, q_3\}$,	$\pi^{u_1 r_1} = [\frac{1}{2}, 0, \frac{1}{2}, 0]$	$I^{u_1 r_1} = \frac{1}{2}(m_{p_1} + m_{p_3})$
$u_1(q_4) = p_2$	$r_2 = \{q_2, q_4\}$	$\pi^{u_1 r_2} = [0, \frac{1}{2}, 0, \frac{1}{2}]$	$I^{u_1 r_2} = \frac{1}{2}(m_{p_2} + m_{p_4})$
$u_2(q_3) = p_1$			
$u_2(q_4) = p_5$	r_1	$\pi^{u_1 r_1}$	$I^{u_1 r_1}$
$u_3(q_3) = p_5$			
$u_3(q_4) = p_2$	r_2	$\pi^{u_1 r_2}$	$I^{u_1 r_2}$
$u_4(q_3) = p_5$			
$u_4(q_4) = p_5$	$r_3 = \{q_1, q_2, q_3, q_4\}$	$\pi^{u_4 r_3} = [\frac{1}{4}, \frac{1}{4}, \frac{1}{4}, \frac{1}{4}]$	$I^{u_4 r_3} = \frac{1}{4}(m_{p_3} + m_{p_2} + m_{p_5})$

Finally, we indicate how the above results can be extended to CTPN with potential. With a feedback policy u we associate the matrix R^u: $R_{qq'}^u = \mu_{qu(q)}\rho_{qq'}^{u(q)}\mu_{u(q)q'}$ if $q' \in u(q)^{\text{in}}$ ($R_{qq'}^u = 0$ otherwise); we denote by $\mathcal{R}(u)$ the set of final classes of R^u; with each final class r we associate a left eigenvector of R^u: $\pi^{ru} = \pi^{ru}R^u$ with support r; and we define ν^u, τ^u as in Theorem 5.7. We denote by diag v the diagonal matrix with diagonal entries $(\text{diag}\,v)_{qq} = v_q$. Then the following formula is an immediate consequence of Theorem 5.7.

Corollary 5.11 *For a strongly connected CTPN with potential v, we have*

$$\lim_{t\to\infty} \frac{1}{t}Z_q(t) = \lambda_q, \quad \text{where} \quad v_q^{-1}\lambda_q = \min_u \min_{r\in\mathcal{R}(u)} \frac{\pi^{ru}\nu^u}{\pi^{ru}(\text{diag}\,v)\tau^u}. \tag{5.20}$$

The terms $\pi^{ru}\nu^u$ which determine the throughput are of course *invariants* of the net. More generally, it follows from standard dynamic programming results that the counter functions of α-discounted CTPN exhibit a geometric growth (or convergence) with rate α. The geometric growth of other classes of CTPN could be obtained by transferring existing results about nonnormalized dynamic programming inductions [29].

References

[1] F. Baccelli, G. Cohen, and B. Gaujal. Recursive equations and basic properties of timed Petri nets. *J. of Discrete Event Dynamic Systems*, 1(4):415–439, 1992.

[2] F. Baccelli, G. Cohen, G.J. Olsder, and J.P. Quadrat. *Synchronization and Linearity*. Wiley, 1992.

[3] F. Baccelli, S. Foss, and B. Gaujal. Structural, temporal and stochastic properties of unbounded free-choice Petri nets. Rapport de recherche, INRIA, 1994.

[4] J. Berstel and C. Reutenauer. *Rational Series and their Languages*. Springer, 1988.

[5] H. Braker. *Algorithms and Applications in Timed Discrete Event Systems*. PhD thesis, Delft University of Technology, December 1993.

[6] H. Braker and J. Resing. Periodicity and critical circuits in a generalized max-algebra setting. *Discrete Event Dynamic Systems*. To appear.

[7] P. Chretienne. *Les Réseaux de Petri Temporisés*. Thèse Université Pierre et Marie Curie (Paris VI), Paris, 1983.

[8] G. Cohen, D. Dubois, J.P. Quadrat, and M. Viot. Analyse du comportement périodique des systèmes de production par la théorie des dioïdes. Rapport de recherche 191, INRIA, Le Chesnay, France, 1983.

[9] G. Cohen, P. Moller, J.P. Quadrat, and M. Viot. Algebraic tools for the performance evaluation of discrete event systems. *IEEE Proceedings: Special issue on Discrete Event Systems*, 77(1), Jan. 1989.

[10] P. Dudnikov and S. Samborskiĭ. Endomorphisms of finitely generated free semimodules. In V. Maslov and S. Samborskiĭ, editors, *Idempotent analysis*, volume 13 of *Adv. in Sov. Math.* AMS, RI, 1992.

[11] E. Sontag. *Polynomial response maps*. Lecture Notes in Control and Information Sciences. Springer, 1979.

[12] S. Gaubert. *Théorie des systèmes linéaires dans les dioïdes*. Thèse, École des Mines de Paris, July 1992.

[13] S. Gaubert. Rational series over dioids and discrete event systems. In *Proc. of the 11th Conf. on Anal. and Opt. of Systems: Discrete Event Systems*, number 199 in Lecture Notes in Control and Information Sciences, Sophia Antipolis, June 1994. Springer.

[14] S. Gaubert. Resource optimization and (min,+) spectral theory. IEEE-TAC, 1995, to appear.

[15] S. Gaubert. Rational series over the (max,+) semiring, discrete event systems and Bellman processes. In preparation, 1995.

[16] G. Gierz, K.H. Hofmann, K. Keimel, J.D Lawson, M. Mislove, and D.S. Scott. *A Compendium of Continuous Lattices*. Springer, 1980.

[17] M. Gondran and M. Minoux. *Graphes et algorithmes*. Eyrolles, Paris, 1979. Engl. transl. *Graphs and Algorithms*, Wiley, 1984.

[18] H.P. Hillion and J.M. Proth. Performance evaluation of job-shop systems using timed event-graphs. *IEEE Trans. on Automatic Control*, 34(1):3–9, January 1989.

[19] D. Krob. Quelques exemples de séries formelles utilisées en algèbre non commutative. Rapport de recherche 90-2, Université de Paris 7, LITP, January 1990.

[20] D. Krob and A. Bonnier-Rigny. A complete system of identities for one letter rational expressions with multiplicities in the tropical semiring. Rapport de recherche 93.07, Université de Paris 7, LITP, 1993.

[21] S. Laftit, J.M. Proth, and X.L. Xie. Optimization of invariant criteria for event graphs. *IEEE Trans. on Automatic Control*, 37(6):547–555, 1992.

[22] P. Moller. *Théorie algébrique des Systèmes à Événements Discrets*. Thèse, École des Mines de Paris, 1988.

[23] A. Munier. Régime asymptotique optimal d'un graphe d'événements temporisé généralisé: application à un problème d'assemblage. *APII*, 27(5):487–513, 1993.

[24] M. Plus. A linear system theory for systems subject to synchronization and saturation constraints. In *Proceedings of the first European Control Conference*, Grenoble, July 1991.

[25] I.V. Romanovskiĭ. Optimization and stationary control of discrete deterministic process in dynamic programming. *Kibernetika*, 2:66–78, 1967. Engl. transl. in Cybernetics 3 (1967).

[26] L.H. Rowen. *Polynomial Identities in Ring Theory*. Academic Press, 1980.

[27] P.J. Schweitzer and A. Federgruen. Geometric convergence of value-iteration in multichain Markov decision problems. *Adv. Appl. Prob.*, 11:188–217, 1979.

[28] P. Whittle. *Optimization over Time*, volume II. Wiley, 1986.

[29] W.H.M. Zijm. Generalized eigenvectors and sets of nonnegative matrices. *Lin. Alg. and Appl.*, 59:91–113, 1984.

Ergodic Theorems for Stochastic Operators and Discrete Event Networks

François Baccelli and Jean Mairesse

Abstract

We present a survey of the main ergodic theory techniques which are used in the study of iterates of monotone and homogeneous stochastic operators. It is shown that ergodic theorems on discrete event networks (queueing networks and/or Petri nets) are a generalization of these stochastic operator theorems. Kingman's subadditive ergodic theorem is the key tool for deriving what we call first order ergodic results. We also show how to use backward constructions (also called Loynes schemes in network theory) in order to obtain second order ergodic results. We present a review of systems within this framework, concentrating on two models, precedence constraint networks and Jackson type networks.

Introduction Many systems appearing in manufacturing, communication or computer science accept a description in terms of discrete event systems. A usual characteristic of these systems is the existence of some sources of randomness affecting their behaviour. Hence a natural framework to study them is the one of stochastic discrete event systems.

In this survey paper, we are concerned with two different types of models. First, we consider the study of the iterates $T_n \circ T_{n-1} \circ \cdots \circ T_0$, where $T_i : \mathbb{R}^k \times \Omega \to \mathbb{R}^k$ is a random monotone and homogeneous operator. Second, we introduce and study stochastic discrete event networks entering the so-called monotone-separable framework. A subclass of interest is that of stochastic open discrete event networks.

It will appear that these models, although they have been studied quite independently in past years, have a lot of common points. They share the same kind of assumptions and properties: monotonicity, homogeneity and non-expansiveness. In fact, we are going to show that monotone-separable discrete event networks are a generalization of monotone-homogeneous operators. However, when a system can be modelled as an operator, it provides a more precise description and stronger results.

In both types of model, we are working with daters. Typically, we have to study a random process $X(n) \in \mathbb{R}^k$, where $X(n)_i$ represents the nth occurrence of some event in the system. We are going to present two types of asymptotic results:

1. First order results, concerning the asymptotic rates $\lim_n X(n)_i/n$.

2. Second order results, concerning the asymptotic behaviour of differences such as $X(n)_i - X(n)_j$.

The main references for the results presented in the paper are the following. First order results for operators appear in Vincent [43]. Second order results for operators are new. First and second order results for open discrete event networks are proved in Baccelli and Foss [5]. First order results for general discrete event networks are new. A more complete presentation will be given in a forthcoming paper [7].

The paper is organized as follows. In Part I we treat first order results, and in Part II second order ones. In each part, we consider operators and discrete event networks separately. In the final section, we present a review of systems within the framework, concentrating on two models, precedence constraints networks and Jackson type networks.

We aim at emphasizing how theorems on stochastic systems are obtained as an interaction between structural properties of deterministic systems and probabilistic tools. In order to do so, we first introduce the probabilistic tools (§1 and 5). Then we present some properties of deterministic systems. Finally, we prove the main theorem for stochastic systems.

Part I

First Order Ergodic Results

1   Probabilistic Tools

We consider a probability space $(\Omega, \mathcal{F}, P)$. We consider a bijective and bi-measurable shift operator $\theta : \Omega \to \Omega$. We assume that θ is stationary and ergodic with respect to the probability P.

Lemma 1.1 (Ergodic lemma) *If $\mathcal{A} \in \mathcal{F}$ is such that $\theta(\mathcal{A}) \subset \mathcal{A}$ then $P\{\mathcal{A}\} = 0$ or 1.*

Theorem 1.2 (Kingman's subadditive ergodic theorem [32]) *Let us consider $X_{l,n}, l < n \in \mathbb{Z}$, a doubly-indexed sequence of integrable random variables such that*

- **stationarity** $X_{n,n+p} = X_{0,p} \circ \theta^n$, $\forall n, p,\ p > 0$.

- **boundedness** $E[X_{0,n}] \geq -Cn$, $\forall n > 0$, *for some finite constant $C > 0$.*

- **subadditivity** $X_{l,n} \leq X_{l,m} + X_{m,n}$, $\forall l < m < n$.

Then there exists a constant γ such that the following convergence holds both in expectation and a.s.

$$\lim_{n\to\infty} \frac{E[X_{0,n}]}{n} = \gamma, \quad \lim_{n\to\infty} \frac{X_{0,n}}{n} = \gamma \quad P \text{ a.s.} \tag{1.1}$$

Remark 1.1 The convergence in expectation is straightforward. In fact, we have by subadditivity, $E(X_{0,n}) \leqslant E(X_{0,m}) + E(X_{m,n})$. By stationarity, this implies $E(X_{0,n}) \leqslant E(X_{0,m}) + E(X_{0,n-m})$. The real sequence $u_n = \{E(X_{0,n})\}$ is subadditive, hence u_n/n converges in $\mathbb{R} \cup \{-\infty\}$. Because of the boundedness assumption, we conclude that the limit is finite.

Remark 1.2 If we have additivity instead of subadditivity, then the previous theorem reduces to the following result:

$$\lim_{n\to\infty} \frac{\sum_{i=0}^{n} X_{i,i+1}}{n} \xrightarrow{n\to\infty} E(X_{0,1}) \quad P \text{ a.s.}$$

When the sequence $\{X_{n,n+1}, n \in \mathbb{N}\}$ is i.i.d., this is simply the Strong Law of Large Numbers. More generally, when the sequence $\{X_{n,n+1}, n \in \mathbb{N}\}$ is stationary ergodic (i.e. $X_{n,n+1} = X_{0,1} \circ \theta^n$), it is Birkhoff's ergodic theorem.

2   Application to Operators

2.1   Subadditivity

By a (deterministic) operator we mean a map $T : \mathbb{R}^k \to \mathbb{R}^k$ which is measurable with respect to $\mathcal{B}$, the Borel σ-field of $\mathbb{R}^k$. Let $\{T_n, n \in \mathbb{N}\}$ be a sequence of operators. We associate with it and an initial condition $x_0 \in \mathbb{R}^k$, a sequence on $\mathbb{R}^k$:

$$\begin{cases} x(n+1) = T_n(x(n)) = T_n \circ \cdots \circ T_0(x(0)) \\ \quad x(0) = x_0 . \end{cases} \tag{2.1}$$

We will sometimes use the notation $x(n, x_0)$ to emphasize the value of the initial condition.

We consider a probability space $(\Omega, \mathcal{F}, P, \theta)$ as defined above. By a random (or stochastic) operator we mean a map $T : \mathbb{R}^k \times \Omega \to \mathbb{R}^k$ which is measurable with respect to $\mathcal{B} \times \mathcal{F}$. As usual, we often write $T(x)$ for $T(x,\omega), x \in \mathbb{R}^k, \omega \in \Omega$. A stationary and ergodic sequence of random operators is a sequence $\{T_n, n \in \mathbb{N}\}$ satisfying $T_n(x,\omega) = T_0(x, \theta^n\omega)$. In the same way as in Equation (2.1), we associate with $\{T_n, n \in \mathbb{N}\}$ and a (possibly random) initial condition x_0, a random process $\{x(n), n \in \mathbb{N}\}$ taking its values in $\mathbb{R}^k$.

In what follows, definitions apply to deterministic *and* random operators. For random operators, the properties have to be satisfied with probability 1.

Definition 2.1

1. **Homogeneity** T *is homogeneous if for all* $x \in \mathbb{R}^k$ *and* λ *in* $\mathbb{R}$, $T(x + \lambda \vec{1}) = \lambda \vec{1} + T(x)$, *where* $\vec{1}$ *is the vector of* $\mathbb{R}^k$ *with all its coordinates equal to* 1.

2. **Monotonicity** T *is monotone if* $x \leqslant y$ *implies* $T(x) \leqslant T(y)$ *coordinatewise.*

For a "physical" interpretation of these conditions, see Remark 2.1. The next theorem is a key tool in understanding the importance of homogeneity and monotonicity in what follows.

Theorem 2.2 (Crandall and Tartar [19]) *We consider an operator* $T : \mathbb{R}^k \to \mathbb{R}^k$ *and the following properties* **H**: T *is homogeneous;* **M**: T *is monotone;* **NE**: T *is non-expansive with respect to the sup-norm, i.e* $\forall x, y \in \mathbb{R}^k$, *we have* $\|T(x) - T(y)\|_\infty \leqslant \|x - y\|_\infty$. *If* **H** *holds, then there is equivalence between* **M** *and* **NE**. *Such operators will be referred to as monotone-homogeneous operators.*

Corollary 2.3 *Let us consider a sequence* $T_n : \mathbb{R}^k \to \mathbb{R}^k$, $n \in \mathbb{N}$, *of monotone-homogeneous operators. If* $\exists x \in \mathbb{R}^k$, $\exists i \in \{1, \ldots, k\}$ *such that* $\lim_n T_n \circ \cdots \circ T_0(x)_i / n$ *exists then*

$$\forall y \in \mathbb{R}^k, \; \lim_n \frac{T_n \circ \cdots \circ T_0(y)_i}{n} = \lim_n \frac{T_n \circ \cdots \circ T_0(x)_i}{n} . \qquad (2.2)$$

Proposition 2.4 *Let* $T_n : \mathbb{R}^k \to \mathbb{R}^k$ *be a sequence of monotone-homogeneous operators. We define* $e = (0, \ldots, 0)'$ *and for* $l < n$, $x_{l,n} = T_{n-1} \circ \cdots \circ T_l(e)$. *The maximal (resp. minimal) coordinate of* $x_{l,n}$ *forms a subadditive (resp. superadditive) process, i.e.* $\forall l < m < n \in \mathbb{N}$,

$$\max_i (x_{l,n})_i \leqslant \max_i (x_{l,m})_i + \max_i (x_{m,n})_i, \; \min_i (x_{l,n})_i \geqslant \min_i (x_{l,m})_i + \min_i (x_{m,n})_i .$$

Proof We have $\forall l < m < n \in \mathbb{N}$,

$$\begin{aligned}
x_{l,n} &= T_{n-1} \circ \cdots \circ T_m \circ T_{m-1} \circ \cdots \circ T_l(e) = T_{n-1} \circ \cdots \circ T_m (x_{l,m}) \\
&\leqslant T_{n-1} \circ \cdots \circ T_m \left(e + (\max_i (x_{l,m})_i) \vec{1} \right) \quad \text{(monotonicity)} \\
&\leqslant T_{n-1} \circ \cdots \circ T_m(e) + (\max_i (x_{l,m})_i) \vec{1} \quad \text{(homogeneity)}.
\end{aligned}$$

Therefore,

$$\max_i (x_{l,n})_i \leqslant \max_i (x_{l,m})_i + \max_i (x_{m,n})_i.$$

The proof of the superadditivity of the minimal coordinate is equivalent. $\square$

We are now ready to prove the following theorem on stochastic operators.

Theorem 2.5 (Vincent [43]) *Let $\{T_n, n \in \mathbb{N}\}$ be a stationary ergodic sequence of monotone-homogeneous random operators. We define the process $x(n, y), y \in \mathbb{R}^k$, as in Equation (2.1). If, for all n, the random variable $T_n \circ \cdots \circ T_1(0)$ is integrable and such that $E(T_n \circ \cdots \circ T_1(0)) > -Cn$, for some positive C, then $\exists \overline{\gamma}, \underline{\gamma} \in \mathbb{R}$ such that $\forall y \in \mathbb{R}^k$,*

$$\lim_n \frac{\max_i x(n, y)_i}{n} = \overline{\gamma} \ P \text{ a.s.}, \quad \lim_n \frac{E(\max_i x(n, y)_i)}{n} = \overline{\gamma} \qquad (2.3)$$

$$\lim_n \frac{\min_i x(n, y)_i}{n} = \underline{\gamma} \ P \text{ a.s.}, \quad \lim_n \frac{E(\min_i x(n, y)_i)}{n} = \underline{\gamma} \qquad (2.4)$$

Proof We define as previously the doubly-indexed sequence $x_{l,n} = T_{n-1} \circ \cdots \circ T_l(e)_i$, $l < n$. Using Proposition 2.4, the sequences $\max_i(x_{n,m})_i$ and $-\min_i(x_{n,m})_i$ are subadditive. Hence they satisfy the conditions of Theorem 1.2. So Equation (2.3) holds for $y = e = (0, \ldots, 0)'$. For any other initial condition y, we obtain $\lim_n x(n, y)/n = \lim_n x(n, e)/n$ using the non-expansiveness as in Corollary 2.3. $\qquad \square$

The convergence for the maximal and minimal rates does not imply that of the coordinates. Here is a counterexample borrowed from [43].

Example 2.6 We consider a random operator $T_0 : \mathbb{R}^3 \to \mathbb{R}^3$ satisfying:

$$x = (x_1, x_2, x_3)', \ T_0(x) = (x_1 + 1, x_2 + 2, U_0 x_1 + (1 - U_0) x_2)',$$

where U_0 is a $[0, 1]$-uniform random variable. We have $\liminf(T_n \ldots T_0(x)_3)/n = 1$ and $\limsup(T_n \ldots T_0(x)_3)/n = 2$.

Here is another example of the same kind:

$$\begin{aligned} T_0(x) = \ & (\delta_0(\max(x_1, x_2) + 2) + (1 - \delta_0)(\min(x_1, x_2) + 1), (1 - \delta_0) \\ & (\max(x_1, x_2) + 2) + \delta_0(\min(x_1, x_2) + 1), U_0 x_1 + (1 - U_0) x_2)', \end{aligned}$$

where U_0 is a $[0, 1]$-uniform random variable and δ_0 is a $(0, 1)$ Bernoulli random variable. The random variables U_0 and δ_0 are independent.

2.2 Projective boundedness

In order to complete Proposition 2.4 or Theorem 2.5, the main questions are:

(i) Does a limit exist for $(T_n \circ \cdots \circ T_0(y)_1/n, \ldots, T_n \circ \cdots \circ T_0(y)_k/n)$?

(ii) Is this limit equal to a constant $(\gamma, \ldots, \gamma)$?

The general answers to these questions are not known (even for deterministic operators). We are going to state a sufficient condition to answer (i) and (ii) positively. Let us introduce some definitions.

Definition 2.7 ($\mathbb{PR}^k$) *We consider the parallelism relation: for $u, v \in \mathbb{R}^k$ $u \simeq v \Leftrightarrow \exists a \in \mathbb{R}$ such that $\forall i$, $u_i = a + v_i$. We define the projective space $\mathbb{PR}^k$ as the quotient of $\mathbb{R}^k$ by this parallelism relation. Let π be the canonical projection of $\mathbb{R}^k$ into $\mathbb{PR}^k$.*

Definition 2.8 *Let T be an operator of $\mathbb{R}^k$ into $\mathbb{R}^k$.*

1. *T is projectively bounded if $\exists K$ a compact of $\mathbb{PR}^k$ such that the image of T is included in K, i.e. $\pi(\text{Im}(T)) \subset K$.*

2. *T has a generalized fixed point if $\exists \gamma \in \mathbb{R}, x_0 \in \mathbb{R}^k$ such that $T(x_0) = \gamma\vec{1}+x_0$. This is equivalent to saying that T has a fixed point in the projective space (see Definition 2.7).*

Proposition 2.9 *Let us consider $T : \mathbb{R}^k \to \mathbb{R}^k$ a monotone-homogeneous operator. Let us consider the following assumptions.*
    A. T is projectively bounded.
    B. T has a generalized fixed point.
    C. $\forall x$, $\lim_n T^n(x)/n = (\gamma, \ldots, \gamma)'$.
The following implications hold : $A \Rightarrow B \Rightarrow C$. The other implications are false, $C \not\Rightarrow B \not\Rightarrow A$ and $C \not\Rightarrow A$.

Proof $A \Rightarrow B$. Let K be a compact of $\mathbb{PR}^k$ such that $\pi(T(\mathbb{R}^k)) \subset K$. This implies that $\pi(T) : K \to K$. Hence $\pi(T)$ is continuous on a compact and has a fixed point by application of Brouwer's Theorem.

$B \Rightarrow C$. Let $x \in \mathbb{R}^k$ be a generalized fixed point of T, i.e. $T(x) = \gamma\vec{1} + x$. This implies $T^n(x) = n\gamma\vec{1}+x$ and $\lim_n T^n(x)/n = (\gamma, \ldots, \gamma)'$. From Corollary 2.3, we have $\forall y \in \mathbb{R}^k, \lim_n T^n(y)/n = (\gamma, \ldots, \gamma)'$.

$B \not\Rightarrow A$ and $C \not\Rightarrow A$. An easy counterexample is obtained by considering the identity operator $I : \mathbb{R}^k \to \mathbb{R}^k$, $I(x) = x$.

$C \not\Rightarrow B$ There exist counterexamples of dimension 2, [27]. $\square$

This proposition has an interesting application for stochastic operators.

Theorem 2.10 *Let $\{T_n, n \in \mathbb{N}\}$ be a stationary and ergodic sequence of random operators. We assume that there exist $l \in \mathbb{N}$ and K a compact of $\mathbb{PR}^k$ such that:*

$$\mathcal{E} = \{\pi(\text{Im}(T_{l-1} \circ \cdots \circ T_0)) \subset K\}, \quad P(\mathcal{E}) > 0. \tag{2.5}$$

Then $\exists \gamma \in \mathbb{R}$, such that $\forall x \in \mathbb{R}^k$, $\lim_n (1/n)T_n \circ \cdots \circ T_0(x) = (\gamma, \ldots, \gamma)'$.

Proof We define recursively the random variables

$$N_1 = \min\{n \in \mathbb{N} \mid T_{n+l-1} \circ \cdots \circ T_n \in \mathcal{E}\},$$
$$N_{i+1} = \min\{n \in \mathbb{N} \mid n \geqslant N_i + l, \ T_{n+l-1} \circ \cdots \circ T_n \in \mathcal{E}\}.$$

First of all, let us prove that the random variables N_i are almost surely finite. Let us consider the event $\mathcal{A}_1 = \{N_1 < +\infty\}$. It is easy to see that $\mathcal{A}_1$ is invariant by the shift θ. In fact $N_1(\theta^{-1}\omega) = N_1(\omega) + 1$ or 0. Hence $\{N_1(\omega) < +\infty\} \Rightarrow \{N_1(\theta^{-1}\omega) < +\infty\}$, i.e. $\theta(\mathcal{A}_1) \subset \mathcal{A}_1$. By Lemma 1.1, this implies that $\mathcal{A}_1$ has probability 0 or 1. But $(\{N_1 = 0\} = \mathcal{E}) \subset \mathcal{A}_1$ and by assumption $P(\mathcal{E}) > 0$. We conclude that $P(\mathcal{A}_1) = 1$. A similar argument can now be applied to N_2. For $\mathcal{A} \in \mathcal{F}$, we define the indicator function $\mathbf{1}_\mathcal{A} : \Omega \to \Omega$, $\mathbf{1}_\mathcal{A}(\omega) = 1$ iff $\omega \in \mathcal{A}$. We have

$$
\begin{aligned}
P(N_2 < +\infty) &= E(\mathbf{1}_{\{N_2 < +\infty\}}) = E(\sum_k \mathbf{1}_{\{N_1=k\}} \mathbf{1}_{\{N_2 < +\infty\}}) \\
&= E(\sum_k \mathbf{1}_{\{N_1=k\}} \mathbf{1}_{\{N_1 \circ \theta^{k+l} < +\infty\}}) = E(\sum_k \mathbf{1}_{\{N_1=k\}}) = 1 .
\end{aligned}
$$

We conclude the proof by induction.

Let $\overline{\gamma}$ and γ be the maximal and minimal rates as defined in Proposition 2.4. Let us assume that $\overline{\gamma} \neq \gamma$. This implies, $\forall x \in \mathbb{R}^k$,

$$\liminf_n (\max_i x(n)_i - \min_i x(n)_i) = +\infty . \tag{2.6}$$

But we also have that $\forall i \in \mathbb{N}, \pi(x(N_i + l)) \subset K$. Hence $(\max_j x(N_i + l)_j - \min_j x(N_i + l)_j) \subset K'$ where K' is a compact of $\mathbb{R}$. Hence there exists a subsequence $N_{\sigma(i)}$ such that $(\max_j x(N_{\sigma(i)} + l)_j - \min_j x(N_{\sigma(i)} + l)_j)$ converges to a finite limit. This is in contradiction with (2.6). $\qquad\square$

Remark 2.1 In many applications, the operator will be applied to a vector of dates for a physical system. The vectors $x(n)$ and $x(n+1) = T_n(x(n))$ will represent the dates of the nth and $(n + 1)$th occurrences of some events in a system. In such a case, the homogeneity property can be interpreted as the fact that changing the absolute origin of time does not modify the dynamics of the system. Hence it becomes a very natural assumption. The monotonicity is interpreted as the fact that delaying an event delays all following events.

3 Application to Discrete Event Networks

3.1 Discrete event networks

A *discrete event network* is characterized by

1. A sequence
$$N = N_{[-\infty,\infty]} = \{\sigma(k), M(k), \ k \in \mathbb{Z}\},$$

   where $\sigma(k) \in \mathbb{R}^+$ and $\{M(k)\}$ is a sequence of F-valued variables, where F is some measurable space. With N and $n \leqslant m \in \mathbb{Z}$, we associate the sequence $N_{[n,m]}$ defined by

$$N_{[n,m]} \stackrel{\text{def}}{=} \{\sigma_{[n,m]}(k), M(n+k), \ k \in \mathbb{N}\},$$

where $\sigma_{[n,m]}(k) \stackrel{\text{def}}{=} \sigma(n+k)$, for $0 \leqslant k \leqslant m-n$, and $\sigma_{[n,m]}(k) \stackrel{\text{def}}{=} \infty$, for $k > m-n$.

2. Measurable operators $\Phi(k,.)$ and $\Psi(.)$: $(\mathbb{R}^+ \times F)^{\mathbb{N}} \to \mathbb{R} \cup \{\infty\}$, $k \in \mathbb{N}^*$, through which are defined

$$X_{[n,m]} = \Psi(N_{[n,m]}), \quad n \leqslant m, \quad X_{[n,m]}^-(k) = \Phi(k, N_{[n,m]}), \quad k \geqslant 1.$$

Remark These variables receive the following interpretations: $X_{[n,m]}^-(k)$ is the initiation date of the kth event on some reference node, for the *driving sequence* $N_{[n,m]}$.

$$X_{[n,m]}^+(k) \stackrel{\text{def}}{=} X_{[n,m]}^-(k) + \sigma_{[n,m]}(k), \quad n \leqslant m, \ k \geqslant 0$$

is the completion date of this event. $X_{[n,m]}^+(k)$ and $X_{[n,m]}^-(k)$ are called *internal daters*. $X_{[n,m]}$ is the *maximal dater*, i.e. the date of the last event in the network, for the sequence $N_{[n,m]}$.

3.2   The monotone-separable framework

Let N and $\widetilde{N}$ be two driving sequences such that $\sigma(k) \leqslant \widetilde{\sigma}(k) < \infty$ for all k, and with $M(k) = \widetilde{M}(k)$ for all k. We denote by $X_{[1,m]}^-(k)$, $X_{[1,m]}^+(k)$ and $X_{[1,m]}$ the daters associated with $N_{[1,m]}$, and by $\widetilde{X}_{[1,m]}^-(k)$ etc. those associated with $\widetilde{N}_{[1,m]}$.

A network is said to be *monotone-separable* if it satisfies the following properties for all $m \geqslant 1$, $k \geqslant 1$ and for all N and $\widetilde{N}$ as above:

- **causality** $X_{[1,m]}^-(m+1) \leqslant X_{[1,m]} < \infty$.

- **monotonicity** $X_{[1,m]}^-(k) \leqslant \widetilde{X}_{[1,m]}^-(k)$ and $X_{[1,m]} \leqslant \widetilde{X}_{[1,m]}$.

- **non-expansiveness**[1] $\widetilde{X}_{[1,m]}^-(k) - X_{[1,m]}^-(k) \leqslant x$ and $\widetilde{X}_{[1,m]} - X_{[1,m]} \leqslant x$, if $\widetilde{\sigma}(k) = \sigma(k)$ for all $k \neq l$, and $\widetilde{\sigma}(l) = \sigma(l) + x$, $x > 0$.

---

[1]If one regards $(\Psi(.), \Phi(k,.), \ k \geqslant 1)$ as an operator: $(\mathbb{R}^+)^{\mathbb{N}} \to (\mathbb{R} \cup \{\infty\})^{\mathbb{N}}$ – the sequence $\{M(k)\}$ being fixed – this is indeed non-expansiveness when taking an L^1 norm on $(\mathbb{R}^+)^{\mathbb{N}}$ and an L^{∞} norm on $(\mathbb{R} \cup \{\infty\})^{\mathbb{N}}$.

- **separability** For $1 \leqslant l < m$, if $X_{[1,l]} \leqslant X^+_{[1,m]}(l+1)$ then $X_{[1,m]} \leqslant X^-_{[1,m]}(l+1) + X_{[l+1,m]}$.

Proposition 3.1 *Under the above assumptions, the sequence $X_{[m,n]}$ satisfies the subadditive inequality $X_{[m,n]} \leqslant X_{[m,l]} + X_{[l+1,n]}$, $\forall m \leqslant l < n$.*

Proof It is enough to prove the property for $m = 1$, since the general relation will then be obtained by applying the relation for $m = 1$ to the variables associated with some suitable sequence. Let $1 \leqslant l < n$. There are two cases:

Case 1 $X_{[1,l]} \leqslant X^+_{[1,n]}(l+1)$. Then, in view of separability

$$X_{[1,n]} \leqslant X^-_{[1,n]}(l+1) + X_{[l+1,n]} \leqslant X_{[1,l]} + X_{[l+1,n]},$$

where we have used the fact that $X_{[1,l]} \geqslant X^-_{[1,l]}(l+1) \geqslant X^-_{[1,n]}(l+1)$, which follows from causality and monotonicity ($X^-_{[1,l]}(l+1) = \widetilde{X}^-_{[1,n]}(l+1)$ with $\widetilde{\sigma}(k) = \sigma(k)$ for $1 \leqslant k \leqslant l$ and $\widetilde{\sigma}(k) = \infty$ for $k > l$).

Case 2 $X_{[1,l]} > X^+_{[1,n]}(l+1)$. Consider the two sequences $\{\sigma(k)\}$ and $\{\widetilde{\sigma}(k)\}$, which differ only in their $(l+1)$st coordinate, for which we take $\widetilde{\sigma}(l+1) = \sigma(l+1) + x$, $x > 0$. In view of monotonicity, $X_{[1,n]} \leqslant \widetilde{X}_{[1,n]}$. In particular, if we take $x = x^*$ with $x^* = X_{[1,l]} - X^+_{[1,n]}(l+1) > 0$, then

$$
\begin{aligned}
\widetilde{X}^+_{[1,n]}(l+1) &= \widetilde{X}^-_{[1,n]}(l+1) + \sigma(l+1) + x^* \\
&= \widetilde{X}^-_{[1,n]}(l+1) + \sigma(l+1) + X_{[1,l]} - X^+_{[1,n]}(l+1) \\
&= X_{[1,l]} + \widetilde{X}^-_{[1,n]}(l+1) - X^-_{[1,n]}(l+1). \quad (3.1)
\end{aligned}
$$

But $X_{[1,l]}$ does not depend on $\sigma(l+1)$, and so $X_{[1,l]} = \widetilde{X}_{[1,l]}$. Therefore

$$\widetilde{X}^+_{[1,n]}(l+1) = \widetilde{X}_{[1,l]} + \widetilde{X}^-_{[1,n]}(l+1) - X^-_{[1,n]}(l+1) \geqslant \widetilde{X}_{[1,l]} \text{ (monot.).}$$

We finally obtain that, for $x = x^*$

$$
\begin{aligned}
\widetilde{X}_{[1,n]} &\leqslant \widetilde{X}^-_{[1,n]}(l+1) + \widetilde{X}_{[l+1,n]} \quad \text{(separability)} \\
&= \widetilde{X}^+_{[1,n]}(l+1) + X^-_{[1,n]}(l+1) - X_{[1,l]} + \widetilde{X}_{[l+1,n]} \quad \text{(Equation (3.1))} \\
&\leqslant \widetilde{X}^+_{[1,n]}(l+1) + X^-_{[1,n]}(l+1) - X_{[1,l]} + x^* + X_{[l+1,n]} \quad \text{(non-exp.)} \\
&= \widetilde{X}^+_{[1,n]}(l+1) + X^-_{[1,n]}(l+1) - X_{[1,l]} + X_{[1,l]} - X^+_{[1,n]}(l+1) + X_{[l+1,n]} \\
&= \widetilde{X}^+_{[1,n]}(l+1) - X^+_{[1,n]}(l+1) + X^-_{[1,n]}(l+1) + X_{[l+1,n]} \\
&\leqslant x^* + X^-_{[1,n]}(l+1) + X_{[l+1,n]} \quad \text{(non-exp.)} \\
&= X^-_{[1,n]}(l+1) - X^+_{[1,n]}(l+1) + X_{[1,l]} + X_{[l+1,n]} \\
&\leqslant X_{[1,l]} + X_{[l+1,n]}.
\end{aligned}
$$

$\square$

Remark 3.1 Under the additional assumption that $X^-_{[1,m]}(l+1)$ is a function of $\{\sigma(k),\ 1 \leqslant k \leqslant l$ and $M(p),\ 1 \leqslant p \leqslant m\}$ only, non-expansiveness can be replaced by the following property:

- **sub-homogeneity** $\widetilde{X}_{[1,m]} \leqslant X_{[1,m]} + \lambda$, if $\widetilde{\sigma}(1) = \sigma(1) + \lambda$ and $\widetilde{\sigma}(k) = \sigma(k)$ for all $k > 1$, $\lambda > 0$ and $m \geqslant 1$.

The proof is exactly the same for case 1. For case 2, taking x^* as in the proof of Proposition 3.1 gives $\widetilde{X}^+_{[1,n]}(l+1) = X_{[1,l]}$ and

$$
\begin{aligned}
\widetilde{X}_{[1,n]} &\leqslant \widetilde{X}^-_{[1,n]}(l+1) + \widetilde{X}_{[l+1,n]} \quad \text{(separability)} \\
&= X^-_{[1,n]}(l+1) + \widetilde{X}_{[l+1,n]} \\
&\leqslant X^-_{[1,n]}(l+1) + X_{[l+1,n]} + X_{[1,l]} - X^+_{[1,n]}(l+1) \quad \text{(sub-homog.)} \\
&\leqslant X_{[1,l]} + X_{[l+1,n]}.
\end{aligned}
$$

Remark 3.2 Some generalizations of the framework, with internal daters, are presented in [7]. See also the Jackson network example of §8.2. The comments on the physical interpretation of homogeneity or monotonicity made in Remark 2.1 also apply to discrete event networks.

3.3   Open discrete event networks

A discrete event network is said to be *open* if the following additional assumption holds for all $m \geqslant 1$:

$$
\forall 1 \leqslant k \leqslant m,\ X^-_{[1,m]}(k+1) = X^-_{[1,\infty]}(k+1) = X^+_{[1,m]}(k), \text{ and } X^-_{[1,m]}(1) = 0.
$$

One can then define a point process $\{A_k\}_{k \geqslant 1}$ by $A_k = A_1 + X^-_{[1,\infty]}(k)$. The origin of this point process is arbitrary. It is then possible to interpret $\{A_k\}$ as an external arrival process, the inter-arrival times being the sequence $\{\sigma(k)\}$. To summarize, an open discrete network is described by a sequence $N = N_{[-\infty,\infty]} = \{A_k, M(k),\ k \in \mathbb{Z}\}$.

The conditions of the monotone separable framework take the following form for an open network (which corresponds to the conditions of [5]): for all $m \geqslant 1$, the following properties hold:

- **causality** $A_m \leqslant A_1 + X_{[1,m]} < \infty$.

- **monotonicity** $\widetilde{X}_{[1,m]} \geqslant X_{[1,m]}$, for $\widetilde{N}$ and N with $\widetilde{\sigma}(k) \geqslant \sigma(k)\ \forall k$.

- **homogeneity** Let $\widetilde{N}$ be the point process obtained by shifting the points of N A_k, $k \geqslant 1$, by $\lambda > 0$ to the right. Then $\widetilde{X}_{[1,n]} = X_{[1,n]}$.

- **separability** $A_1 + X_{[1,m]} \leqslant A_{l+1} + X_{[l+1,m]}$ for all $1 \leqslant l < m$ such that $A_1 + X_{[1,l]} \leqslant A_{l+1}$.

For an open network, monotonicity can be interpreted as the fact that delaying an arrival delays all forthcoming events in the network. For a possible interpretation of separability, see Remark 6.1.

3.4 Stochastic discrete event networks

We consider a probability space $(\Omega, \mathcal{F}, P, \theta)$ as in §1. The following stochastic assumptions are made:

- **compatibility** $(\sigma(k), M(k)) = (\sigma(0), M(0)) \circ \theta^k$ for all $k \in \mathbb{Z}$.

- **integrability** $\exists C > 0, \ -Cm \leqslant E[X_{[1,m]}] < \infty$ for all $m \geqslant 0$.

Theorem 3.2 *For every discrete event network which satisfies the monotone-separable assumptions and the above stochastic assumptions, we have*

$$\lim_{n \to \infty} \frac{X_{[1,n]}}{n} = \gamma \quad \text{a.s.} \quad \text{and} \quad \lim_{n \to \infty} \frac{E[X_{[1,n]}]}{n} = \gamma \qquad (3.2)$$

for some finite constant γ.

Proof We have $X_{[m,m+p]} = X_{[0,p]} \circ \theta^m$, for all $m \in \mathbb{Z}$ and $p \geqslant 0$. For $m \leqslant n$, define $Y_{[m,n+1]} = X_{[m,n]}$. From Proposition 3.1, for all $m \leqslant l < n$, $Y_{[m,n+1]} \leqslant Y_{[m,l+1]} + Y_{[l+1,n+1]}$. So $\{Y_{[m,n]}\}$, $m < n$, satisfies all the assumptions of Theorem 1.2. $\qquad\square$

4 Relations Between Operators and Networks

Let us investigate the relation between the operator framework considered in §2 and the monotone-separable framework considered above. Let $\{T_n\}$ be a sequence of monotone-homogeneous operators. Let $\sigma(n) \equiv 0$ and $M(n) = T_n$. Let $x(n, 0)$ be the variables associated with the operator recurrence equation (2.1) with initial condition $x_0 = 0$. With these variables, we associate

$$X^-_{[0,n]}(k) = X^+_{[0,n]}(k) = \max_i x(k-1, 0)_i, \quad k \geqslant 1, \quad \text{and} \quad X_{[0,n]} = \max_i x(n, 0)_i.$$

Note that these variables are functions of $\{M(l)\}$. We have

- $X^-_{[0,n]}(n + 1) = X_{[0,n]} < \infty$, so that causality holds.

- Monotonicity and non-expansiveness trivially hold as neither $X^-_{[0,n]}(k)$ nor $X_{[0,n]}$ depends upon $\{\sigma(l)\}$.

- Separability holds as it is always true that $X_{[0,l]} = X^-_{[0,m]}(l+1)$ and

$$
\begin{aligned}
X_{[0,m]} &= \max_i \left(T_m \circ \cdots \circ T_{l+1}(x(l,0)) \right)_i \\
&= \max_i \left(T_m \circ \cdots \circ T_{l+1}(x(l,0) + (X_{[0,l]} - X_{[0,l]})\vec{1}) \right)_i \\
&= X_{[0,l]} + \max_i \left(T_m \circ \cdots \circ T_{l+1}(x(l,0) - X_{[0,l]}\vec{1}) \right)_i, \quad \text{(homog.)} \\
&\leqslant X_{[0,l]} + \max_i \left(T_m \circ \cdots \circ T_{l+1}(0) \right)_i, \quad \text{(monotonicity)} \\
&= X^-_{[0,m]}(l+1) + X_{[l+1,m]}.
\end{aligned}
$$

Hence, monotone-separable operators are a special case of monotone-separable discrete event networks. On the other hand, it should be remarked that an operator can *not* be represented as an *open* discrete event network. A representation in terms of operators is interesting as it is more precise than the corresponding one in terms of discrete event networks. In particular, we will see that we are able to obtain second order results for operators, §7, and not for non-open discrete event networks, §6.2.

Part II

Second Order Ergodic Results

We will introduce a construction which is known as the Loynes scheme. This type of construction will be used for both types of models, discrete event networks and operators, but in a rather different way.

5 Basic Example and Probabilistic Tools

The basic construction was introduced by Loynes in [34] to study the stability of the $G/G/1/\infty$ queue. A G/G arrival process is a stationary and ergodic marked point process $N = \{(\tau_n, \sigma_n), n \in \mathbb{Z}\}$, where $\sigma_n \in \mathbb{R}^+$ is the service time required by customer n and $\tau_n = A_{n+1} - A_n$, the inter-arrival time between customers n and $n+1$. The $1/\infty$ part describes the queueing mechanism. There is a single server and an infinite waiting room or buffer. Upon arrival at instant A_n, customer n is served immediately if the server is idle at A_n^- and is queued in the buffer otherwise. The server operates at unit rate until all customers present in the buffer have been served. Let $X_{[l,n]}$ be the time of last activity in the system, i.e. the departure of the last customer, for the restriction $N_{[l,n]}$. The system can be described in two equivalent ways:

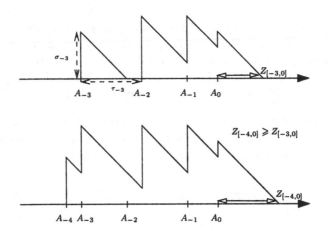

Figure 1: Loynes scheme for the $G/G/1/\infty$ queue.

- As a stochastic operator,

$$\begin{pmatrix} A_{n+1} \\ X_{[l,n+1]} \end{pmatrix} = \begin{pmatrix} \tau_n + A_n \\ \max(\tau_n + \sigma_{n+1} + A_n\,,\,\sigma_{n+1} + X_{[l,n]}) \end{pmatrix} \qquad (5.1)$$

$$= \begin{pmatrix} \tau_n & \varepsilon \\ \tau_n \otimes \sigma_{n+1} & \sigma_{n+1} \end{pmatrix} \otimes \begin{pmatrix} A_n \\ X_{[l,n]} \end{pmatrix}. \qquad (5.2)$$

Equation (5.1) can be written $X_{[l,n+1]} = \max(A_{n+1}, X_{[l,n]}) + \sigma_{n+1}$. The meaning is that the server starts working on customer $n+1$ as soon as this customer has arrived (A_{n+1}) *and* the server has completed the services of the previous customers ($X_{[l,n]}$). Equation (5.2) is just a re-writing using the (max,+) notation, see also §8.1. It is easy to verify that this operator is monotone and homogeneous.

- As an open network, by means of the function Ψ of §3.

$$X_{[l,n]} = \Psi(\tau_i, \sigma_i, i \in \{l, \ldots, n\})$$

$$= (A_n - A_l) + \sigma_n + \max(0, \max_{k=1}^{n-l} \sum_{i=1}^{k}(\sigma_{n-i} - \tau_{n-i})). \qquad (5.3)$$

The easiest way to understand Equation (5.3) is to look at Figure 1. The function Ψ is monotone, homogeneous and separable.

Let us consider the sequence of variables $\{Z_{[l,n]},\ l \leqslant n \in \mathbb{Z}\}$ defined by $Z_{[l,n]} = X_{[l,n]} - (A_n - A_l)$. The variables $Z_{[l,n]}$ satisfy Lindley's equation[2] $Z_{[l,n+1]} = (Z_{[l,n]} - \tau_n)^+ + \sigma_{n+1}$.

---

[2]It is more classical, but equivalent, to work with the workload variable $W_n = X_{[0,n]} - A_n - \sigma(n)$, yielding the equation $W_{n+1} = (W_n + \sigma_n - \tau_n)^+$.

Theorem 5.1 (Loynes [34]) *The sequence $Z_{[-n,0]}$ is increasing in n, i.e.*
$Z_{[-n-1,0]} \geqslant Z_{[-n,0]}$. *The limit $Z = \lim_n Z_{[-n,0]}$ satisfies $P\{Z < +\infty\} = 0$ or*
1. Furthermore Z is a stationary solution of Lindley's equation, i.e. $Z(\theta\omega) = (Z(\omega) - \tau_0)^+ + \sigma_1$. When $P\{Z < +\infty\} = 1$, the sequence $\{Z_{[0,n]}, n \in \mathbb{N}\}$
couples in finite time with the stationary sequence $\{Z \circ \theta^n\}$.

Proof The monotonicity of $Z_{[-n,0]}$ is easy to obtain from Equation (5.3). It
is also illustrated in Figure 1. Hence the limit $Z = \lim_n Z_{[-n,0]}$ exists. Let us
denote $\mathcal{A} = \{Z < +\infty\}$. From $Z_{[-n,1]} = (Z_{[-n,0]} - \tau_0)^+ + \sigma_1$ and the fact that
σ_1 is a.s. finite, we obtain

$$Z(\omega) < +\infty \Leftrightarrow \exists K \; \forall n, \; Z_{[-n,0]}(\omega) < K \Rightarrow \exists K' \; \forall n, \; Z_{[-n,1]}(\omega) < K'.$$

But we also have

$$Z_{[-n,1]}(\omega) = Z_{[-n-1,0]}(\theta\omega) . \tag{5.4}$$

We conclude that $Z(\theta\omega) < +\infty$. We have proved that $\theta(\mathcal{A}) \subset \mathcal{A}$ which
implies, by Lemma 1.1, that $P\{\mathcal{A}\} = 0$ or 1. From Equation (5.4), letting
n go to ∞, we deduce that $Z(\theta\omega) = (Z(\omega) - \tau_0)^+ + \sigma_1$. For a proof of the
remaining point, see [34] or [1].   □

The limit Z is usually referred to as the Loynes variable. We can obtain,
using Equation (5.3), $P\{Z < +\infty\} = 1 \Leftrightarrow E(\sigma) < E(\tau)$. The condition
$E(\sigma) < E(\tau)$ is called the stability condition and is usually written in the
form $\rho = E(\sigma)/E(\tau) < 1$. We will see a similar type of stability condition in
Theorem 6.2.

6 Application to Discrete Event Networks

6.1 Open discrete event networks

The assumptions and notations are those of §3.3 but we replace the separability assumption by

- **strong separability** For $1 \leqslant l < m$, if $A_1 + X_{[1,l]} \leqslant A_{l+1}$ then $A_1 + X_{[1,m]} = A_{l+1} + X_{[l+1,m]}$.

Remark 6.1 Strong separability can be interpreted as follows. If the arrival
of customer $l+1$ takes place later than the last activity for the arrival process
$[1, l]$, then the evolution of the network after time A_{l+1} is the same as in the
network which starts "empty" at this time.

We define $\lambda = E(A_{n+1} - A_n)^{-1}$ interpreted as the arrival rate and

$$Z_{[l,n]} = X_{[l,n]} - (A_n - A_l), \quad l \leqslant n . \tag{6.1}$$

Proposition 6.1 (Internal monotonicity) *Under the above assumptions, we have* $Z_{[l-1,n]} \geqslant Z_{[l,n]}$, $l \leqslant n$.

Proof Consider the point process $\widetilde{N}$ with $\widetilde{\sigma}(l-1) = \sigma(l-1) + Z_{[l-1,l-1]}$ and $\widetilde{\sigma}(k) = \sigma(k)$ everywhere else. For $\widetilde{N}_{[l-1,n]}$, we have separability in l so that

$$
\begin{aligned}
\widetilde{X}_{[l-1,n]} &= \widetilde{X}_{[l,n]} + \widetilde{A}_l - \widetilde{A}_{l-1} \\
&= X_{[l,n]} + \widetilde{A}_l - \widetilde{A}_{l-1} \quad \text{(strong separability)} \\
&= X_{[l,n]} + A_l - A_{l-1} + Z_{[l-1,l-1]} .
\end{aligned}
\tag{6.2}
$$

Therefore

$$
\begin{aligned}
Z_{[l-1,n]} &= X_{[l-1,n]} - (A_n - A_{l-1}) \\
&= X_{[l-1,n]} - (A_n - A_l) - (A_l - A_{l-1}) \\
&= X_{[l-1,n]} - (A_n - A_l) + X_{[l,n]} - \widetilde{X}_{[l-1,n]} + Z_{[l-1,l-1]} \quad \text{(by (6.2))} \\
&= Z_{[l,n]} + X_{[l-1,n]} - \widetilde{X}_{[l-1,n]} + Z_{[l-1,l-1]} \\
&\geqslant Z_{[l,n]}, \quad \text{(non-expansiveness).}
\end{aligned}
$$

$\square$

Let $Z = \lim_n Z_{[-n,0]}(N)$, which exists by internal monotonicity of $Z_{[-n,0]}(N)$. We define a c-scaling of the arrival point process N in the following way:

$$
0 \leqslant c < +\infty, \quad cN = \{cA_n, M(n), n \in \mathbb{Z}\}.
$$

From Equation (6.1) and Proposition 3.1, we obtain that $Z_{[1,n]}$ is subadditive. Applying Theorem 3.2, we obtain the existence of the limits

$$
\lim_n \frac{Z_{[1,n]}(cN)}{n} = \lim_n \frac{Z_{[-n,0]}(cN)}{n} = \gamma(c) .
$$

From Equation (6.1), we obtain

$$
\lim_n \frac{X_{[1,n]}(cN)}{n} = \lim_n \frac{X_{[-n,0]}(cN)}{n} = \gamma(c) + \frac{c}{\lambda} .
$$

For $c \geqslant \tilde{c}$, we have $cN \geqslant \tilde{c}N$. We obtain by internal monotonicity and monotonicity respectively:

1. $Z_{[-n,0]}(cN)$ is decreasing in $c \Rightarrow \gamma(c)$ is decreasing in c.

2. $X_{[0,n]}(cN)$ is increasing in $c \Rightarrow \gamma(c) + c/\lambda$ is increasing in c.

We deduce the existence of a constant $\gamma(0)$ defined by :

$$\lim_{c \to 0} \searrow \gamma(c) + \frac{c}{\lambda} = \gamma(0) = \lim_{c \to 0} \nearrow \gamma(c) . \qquad (6.3)$$

The intuitive interpretation is that $\gamma(0)^{-1}$ is the throughput of the network when we saturate the input, i.e. when $A_n = 0, \forall n$. It is the maximal possible throughput.

Theorem 6.2 *Let $N = \{A_n, M_n, n \in \mathbb{Z}\}$ be a stationary ergodic point process. We set $\rho = \lambda\gamma(0)$. If $\rho > 1$, then $P(Z = +\infty) = 1$. If $\rho < 1$, then $P(Z < +\infty) = 1$ and $\{Z_{[0,n]}, n \in \mathbb{N}\}$ couples in finite time with the stationary sequence $\{Z \circ \theta^n\}$.*

Proof The first part of the theorem is immediate. In fact relation (6.3) implies $\gamma(1) + 1/\lambda \geqslant \gamma(0)$. We have :

$$\left(\lim_n \frac{Z_{[-n,0]}}{n} = \gamma(1) \right) \geqslant \left(\gamma(0) - \frac{1}{\lambda} = \frac{\rho - 1}{\lambda} \right) .$$

Therefore $\rho - 1 > 0$ implies $P(Z = +\infty) = 1$. For a complete proof of the result, the reader is referred to [5]. $\qquad\qquad\qquad\square$

Remark 6.2 For $\rho < 1$, Z is the smallest stationary regime for the response time of the system (which is defined as the time to the last activity under the restriction $[-\infty, 0]$ of N). Intuitively it is the stationary regime corresponding to an "empty" initial condition, as it is the limit of the systems starting "empty" and fed up with the restrictions $[-n, 0]$ of N. In many cases, there will be multiple stationary regimes depending on the initial condition. A simple example of a monotone and separable open network having multiple stationary regimes is proposed in [1, p. 83.], It is a $G/G/2/\infty$ queue with a "shortest workload" allocation rule (see also Theorem 7.5).

6.2 General discrete event networks

For discrete event networks which are not open, there are no general results. The reason is the absence of internal monotonicity of the variables $Z_{[-n,0]} = X_{[-n,0]} - X_{[-n,0]}^-$. We illustrate the phenomenon in Figure 2 where we compare the case of a general network and the case of an open network. For open and general networks, we consider successively the restrictions $[-n, 0]$ and $[-n-1, 0]$. In the open case, the internal monotonicity has been illustrated in Figure 2. In the general case, the variables X^- are internal variables, hence their values are modified when we go from the restriction $[-n, 0]$ to $[-n-1, 0]$. As a consequence, there is no internal monotonicity. In Figure 2, for ease of comparison, we have assumed that $X_{[-n-1,0]}^-(-n) = X_{[-n-1,0]}^-(-n)$ (these quantities are defined up to an additive constant).

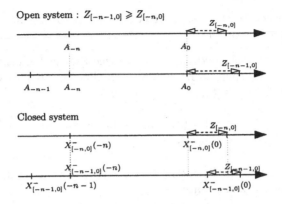

Figure 2: Loynes scheme for monotone-separable networks.

7   Application to Operators

We discuss in Sections 7.1 and 7.2 two very different approaches. They correspond to two different types of operators, see Remark 7.1. The first approach is directly based on the Loynes scheme. The second one uses fixed point results.

7.1   Monotonicity

Definition 7.1 *We say that the operator $T : \mathbb{R}^k \to \mathbb{R}^k$ has a minimal value if there exists $x_0 \in \mathbb{R}^k$ such that $\forall y \geqslant x_0$, $T(y) \geqslant x_0$.*

Let us consider a sequence of monotone operators $\{T_n, n \in \mathbb{Z}\}$. If all the operators have a common minimal value x_0, then we are able to construct a Loynes scheme, in the same way as in §5. In fact, we have $T_0(x_0) \geqslant x_0$ and $T_0 \circ T_{-1}(x_0) \geqslant T_0(x_0) \geqslant x_0$ using monotonicity. We obtain that

$$\exists Z \in (\mathbb{R} \cup \{+\infty\})^k, \ \lim_n T_0 \circ T_{-1} \circ \cdots \circ T_{-n}(x_0) = Z . \qquad (7.1)$$

The main question is whether the limit Z is finite or not, the finite case being the interesting one. In particular, if we consider a sequence of monotone-homogeneous operators, then the limits $\overline{\gamma}$ and γ defined in Theorem 2.5 exist. Because of the existence of the minimal value x_0, we have $\overline{\gamma} \geqslant \gamma \geqslant 0$. If $\overline{\gamma} > 0$ then there exists i such that $Z_i = +\infty$ (the proof is immediate).

For this reason, it is usually not interesting to construct a Loynes scheme directly on the sequence of operators T_n. For example, in the case of the operator of the G/G/1 queue, see Equation (5.1), the Loynes scheme was not built on $(A_n, X_{[l,n]})'$ but on the differences $Z_{[l,n]} = X_{[l,n]} - A_n$. In order to generalize the construction, the correct approach is to consider the operators T_n in a projective space.

We have already defined the projective space $\mathbb{PR}^k$ in Definition 2.7. The space $\mathbb{PR}^k$ is isomorphic to $\mathbb{R}^{k-1}$. There are different possible ways to map $\mathbb{PR}^k$ onto $\mathbb{R}^{k-1}$. For $i \in \{1, \ldots, k\}$, we define

$$\pi_i : \quad \mathbb{R}^k \longrightarrow \mathbb{R}^{k-1}, \quad \pi_i(x) = (x_1 - x_i, \ldots, x_{i-1} - x_i, x_{i+1} - x_i, \ldots, x_k - x_i)'$$
$$\phi_i : \quad \mathbb{PR}^k \to \mathbb{R}^{k-1}, \quad \phi_i = \pi_i \circ \pi^{-1},$$

where π was defined in Definition 2.7. It is easy to verify that ϕ_i is defined without ambiguity and is bijective.

Definition 7.2 Let $x \in \mathbb{R}^k$. We define $|x|_\mathcal{P} = \max_i x_i - \min_i x_i$. Let $u \in \mathbb{PR}^k$ (resp. $u \in \mathbb{R}^{k-1}$) and x be a representative of u, i.e. $\pi(x) = u$ (resp. $\pi_i(x) = u$) We define $|u|_\mathcal{P} = \max_i x_i - \min_i x_i$.

The function $|.|_\mathcal{P}$ is a semi-norm on $\mathbb{R}^k$ as $|x|_\mathcal{P} = 0 \Rightarrow x_i = \lambda, \forall i$. On the other hand, it defines a norm on $\mathbb{PR}^k$ or $\mathbb{R}^{k-1}$. We call it the projective norm. We use the same notation for the semi-norm on $\mathbb{R}^k$ and the norms on $\mathbb{PR}^k$ and $\mathbb{R}^{k-1}$ in order not to carry too many notations.

From now on, we are going to work on $\mathbb{R}^{k-1}$ equipped with the projective norm. Without loss of generality, we will restrict our attention to π_1, ϕ_1. Working on $\mathbb{R}^{k-1}$ rather than on $\mathbb{PR}^k$ enables us to have a natural partial order. The projective norm is indeed compatible with the coordinatewise partial ordering on $\mathbb{R}^{k-1}$, i.e. for $u, v \in \mathbb{R}^{k-1}, u \geqslant v \Rightarrow |u|_\mathcal{P} \geqslant |v|_\mathcal{P}$.

Let $T : \mathbb{R}^k \to \mathbb{R}^k$ be a homogeneous operator. We define

$$\tilde{T} : \quad \mathbb{R}^{k-1} \to \mathbb{R}^{k-1} \quad \tilde{T}(u) = \pi_1(T(x)), \quad x \in \pi_1^{-1}(u) \,.$$

Because of homogeneity, $\tilde{T}(u)$ is unambiguously defined. We can write with abbreviated notations $\tilde{T} = \pi_1 \circ T \circ \pi_1^{-1}$.

Lemma 7.3 We consider a homogeneous operator $T : \mathbb{R}^k \to \mathbb{R}^k$ and the associated operator $\tilde{T} : \mathbb{R}^{k-1} \to \mathbb{R}^{k-1}$, satisfying the following assumptions:

A. T is monotone.

B. $T(x)_1 - x_1$ is independent of $x \in \mathbb{R}^k$.

C. $\exists x_0$ such that $T(x_0)_1 - (x_0)_1 = \min_i(T(x_0)_i - (x_0)_i)$.

Under Assumption A, $\tilde{T}$ is non-expansive. Under Assumptions $A + B$, $\tilde{T}$ is monotone. Under Assumptions $A+Brm +C$, $\tilde{T}$ has minimal value $\tilde{x}_0 = \pi_1(x_0)$.

Proof We consider $u, v \in \mathbb{R}^{k-1}$ satisfying $u \geqslant v$. Let $x, y \in \mathbb{R}^k$ be such that $\pi_1(x) = u, \pi_1(y) = v$ and $x_1 = y_1$.

1. $A \Rightarrow \tilde{T}$ is non-expansive. The representatives x and y are such that $|u - v|_{\mathcal{P}} = |x - y|_{\mathcal{P}} = ||x - y||_\infty$. By monotonicity of T, we have $T(x) \geqslant T(y)$, hence $|T(x) - T(y)|_{\mathcal{P}} \leqslant ||T(x) - T(y)||_\infty$. By non-expansiveness of T (Theorem 2.2), we have $||T(x) - T(y)||_\infty \leqslant ||x - y||_\infty$. We conclude that :

$$\begin{aligned} |\tilde{T}(u) - \tilde{T}(v)|_{\mathcal{P}} = |T(x) - T(y)|_{\mathcal{P}} &\leqslant ||T(x) - T(y)||_\infty \\ &\leqslant ||x - y||_\infty = |u - v|_{\mathcal{P}} \, . \end{aligned}$$

2. $A + B \Rightarrow \tilde{T}$ is monotone. Let the representatives x and y satisfy $x_1 = y_1$. Hence by Assumption B, we have $T(x)_1 = T(y)_1$. We conclude that $T(x) \geqslant T(y) \Rightarrow \tilde{T}(u) \geqslant \tilde{T}(v)$.

3. $A + B + C \Rightarrow \tilde{T}$ has minimal value $\tilde{x}_0 = \pi_1(x_0)$. We have

$$\begin{aligned} \tilde{T}_0(\tilde{x}_0)_i &= T(x_0)_i - T(x_0)_1 = T(x_0)_i - (x_0)_i + (x_0)_i - T(x_0)_1 \\ &\geqslant T(x_0)_1 - (x_0)_1 + (x_0)_i - T(x_0)_1 = (\tilde{x}_0)_i \, . \end{aligned}$$

From the monotonicity of $\tilde{T}$, $\forall y \in \mathbb{R}^{k-1}, y \geqslant \tilde{x}_0 \Rightarrow \tilde{T}(y) \geqslant \tilde{x}_0$.   □

The operator $\tilde{T}$ is not homogeneous in general. Hence the conditions ensuring monotonicity and non-expansiveness are not the same (compare Theorem 2.2).

Remark 7.1 Assumption B can easily be weakened and replaced by:

B'. $\forall x, y \in \mathbb{R}^k$, $x_1 - y_1 = \min_i x_i - y_i \Rightarrow T(x)_1 - T(y)_1 = \min_i T(x)_i - T(y)_i$.

In Lemma 7.3, we have presented the assumptions which appear naturally in physical systems. In particular, Assumption B is satisfied when the first coordinate of T is the dater of an exogeneous arrival process. Assumption C is satisfied if the other coordinates of T correspond to events which are induced by the arrivals (hence occur later on). This was the case for the operator associated with the $G/G/1/\infty$ queue, see Equation (5.1). In that example, the minimal value was $e = (0, \ldots, 0)'$.

These assumptions are of course restrictive. Roughly speaking, they will apply only to some operators associated with "open systemsC". For operators associated with "closed systems", the conditions and results of Section 7.2 are more appropriate.

Theorem 7.4 *Let $\{T_n, n \in \mathbb{N}\}$ be a stationary and ergodic sequence of homogeneous random operators on $\mathbb{R}^k$ and let $\{\tilde{T}_n, n \in \mathbb{N}\}$ be the associated sequence on $\mathbb{R}^{k-1}$. We assume that Assumptions A, B and C of Lemma*

7.3 are satisfied with probability 1 by the operators $\{T_n\}$ (in particular they have a constant minimal value x_0). We set $\tilde{x}_0 = \pi_1(x_0)$. Then the limit $Z = \lim_n \tilde{T}_0 \circ \cdots \circ \tilde{T}-n(\tilde{x}_0)$ exists and satisfies $P\{Z < +\infty\} = 0$ or 1. Furthermore Z is a stationary solution, i.e. $Z(\theta\omega) = \tilde{T}_1(Z(\omega))$. When $P\{Z < +\infty\} = 1$, the sequence $\{T_n \circ \cdots \circ T_1(x_0)\}$ couples in finite time with the stationary sequence $\{Z \circ \theta^n\}$.

Proof This is exactly similar to the proof of Loynes' Theorem 5.1. $\square$

The main difficulty is often to prove the finiteness of Z. Moreover, when finite, Z is usually not the unique stationary solution. Indeed, we have that $\forall \lambda \in \mathbb{R}$, $\tilde{x}_0 + \lambda\vec{1}$ is a minimal value for the operators $\tilde{T}_n$. Hence by Theorem 7.4, the limits

$$Z^\lambda = \lim_n \tilde{T}_0 \circ \cdots \circ \tilde{T}-n(\tilde{x}_0 + \lambda\vec{1})$$

exist and are stationary solutions. The variables Z^λ are increasing in λ by monotonicity of $\tilde{T}_n$. Hence we can define the limit

$$Z^\infty = \lim_{\lambda \to +\infty} Z^\lambda. \tag{7.2}$$

The next theorem was originally proved by Brandt for a special operator associated with the $G/G/k/\infty$ queue.

Theorem 7.5 (Brandt [14]) *We have $P\{Z^\infty < +\infty\} = 0$ or 1. If we have $P\{Z^\infty < +\infty\} = 1$, then Z^∞ is the maximal finite stationary solution, i.e. $Z(\theta\omega) = \tilde{T}_1(Z(\omega))$ and*

$$Y(\theta\omega) = \tilde{T}_1(Y(\omega)), \ P\{Y < +\infty\} = 1 \Rightarrow P\{Z^\infty \geqslant Y\} = 1.$$

Proof The essential ingredient is the non-expansiveness of $\tilde{T}_n$. For more details, the reader is referred to [14] or [15, Theorem 1.3.2]. $\square$

Remark 7.2 The results presented in this section §7.1 are just a specialization to operators of finite dimension of more general results. Let $(E, \mathcal{E})$ be a Polish space (complete separable metric space) equipped with its Borel σ-field. We consider $\{\phi_n, n \in \mathbb{Z}\}$ a stationary and ergodic sequence of measurable random functions $\phi_n : E \times \Omega \to E$. The recursive equations $x(n+1) = \phi_n(x(n))$, $x(0) = x_0$ define a stochastic recursive sequence, following the terminology of Borovkov [12]. If the functions ϕ_n are monotone and satisfy $\phi_n(x_0) \geqslant x_0$ then the results of Theorem 7.4 hold (just replace T_n by ϕ_n). If we assume furthermore that the functions ϕ_n are non-expansive (with respect to the metric of E) then the results of Theorem 7.5 hold. For a detailed presentation of this framework, see [15] or [13].

7.2   Fixed point

We will see, in this section, a rather different use of Loynes' backward construction.

The next result generalizes Proposition 2.9. The proof of $A \Rightarrow B$ in Proposition 2.9 used only the continuity of the operator T. In fact, using the non-expansiveness of T, we can get stronger results.

Theorem 7.6 (Weller [44], Sine [42]) *Let C be a compact of $\mathbb{R}^k$. We consider an operator $T : C \to C$, non-expansive with respect to the sup-norm $\|.\|_\infty$. Then we have :*

$$\forall x \in C, \exists p \in \mathbb{N}, \exists u \in C : \lim_{n \to \infty} T^{np}(x) = u \text{ and } T^p(u) = u.$$
(7.3)

The following corollary is the essential result in what follows.

Corollary 7.7 *Let T be defined as in Theorem 7.6. We assume that $\forall n \geq 1$, T^n has a unique fixed point u. Then $\forall \varepsilon > 0, \exists N \in \mathbb{N}$ such that*

$$\forall n \geq N : \sup_{x \in C} \|T^n(x) - u\|_\infty \leq \varepsilon.$$
(7.4)

In other words, there is uniform convergence of T^n to u.

Proof Let us prove first that T^n converges simply to u. Let x belong to C. As u is the unique fixed point of the powers of T, we obtain by application of Theorem 7.6:

$$\forall x \in C, \exists p \in \mathbb{N}, \forall \varepsilon > 0, \exists M(x, \varepsilon) \in \mathbb{N}, \forall n \geq M(x, \varepsilon) : \|T^{np}(x) - u\|_\infty \leq \varepsilon.$$

By non-expansiveness, we have

$$\|T \circ T^{np}(x) - T(u)\|_\infty = \|T^{np+1}(x) - u\|_\infty \leq \|T^{np}(x) - u\|_\infty$$

and by induction, $\forall q \in \mathbb{N}$, $\|T^{np+q}(x) - u\|_\infty \leq \|T^{np}(x) - u\|_\infty$. This implies

$$\forall x \in C, \forall \varepsilon > 0, \exists N(x, \varepsilon) \in \mathbb{N}, \forall n > N(x, \varepsilon), \|T^n(x) - u\|_\infty \leq \varepsilon.$$

We are now ready to prove that the convergence is uniform. Let us denote by $\mathcal{B}(x, \varepsilon)$ the open ball of centre x and radius ε for the sup-norm. Using non-expansiveness, we have that $\forall y \in \mathcal{B}(x, \varepsilon), \forall n \geq N(x, \varepsilon)$,

$$\|T^n(y) - u\|_\infty \leq \|T^n(y) - T^n(x)\|_\infty + \|T^n(x) - u\|_\infty \leq 2\varepsilon.$$
(7.5)

Using the Borel–Lebesgue characterization of compact sets, there exists a finite number of points x_i such that $C \subset \bigcup_i \mathcal{B}(x_i, \varepsilon)$. Using Equation (7.5), we obtain: $\forall \varepsilon > 0, \forall n \geq \max_i N(x_i, \varepsilon), \forall x \in C : \|T^n(x) - u\|_\infty \leq 2\varepsilon.$ □

We are now ready to prove the main theorem of this section, which generalizes the approach proposed in [35] and [36].

Theorem 7.8 *Let $\{T_n, n \in \mathbb{N}\}$ be a stationary ergodic sequence of monotone-homogeneous random operators on $\mathbb{R}^k$ and let $\{\tilde{T}_n\}$ be the associated sequence on $\mathbb{R}^{k-1}$. We assume that there exists a deterministic monotone-homogeneous operator S on $\mathbb{R}^k$ ($\tilde{S}$ on $\mathbb{R}^{k-1}$) such that*

(i) *$\tilde{S}$ is bounded, i.e. there exists a compact K of $\mathbb{R}^{k-1}$ such that $\mathrm{Im}(\tilde{S}) \subset K$.*

(ii) *$\forall n \geqslant 1$, $\tilde{S}^n$ has a unique fixed point.*

(iii) *There exists a deterministic constant l such that $\tilde{S}$ belongs to the support of the random operator $\tilde{T}_l \circ \cdots \circ \tilde{T}_1$ and $\forall n > 0, \tilde{S}^n$ belongs to the support of $\tilde{T}_{nl} \circ \cdots \circ \tilde{T}_1$, with the following precise meaning:*

$$\forall \varepsilon > 0, \quad P\{ \sup_{x \in \mathbb{R}^{k-1}} |\tilde{T}_l \circ \cdots \circ \tilde{T}_1(x) - \tilde{S}(x)|_{\mathcal{P}} \leqslant \varepsilon \} > 0,$$

$$P\{ \sup_{x \in \mathbb{R}^{k-1}} |\tilde{T}_{nl} \circ \cdots \circ \tilde{T}_1(x) - \tilde{S}^n(x)|_{\mathcal{P}} \leqslant \varepsilon \} > 0 \,.$$

Then $\forall x \in \mathbb{R}^{k-1}$, $\tilde{x}(n) = \tilde{T}_{n-1} \circ \cdots \circ \tilde{T}_0(x)$ converges weakly to a unique stationary distribution.

Proof We first prove the theorem when assumption (iii) is replaced by the stronger assumption

(iv) $\exists l$ s.t. $P\{\tilde{T}_l \circ \cdots \circ \tilde{T}_1 = \tilde{S}\} > 0$ and $\forall n > 0, P\{\tilde{T}_{nl} \circ \cdots \circ \tilde{T}_1 = \tilde{S}^n\} > 0.$

For $x \in \mathbb{R}^{k-1}$, we define the variables

$$Z_{-n,0}(x) = \tilde{T}_0 \circ \cdots \circ \tilde{T}_{-n}(x) = \tilde{x}(n, x) \circ \theta^{-n} \,. \tag{7.6}$$

We now prove that $Z_{-n,0}(x)$ admits P.-a.s. a limit which is independent of x.

The compact K of assumption (i) is stable under $\tilde{S}$, and from assumption (ii), there is a unique fixed point $u \in \mathbb{R}^{k-1}$ for the powers of $\tilde{S}$. From Lemma 7.3, $\tilde{S}$ is non-expansive with respect to the projective norm. Hence Corollary 7.7 can be applied to $\tilde{S}$ on $(\mathbb{R}^{k-1}, |.|_{\mathcal{P}})$. This implies

$$\forall \varepsilon > 0, \ \exists N(\varepsilon), \ \forall n \geqslant N(\varepsilon), \ \forall x \in \mathbb{R}^{k-1}, \ |\tilde{S}^n(x) - u|_{\mathcal{P}} \leqslant \varepsilon. \tag{7.7}$$

We define the random variables

$$\forall \varepsilon > 0, \ M(\varepsilon) = \min\{n \geqslant N(\varepsilon)l \mid \tilde{T}_{-n} \circ \cdots \circ \tilde{T}_{-n-N(\varepsilon)l+1} = \tilde{S}^{N(\varepsilon)}\} \,, \tag{7.8}$$

where $N(\varepsilon)$ and l are defined in Equation (7.7) and in assumption (iv) respectively. Assumption (iv) also implies that $P\{M(\varepsilon) < +\infty\} > 0$. We obtain

$$P\{M(\varepsilon) < +\infty\} = 1, \qquad (7.9)$$

in exactly the same way as we obtained $P\{N_1 < +\infty\} = 1$ in the proof of Theorem 2.10.

Let us fix $\varepsilon = 1$. We define the events $\mathcal{A}_n = \{M(1, \omega) = n\}$, which form a countable partition of Ω.

Let us work for a moment on the event $\mathcal{A} = \mathcal{A}_m$ for a given integer m. Let us consider the variables $Z_{-n,-m}(x) = \tilde{T}_{-m} \circ \cdots \circ \tilde{T}_{-n}(x)$, $n > m$. We have

$$\forall n \geqslant m + N(1)l, \ Z_{-n,-m}(x) = \tilde{S}^{N(1)}(\tilde{T}_{-m-N(1)l} \circ \cdots \circ \tilde{T}_{-n}(x)) .$$
$$(7.10)$$

Hence on $\mathcal{A}_m$, the image of $Z_{-n,-m}$ is included in the closed ball of center u and radius $\varepsilon = 1$ (Equation (7.7)), which we denote by $K(1)$,

$$\forall n \geqslant m + N(1)l, \ \mathrm{Im}(Z_{-n,-m}) \subset K(1). \qquad (7.11)$$

We consider the sequence of random variables $\{M(1/i), i \in \mathbb{N}\}$. By definition of the variables $M(\varepsilon)$, (7.8), the sequence $M(1/i)$ is increasing in i; in particular $M(1/i) \geqslant M(1)$. We have, for all $n \geqslant M(1/i) + N(1/i)l$ (note that we consider the variables Z with respect to an unchanged ending point $-m$),

$$Z_{-n,-m}(x) = \tilde{T}_{-m} \circ \cdots \circ \tilde{T}_{-n}(x)$$
$$= \tilde{T}_{-m} \circ \cdots \circ \tilde{T}_{-M(1/i)+1} \circ \tilde{S}^{N(1/i)} \circ \tilde{T}_{-M(1/i)-N(1/i)l} \circ \cdots \circ \tilde{T}_{-n}(x) .$$

Using Equation (7.7), we have that $\tilde{S}^{N(1/i)} \circ \tilde{T}_{-M(1/i)-N(1/i)l} \circ \cdots \circ \tilde{T}_{-n}(x)$ is included in the closed ball of centre u and radius $1/i$. Using the non-expansiveness of the operators, we obtain the existence of a compact set, denoted $K(1/i)$, such that

$$\forall n \geqslant M(\tfrac{1}{i}) + N(\tfrac{1}{i})l, \ \mathrm{Im}(Z_{-n,-m}) \subset K(\tfrac{1}{i}). \qquad (7.12)$$

We have built a decreasing sequence of compact sets $K(1/i)$ whose radius goes to zero. By a classical theorem on decreasing sequences of compact sets (Borel–Lebesgue Theorem), the intersection of the sets $K(1/i)$ is a single point. This means precisely that the limit of $Z_{-n,-m}(x)$, $n \to +\infty$, exists and is independent of x. We define the following notations:

$$\forall \omega \in \mathcal{A}_m, \ \forall x \in \mathbb{R}^{k-1}, \ \lim_{n \to \infty} Z_{-n,-m}(x) = Z_{\infty,m}, \ Z = \tilde{T}_0 \circ \cdots \circ \tilde{T}_{-m+1}(Z_{\infty,m}) .$$

It is straightforward to prove that $Z = \lim_{n \to +\infty} Z_{-n,0}(x)$. By applying the same construction to all the events $\mathcal{A}_m, m \in \mathbb{N}$, we prove the a.s. existence of $Z = \lim_n Z_{-n,0}(x)$, the limit being independent of x. By analogy with §5, we call Z the Loynes variable.

We are now going to prove the existence of the Loynes variable Z under the weaker assumption (iii).

We define the random variables $N(\varepsilon)$ as previously, (7.7). On the other hand, the definition of the variables $M(\varepsilon)$ is modified:

$$\forall \varepsilon, \ M(\varepsilon) = \min\{n \geqslant N(\varepsilon)l \mid \sup_{x \in \mathbb{R}^{k-1}} |\tilde{T}_{-n+N(\varepsilon)l} \ldots \tilde{T}_{-n}(x) - \tilde{S}^{N(\varepsilon)}(x)|_{\mathcal{P}} \leqslant \varepsilon\}. \tag{7.13}$$

From assumption (iii) and the Ergodic Lemma 1.1, we obtain $P\{M(\varepsilon) < +\infty\} = 1$.

We define the variable $M(1)$, then the partition $\mathcal{A}_n$, the event $\mathcal{A}$ and the variables $M(1/i)$ as before. We define the variables

$$\hat{Z}^i_{-n,-m}(x) =$$
$$\tilde{T}_{-m} \circ \cdots \circ \tilde{T}_{-M(1/i)+1} \circ \left(\tilde{S}^{N(1/i)}\right) \circ \tilde{T}_{-M(1/i)-1-N(1/i)l} \circ \cdots \circ \tilde{T}_{-n}(x) \tag{7.14}$$

There exists a sequence of compacts $\hat{K}(1/i)$ of radius $1/i$ such that (see the first part of the proof)

$$\forall n \geqslant M(1/i) + N(1/i)l, \ \mathrm{Im}(\hat{Z}^i_{-n,-m}) \subset \hat{K}(1/i). \tag{7.15}$$

From the definition of $M(1/i)$, Equation (7.13), we get

$$\forall n \geqslant M(1/i) + N(1/i)l, \forall x \in \mathbb{R}^{k-1}, \ |Z_{-n,-M(1/i)}(x) - \hat{Z}^i_{-n,-M(1/i)}(x)|_{\mathcal{P}} \leqslant \frac{1}{i}.$$

Using non-expansiveness, we obtain

$$\forall n \geqslant M(1/i) + N(1/i)l, \forall x \in \mathbb{R}^{k-1}, \ |Z_{-n,-m}(x) - \hat{Z}^i_{-n,-m}(x)|_{\mathcal{P}} \leqslant \frac{1}{i}.$$

We conclude that $\forall n \geqslant M(1/i) + N(1/i)l, \forall x, y \in \mathbb{R}^{k-1}$

$$|Z_{-n,-m}(x) - Z_{-n,-m}(y)|_{\mathcal{P}} \leqslant |Z_{-n,-m}(x) - \hat{Z}^i_{-n,-m}(x)|_{\mathcal{P}} +$$
$$|\hat{Z}^i_{-n,-m}(x) - \hat{Z}^i_{-n,-m}(y)|_{\mathcal{P}} + |\hat{Z}^i_{-n,-m}(y) - Z_{-n,-m}(y)|_{\mathcal{P}} \leqslant \frac{3}{i}.$$

Hence there exists a sequence of compacts $K(1/i)$ of radius $3/i$ such that $\forall n \geqslant M(1/i) + N(1/i)l, \ \mathrm{Im}(Z_{-n,-m}) \subset K(1/i)$. We conclude as in the first part of the proof.

Our aim is now to prove that we have weak convergence of the process $\tilde{x}(n) = \tilde{T}_n \circ \cdots \circ \tilde{T}_0(x(0))$ to the stationary distribution of Z. We consider a function $f : \mathbb{R}^k \to \mathbb{R}$, continuous and bounded. We have, using the stationarity of $\{\tilde{T}_n\}$,

$$
\begin{aligned}
E\left(f(x(n, x(0))) \right) &= E\left(f(\tilde{T}_{n-1} \circ \cdots \circ \tilde{T}_0(x(0))) \right) \\
&= E\left(f(\tilde{T}_0 \circ \cdots \circ \tilde{T}_{-n+1}(x(0))) \right) \xrightarrow{n} E\, f(Z) \quad (7.16)
\end{aligned}
$$

The convergence in (7.16) is obtained from Lebesgue's dominated convergence theorem (f is bounded). This proves weak convergence. $\qquad\square$

Remark 7.3 It would be nice to replace assumption (iii) by the following weaker assumption

(v) $\quad \forall \varepsilon > 0,\ \forall K$ compact, $\quad P\{\sup_{x \in K} |\tilde{T}_l \circ \cdots \circ \tilde{T}_1(x) - \tilde{S}(x)|_p \leqslant \varepsilon\} > 0,$

$$
P\{\sup_{x \in K} |\tilde{T}_{nl} \circ \cdots \circ \tilde{T}_1(x) - \tilde{S}^n(x)|_p \leqslant \varepsilon\} > 0 \,.
$$

Assumption (v) means precisely that $\tilde{S}$ is in the support of $\tilde{T}_0$ for the topology of weak convergence on the functional space $C_0(\mathbb{R}^{k-1}, \mathbb{R}^{k-1})$ (continuous functions of $\mathbb{R}^{k-1}$).

However, Theorem 7.8 is not true under assumption (v) Here is a counterexample. We consider $a, b \in \mathbb{R}^+$ and we define the monotone homogeneous operators on $\mathbb{R}^2$:

$$
T_A(x) = \begin{pmatrix} x_1 \\ x_2 + a \end{pmatrix}, \quad \forall i \in \mathbb{N}^+,\ T_{B_i}(x) = \begin{pmatrix} x_1 \\ \max(x_2 - ib, x_1) \end{pmatrix}. \quad (7.17)
$$

We consider a sequence of i.i.d. random operators $\{T_n, n \in \mathbb{N}\}$ with the following distribution:

$$
P\{T_0 = T_A\} = \frac{1}{2}, \quad P\{T_0 = T_{B_i}\} = \frac{1}{2^{i+1}}, i \in \mathbb{N}^+ \,.
$$

We define the monotone homogeneous operator $S : \mathbb{R}^2 \to \mathbb{R}^2$, $S(x) = (x_1, x_1)'$. It is clear that $\tilde{S}$ satisfies assumptions (i) and (ii) as $\tilde{S}$ is constant. Let K be a compact set of $\mathbb{R}$ and n be such that $K \subset [-n, n]$. We obtain immediately that $\forall x \in K$, $\tilde{T}_{B_i}(x) = \tilde{S}(x)$ as soon as $ib \geqslant n$. Hence $\tilde{S}$ verifies also assumption (v).

The description of the process $\tilde{x}(n) = \tilde{T}_{n-1} \circ \cdots \circ \tilde{T}_0(0)$ is very easy. It is a random walk on the real line with an absorbing barrier at 0. The drift of the random walk is

$$
\delta = \frac{a}{2} - \sum_{i=1}^{\infty} \frac{ib}{2^{i+1}} = \frac{a}{2} - b \,.
$$

We conclude that the process $\tilde{x}(n)$ is transient if $a > 2b$, which provides the announced counter-example.

Practically speaking, the main difficulty consists in finding a deterministic operator S satisfying the assumptions of Theorem 7.8. We discuss this point for some specific models in §8.1.

8 Models Within the Framework

8.1 Operators

Let $\mathcal{A}$ and $\mathcal{B}$ be two arbitrary sets. We define applications ($\mathbb{M}_k$ denotes the set of matrices of dimension $k \times k$)

$$P : \mathcal{A} \times \mathcal{B} \to \mathbb{M}_k(\mathbb{R}), \ A : \mathcal{A} \times \mathcal{B} \to \mathbb{M}_k(\mathbb{R} \cup \{-\infty, +\infty\}),$$

where the matrices $P(\alpha, \beta)$ are "Markovian", i.e. satisfy

$$\forall i \in \{1, \ldots, k\}, \ p_{ij}(\alpha, \beta) \geqslant 0, \ \sum_{j=1}^{k} p_{ij}(\alpha, \beta) = 1. \tag{8.1}$$

Let us consider the following "(min,max,+,×)" function:

$$x \in \mathbb{R}^k, i \in \{1, \ldots, k\}, \ T(x)_i = \inf_{\alpha \in \mathcal{A}} \sup_{\beta \in \mathcal{B}} \sum_{j=1}^{k} p_{ij}(\alpha, \beta) \, (x_j + a_{ij}(\alpha, \beta)). \tag{8.2}$$

Equation (8.2) arises in stochastic control of dynamic games, see [11]. If $T(x)_i$ is finite ($\forall x \forall i$) then it defines a monotone-homogeneous operator. For example, let us prove homogeneity. We have for $x \in T^k, \lambda \in \mathbb{R}$,

$$T(x + \lambda \vec{1})_i = \inf_{\alpha} \sup_{\beta} \sum_{j=1}^{k} p_{ij}(\alpha, \beta) \, (x_j + \lambda + a_{ij}(\alpha, \beta))$$

$$= \inf_{\alpha} \sup_{\beta} (\sum_{j=1}^{k} p_{ij}(\alpha, \beta)\lambda) + \sum_{j=1}^{k} p_{ij}(\alpha, \beta) \, (x_j + a_{ij}(\alpha, \beta)) = \lambda + T(x)_i.$$

The following representation theorem provides a precise idea of the degree of generality of the class of monotone-homogeneous operators.

Theorem 8.1 (Kolokoltsov [33]) *Let* $T : \mathbb{R}^k \to \mathbb{R}^k$ *be a monotone and homogeneous operator. Then it can be represented in the form (8.2).*

The next lemma, which is based on this representation, is proved in [33]. It can be coupled with Theorem 7.8 to obtain second order results for some stochastic operators.

Lemma 8.2 *Let $T : \mathbb{R}^k \to \mathbb{R}^k$ be a monotone-homogeneous operator, written in the form of Equation (8.2). Let us assume that*

$$\exists \eta > 0 : \forall i, j \exists l : \forall \alpha, \beta, \ p_{il}(\alpha, \beta) > \eta, p_{jl}(\alpha, \beta) > \eta \,.$$

Then the operators $T^n, n \in \mathbb{N}$ have a unique generalized fixed point.

From the point of view of applications, the interesting case is when the sets $\mathcal{A}$ and $\mathcal{B}$ are finite. Here are some specializations of Equation (8.2).

$(+,\times)$ linear systems The operator T is just a Markovian matrix P, see Equation (8.1). We have $T(x) = Px$ (matrix–vector multiplication in the usual algebra). The matrix P can be interpreted as the matrix of transition probabilities of a Markov chain (MC) having state space $\{1, \ldots, k\}$. The most interesting operator for a MC is $S(y) = yP$ where y is a row vector. It is well known that the limit of $S^n(y), y \geqslant 0, \sum_i y_i = 1$ is the stationary distribution of the MC. But the operator $T(x) = Px$ is also interesting from the point of view of applications. It appeared in [21] to model the problem of reaching agreement on subjective opinions. More generally, it has been studied as a special case of the general theory of products of non-negative matrices; see for example [41, Chapter 4.6].

For any Markovian matrix P, we have $T(\vec{1}) = P\vec{1} = \vec{1}$. Hence the vector $\vec{1}$ is a generalized fixed point (Definition 2.8) of the operator T. By application of the Perron–Frobenius Theorem, it is the only one. Hence, applying the ergodic results of this paper to a stochastic sequence of matrices P_n is going to yield trivial results (the convergence of $\pi(P_n \ldots P_0 x)$ to $\pi(\vec{1})$). In fact much stronger results are known for such models. The necessary and sufficient conditions for convergence of $\pi(P_n \ldots P_0 x)$ to $\pi(\vec{1})$ are known for a general sequence of matrices P_n, without any stochastic assumptions, see [41, Theorem 4.18].

(max,+) linear systems Such operators have the form

$$x \in \mathbb{R}^k, i \in \{1, \ldots, k\}, \quad T(x)_i = \max_j (x_j + a_{ij}) \,, \tag{8.3}$$

$$T(x) = A \otimes x \,. \tag{8.4}$$

Equation (8.3) can be interpreted as a matrix–vector product in the (max,+) algebra. Equation (8.4) is simply a rewriting of Equation (8.3) using (max,+) notations. The (min,+) linear case reduces to the (max,+) case by switching to the operator $-T$.

Such systems appear in many domains of applications, under various forms. For example (the list is far from complete)

- Computer science: parallel algorithms, shared memory systems, PERT graphs, see [43] or [23].

- Queueing theory: $G/G/1/\infty$ queue (see §5), queues in series, queues in series with blocking, fork–join networks [3].

- Operations research and manufacturing: job-shop models, event graphs (a subclass of Petri nets), see [17], [28] and [3].

- Economy or control theory: dynamic optimization, see [46].

- Physics of crystal structures: Frenkel–Kontorova model, see [24].

Among the very large and complete literature on the theoretical aspects of deterministic (max,+) systems, let us quote only [3], [37] and the references therein. As far as we know, the first references to stochastic (max,+) linear systems are [18] and [39]. Thanks to the rich deterministic theory, Theorems 2.5, 7.8 become very operational for (max,+) systems. The different assumptions in these theorems can be interpreted as properties of the underlying graph structure of the model. For more details, see [35].

(min,max,+) linear systems   These systems can be represented in one of the following dual forms. We use the symbol $\otimes$ for the (max,+) matrix–vector product, see (8.4), and the symbol $\odot$ for the (min,+) matrix–vector product.

$$x \in \mathbb{R}^k, \quad T(x) = \min\left(A_1 \otimes x, A_2 \otimes x, \ldots, A_l \otimes x\right),$$
$$T(x) = \max\left(B_1 \odot x, B_2 \odot x, \ldots, B_p \odot x\right).$$

Here are some domains of application where such systems appear:

- Minimax control in dynamic game theory, see [11].

- Study of timed digital circuits, see [26]. The (min,max) structure arises from the (and,or) operations of logical circuits.

- Queueing theory. $G/G/s/\infty$ file, resequencing file, see for example [1]. Parallel processing systems [9]: there are k processors, and a customer requires p out of the k processors to be executed concurrently .

- Motion of interfaces in particle systems [22]. As an illustration, let us describe a little more precisely a special case known as the marching soldier model. There is a row of k soldiers who advance in the same direction. In order to try to keep a common pace, they adopt the following strategy. At regular instants of time, each soldier checks the position of his right and left neighbours. He advances by 1 if they are both ahead of him and stays at the same position otherwise. Let $x \in \mathbb{R}^k$

denote the position of the soldiers at instant 0. Their position at instant 1 will be (with the convention $x_0 = x_{k+1} = +\infty$)

$$T(x)_i = \max\left(\min(x_{i-1}, x_i, x_{i+1}) + 1, x_i\right) .$$

The study of deterministic (min,max,+) systems (existence of generalized fixed points, projective boundedness, ...) has been considered in several papers [38], [25]. However, it is far from being complete. For this reason, the only references to stochastic (min,max,+) systems concern first order results [22], [30].

(max,+,×) linear systems   These systems can be represented in the form

$$x \in \mathbb{R}^k, \ T(x)_i = \max_{\alpha \in \mathcal{A}} \sum_{j=1}^{k} p_{ij}(\alpha) \left(x_j + a_i(\alpha)\right) . \tag{8.5}$$

Equation (8.5) appears in many domains of applications like operational research, management science and engineering. It is in fact one of the optimality equation of stochastic[3] dynamic programming in discrete time, on a finite state space and with undiscounted rewards. A controller observes a system which evolves in a state space $\{1, \ldots, k\}$. The set of possible decisions for the controller is $\mathcal{A}$. Under the decision $\alpha \in \mathcal{A}$, the system evolves from a state i to a state j according to the transition probabilities $p_{ij}(\alpha)$. Also, under the decision $\alpha \in \mathcal{A}$, there is an immediate reward for being originally in state i which is $a_i(\alpha)$. It is well known that the optimal decision and the reward vector are obtained as $\lim_n T^n(x)$; see for example [45, Chapter 3.2.].

There is a very important literature on deterministic operators of type (8.5), see [40] or [45] and the references there. The next theorem is classical; for a proof see for example [45, Chapter 4.3]

Theorem 8.3 *Let T be an operator satisfying Equation (8.5). A sufficient condition for the existence of a unique generalized fixed point for T is that $\forall \alpha \in \mathcal{A}$, the matrix $P(\alpha)$ is ergodic, i.e. the graph of the non-zero terms of $P(\alpha)$ is strongly connected and aperiodic.*

Remark 8.1 A (max,+) system can be viewed as a (max,+,×) system with $\mathcal{A} = \{1, \ldots, k\}$ and $P(\alpha)$ defined by $P_{ij}(\alpha) = 1$ if $j = \alpha$ and $P_{ij}(\alpha) = 0$ otherwise. Such matrices do not satisfy the assumption of Theorem 8.3.

The theorems presented in this paper, when coupled with results like Theorem 8.3, can be used in an efficient way for systems satisfying (8.5) when the rewards $a(\alpha)$ and/or the transition matrices $P(\alpha)$ become random. The authors do not know of any reference on the subject.

---

[3]The term stochastic refers here to the Markovian interpretation of the matrices $P(\alpha)$. According to our terminology, Equation (8.5) is that of a deterministic operator.

8.2   Discrete event networks

We are now going to review some classes of discrete event networks. We restrict our attention to systems which cannot be modelled as monotone-homogeneous operators. The references that are quoted are only the ones using the monotone-separable framework or similar approaches.

- Precedence constraint models. Their study has been motivated by database systems. Different variations are considered in [8], [10], [20].

- Polling models. A wide class of polling models with general routing policies and stationary ergodic inputs enters the monotone-separable framework, see [16].

- Free choice Petri nets. Event graphs, which are represented as (max,+) linear operators, see §8.1, or Jackson networks, see below, are subclasses of free choice Petri nets. Free choice Petri nets belong to the monotone-separable framework, see [6], [4], [7].

Let us detail two of these models. We present firstly a simple example of precedence constraint systems, and secondly of Jackson networks.

Precedence constraint models   There is a stream of customers $j(n)$, $n \in \mathbb{N}$. Each customer $j(n)$ has a service time requirement $t(n)$ and precedence constraints under the form of a list $L(n)$ of customers. More precisely, we have $L(n) = \{j(i_1), j(i_2), \ldots, j(i_{l_n})\}$ with $n > i_1 > i_2 > \cdots > i_{l_n} \geqslant 0$. Job j_n starts its execution as soon as all the customers of the list $L(n)$ have completed their execution. The execution of customer $j(n)$ takes $t(n)$ units of time.

Let us distinguish two cases.

1. We assume that the length of the precedence list is uniformly bounded by k, i.e. $\forall n \in \mathbb{N}$, $l_n \leqslant k$. We define the vector $x(n) \in \mathbb{R}^k$ such that $x(n)_i$ is the instant of completion of customer $j(n-i)$. From the dynamic described above, we have $x(n+1) = T_n(x(n))$, where the operator $T_n : \mathbb{R}^k \to \mathbb{R}^k$ is defined by

$$\begin{cases} T_n(x)_1 = \max_{\{i \mid j(n-i) \in L(n)\}} x_i + t(n) \\ T_n(x)_i = x_{i-1}, \ i = \{2, \ldots, k\}. \end{cases}$$

This operator is monotone-homogeneous. It is in fact a (max,+) linear system, see §8.1.

2. Let us assume now that the length l_n is not uniformly bounded. It is not possible to describe the system as an operator of finite dimension. Let $X_{[1,n]}$ be the last instant of completion of one of the customers $j(i), i \in \{1, \ldots, n\}$. It is easy to verify that $X_{[1,n]}$ satisfies the properties of the monotone-separable framework for discrete event networks, see §3.

In both cases, when $\{t(n), L(n), n \in \mathbb{N}\}$ forms a stationary ergodic sequence of random variables, we can apply the ergodic theorems presented in this paper.

Jackson networks   A Jackson network (introduced in [29]) is a queueing network with I nodes, where each node is a single server FIFO queue (cf. §5). Customers move from node to node in order to receive some service there. The data are $2I$ sequences

$$\{\sigma^i(n), n \in \mathbb{N}\}, \quad \{\nu^i(n), n \in \mathbb{N}\}, \quad i \in \{1, \dots, I\},$$

where $\sigma^i(n) \in \mathbb{R}^+$ and $\nu^i(n) \in \{1, \dots, I, I+1\}$.

In the nominal network, the nth, $n \geqslant 1$, customer to be served by node i after the origin of time requires a service time $\sigma^i(n)$; after completion of its service there, it moves to node $\nu^i(n)$, where $I+1$ is the exit. We say that $\nu^i(n)$ is the nth routing variable on node i.

We are going to describe the closed (resp. open) Jackson network as a discrete event network (resp. open discrete event network), using the notations of §3.

- **Closed case** The state at the origin of time is that all customers are in node 1, and service 1 is just starting on node 1. There are no external arrivals and $\nu^i(n) \in \{1, \dots, I\}$, for all i and n. The total number of customers in the network is then a constant. We take

$$\sigma(n) \stackrel{\text{def}}{=} \sigma^1(n).$$

  The *internal daters* $X_{[1,\infty]}^{i-}(n)$ and $X_{[1,\infty]}^{i+}(n)$, $n \geqslant 1$, $i \in \{1, \dots, I\}$, are the initiation and completion instants of the nth service on node i. We take

$$X_{[1,\infty]}^{-}(n) \stackrel{\text{def}}{=} X_{[1,\infty]}^{1-}(n),$$

  so that $X_{[1,\infty]}^{-}(1) = 0$.

- **Open case** The state at the origin of time is that all queues are empty and a customer is just arriving in the network. There is an external arrival point process $\{A_n, n \geqslant 1\}$, with $A_1 = 0$, or equivalently an additional saturated node (numbered 0), which produces customers with inter-arrival times $\sigma^0(n) = A_{n+1} - A_n$, $n \geqslant 1$, regardless of the state of the network. The nth external arrival is routed to node $\nu^0(n) \in \{1, \dots, I\}$. We take

$$\sigma(n) \stackrel{\text{def}}{=} \sigma^0(n).$$

We can extend the definition of internal daters, which is the same as above, to $i = 0$ by taking $X_{[1,\infty]}^{0-}(n) = A_n$ and $X_{[1,\infty]}^{0+}(n) = A_n + \sigma(n) = A_{n+1}$. We take

$$X_{[1,\infty]}^{-}(n) \overset{\text{def}}{=} X_{[1,\infty]}^{0-}(n),$$

so that $X_{[1,\infty]}^{-}(1) = 0$.

In both cases, the restrictions $[1, m]$ of the process are obtained by modifying the $\{\sigma(n), n \in \mathbb{N}\}$ sequence in the following way:

$$\sigma_{[1,m]}^i(n) = \begin{cases} \sigma^i(n) & \text{for all } n \geqslant 1 \text{ and } i \neq 1 \text{ (resp. } i \neq 0); \\ \sigma^i(n) & \text{for all } 1 \leqslant n \leqslant m \text{ and } i = 1 \text{ (resp. } i = 0); \\ \infty & \text{for all } n > m \text{ and } i = 1 \text{ (resp. } i = 0). \end{cases}$$

The corresponding variables are denoted $X_{[1,m]}^{-}(n), X_{[1,m]}^{+}(n)$. In both cases, the maximal dater is defined as

$$X_{[1,m]} = \max \left(\sup_{i,n} \left\{ X_{[1,m]}^{i-}(n) \text{ s.t. } X_{[1,m]}^{i-}(n) < \infty \right\}, \right.$$
$$\left. \sup_{i,n} \left\{ X_{[1,m]}^{i+}(n) \text{ s.t. } X_{[1,m]}^{i+}(n) < \infty \right\} \right),$$

where the supremum is taken over $n \geqslant 1$ and $i \in \{1, \ldots, I\}$ (resp. $i \in \{0, \ldots, I\}$) in the closed (resp. open) case.

The following lemma follows from results proved in [2].

Lemma 8.4 *For all $i \in \{1, \ldots, I\}$ and $l \geqslant 1$, there exist finite sets $\mathcal{A}(i, l) \subset \mathbb{N}$, $\mathcal{B}(i, l, p) \subset \mathbb{N}$ where $p \in \mathcal{A}(i, l)$, and $\mathcal{C}(i, l, p, q) \subset \mathbb{N} \times \mathbb{N}$ where $q \in \mathcal{B}(i, l, p)$, which depend on the routing sequences only (not on the service sequences). These sets are such that*

$$\forall m, n \geqslant 1, \ X_{[1,m]}^{i-}(n) = \inf_{l \in \mathcal{A}(i,n)} \max_{p \in \mathcal{B}(i,n,l)} \sum_{(i_q, n_q) \in \mathcal{C}(i,n,l,p)} \sigma_{[1,m]}^{i_q}(n_q). \tag{8.6}$$

A pair (i, n) appears at most once in each set $\mathcal{C}(i, n, l, p)$.

This lemma has to be interpreted as the fact that Jackson networks have a (min,max,+) structure, although a very complicated one. Hence, it should come as no surprise that they belong to the monotone-separable framework. Let us prove it.

Causality In both cases, the assumption is that $X_{[1,m]}$ is a.s. finite for all m. Note that this implies *causality* as defined in §3.

Lemma 8.5 *Causality is satisfied whenever the routing sequences $\{\nu^i(n)\}_{n \in \mathbb{N}}$ are i.i.d. and independent of the service times, and the routing matrix*

$$\mathbb{P} = (p_{ij}), \quad p_{ij} = P(\nu^i(1) = j), \quad i, j \in \{1, \dots, I\}$$

is without capture in the open case, and irreducible in the closed case.

Proof The proof is based on the following coupling idea: consider a Kelly network (i.e. a route is attached to a customer, see [31]) where the routes are independent and sampled according to the stopped Markov chain with transition matrix $\mathbb{P}$. By this we mean that in the $[1,m]$-network, the route of the first customer to leave node 1 (resp. 0) is

$$\{N_0 = 1, N_1, \dots, N_{U_1}\} \quad \text{in the closed case}$$
$$\{N_0 = D, N_1, \dots, N_{U_{I+1}}\} \quad \text{in the open case},$$

where $\{N_p\}$ is a path of the Markov chain $\mathbb{P}$, U_i is the return time to state i, and D is an independent random variable on $\{1, \dots, I\}$, with distribution $\pi(i) = P(\nu^0(1) = i)$. The routes of the first m customers to be served at node 1 (resp. to arrive from node 0) are assumed to be independent and identically distributed. In this Kelly network, the routes of these m customers are not affected by the service times (in contrast with what happens in the initial network). Thus, in the closed (resp. open) case, all m customers eventually return to node 1 (resp. leave) provided $\mathbb{P}$ is irreducible (resp. $\mathbb{P}$ is without capture). In addition, such a Kelly network is identical in law to the $[1,m]$ restriction of the original network. So $P(X_{[1,m]} < \infty) = 1$. $\square$

In what follows, we will adopt the assumptions of Lemma 8.5 and assume in addition that the service times are integrable.

Monotonicity As an immediate corollary of Lemma 8.4, for all fixed routing sequences, for all $m, n \geqslant 1$ and i, the variable $X_{[1,m]}^{i-}(n)$ (and therefore $X_{[1,m]}^{i+}(n)$ as well) is a monotone non-decreasing function of $\{\sigma^j(n), j \in \{2, \dots, I\}, n \geqslant 1, \sigma^1(n), 1 \leqslant n \leqslant m\}$ (resp. $\{\sigma^j(n), j \in \{1, \dots, I\}, n \geqslant 1, \sigma^0(n), 1 \leqslant n \leqslant m\}$). This monotonicity extends to the maximal dater as well.

Non-expansiveness Let $j \leqslant I$ and $l \geqslant 1$ be fixed. Consider $\sigma^j(l)$ as a variable and all other service times as constants. Then, it follows from Lemma 8.4 that $X_{[1,m]}^{i-}(n)$ is a (min, max) function of $\sigma^j(l)$. Thus non-expansiveness as defined in §3 holds.

Separability Let $\varphi^i_{[1,m]} = \sup\{n \geqslant 1 \mid X^{i+}_{[1,m]}(n) < \infty\}$, $m \geqslant 1$ (the total number of events which ever complete on station i in the $[1, m]$-network). Of course $\varphi^1_{[1,m]} = m$ in the closed case, and $\varphi^0_{[1,m]} = m$ in the open case. The following two properties hold:

1. For all i and m, $\varphi^i_{[1,m]}$ does not depend on the (finite) values of the variables $\{\sigma^j(n), \; j \in \{2, \ldots, I\}, \; n \geqslant 1, \; \sigma^1(n), \; 1 \leqslant n \leqslant m\}$ (resp. $\{\sigma^j(n), \; j \in \{1, \ldots, I\}, \; n \geqslant 1, \; \sigma^0(n), \; 1 \leqslant n \leqslant m\}$) – this follows from Lemma 8.4.

2. For all $m \geqslant 1$, the random variables $\{\varphi^i_{[1,m]}, \; i \leqslant I\}$ form a stopping time of the sequences $\{\nu^i(n), \; i \leqslant I, \; n \geqslant 1\}$ in the sense that

$$\{\varphi^i_{[1,m]} \leqslant n^i, \; i \leqslant I\} \in \mathcal{F}\{\nu^i(l), \; l \leqslant n^i, \; i \leqslant I\},$$

    where $\mathcal{F}(u)$ denotes the σ-algebra generated by the random variable u.

We are now in a position to complete the definition of $N = \{\sigma(n), M(n), \; n \in \mathbb{N}^*\}$ (see §3) for this network, by taking

$$M(n) \overset{\text{def}}{=} \{\sigma^i(l), \; \nu^i(l), \; l = \varphi^i_{[1,n-1]} + 1, \ldots, \varphi^i_{[1,n]}, \; i \leqslant I\}, \quad n \geqslant 1,$$

with the convention $\varphi^i_{[1,0]} = 0$.

With this definition, the $[m, \infty]$-network, $1 \leqslant m$, is a Jackson network as defined above, but with the driving sequences

$$\sigma^i_{[m,\infty]}(n) = \sigma^i(n + \varphi^i_{[1,m-1]}), \quad \nu^i_{[m,\infty]}(n) = \nu^i(n + \varphi^i_{[1,m-1]}), \quad n \geqslant 1.$$

From the i.i.d. assumptions on the sequences $\{\sigma^i(n), \nu^i(n), n \in \mathbb{N}\}$ and the fact that the random variables $\varphi^i_{[1,m-1]}$ are stopping times, we obtain that the $[m, \infty]$-network is equal in distribution to the original $[1, \infty]$-network. Separability is now clear:

- **Open case** If $A_{l+1} \geqslant A_1 + X_{[1,l]}$, then from monotonicity, for all i,

$$A_{l+1} \geqslant A_1 + X^{i+}_{[1,l]}(\varphi^i_{[1,l]}) \geqslant A_1 + X^{i+}_{[1,n]}(\varphi^i_{[1,l]}),$$

  and so the $(l + 1)$st external arrival finds an empty network (we know that if there are l external arrivals and $\varphi^i_{[1,l]}$ departures from node i, then the network is empty). In addition, the next customer to be served on node i is that with index $\varphi^i_{[1,l]} + 1$, $i \leqslant I$. Thus $A_1 + X_{[1,m]} = A_{l+1} + X_{[l+1,m]}$.

- **Closed case** If $X^{+}_{[1,m]}(l+1) \geqslant X_{[1,l]}$, then

$$X^{+}_{[1,m]}(l+1) \geqslant X^{i+}_{[1,l]}(\varphi^{i}_{[1,l]}) \geqslant X^{i+}_{[1,m]}(\varphi^{i}_{[1,l]}),$$

and so, by the same argument as above, when the $(l+1)$st service ends on node 1, all customers are present in node 1. Separability follows in a way which is similar to that of the previous case.

First order ergodic theorem

Compatibility is immediate from property 2 of $\{\varphi^{i}_{[1,m]}\}$. To prove integrability, it is enough to prove that $X_{[1,1]}$ is integrable. This follows from the fact that the stopping times U_1 (resp. U_{I+1}) of $\mathbb{P}$ are integrable and from the assumption that service times are integrable.

Therefore, Theorem 3.2 applies and $\lim_{m\to\infty} (1/m)X_{[1,m]} = \gamma$, a.s. for some positive and finite constant, both in the open and closed case. More generally, it can be shown that the above limit implies that there exist finite constants rates γ^i such that $\lim (1/m)X^{i-}_{[1,m]} = \gamma^i$, a.s., $i \leqslant I$, both in the open and closed cases. For more details on the computation of these rates see [4] and [7].

Acknowledgment The authors would like to point out the importance for the maturation of this paper of the *Idempotency* workshop organized by Jeremy Gunawardena. The second author would also like to thank Jean-Marc Vincent for fruitful discussions on the subject.

References

[1] F. Baccelli and P. Brémaud. *Elements of Queueing Theory.* Number 26 in Applications of Mathematics. Springer Verlag, Berlin, 1994.

[2] F. Baccelli, G. Cohen, and B. Gaujal. Recursive equations and basic properties of timed Petri nets. *J. of Discrete Event Dynamic Systems*, 1(4):415–439, 1992.

[3] F. Baccelli, G. Cohen, G.J. Olsder, and J.P. Quadrat. *Synchronization and Linearity.* John Wiley & Sons, New York, 1992.

[4] F. Baccelli and S. Foss. Ergodicity of Jackson-type queueing networks. *Queueing Systems*, 17:5–72, 1994.

[5] F. Baccelli and S. Foss. On the saturation rule for the stability of queues. *J. Appl. Prob.*, 32(2):494–507, 1995.

[6] F. Baccelli, S. Foss, and B. Gaujal. Structural, temporal and stochastic properties of unbounded free-choice Petri nets. Technical Report 2411, INRIA, Sophia Antipolis, France, 1994. To appear in IEEE TAC.

[7] F. Baccelli, S. Foss, and J. Mairesse. Closed Jackson networks under stationary ergodic assumptions. In preparation, 1995.

[8] F. Baccelli and Z. Liu. On the stability condition of a precedence-based queuing discipline. *Adv. Appl. Prob.*, 21:883–898, 1989.

[9] N. Bambos and J.Walrand. Scheduling and stability aspects of a general class of parallel processing systems. *Adv. Appl. Prob.*, 25:176–202, 1993.

[10] N. Bambos and J. Walrand. On stability and performance of parallel processing systems. *J. Ass. for Comp. Machinery*, 38(2):429–452, 1991.

[11] T. Başar and G.J. Olsder. *Dynamic Non-cooperative Game Theory*. Chapman and Hill, 1995. 2nd edition.

[12] A. Borovkov. *Asymptotic Methods in Queueing Theory*. John Wiley & Sons, New York, 1984.

[13] A. Borovkov and S. Foss. Stochastically recursive sequences and their generalizations. *Siberian Adv. in Math.*, 2:16–81, 1992.

[14] A. Brandt. On stationary waiting time and limiting behaviour of queues with many servers I: the general $G/G/m/\infty$ case. *Elektron. Inform. Kybern.*, 21:47–64, 1985.

[15] A. Brandt, P. Franken, and B. Lisek. *Stationary Stochastic Models*. Prob. and Math. Stat. Wiley, New York, 1990.

[16] N. Chernova and S. Foss. Ergodic properties of polling systems. Technical report, Novosibirsk University, 1994. Submitted to Problems of Information Transmission.

[17] G. Cohen, D. Dubois, J.P. Quadrat, and M. Viot. A linear system-theoretic view of discrete-event processes and its use for performance evaluation in manufacturing. *IEEE Trans. Automatic Control*, AC-30:210–220, 1985.

[18] J.E. Cohen. Subadditivity, generalized product of random matrices and operations research. *SIAM Review*, 30(1):69–86, 1988.

[19] M. Crandall and L. Tartar. Some relations between nonexpansive and order preserving mappings. *Proceedings of the AMS*, 78(3):385–390, 1980.

[20] H. Daduna. On the stability of queuing systems under precedence restrictions for the service of customers. *Queueing Systems*, 17:73–88, 1994.

[21] M. de Groot. Reaching a consensus. *J. Amer. Stat. Assoc.*, 69:118–121, 1974.

[22] M. Ekhaus and L. Gray. Convergence to equilibrium and a strong law for the motion of restricted interfaces. Working paper.

[23] S. Gaubert and J. Mairesse. Task resource models and (max,+) automata. In J. Gunawardena, editor, *Idempotency*. Cambridge University Press, 1995.

[24] R. Griffiths. Frenkel-Kontorova models of commensurate-incommensurate phase transitions. In H. van Beijeren, editor, *Fundamental problems in statistical mechanics VII*. Elsevier Science Publishers, 1990.

[25] J. Gunawardena. Min-max functions. Technical Report STAN-CS-93-1474, Stanford University, 1993. To appear in Discrete Event Dynamic Systems.

[26] J. Gunawardena. Timing analysis of digital circuits and the theory of min-max functions. In *TAU' 93, ACM Int. Workshop on Timing Issues in the Specification and Synthesis of Digital Systems*, 1993.

[27] J. Gunawardena and M. Keane. Private communication, 1995.

[28] C. Hanen and A. Munier. A study of the cyclic scheduling problem on parallel processors. *Discrete Appl. Math.*, 57:167–192, 1995.

[29] J.R. Jackson. Jobshop-like queueing systems. *Manag. Science*, 10:131–142, 1963.

[30] A. Jean-Marie and G.J. Olsder. Analysis of stochastic min-max systems : results and conjectures. Technical Report 93-94, Faculty Technical Math. and Informatics, Delft University of Technology, 1993.

[31] F. Kelly. *Reversibility and Stochastic Networks*. Wiley, New-York, 1979.

[32] J. Kingman. Subadditive ergodic theory. *Annals of Probability*, 1:883–909, 1973.

[33] V. Kolokoltsov. On linear, additive and homogeneous operators in idempotent analysis. In V. Maslov and S. Samborskiĭ, editors, *Idempotent Analysis*, volume 13 of *Adv. in Sov. Math.* AMS, 1992.

[34] R. Loynes. The stability of a queue with non-independent interarrival and service times. *Proc. Camb. Philos. Soc.*, 58:497–520, 1962.

[35] J. Mairesse. Products of irreducible random matrices in the (max,+) algebra. Technical Report RR-1939, INRIA, Sophia Antipolis, France, 1993. To appear in *Adv. Applied Prob.*

[36] J. Mairesse. *Stability of Stochastic Discrete Event Systems. Algebraic Approach*. PhD thesis, Ecole Polytechnique, Paris, 1995.

[37] V. Maslov and S. Samborskiĭ, editors. *Idempotent analysis*, volume 13 of *Adv. in Sov. Math.* AMS, 1992.

[38] G.J. Olsder. Eigenvalues of dynamic min-max systems. *JDEDS*, 1:177–207, 1991.

[39] G.J. Olsder, J. Resing, R. de Vries, M. Keane, and G. Hooghiemstra. Discrete event systems with stochastic processing times. *IEEE Trans. on Automatic Control*, 35(3):299–302, 1990.

[40] S. Ross. *Introduction to Stochastic Dynamic Programming*. Academic Press, New York, 1983.

[41] E. Seneta. *Non-negative Matrices and Markov Chains*. Springer series in statistics. Springer Verlag, Berlin, 1981.

[42] R. Sine. A non-linear Perron-Frobenius theorem. *Proc. AMS*, 109(2):331–336, 1990.

[43] J.M. Vincent. Some ergodic results on stochastic iterative DEDS. Technical Report 4, Apache, IMAG, Grenoble, 1994. To appear in JDEDS.

[44] D. Weller. *Hilbert's metric, part metric and self-mappings of a cone*. PhD thesis, University of Bremen, Germany, 1987.

[45] D.J. White. *Markov Decision Processes*. Wiley, New York, 1993.

[46] S. Yakovenko and L. Kontorer. Nonlinear semigroups and infinite horizon optimization. In V. Maslov and S. Samborskiĭ, editors, *Idempotent analysis*, volume 13 of *Adv. in Sov. Math*. AMS, 1992.

Computational Issues in Recursive Stochastic Systems

Bruno Gaujal and Alain Jean-Marie

Abstract

Estimation of asymptotic quantities in stochastic recursive systems can be performed by simulation or exact analysis. In this paper, we show how to represent a system in order to make computation procedures more efficient. A first part of this paper is devoted to parallel algorithms for the simulation of linear systems over an arbitrary semiring. Starting from a linear recursive system of order m, we construct an equivalent system of order 1 which minimizes the complexity of the computations. A second part discusses the evaluation of general recursive systems using Markovian techniques.

1   Introduction

Stochastic recursive systems may be used to model many discrete event systems, such as stochastic event graphs [16, 9, 4], PERT networks, timed automata [10] or min-max systems [15]. Qualitative theorems characterizing the asymptotic behavior of the system have been proved recently [2, 22] but efficient quantitative methods are still to be found. We investigate two approaches to estimate the behavior of recursive systems: parallel simulation and exact Markovian analysis. If we consider a linear recursive system of order m, it is essential for both approaches to provide a standard representation of the system that yields a minimum "cost". A standard representation is a larger system of order 1 which includes the original one, path-wise. The cost is different according to the technique used.

In the first part, we present two algorithms: a *space parallel* and a *time parallel* simulation of linear recursive systems. For both of them, we construct an optimal standard representation. This is done by modifying the *marking* of the associated *reduced graph*.

In the second part, we show how *Markov additive processes* can be used to compute performance measures of recursive systems of order 1.

This paper is organized as follows. In §2, we introduce the definitions of linear systems over semirings and we define the reduced and developed graphs associated with the linear system. In §3, we present the parallel algorithms and evaluate their complexity. In §4, we present several optimizations of the initial marking applying to these algorithms. In §5 we show how Markov additive processes arise naturally in the framework of recursive stochastic systems.

2   Linear Systems Over a Semiring.

In this section, we describe the basic concepts used in the paper.

Definition 1 (semiring) *A semiring is a set $\mathcal{D}$ equipped with an associative and commutative operation $\oplus$ with null element ε, and with an associative operation $\otimes$ with null element e, distributive with respect to $\oplus$.*

We also define the product spaces, $\mathcal{D}^{n \times p}$ with the following operations:

$$\forall A, B \in \mathcal{D}^{n \times p}, (A \oplus B)_{i,j} = A_{i,j} \oplus B_{i,j} ,$$

and $\forall A \in \mathcal{D}^{n \times p}, B \in \mathcal{D}^{p \times m}, (A \otimes B)$ is in $\mathcal{D}^{n \times m}$ with

$$(A \otimes B)_{i,j} = \bigoplus_{k=1}^{p} A_{i,k} \otimes B_{k,j}.$$

Definition 2 (linear system) *A linear system $\mathcal{S}$ of dimension T and of order m_0 is a system with state variables $Y(n) \in \mathcal{D}^T$, $n \in \mathbb{N}$ satisfying the evolution equation:*

$$Y(n) = A(n,0) \otimes Y(n) \oplus A(n,1) \otimes Y(n-1) \oplus \cdots \oplus A(n,m_0) \otimes Y(n-m_0), \quad (2.1)$$

where $A(n,i), i = 0, \ldots, m_0$ are matrices in $\mathcal{D}$ of size $T \times T$ and the initial conditions are $Y(-1), \ldots, Y(-m_0)$.

The linear system is *stochastic* if $A(n,k)$ are random matrices.

Structural assumptions   The following assumptions will be made throughout the paper.

$\mathcal{A}_1$: $A(n,i)$ has a fixed support,

$\mathcal{A}_2$: $A(n,0)$ is acyclic,

$\mathcal{A}_3$: $A(n,1)$ has a non-ε diagonal.

The matrices $A(n,i)$, $n \in \mathbb{N}$ are said to have the same support, if $\forall n_1, n_2 \in \mathbb{N}$, $A(n_1, i) = \varepsilon \Leftrightarrow A(n_2, i) = \varepsilon$. In other words, the dependency structure of the variables $Y_i(n)$ is independent of n.

Assumption $\mathcal{A}_3$ will be satisfied in practice for most FIFO systems.

Assumption $\mathcal{A}_2$ ensures that the system (2.1) has a unique solution $Y(n)$ [14]. In this case we can define

$$A(n,0)^\star = \bigoplus_{k=0}^{\infty} A(n,0)^k,$$

and (2.1) becomes

$$Y(n) = \bigoplus_{i=1}^{m_0} A(n,0)^\star \otimes A(n,i) \otimes Y(n-i). \qquad (2.2)$$

2.1 Dependency graph

We introduce another structure, which is a graph $\delta(\mathcal{S})$ based on the dependency relations in the linear system $\mathcal{S}$. It belongs to the class of task graphs called PERT graphs (or activity networks). It is also called the developed graph and is presented in [6] for example. Basically, this graph is constructed using the variables $Y_i(n), i \leq T, n \in \mathbb{N}$ as nodes with an arc between node $Y_j(m)$ and node $Y_i(n)$ if the computation of $Y_i(n)$ involves $Y_j(m)$. Furthermore we put a weight on this arc, which is the matrix coefficient that relates $Y_i(n)$ and $Y_j(m)$.

A more formal definition of the dependency graph is:

Definition 3 (dependency graph) *The dependency graph associated with the linear system $\mathcal{S}$ is a weighted graph $\delta(\mathcal{S})$.*

- *The set of vertices of $\delta(\mathcal{S})$ is $\{(t,n), \; 1 \leq t \leq T, \; n \in \mathbb{N}\} \cup \{\bot\}$.*

- *The set of arcs of $\delta(\mathcal{S})$ is*

   - *$\{(t,n),(t',n') \; : \; [A(\cdot, n'-n)]_{t',t} \neq \varepsilon\}$. This arc has a weight $[A(n', n'-n)]_{t',t}$.*
   - *$\{(\bot),(t,k) \; : \; t \in [0,T], \; k \leq M(t)\}$ where $M(t) = \max\{0 \leq m \leq m_0 \; : \; \exists i, [A(\cdot, m+1)]_{t,i} \neq \varepsilon\}$. The arc $(\bot),(t,k)$ has a weight $[A(k, k+1) \otimes Y(-1) \oplus \cdots \oplus A(k, m_0) \otimes Y(k-m_0)]_t$.*

The node $\bot$ represents the "initiation" of the dependency graph. Assumption $\mathcal{A}_1$ implies that $\delta(\mathcal{S})$ is periodic (i.e. if $(t,n),(t',n')$ is an arc of $\delta(\mathcal{S})$, then $(t, n+1),(t', n'+1)$ is also an arc in $\delta(\mathcal{S})$); assumption $\mathcal{A}_2$ implies that $\delta(\mathcal{S})$ is acyclic and assumption $\mathcal{A}_3$ implies that $(t,n),(t,n+1)$ is an arc in $\delta(\mathcal{S})$ for all t and n.

To illustrate the definition of $\delta(\mathcal{S})$, let $\mathcal{S}$ be the system:

$$Y(0) = Y(-1) = (0,0,0,0)^t,$$

$$
\begin{pmatrix} Y_1(n) \\ Y_2(n) \\ Y_3(n) \\ Y_4(n) \end{pmatrix} = \begin{pmatrix} \varepsilon & \varepsilon & \varepsilon & \varepsilon \\ a_{2,1}(n) & \varepsilon & \varepsilon & \varepsilon \\ \varepsilon & a_{3,2}(n) & \varepsilon & \varepsilon \\ \varepsilon & \varepsilon & a_{4,3}(n) & \varepsilon \end{pmatrix} \otimes \begin{pmatrix} Y_1(n) \\ Y_2(n) \\ Y_3(n) \\ Y_4(n) \end{pmatrix}
$$

$$
\oplus \begin{pmatrix} b_{1,1}(n) & \varepsilon & \varepsilon & \varepsilon \\ \varepsilon & b_{2,2}(n) & \varepsilon & \varepsilon \\ \varepsilon & \varepsilon & b_{3,3}(n) & \varepsilon \\ \varepsilon & \varepsilon & \varepsilon & b_{4,4}(n) \end{pmatrix} \otimes \begin{pmatrix} Y_1(n-1) \\ Y_2(n-1) \\ Y_3(n-1) \\ Y_4(n-1) \end{pmatrix}
$$

$$
\oplus \begin{pmatrix} \varepsilon & \varepsilon & \varepsilon & c_{1,4}(n) \\ \varepsilon & \varepsilon & \varepsilon & \varepsilon \\ \varepsilon & \varepsilon & \varepsilon & \varepsilon \\ \varepsilon & \varepsilon & \varepsilon & \varepsilon \end{pmatrix} \otimes \begin{pmatrix} Y_1(n-2) \\ Y_2(n-2) \\ Y_3(n-2) \\ Y_4(n-2) \end{pmatrix}. \qquad (2.3)
$$

Figure 1 depicts the initial part of $\delta(\mathcal{S})$, with some of its weights.

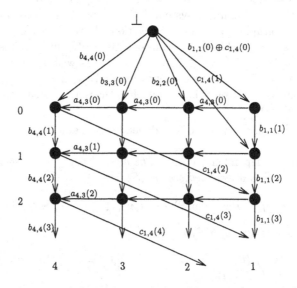

Figure 1: Associated dependency graph of the linear system (2.3).

2.2 Reduced graph

We can also construct a marked reduced graph $\rho(\mathcal{S})$ associated with the system $\mathcal{S}$, which is a finite representation of $\delta(\mathcal{S})$. It has T nodes and if there is an arc between nodes (t, n) and $(t', n+m)$ for all $n \in \mathbb{N}$, in $\delta(\mathcal{S})$, there is an arc between nodes t and t' in $\rho(\mathcal{S})$ with a mark equal to m and a sequence of weights equal to the weights of all the arcs $(t, n), (t', n+m)$ in $\delta(\mathcal{S})$.

More precisely, we have the following definition.

Definition 4 (reduced graph) *The reduced graph associated with the linear system S is an oriented multi-graph $\rho(S) = (T, P, M_0, \Sigma, \Xi)$ where*

- $T = \{1, \ldots, T\}$ *is the set of nodes,*

- P *is the set of arcs. An arc p with ends (t, t') is in P if there exists $m, 0 \le m \le m_0$ such that $[A(\cdot, m)]_{t',t} \ne \varepsilon$. In that case:*

- $M_0(p) \overset{\text{def}}{=} m \in \mathbb{Z}$ *is the marking of p,*

- $\Sigma(p) \overset{\text{def}}{=} \{\sigma_p(n), n \in \mathbb{N}\} \overset{\text{def}}{=} \{[A(n, M_0(p))]_{t',t}, n \in \mathbb{N}\}$ *is the sequence of weights of arc p and*

- $\Xi(p) \overset{\text{def}}{=} \{\xi_p(k), 0 < k \le M_0(p)\} \overset{\text{def}}{=} \{[A(k,m)]_{t',t} \otimes Y_t(k - M_0(p)), \ 0 \le k < M_0(p)\}$ *are the initial weights of arc p.*

The cardinal of the set of edges of the reduced graph P is denoted P. Note that $m_0 = \max_{p \in P} M_0(p)$. Note also that negative markings are allowed.

The reduced graph associated with the system given in (2.3) is depicted in Figure 2. The marks associated with the arcs are depicted in boxes by the arcs. The weights and the initial weights are not displayed in the figure.

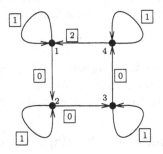

Figure 2: Reduced graph associated with the linear system (2.3).

Remark Conversely a linear system can be defined by the reduced graph associated with it. Indeed, If $G = (T, P, M_0, \Sigma, \Xi)$ is given, this defines uniquely a linear system S such that $G = \rho(S)$. Therefore, in the following, a linear system will often be defined by its reduced graph. If G is a reduced graph, we will for simplicity denote by $\delta(G)$ the dependency graph of the linear system defined by G.

Remark The notion of reduced graph is equivalent to that of Timed Event Graphs (TEG), studied in [4]. TEG are a class of Petri nets whose dynamic behavior is described a by linear system in the (max, +) semiring. We might have used this formalism instead of introducing a seemingly new concept, but we felt that for linear systems over other semirings, and above all for the more general recurrence equations to which our modeling applies (see §4.3.5), a Petri net interpretation of the equations might disturb a reader not familiar with this formalism. Readers who are familiar with TEG analysis will easily interpret the following results in this context.

In the following, $t^{\bullet}$ will denote the set of arcs in $\mathcal{P}$ with starting node t and $^{\bullet}t$ the set of arcs in $\mathcal{P}$ with ending node t.

Definition 5 (incidence matrix) *The incidence matrix I of a reduced graph G is a matrix of size $P \times T$ defined by $I = I_+ - I_-$ where the matrices I_+ and I_- are respectively*

$$[I_+]_{p,t} = \begin{cases} 1 & \text{if arc } p \in t^{\bullet} \\ 0 & \text{otherwise,} \end{cases} \qquad [I_-]_{p,t} = \begin{cases} 1 & \text{if arc } p \in {}^{\bullet}t \\ 0 & \text{otherwise.} \end{cases}$$

Note that the matrix I has exactly one element equal to 1 and one element equal to -1 per line. It is not difficult to see that if G is connected, I is of rank $T - 1$ and the eigenvector of eigenvalue 0 is $(1, 1, \ldots, 1)$.

Let M be an integer valued vector of size P. We say that M is *structurally reachable* from M_0 [23] if and only if there exists a non-negative integer vector x (x is a vector of dimension T, called a firing vector), such that

$$M = M_0 + I \cdot x.$$

Starting with the marking M_0, we denote by $R(M_0)$ the set of all the markings which are structurally reachable from M_0, and we denote by $R^+(M_0)$ the subset of $R(M_0)$ consisting of non-negative markings.

2.3   Relations between the dependency graph and the reduced graph

These two graphs are strongly related. Here we present only a few properties which will be useful in the following. For more on this, see [11]. Let G be a reduced graph.

Definition 6 (level, column) *The level n is the set $\{(t, n), t \in T\}$ of vertices in $\delta(G)$. The column t is the set $\{(t, n), n \in \mathbb{N}\}$ of vertices in $\delta(S)$.*

Definition 7 (section) *A section S is a set of nodes in $\delta(G)$, with exactly one node per column. $S = \{(i, n_i), \ i \in T\}$.*

Note that a level is a section where the n_i are all equal. A section $S = \{(i, n_i), \ i \in T\}$ in $\delta(S)$ is associated with the marking M in G defined by $M =$

$M_0 + I(n_1, \cdots, n_T)^t$. This marking may be negative for some arcs as in Figure 3. Conversely, M is associated with all sections of the form $\{(i, n_i + k), i \in T\}$, with $k \in \mathbb{Z}$ such that $n_i + k \geq 0$, and with no other. We shall denote such a section by $S(M)$. Figure 3 depicts the sets $S(M)$ and $C(M)$ (defined later in §4.3.3) for a (non-positive) marking $M = M_0 + I.(1, 0, 0, 1)^t = (1, 0, -1, 2)$ in the reduced graph depicted in Figure 2.

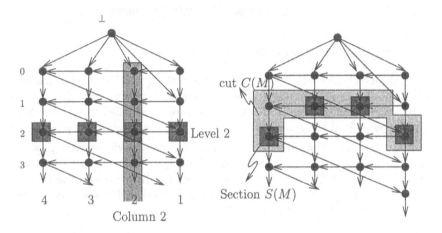

Figure 3: the sets $S(M)$ and $C(M)$, with $M = (1, 0, -1, 2)$.

2.4 Standard representation

Definition 8 (standard representation) *A standard representation of a linear system S of order m_0 on the variables $Y(n)$ is a linear system of order 1 and dimension $\tilde{T} \geq T$:*

$$\tilde{Y}(n) = A(n) \otimes \tilde{Y}(n-1), \tag{2.4}$$

such that there exist an integer n_0 and integers θ_i, $i = 1, \ldots, T$ satisfying

$$\forall n > n_0, \ \forall i \leq T, \ Y_i(n) = \tilde{Y}_i(n + \theta_i).$$

The usual way to construct a standard representation of a linear system (see e.g. [4, p. 82]) is to extend the state variable $Y(n)$ into the "past" by adding to it $Y(n-1), \ldots, Y(n-m_0)$. It is easy to see that this leads to a standard representation of dimension $m_0 T$, for which all $\theta_i = 0$.

However, there are cheaper ways to transform the equation of order m_0 into an equation of order 1. Such techniques will be shown in §4.3.2 and constitute one of the main points of this paper.

3   Parallel Simulation Algorithms

In this section we present two dual algorithms to simulate a linear system, both based on a standard representation (2.4). The simulation simply consists in computing all the state variables $\tilde{Y}_i(n)$, from which the statistics of the system can be estimated. The first algorithm uses a classical matrix–vector parallel multiplication while the second uses the associativity of the product of matrices in the semiring and a parallel prefix procedure.

3.1   Space parallel algorithm

The first idea consists in computing the product $A(n) \otimes \tilde{Y}(n-1)$ in parallel using conventional linear algebra parallel algorithms. This algorithm can be efficiently implemented on a SIMD architecture. It is called "space parallel" because for a fixed n the variables $\tilde{Y}_i(n)$ are computed by different processors.

For computation purposes, the matrix $A(n)$ can be viewed as a product $A(n) = A_0(n)^\star \otimes A_1(n)$, where the matrices $A_0(n)$ and $A_1(n)$ are matrices of size $\tilde{T} \times \tilde{T}$ (see §4.3.2). The computation of $A_0(n)^\star$ can be done in time $O(\tilde{T}^3)$ using one processor, but on a SIMD machine the following construction is more efficient. First, let us note that

$$A_0(n)^\star \;=\; E \oplus A_0(n) \oplus A_0(n)^2 \oplus \cdots A_0(n)^L,$$

where L is the length of the longest path with no marks in the reduced graph. If $\oplus$ is idempotent we have

$$A_0(n)^\star \;=\; (E \oplus A_0(n))^L .$$

The evolution equation is now $Y(n) = (E + A_0(n))^L \otimes A_1(n) \otimes \tilde{Y}(n-1)$. From this last equation, one can derive a parallel algorithm to compute $\tilde{Y}(n)$ by performing only matrix–vector multiplications, see [3].

Complexity   We use the PRAM (Parallel Random Access Machine) model to evaluate the complexity of our parallel algorithms. Assuming that the time parallel algorithm is run on a SIMD machine with K processors, K being less than $\tilde{T}^2$, computing $\tilde{Y}(N)$ will have a complexity of

$$O\!\left((N/K)\tilde{T}^2 (L \log \tilde{T})\right).$$

Using sparse matrix techniques based on D (the maximal degree of a node) for the multiplications gives a PRAM complexity of

$$O\!\left((N/K)D\tilde{T}(L \log \tilde{T})\right). \tag{3.1}$$

3.2 Time parallel algorithm

A dual way to use parallelism to simulate the system (2.1) is to write one of its standard representations (2.4) in the form

$$\tilde{Y}(N) = A(N) \otimes \ldots \otimes A(1) \otimes \tilde{Y}(0),$$

and to use the associativity of the matrix operation $\otimes$ to design a classical parallel prefix algorithm [5] to compute all vectors $\tilde{Y}(N), \cdots, \tilde{Y}(1)$.

This algorithm is "time parallel" because the computation of these vectors, which represent points in time, is carried out in parallel.

Complexity Processors have to compute sequentially the product of N/K matrices. Each can be done in a time $O(\tilde{T}^2 D)$. Adding the complexity of the parallel prefix algorithm on K processors gives an overall asymptotic complexity of

$$O\Big((N/K)\tilde{T}^2 D + \tilde{T}^3 \log K\Big). \tag{3.2}$$

3.3 Comparison

For both algorithms, the PRAM complexity depends on the initial marking of the reduced graph, through $\tilde{T}$ and/or L. Therefore, modifying the initial marking in a suitable way *before* deriving the evolution equations may improve these algorithms. This is the purpose of the next section.

Note however that PRAM complexity does not allow comparison of precise practical implementations of parallel algorithms. On the other hand, it does give a general idea of their actual efficiency. The optimizations of the PRAM complexity which are presented in the following also have the advantage of minimizing the overall number of communications and the memory requirement of the program. It turns out that the space algorithm is more suitable for simulating very large systems over a moderate period of time while the time algorithm is more suitable for simulating moderate size systems over a long period of time.

4 Minimal Standard Representations

In this section we compute initial markings which are minimal in the sense that they minimize the overall complexity of the algorithms, given in (3.1) and (3.2).

In §4.1, we show under which conditions changing the marking should be done in order to preserve the behavior of the linear system it represents (Theorem 9). We then give characterizations of the markings that optimize the space parallel and time parallel algorithms, as well as algorithms to find them. In particular, it is shown in Theorem 12, Lemma 13 and Corollary 21 that these markings can be computed in polynomial time.

4.1 Modification of the initial marking

The following property states that by modifying properly the marks of a reduced graph, it is possible to change a marking into a new one structurally reachable from the first. Its proof is straightforward.

Theorem 9 *Let $G^1 = (\mathcal{P}, \mathcal{T}, M^1, \Sigma^1, \Xi^1)$ be the reduced graph associated with a linear system. Let $G^2 = (\mathcal{P}, \mathcal{T}, M^2, \Sigma^2, \Xi^2)$ be another reduced graph with the same topology and $M^2 \in R(M^1)$. Let $S(M^2)$ be a section associated with M^2 in $\tau(G^1)$; $S(M^2) = \{(1, n_1), \ldots, (T, n_T)\}$. If the conditions*

  (i) $\sigma^2_{j,i}(n) = \sigma^1_{j,i}(n + n_j)$,

  (ii) $\xi^2_{j,i}(n) = \sigma^1_{j,i}(n + n_j) \otimes Y^1_i(n + n_j - M^1(j, i))$, *for* $1 \le n \le M^1(j, i)$

are satisfied, then $Y^2_j(n) = Y^1_j(n + n_j)$, $\forall n \in \mathbb{N}$, $j \in \mathcal{T}$.

    This result can be illustrated on the dependency graphs associated with G^1 and G^2 by recalling that $Y_i(n)$ is the longest path from $\perp$ to (i, n). Since $\delta(G^2)$ is constructed from $\delta(G^1)$ by merging all the nodes between $\perp$ and $S(M^2)$ into $\perp$, the weights that must be put on the arcs of $\delta(G^2)$ are easy to guess. See Figure 4 for an illustration of Theorem 9 from this viewpoint.

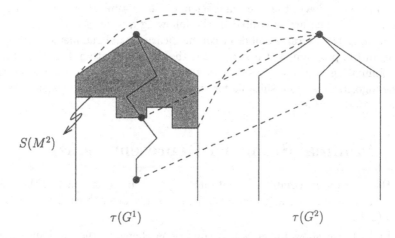

$$S(M^2)$$

$$\tau(G^1) \qquad\qquad\qquad \tau(G^2)$$

Figure 4: How to convert the weights in $\delta(G^1)$ into weights in $\delta(G^2)$.

4.2 Optimization for space parallelism

In this section, we will provide a computation of an initial markings that minimizes the parameter L appearing in the complexity formula (3.1) of the space algorithm. Since L depends on the marking M_0, we will denote it by $L(M_0)$. In this section, we will find a marking $M^{(L)}$ that minimizes L.

Definition 10 (L^* and $M^{(L)}$) $L^* = \min\{L(M) \ : \ M \in R^+(M_0)\}$ and $M^{(L)}$ is a marking such that $L(M^{(L)}) = L^*$.

Let S be a linear system, $\delta(S)$ is its associated dependency graph. Let $\mu(n)$ be the length of the longest path in $\delta(S)$ from any node in level 1 to any node in level n.

Definition 11 (sequentiality) *The sequentiality of a linear system S is*

$$\mu = \lim_{n\to\infty} \mu(n)/n.$$

This limit μ is equal to the maximal average marking of the cycles of the reduced graph (see [4]).

The sequentiality measures the asymptotic longest causal chain in the dependency graph. It is equal to the time it takes on a PRAM with any number of processors to compute $Y(n)$. This time can be viewed indeed as the degree of intrinsic "sequentiality" contained in the system.

Theorem 12 $L^* = \lceil \mu \rceil - 1$.

In other words, L^* is the integer approximation of the sequentiality of the system. This remark gives an insight into why the complexity of the parallel simulation of a linear system is linear in L^* in the best case.

In [11, 12], a "greedy algorithm" that computes a marking attaining L^* is provided. It consists in repeatedly firing all transitions which are at the head of the longest paths with no marks in the reduced graph. It can be proved that this algorithm converges to a marking M with $L(M) = L^*$, and that its complexity is polynomial.

In [21], another algorithm to compute L^* and a marking realizing it is provided. The "greedy" algorithm gives further insight on the relations that can be found between the local optimization of the longest path with no mark and the asymptotic sequentiality of the whole system (Theorem 12).

4.3 Optimization for time parallelism

4.3.1 Minimization of m_0

As mentioned briefly in §2.4, one can build a standard representation of a linear system of size $m_0 T$. In this case, one may want to minimize m_0, the maximal number of marks present in one arc for the initial marking. For a given marking M, we will write $\overline{M} \stackrel{\text{def}}{=} \max_{p \in \mathcal{P}} M(p)$. Note that $m_0 = \overline{M_0}$.

Lemma 13 *If (G, M_0) is a linear system, then the computation of a marking $M^{(B)}$ such that $\overline{M^{(B)}} = \min_{M \in R^+(M_0)} \overline{M}$ is polynomial in the size of G.*

The proof of this lemma is not difficult and uses a linear program and an integer approximation of its solution.

4.3.2 Tree expansion

As claimed in §2.4, it is possible to construct standard representations of a linear system without using the usual state space extension. The principle we apply in this section is that it suffices to transform the reduced graph of the original system into an equivalent one, with at most one mark per arc.

In order to do this, we apply a local transformation to all the nodes in the graph. Let $G = (\mathcal{P}, \mathcal{T}, M_0, \Phi, \Sigma, \Xi)$ be a reduced graph associated with a linear system. We construct a *down-tree* linear system $G^d = (\mathcal{P}^d, \mathcal{T}^d, M^d, \Phi^d, \Sigma^d, \Xi^d)$ in the following way. For each node t in $\mathcal{T}$, we add $m = \max\{M_0(t^\bullet)\} - 1$ nodes $a_1, \ldots, a_m$ and m arcs $b_1, \ldots, b_m$ with ε weights, ε initial weights and one mark. The original arcs $p \in t^\bullet$ have one mark in G^d if not originally marked. Their weights and initial weights are modified: $\sigma_p^d(n) = \sigma_p(n - M_0(p) + 1)$ for $n \geq M(p)$ and $\sigma_p^d(n) = \xi_p(n + 1)$ otherwise. As for the initial weights, $\xi_p^d(1) = \xi_p(1)$ if p is marked.

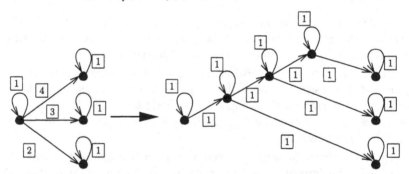

Figure 5: Construction of a new reduced graph with at most one mark per arc from the original reduced graph

This transformation is depicted in Figure 5. Note that the number of nodes of G^d is

$$T^d = \sum_{t \in \mathcal{T}} \max_{p \in t^\bullet} M_0(p). \tag{4.1}$$

The linear system associated with G^d is thus of order 1 and dimension T^d. A standard representation with the same dimension is obtained as in (2.2).

In the following, we shall compute a marking for which the dimension of this system is minimized.

4.3.3 Dependency graph interpretation

We first give an interpretation of T^d in the dependency graph. Let M be an arbitrary marking in $R(M_0)$. Let $T^d(M)$ be the size of the down-tree expansion of G obtained with this marking.

Definition 14 (cut) *A cut in $\delta(S)$ associated with a reachable marking M is the set $C(M)$ defined by*

$$(i, n) \in C(M) \quad \text{iff} \quad (i, n) \in S(M) \text{ or}$$
$$\exists (j, m) \in S(M) \text{ and } M(p) > m - n,$$

where $M(p)$ is the marking in $\rho(S)$ on an arc p between nodes i and j.

In other words, $C(M)$ comprises $S(M)$ and all nodes "above" $S(M)$ that have a successor strictly below it. Hence, $C(M)$ is indeed a cut of the graph $\delta(S)$. Figure 3 gives an example of a cut $C(M)$, with $M = (1, 0, -1, 2)$.

By definition of $C(M)$ and with (4.1), one has:

Lemma 15 *The size of the cut $C(M)$ is $T^d(M)$.*

Let $A_M^d(n)$ be the sequence of matrices of the standard representation associated with G^d. So far, the marking M we consider does not need to be positive. However, in order to apply the parallel prefix algorithm (§4.3) to compute the product $A_M^d(n) \otimes \cdots \otimes A_M^d(1)$, the matrices need to be generated in parallel, which means that they must not share any variable. We call matrices with this property *disjoint* matrices.

Definition 16 (compatible cut) *A cut $C(M)$ is compatible if it is associated with a positive marking.*

Lemma 17 *The matrices $A_M^d(n)$ are disjoint if and only if the marking M is positive (i.e. the cut $C(M)$ is compatible).*

Proof In the dependency graph $\delta(G_M)$, the element $A_M^d(n)_{i,j}$ is the longest path from node $(j, n-1)$ to node (i, n). If M is positive, the longest path from $(j, n-1)$ to (i, n) does not go through nodes below $S(M)$, which means that the matrices $A_M^d(n)$ and $A_M^d(n-1)$ do not share any variable. □

Definition 18 $(H^*, M^{(H)})$ *Let H^* be the minimum size of a compatible cut:*

$$H^* = \min_{M \in R^+(M_0)} H(M),$$

where $H(M) = |C(M)|$. We denote by $M^{(H)}$ a marking minimizing H, i.e. $H(M^{(H)}) = H^*$.

Theorem 19 *The system of order 1,*

$$Y^d(n) = A^d_{M^{(H)}}(n) \otimes Y^d(n-1), \qquad (4.2)$$

is a standard representation of the original system with disjoint matrices of minimal size.

Proof We only give a sketch of the proof. First, it should be clear by applying Theorem 9 that the system (4.2) is a standard representation of the original system. Second, any standard representation with disjoint matrices can be associated with a compatible cut in the original developed graph (see Lemma 17), and the marking $M^{(H)}$ generates the smallest compatible cut by definition.   □

4.3.4   Computation of H^*

The problem of computing H^* may be stated in the form of a linear program:

$$\{\min \sum_{i \in T} N_i \; : \; M = M_0 + I.x, \; x \geq 0, \; M \geq 0, \; N_i \geq M(p), \; \forall p \in i^\bullet\} \, .$$

Theorem 20 *This linear program is totally unimodular.*

Corollary 21 *The computation of $M^{(H)}$ is polynomial in the size of the net.*

The proof of Theorem 20 is technical and is not given here. A complete proof can be found in [12]. Corollary 21 is an immediate consequence of Theorem 20 using the fact that the basic optimal solution of a linear program whose constraint matrix is totally unimodular is an integer-valued vector [17].

4.3.5   Generalizations

This optimization can be applied to a more general framework, namely the *uniform recurrence equations*, arising for instance in cyclic computation circuits [13].

   Let $\mathcal{G}$ be an oriented graph with T nodes. We attach to each node i a sequence of values $V_i(n)$ to be computed and an operator F_i. If nodes $j_1, \ldots, j_l$ are the predecessors of node i, there exists a sequence $k_{1,i}, \ldots k_{l,i}$ such that $V_i(n)$ is given by

$$V_i(n) = F_i(V_{j_1}(\gamma^{k_{1,i}}(n)), \ldots, V_{j_l}(\gamma^{k_{l,i}}(n))),$$

where γ is a time shift operator, i.e. $\gamma(n) = n - 1$ and all the F_i are γ homogeneous.

We denote by $k_i = \max_j k_{i,j}$ the maximal delay between node i and its successors. If we want to carry out the computation of $V_i(n)$ for $n \in \mathbb{N}$, we need registers for each node to store the values of $V_i(\gamma^{k_i}(n)), \ldots, V_i(\gamma(n)), V_i(n)$.

Since F_j are all γ homogeneous, we can carry out the computation by moving the delays around (which corresponds to changing the initial marking). The optimal marking $M^{(H)}$ corresponds to a position of the delays such that the total number of registers necessary to carry the computation is as small as possible.

Remark If non-positive values for the delays are acceptable, the minimal representation procedure gives an even smaller standard representation of the system. It is minimum among all the possible representations of the system, including those where the transformations required to retrieve the initial values are not restricted to translations as in Definition 8 but can be arbitrary maps. For more on this, see [13].

5   Markovian Analysis of Recursive Systems

In this section, we address the problem of computing stationary or transient performance measures for dynamical systems described by stochastic recurrences.

The class of systems we have particularly in mind is that of stochastic $(\max, +)$ systems, with evolution equations $Y(n+1) = A(n) \otimes Y(n)$. Indeed, as mentioned in the introduction, this class of equations models stochastic event graphs, and certain classes of queuing networks. However, it appears that some techniques initially applied to $(\max, +)$ systems have a wider range of applications.

Statistics of interest for such systems are the asymptotic growth rate of the sequence $\mathbf{E}Y(n)$ (sometimes called the Lyapunov exponent, cycle time, or first order statistic [4] of the system), and limit behavior of the increments $Y_i(n) - Y_j(n - m)$ (second order statistics), to quote the most basic ones.

The purpose of this section is to show how, under natural assumptions on the random driving sequences, one may effectively compute these statistics.

Most of the results stated below will be left without proof, as they are either "well known" in the literature, or straightforward to derive using standard arguments.

All the methods discussed below are ultimately based on the existence of an underlying Markov chain. One may however distinguish two different classes of Markovian models, which differ by the state space they use. The first approach (§5.1) uses as a state space a direct representation of the system. The second (§5.2) uses as a state space the instants at which events occur.

5.1 Markov chains based on system states

We briefly discuss here the most usual way to handle performance problems in random DEDS. It consists in considering that the system evolves in a Markovian way on a state space which describes it. Typically, this state will be the number of customers (in a queuing network interpretation) or tokens (in a Petri net interpretation) of different types in different locations of the system, plus possibly additional variables keeping track of the "phase" of the random variables driving the system. This modeling technique is described in numerous references: for instance [20] for queues, and [1] for Petri nets.

Among the drawbacks attached to this method is the constraint that the distributions of the random variables involved must have a Coxian law, at best, or an exponential law, more usually.

The method presented in the next paragraph somewhat relaxes this restriction.

5.2 Markov chains based on evolution equations

This second approach uses the description of the system which uses recurrences on the instants at which events happen in the system. It was used by Resing et al. in [24] for solving (max, +) systems. To illustrate this approach, consider the linear (max, +) system of order 1 and dimension 2:

$$\begin{cases} Y_1(n+1) &=& \max\{a_{11}(n) + Y_1(n), a_{12}(n) + Y_2(n)\} \\ Y_2(n+1) &=& \max\{a_{21}(n) + Y_1(n), a_{22}(n) + Y_2(n)\} \, , \end{cases} \tag{5.1}$$

starting from some $(Y_1(0), Y_2(0))$. Introducing the variable $Z(n) = Y_2(n) - Y_1(n)$, we obtain:

$$\begin{cases} Z(n+1) &=& \max\{a_{21}(n), a_{22}(n) + Z(n)\} - \max\{a_{11}(n), a_{12}(n) + Z(n)\} \\ Y_1(n+1) &=& Y_1(n) + \max\{a_{11}(n), a_{12}(n) + Z(n)\} \, , \\ Y_2(n+1) &=& Y_2(n) + \max\{a_{21}(n) - Z(n), a_{22}(n)\} \, . \end{cases}$$

Assuming that the sequence $\{a_{ij}(n), i, j = 1, 2, n \in \mathbb{N}\}$ is i.i.d., it is clear that $\mathcal{Z} = \{Z(n), n \in \mathbb{N}\}$ and $\{(Y_1(n), Y_2(n), Z_1(n)), n \in \mathbb{N}\}$ are homogeneous Markov chains. If $\mathcal{Z}$ admits a stationary distribution π, one readily obtains stationary performance measures in terms of π. Of course, the process $\{(Y_1(n), Y_2(n)), n \in \mathbb{N}\}$ is itself a Markov chain, but this chain is transient, and the introduction of $Z(n)$ makes it clear that its growth depends on the behavior of another chain which, in most cases, turns out to have stationary regime. The proper structure to describe this dependency is the class of *Markov Additive Processes* (MAP), discussed in particular in [7, 8]. Roughly speaking, a MAP is a Markov process with two components, such that the increments of the second component are independent given the values of the first component.

The example above is a particular case of the following theorem. Let $\mathbf{1}$ be the vector $(1, \ldots, 1)^t \in \mathbb{R}^m$.

Theorem 22 *Consider the stochastic recurrence*

$$X(n) = \Phi(X(n-1), \xi(n)), \qquad n \geq 1$$

with $\{X(n), n \in \mathbb{N}\}$ *a sequence of* $\mathbb{R}^m$, $\{\xi(n), n \in \mathbb{N}\}$ *a sequence of i.i.d. random variables of* $\mathbb{R}^p$, *and* $\Phi : \mathbb{R}^m \times \mathbb{R}^p \to \mathbb{R}^m$ *a fixed operator such that*

$$\forall (x, y) \in \mathbb{R}^m \times \mathbb{R}^p, \; \forall \lambda \in \mathbb{R}, \quad \Phi(x + \lambda \mathbf{1}, y) = \Phi(x, y) + \lambda \mathbf{1} \; .$$

Then, if $Z(n) = (X_2(n) - X_1(n), \ldots, X_m(n) - X_1(n))$,

(i) *the process* $\{Z(n), n \in \mathbb{N}\}$ *is a homogeneous Markov chain on* $\mathbb{R}^{m-1}$,

(ii) *the processes* $\{(X_i(n), Z(n)), n \in \mathbb{N}\}$ *are discrete time MAPs.*

The proof is immediate, using the homogeneity of the operator with respect to x.

The practical consequences of this theorem follow from the next property. For the sake of simplicity and numerical tractability, we restrict ourselves to the case of discrete random variables.

Theorem 23 *Let* $\{(X(n), Z(n)), n \in \mathbb{N}\}$ *be a discrete time MAP on* $\mathbb{N} \times \mathbb{N}^p$. *Let* $M(z)$ *be the matrix*

$$M(z)_{ij} = \mathbf{E}(z^{X(n+1)-X(n)} 1_{\{Z(n)=j\}} \mid Z(n-1) = i) , \quad i, j \in \mathbb{N}^p, |z| = 1. \tag{5.2}$$

Then the joint generating function of the $X(n)$ *is given by*

$$\mathbf{E}\left(\prod_{n=0}^{N} s_n^{X(n)}\right) = \pi_0 \, M(s_1 \ldots s_N) \, M(s_2 \ldots s_N) \, \ldots \, M(s_{N-1} s_N) \, M(s_N) \, \mathbf{1}, \tag{5.3}$$

where $\pi_0^{(k)} = \mathbf{P}(Z(0) = k)$. *Alternately, let, for all* $i \geq 1$, $d(i) = X(i) - X(i-1)$. *The joint generating function of the* $d(i)$ *is given by*

$$\mathbf{E}\left(\prod_{n=1}^{N} s_n^{d(n)}\right) = \pi_0 \, M(s_1) \, \ldots \, M(s_N) \, \mathbf{1} \; . \tag{5.4}$$

The extension of this theorem when the component X is continuous and multidimensional is immediate; the corresponding definition of M involves Laplace transforms.

The central role played by the matrix $M(z)$ in the transient and asymptotic properties of the chain $X(n)$ has been known in the literature for a long time. See [25] for instance. Notice in particular that for z real and positive, $M(z)$ (where it is defined) satisfies the assumptions of the Perron–Frobenius theorem whenever $M(1)$ (which is the transition matrix of the chain $\mathcal{Z}$) does. This happens in particular when the state space of $\mathcal{Z}$ is finite and the chain satisfies the usual ergodicity conditions. In this case, we shall write: $\mathcal{Z}$ is PF.

Among the corollaries of Theorem 23, we have:

Corollary 24 *The generating function of the distribution of $X(n)$ is given by*

$$\mathbf{E}(s^{X(n)}) = \pi_0 M(s)^n \mathbf{1} .$$

Let $\gamma = \lim_{n \to \infty} \mathbf{E}X(n)/n$ be the asymptotic growth rate of the chain. We have

(i) *If Z is stationary and ergodic with stationary distribution π,*

$$\gamma = \mathbf{E}_\pi X(1) = \pi M'(1)\mathbf{1} . \tag{5.5}$$

(ii) *Assume $\mathcal{Z}$ is PF. Let $\lambda_1(z)$ be the Perron–Frobenius eigenvalue of the matrix $M(z)$, and $P(z, x) = \det(M(z) - xI)$ be the characteristic polynomial of $M(z)$. Then*

$$\gamma = \frac{d\lambda_1}{dz}(1) = -\frac{\partial P}{\partial z}(1, 1) \Big/ \frac{\partial P}{\partial x}(1, 1) . \tag{5.6}$$

The theorems above apply to stochastic recurrences governed by operators which are "time homogeneous" in their variables. Systems whose evolution is governed by $(\max, +)$, $(\min, +)$, $(\min, \max, +)$ recurrences fall in this category.

It is therefore possible to compute the first order measures of these systems using either the stationary distribution with (5.5) or the characteristic polynomial (actually, only partial derivatives of it) with (5.6).

Theorem 23 also provides a way to address the computation of second order quantities, such as the limit behavior of the distribution of $d(n) = Y(n + 1) - Y(n)$ and its moments, as well as autocorrelations of the process $\{d(n), n \in \mathbb{N}\}$ and the central limit behavior of $X(n)$. See [18] for details.

5.3   Examples

5.3.1   A two wimensional $(\max, +)$ system

This example is taken from [18], where it is solved using a different, but strongly related technique.

Consider the recurrence (5.1), where the driving sequences $a_{ij}(n)$ are independent, i.i.d. sequences of Bernoulli variables with respective parameters a, b, c, d.

As noted above, the process $\{(Y_1(n), Z(n)), n \in \mathbb{N}\}$ is a MAP. The chain $\mathcal{Z}$ turns out to live on the state space $\{-1, 0, 1\}$. For notational convenience, introduce $\bar{a} = 1 - a$, $\bar{b} = 1 - b$, $\bar{c} = 1 - c$ and $\bar{d} = 1 - d$. The matrix $M(z)$ is

$$M(z) = \begin{pmatrix} a\bar{c}z & acz + \bar{a}\bar{c} & \bar{a}c \\ (1 - \bar{a}\bar{b})\bar{c}\bar{d}z & (1 - \bar{a}\bar{b})(1 - \bar{c}\bar{d})z + \bar{a}\bar{b}\bar{c}\bar{d} & \bar{a}\bar{b}(1 - \bar{c}\bar{d}) \\ b\bar{d}z^2 & bdz^2 + \bar{b}\bar{d}z & \bar{b}dz \end{pmatrix},$$

from which γ may be computed.

5.3.2   A two dimensional $(\max, \min, +)$ system

This example is taken from [19]. Consider the recurrence:

$$\begin{aligned} Y_1(n+1) &= \max\{a_{11}(n) + Y_1(n), a_{12}(n) + Y_2(n)\} \\ Y_2(n+1) &= \min\{a_{21}(n) + Y_1(n), a_{22}(n) + Y_2(n)\}, \end{aligned}$$

where the driving sequences $a_{ij}(n)$ are independent, i.i.d. sequences of Bernoulli variables with respective parameters a, b, c, d. As above, let $\bar{a} = 1 - a$, $\bar{b} = 1 - b$, $\bar{c} = 1 - c$ and $\bar{d} = 1 - d$.

Unlike the example of §5.3.1, the chain $\mathcal{Z}$ is not bounded, despite the boundedness of the driving random variables. It turns out that only the states $\{1, 0, -1, -2, \ldots\}$ can be recurrent. Fortunately, the regular behavior of the "tail" of the chain allows us to solve for the stationary distribution, and therefore to use equation (5.5).

The probability transition matrix restricted to these states (written down in that order) is:

$$\begin{pmatrix} 0 & b\bar{c} & bc + \bar{b}\bar{c} & \bar{b}c & 0 & \cdots \\ \bar{a}\bar{b}\bar{c}d & \bar{a}c(1 - bd) & \begin{matrix}(1 - \bar{a}\bar{c})(1 - bd) \\ +(1 - \bar{a}\bar{c})bd\end{matrix} & 0 & 0 & \cdots \\ 0 & \bar{a}d & ad + \bar{a}\bar{d} & a\bar{d} & 0 & \cdots \\ 0 & 0 & \bar{a}d & ad + \bar{a}\bar{d} & a\bar{d} & \\ 0 & 0 & 0 & \bar{a}d & ad + \bar{a}\bar{d} & \ddots \\ \vdots & & & & \ddots & \ddots & \ddots \end{pmatrix}. \quad (5.7)$$

The tail of this Markov chain is therefore a birth and death process, and the solution of the equilibrium equations for the stationary distribution π gives

$$\pi(-n) = \pi(-2)\left(\frac{a\bar{d}}{\bar{a}d}\right)^{-n+2}, \quad n \geq 2,$$

as well as the ergodicity condition of the chain: $a(1 - d) < d(1 - a)$, or $a < d$. In particular, $\pi(-3) = a\bar{d}\pi(-2)/\bar{a}d$. The equilibrium equations for π

on the states $\{-2, -1, 0, 1\}$ still have to be solved, and the vector π has to be normalized.

Once this is done, the value of γ is obtained from (5.5). We omit the detailed calculations, which are easily performed symbolically with a computer algebra system.

An interesting point here is that if $a \geq d$, the chain $\mathcal{Z}$ does not have a stationary distribution. Nevertheless, asymptotic growth rates exist for both Y_1 ($\gamma_1 = a$) and Y_2 ($\gamma_2 = d$).

5.4   Conclusions and perspectives

The analysis by means of Markov chains based on the evolution equations provides an alternative way to analyze systems, which allows a complete analysis in some cases where the "standard" approach would fail or require serious complications. It has the further advantage of providing a deep insight into both transient and asymptotic properties.

On the other hand, the technique allows actual computations in a small set of cases.

When the driving random variables have a continuous distribution, the computation goes through with difficulties in only some limited cases [24, 18].

In discrete cases, the combinatorial explosion and the "curse of dimensionality" limits the analysis. In this respect, the optimization technique developed in §4 is likely to prove essential to keep problems numerically tractable.

This approach opens areas for further research in several directions. First, it is necessary to assess the finiteness of the chain, and its ergodicity. In the case of (max, +) systems, the general stability theorems of [4] provide the answers, but for other systems, much remains to be done. Conjectures for (max, min, +) systems can be found in [19].

On the computational side, the use of this approach for solving problems of reasonable size and complexity requires further insight into the structure of the Markov chains involved.

References

[1] M. Ajmone Marsan, G. Balbo, and G. Conte. *Performance Models of Multiprocessor Systems*. MIT Press, Cambridge, Mass., 1986.

[2] F. Baccelli. Ergodic theory of stochastic Petri networks. *Annals of Probability*, 20(1):375–396, 1992.

[3] F. Baccelli and M. Canales. Parallel simulation of stochastic Petri nets. *ACM Transactions on Modeling and Computer Simulation*, 3(1), January 1993.

[4] F. Baccelli, G. Cohen, G.-J. Olsder, and J.-P. Quadrat. *Synchronization and Linearity*. John Wiley & Sons, New York, 1992.

[5] G.E. Blelloch. *Synthesis of Parallel Algorithms* (J.H. Reif, ed.), chapter Prefix Sums and Their Applications. Morgan Kaufmann Publishers, 1993.

[6] P. Chretienne. *Les Réseaux de Petri Temporisés*. PhD thesis, Université Paris VI, Paris, 1983.

[7] E. Çinlar. Markov additive processes I. *Z. Wahrsheinlichktheorie verw. Geb.*, 24:85–93, 1972.

[8] E. Çinlar. Markov additive processes II. *Z. Wahrsheinlichktheorie verw. Geb.*, 24:95–121, 1972.

[9] G. Cohen, D. Dubois, J.-P. Quadrat, and M. Viot. A linear system-theoretic view of discrete-event processes and its use for performance evaluation in manufacturing. *IEEE Trans. Automatic Control*, AC-30:210–220, 1985.

[10] S. Gaubert. Performance evaluation of $(\max, +)$ automata. INRIA Report No. 1922, May 1993. To appear in *IEEE Trans. Automatic Control*, 1995.

[11] B. Gaujal. *Parallélisme et simulation des systèmes à événements discrets*. PhD thesis, University of Nice Sophia Antipolis, June 1994.

[12] B. Gaujal. Initial marking optimization for parallel simulation of marked graphs. Technical report, INRIA, 1995.

[13] B. Gaujal, A. Jean-Marie, and J. Mairesse. Minimal representation of uniform recurrence equations. INRIA Report No 2568, june 1995.

[14] M. Gondran and M. Minoux. *Graphs and Algorithms*. John Wiley & Sons, 1986.

[15] J. Gunawardena. A dynamic approach to timed behaviour. In B. Jonsson and J. Parrow, editors, *CONCUR'94*, number 836 in LNCS, 1994.

[16] H. Hillion and J.M. Proth. Performance evaluation of job shop systems using timed event graphs. *IEEE Trans. Automatic Control*, 34(1):3–9, 1989.

[17] A.J. Hoffman and J.B. Kruskal. Integral boudary points of convex poly-hedra. *Annals of Math. Studies, Princeton*, 38:223, 1956.

[18] A. Jean-Marie. Analytical computation of Lyapunov exponents in stochastic event graphs. In O. Boxma and G. Koole, editors, *Third QMIPS Workshop*. CWI Tracts, 1993.

[19] A. Jean-Marie and G. J. Olsder. Analysis of stochastic min-max systems: results and conjectures. Technical Report 93-94, Delft University of Technology, 1993.

[20] L. Kleinrock. *Queueing Systems, Vol I: Theory*. John Wiley and Sons, 1975.

[21] C.E. Leiserson and J.B. Saxe. Retiming synchronous circuitry. *Algorithmica*, 6:5–35, 1991.

[22] J. Mairesse. Products of irreducible random matrices in the (max,+) algebra, part I. Report 1939, INRIA, Sophia Antipolis, May 1993.

[23] T. Murata. Petri nets: Properties, analysis and applications. *Proceedings of the IEEE*, 77(4):541–580, 1989.

[24] J. Resing, R.E. de Vries, M.S. Keane, G. Hooghiemstra, and G.J. Olsder. Asymptotic behavior of random discrete event systems. *Stochastic Processes and their Applications*, 36:195–216, 1990.

[25] I. S. Volkov. On the distribution of sums of random variables defined on a homogeneous Markov chain with a finite number of states. *SIAM J. Theory Prob. Applications*, 3:384–399, 1958.

Periodic Points of Nonexpansive Maps

Roger D. Nussbaum[1]

1 Introduction

If M is a topological space and $f : M \rightarrow M$ is a continuous map, it is frequently of interest to understand, for $x \in M$, the behaviour of iterates $f^k(x)$ as k approaches infinity. In this generality little can be said; but if f is "nonexpansive with respect to a metric ρ on M", i.e. if $\rho(f(x), f(y)) \leq \rho(x, y)$ for all $x, y \in M$, then a variety of useful theorems, some of recent vintage, can be brought into play. Surprisingly, this simple and powerful observation has often been missed in studying specific classes of nonlinear operators.

In this note we shall state some theorems about nonexpansive maps, describe some conjectures and open questions, and indicate some specific classes of maps for which these observations prove useful.

2 Examples of Maps of Interest

We begin by recalling some terminology and basic facts. Let S denote a compact Hausdorff space, $C(S) = X$, the Banach space of continuous functions on S in the sup norm $\|x\|_\infty = \sup\{|x(s)| \mid s \in S\}$, K the cone of nonnegative functions on S and $\overset{\circ}{K}$ the interior of S. If S is the set of integers i with $1 \leq i \leq n$, we identify $C(S)$ with $\mathbf{R}^n$ in the obvious way and write $K := K^n$, so $K^n := \{x \in \mathbf{R}^n \mid x_i \geq 0 \text{ for } 1 \leq i \leq n\}$. We shall denote by $u \in C(S)$ the constant map defined by $u(s) = 1$ for all $s \in S$. If $S = \{i \in \mathbf{Z} \mid 1 \leq i \leq n\}$, we have $u \in \mathbf{R}^n$ and $u_i = 1$ for $1 \leq i \leq n$.

The set $K \subset C(S)$ induces a partial ordering by $x \leq y$ if and only if $y - x \in K$. More generally, if C is a closed, convex subset of a Banach space Y, C is called a cone (with vertex at 0) if (a) $C \cap (-C) = \{0\}$, where $(-C) = \{-x \mid x \in C\}$ and (b) $tC \subset C$ for all $t \geq 0$, where $tC = \{tx \mid x \in C\}$. A cone C induces a partial ordering on Y by $x \leq y$ if and only if $y - x \in C$. If D is a subset of Y, a map $f : D \rightarrow Y$ is called "order-preserving" if for all $x, y \in D$ with $x \leq y$ one has $f(x) \leq f(y)$. Two elements x and y in C are called "comparable" if there exist positive real numbers r and s such that $rx \leq y$ and $sy \leq x$.

[1]Partially supported by NSF-DMS 9401823

Comparability defines an equivalence relation on C. If $e \in C$, we shall denote by C_e the set of elements of C which are comparable to e. If $\overset{\circ}{C}$ is nonempty and $e \in \overset{\circ}{C}$, then $\overset{\circ}{C} = C_e$. If x and y are nonzero comparable elements of C, we follow Bushell's notation [8] and define $m(y/x)$ and $M(y/x)$ by $m(y/x) := \alpha := \sup\{r > 0 \mid rx \leq y\}$ and $M(y/x) := \beta := \inf\{t > 0 \mid y \leq tx\}$. If $x, y \in C_e$ ($e \neq 0$), we define numbers $\bar{d}(x, y)$ and $d(x, y)$ by $\bar{d}(x, y) = \log(\max\{\alpha^{-1}, \beta\})$ and $d(x, y) = \log(\frac{\beta}{\alpha})$. The function $\bar{d}$ is called "Thompson's metric" or the "part metric" on C_e, and d is called Hilbert's projective metric. It is known that $\bar{d}$ is a metric on C_e, and $(C_e, \bar{d})$ is a complete metric space if C is "normal". (Recall that C is called "normal" if there exists a constant A such that $\|x\| \leq A\|y\|$ whenever $0 \leq x \leq y$. Note that C is necessarily "normal" if Y is finite dimensional.) Similarly, it is known that d satisfies all properties of a metric on C_e except that $d(x, y) = 0$ if and only if $y = tx$ for some $t > 0$.

If $\sum_e$ denotes the intersection of C_e with a hyperplane and if C is normal, then $(\sum_e, d)$ is a complete metric space. Proofs, further details, and references to the literature can be found in Chapter 1 of [26].

Returning to the case $X = C(S)$, define a map $\Phi : X \to \overset{\circ}{K}$ by $(\Phi(x))(s) = \exp(x(s))$, so $\Phi^{-1} : \overset{\circ}{K} \to C(S)$ is given by $(\Phi^{-1}y)(s) = \log(y(s))$. Define a seminorm ω on X by $\omega(x) = \sup\{x(s) \mid s \in S\} - \inf\{x(s) \mid s \in S\}$.

Then one can check that for all $x, y \in C(S)$,

$$\|x - y\|_\infty = \bar{d}(\Phi(x), \Phi(y)) \text{ and } \omega(x - y) = d(\Phi(x), \Phi(y)). \qquad (2.1)$$

Further details can be found in Chapter 1 of [26] and in [33].

If D is a subset of $C(S)$ and $f : D \to C(S)$ is a map, (2.1) implies that $f : D \to C(S)$ is nonexpansive with respect to the metric given by the sup norm if and only if $\Phi f \Phi^{-1} : \Phi(D) \to \overset{\circ}{K}$ is nonexpansive with respect to the part metric $\bar{d}$. Analogous statements hold for ω and d. Furthermore, because Φ and Φ^{-1} are order-preserving, f is order-preserving if and only if $\Phi f \Phi^{-1}$ is order-preserving.

The following theorem, which is a special case of Proposition 2 in [11], is sometimes useful in proving that maps are nonexpansive with respect to the sup norm.

Theorem 2.1 (Crandall and Tartar [11]) *Let S be a compact Hausdorff space and $C(S)$ the Banach space of continuous real-valued functions on S in the sup norm. Let $u \in C(S)$ denote the function which is identically equal to one and let $f : C(S) \to C(S)$ satisfy*

$$f(x + tu) = f(x) + tu \quad \text{for all } x \in C(S) \text{ and } t \in \mathbf{R}. \qquad (2.2)$$

Then f is nonexpansive in the sup norm if and only if f is order-preserving.

We now give some examples of maps of interest. Let S be a compact Hausdorff space and let $a : S \times S \to \mathbf{R}$ be a continuous map. Define maps F_a and G_a of $C(S)$ to $C(S)$ by

$$(F_a x)(s) = \min\{a(s,t) + x(t) \mid t \in S\} \tag{2.3}$$
$$(G_a x)(s) = \max\{a(s,t) + x(t) \mid t \in S\}. \tag{2.4}$$

The map F_a arises in statistical mechanics; see [9,14,15,16,30]. When $S = \{i \in \mathbf{Z} \mid 1 \leq i \leq n\}$, maps like F_a and G_a arise in a variety of contexts, for example, in machine scheduling; see [3,12,13,17,18,19]. If f denotes either F_a or G_a, one can easily see that f is order-preserving and satisfies equation (2.2). Thus Theorem 1.2 implies that F_a and G_a are nonexpansive with respect to the sup norm.

If C is a cone in a Banach space Y, $D \subset C$ satisfies $tD \subset D$ for all $t > 0$, and $g : D \to C$ is a map, g is called homogeneous of degree one if $g(tx) = tg(x)$ for all $x \in D$ and $t > 0$. If $f : C(X) \to C(S)$, note that f satisfies (2.2) if and only if $g = \Phi f \Phi^{-1} \colon \overset{\circ}{K} \to \overset{\circ}{K}$ is homogeneous of degree one. If $f : C(S) \to C(S)$, we follow [9] and say that $v \in C(S)$ is an "additive eigenvector of f" if there exists $\alpha \in \mathbf{R}$ with $f(v) = v + \alpha u$. If C is a cone, D is a subset of C and $g : D \to C$ is a map, we shall say that $w \in D$ is an eigenvector of g if $w \neq 0$ and there exists $\lambda \geq 0$ with $g(w) = \lambda w$. It may happen that $g : C \to C$ has an eigenvector in C but no eigenvector in $\overset{\circ}{C}$. A central and frequently difficult question is whether a map $f : C(S) \to C(S)$ has an additive eigenvector or whether a map g defined on the interior $\overset{\circ}{C}$ of a cone C has an eigenvector in $\overset{\circ}{C}$. We refer, for example, to certain difficult questions about "rescaling" nonnegative integral kernels (so-called DAD theorems). These questions are equivalent to the existence of an eigenvector in the interior of a cone of an associated nonlinear operator: see [6], [32], Chapter 4 of [27] and references there.

In general suppose that $f : C(S) \to C(S)$ is order-preserving and satisfies (2.2). It may happen that $g = \Phi f \Phi^{-1} : \overset{\circ}{K} \to \overset{\circ}{K}$ extends continuously to K and is compact on K and that there exists $w \in K - \{0\}$ and $\delta > 0$ with $g(w) \geq \delta w$. It then follows from general results about nonlinear cone mappings (see [22]) that g has an eigenvector in K. If one can prove that necessarily this eigenvector is in $\overset{\circ}{K}$, then g has an eigenvector in $\overset{\circ}{K}$ and f has an additive eigenvector. By applying this approach to F_a and G_a, one can prove that F_a and G_a have additive eigenvectors. A different approach is given in [10].

If $S = \{i \in \mathbf{Z} \mid 1 \leq i \leq n\}$, it is natural to define more general versions of the maps in (2.3) and (2.4). If T is a set of real numbers and $\delta = \pm 1$, define $\mu_\delta(T) = \inf(T)$ if $\delta = -1$ and $\mu_\delta(T) = \sup(T)$ if $\delta = 1$. Let Γ be a subset of $S = \{i \in \mathbf{Z} \mid 1 \leq i \leq n\}$; possibly Γ is empty. For each i, let Γ_i be a nonempty subset of S; and for each $j \in \Gamma_i$, let a_{ij} be a real number. Define

$\delta_i = -1$ if $i \in \Gamma$ and $\delta_i = +1$ if $i \notin \Gamma, i \in S$. Define a map $f : \mathbf{R}^n \to \mathbf{R}^n$ by $f_i(x) = \mu_{\delta_i}(\{a_{ij} + x_j \mid j \in \Gamma_i\})$, where $f_i(x)$ denotes the ith coordinate of $f(x)$. Let $\mathcal{F}_0$ be the set of functions which arise in this way. More generally, if Γ is any subset of S and $g, h : \mathbf{R}^n \to \mathbf{R}^n$, we can define $\delta_i = -1$ if $i \in \Gamma$ and $\delta_i = 1$ if $i \notin \Gamma$ and define a map $k : \mathbf{R}^n \to \mathbf{R}^n$ by

$$k_i(x) = \mu_{\delta_i}(\{f_i(x), g_i(x)\}). \tag{2.5}$$

We define $\mathcal{F}$ to be the smallest collection of maps $f : \mathbf{R}^n \to \mathbf{R}^n$ which (1) contains $\mathcal{F}_0$, (2) is closed under the operation of taking compositions, (3) is closed under all the operations of the form in (2.5) and (4) is closed in the topology of uniform convergence on compact sets for continuous functions $f : \mathbf{R}^n \to \mathbf{R}^n$. If $f \in \mathcal{F}$, it is easy to show that f is order-preserving and satisfies (2.2) and hence (see Theorem 2.1) is nonexpansive in the sup norm. More generally, suppose that for each $j \in \Gamma_i$, ε_{ij} is a number which equals ± 1. Define $g : \mathbf{R}^n \to \mathbf{R}^n$ by

$$g_i(x) = \mu_{\delta_i}(\{a_{ij} + \varepsilon_{ij}x_j \mid j \in \Gamma_i\}) \tag{2.6}$$

where $g_i(x)$ is the ith coordinate of $g(x)$. Let $\mathcal{G}_0$ denote the class of maps of the form given by (2.6). Let $\mathcal{G}$ denote the smallest collection of maps which contains $\mathcal{G}_0$ and satisfies the properties (2), (3) and (4) listed earlier in this paragraph. Elements of $\mathcal{G}$ are not, in general, order-preserving, so Theorem 2.1 does not apply; but by using Proposition 1.2, on page 1658 in [33], one can easily prove that all elements of $\mathcal{G}$ are nonexpansive in the sup norm.

Other classes of maps are of interest in different contexts. We refer to the class of cone mappings $\mathcal{M}$ defined on page 25 of [27]. A slight generalization of $\mathcal{M}$ can be shown to contain all maps of the form $\Phi f \Phi^{-1}, f \in \mathcal{F}$.

3   Periodic Points of Nonexpansive Maps

Recall that the ℓ_1-norm on $\mathbf{R}^n$ is defined by $\|z\|_1 = \sum_{i=1}^n |z_i|$, where z_i is the i-th coordinate of z. More generally, a norm $\|\cdot\|$ on a finite dimensional vector space V is called a "polyhedral norm" if $\{x \mid \|x\| \leq 1\}$ is a polyhedron. A norm $\|\cdot\|$ is polyhedral if there exist continuous linear functionals $\theta_1, \theta_2, \ldots, \theta_N$ on V such that

$$\|x\| = \max\{|\theta_i(x)| \mid 1 \leq i \leq N\}. \tag{3.1}$$

Theorem 3.1 (Akcoglu and Krengel [1]) *Let D be a compact subset of $\mathbf{R}^n$ and $f : D \to D$ a map which is nonexpansive in the ℓ_1-norm. For each $x \in D$, there is a periodic point $\xi_x = \xi$ of f in D and an integer $p_x = p$ such that*

$$\lim_{k \to \infty} f^{kp}(x) = \xi \quad \text{and} \quad f^p(\xi) = \xi \quad \text{and} \quad f^j(\xi) \neq \xi \quad \text{for } 0 < j < p. \tag{3.2}$$

D. Weller [40] proved that Theorem 3.1 holds for any polyhedral norm. R. Sine [39] introduced important other ideas and indicated the central role played by the sup norm. Other related results have been obtained by A. Blokhuis and H.A. Willbrink [5], S.-K. Lo [23], P. Martus [25], R.N. Lyons and R.D. Nussbaum [24], R.D. Nussbaum [28,29,30,31], and M. Scheutzow [37,38]. These results imply the following theorem, in particular.

Theorem 3.2 *Let D be a compact subset of a finite dimensional vector space V with a polyhedral norm $\|\cdot\|$. Assume that $f : D \to D$ is nonexpansive with respect to $\|\cdot\|$. If $x \in D$, there exists an integer $p_x = p$ and a periodic point $\xi_x = \xi \in D$ of f of minimal period p such that (3.2) holds. If the norm $\|\cdot\|$ satisfies (3.1) and N is as in (3.1) then (see [25]) $p_x \leq 2^N N!$. In particular the minimal period of any periodic point $\xi \in D$ of f is less than or equal to $2^N N!$.*

The estimate above is certainly not best possible, even if $V = \mathbf{R}^n$ and the norm is the sup norm.

Conjecture 3.1 *Suppose $D \subset \mathbf{R}^n$ and $f : D \to \mathbf{R}^n$ is nonexpansive with respect to the sup norm. Suppose that $\xi \in D$ is a periodic point of f of minimal period p (so $f^p(\xi) = \xi$ and $f^j(\xi) \neq \xi$ for $0 < j < p$). Then (?) it follows that $p \leq 2^n$.*

Lyons and Nussbaum [24] have proved Conjecture 3.1 for $n = 1, 2$ and 3. The case $n = 3$ is already nontrivial. The case $n = 2$ was also obtained by S.-K. Lo [23]. Some further evidence for Conjecture 3.1 is given in [24], but it is not even known whether the conjecture is true for $n = 4$ or $n = 5$. As was noted on page 525 of [28], by taking D to be a subset of the vertices of the unit cube $Q = \{x \in \mathbf{R}^n \mid \|x\|_\infty \leq 1\}$ and f any cyclic permutation of the elements of D, one obtains a map $f : D \to D$ which is nonexpansive in the sup norm and has a periodic point of period $p = |D|$. Because the unit cube has 2^n vertices, it follows that for every p with $1 \leq p \leq 2^n$, there is a set D and a map $f : D \to D$ which has a periodic point of period p and is nonexpansive in the sup norm.

It may be worth recalling that a result of Aronszajn and Panitchpakdi [2] implies that if $f : D \to \mathbf{R}^n$ is nonexpansive in the sup norm, then there is an extension $F : \mathbf{R}^n \to \mathbf{R}^n$ which is nonexpansive in the sup norm. Thus one can assume that $D = \mathbf{R}^n$ in Conjecture 3.1. Since the Aronszajn–Panitchpakdi theorem is essentially only an existence result, one might assume that examples of maps $F : \mathbf{R}^n \to \mathbf{R}^n$ which are nonexpansive in the sup norm and possess a periodic point of period 2^n are pathological. However, in some unpublished notes the author has constructed natural and simple examples of such maps.

It frequently happens that one deals with maps which are not only nonexpansive with respect to some norm but are also order-preserving or satisfy other properties. Thus one is naturally led to the following questions.

Question 3.1 Let $\mathcal{F}$ denote the class of maps $f : \mathbf{R}^n \to \mathbf{R}^n$ which are order-preserving and nonexpansive with respect to the sup norm, and suppose that $\xi \in \mathbf{R}^n$ is a periodic point of $f \in \mathcal{F}$ of minimal period $p = p(\xi, f)$. Can one describe the set of periods $\{p(\xi, f) \mid f \in \mathcal{F}, \xi$ is a periodic point of $f\}$ precisely? Can one determine the supremum of this set?

By using results in [24], the author has proved that if $n = 2$ and $f : \mathbf{R}^2 \to \mathbf{R}^2$ is order-preserving and nonexpansive in the sup norm, then any periodic point ξ of f has minimal period $p \leq 2$. It seems likely that by using the analysis in [24], one can also answer Question 3.1 for $n = 3$.

Question 3.2 Let $\mathcal{G}$ denote the class of maps $f : \mathbf{R}^n \to \mathbf{R}^n$ which are order-preserving and satisfy (2.2) (and hence are nonexpansive with respect to the sup norm). Suppose that $\xi \in \mathbf{R}^n$ is a periodic point of $f \in \mathcal{G}$ of minimal period $p = p(\xi, f)$. Can one describe precisely the set of periods $\{p(\xi, f) \mid f \in \mathcal{G}, \xi$ is a periodic point of $f\}$? Can one determine the supremum of this set?

One can ask analogous questions for different norms. The ℓ_1-norm is of particular interest. Thus, for each n, let $\tilde{P}(n)$ denote the set of positive integers p such that there exists a map $f : K^n \to K^n$ which satisfies (a) $f(0) = 0$, (b) f is order-preserving, (c) f is nonexpansive with respect to the ℓ_1-norm and (d) f has a periodic point of minimal period p. Let $P^*(n)$ denote the set of positive integers p such that there exists $f : K^n \to K^n$ which satisfies (a), (c) and (d) above.

In [1], Akcoglu and Krengel proved that $\sup\{p \mid p \in \tilde{P}(n)\} \leq n!$, and M. Scheutzow [37] gave the much sharper result that $\sup\{p \mid p \in P^*(n)\} \leq \mathrm{lcm}(\{i \in \mathbf{Z} \mid 1 \leq i \leq n\})$, where $\mathrm{lcm}(S)$ denotes the least common multiple of a set of positive integers S. In [29,31], Nussbaum gave more precise characterizations of the sets $P^*(n)$ and $\tilde{P}(n)$ and explicitly computed $\varphi(n) = \sup\{p \mid p \in P^*(n)\}$ for $1 \leq n \leq 32$: see [29, p. 365] and [31, p. 969]. In very recent work [34,35,36], Nussbaum, Scheutzow and Verduyn Lunel have defined, for each positive integer n, a set of positive integers $Q(n)$ determined by certain arithmetic and combinatorial constraints. It has been proved that $\tilde{P}(n) = P^*(n) = Q(n)$ for all n, and the set $Q(n)$ has been explicitly computed for $1 \leq n \leq 50$. We refer the reader to [34,35,36] for further details.

A norm $\|\cdot\|$ on $\mathbf{R}^n$ is called strictly monotonic if $\|x\| < \|y\|$ for all $x, y \in \mathbf{R}^n$ with $0 \leq x \leq y$ and $x \neq y$. The ℓ_p-norms on $\mathbf{R}^n$ are strictly monotonic for $1 \leq p < \infty$, but the sup norm is not strictly monotonic. The following theorem follows from Proposition 2.1 and Theorem 2.1 in [31] and from results in [34].

Theorem 3.3 *Let $f : K^n \subset \mathbf{R}^n \to K^n$ be a map which is nonexpansive with respect to a strictly monotonic norm $\|\cdot\|$ on $\mathbf{R}^n$. Assume that $f(0) = 0$ and that f is order-preserving. If $x \in K^n$, there exists a periodic point $\xi = \xi_x \in K^n$ of minimal period $p = p_x \in Q(n)$ such that $\lim_{k \to \infty} f^{kp}(x) = \xi$.*

For a given strictly monotonic norm $\| \cdot \|$ on $\mathbf{R}^n$, what can be said about the set of minimal periods p which arise from maps f as in Theorem 3.3?

4   The Structure of the Fixed Point Set

In this section we shall briefly describe some theorems concerning the structure of the fixed point sets of maps which are nonexpansive with respect to a given metric. For simplicity we shall describe these theorems in the finite dimensional case, but infinite dimensional versions are known: see [7] and Chapter 4 of [26]. The following theorem is a variant of Bruck's basic results [7]; but in the case we consider, it seems most easily proved with the aid of a theorem of Baillon, Bruck and Reich [4]. See also Ishikawa's basic result [20] and the appendix in [21]. Recall that if S is a subset of a topological space X, a continuous map $r : X \to X$ is called a retraction of X onto S if $r(x) \in S$ for all $x \in X$ and $r(x) = x$ for all $x \in S$.

Theorem 4.1 (Compare [7]) *Let* $f : \mathbf{R}^n \to \mathbf{R}^n$ *be nonexpansive with respect to a norm* $\| \cdot \|$ *on* $\mathbf{R}^n$*, and assume that* $S = \{x \in \mathbf{R}^n \mid f(x) = x\}$ *is nonempty. Then there is a nonexpansive retraction* r *of* $\mathbf{R}^n$ *onto* S*. If* f *is order-preserving, the retraction* r *can also be chosen to be order-preserving; and if* f *is order-preserving and satisfies (2.2), the nonexpansive retraction* r *can also be chosen to be order-preserving and to satisfy (2.2).*

Proof Consider the initial value problem

$$x'(t) = f(x(t)) - x(t), \quad x(0) = x_0. \tag{4.1}$$

It follows from [4] that this has a unique solution $x(t; x_0)$ defined for all $t \geq 0$. If we define $T(t) : \mathbf{R}^n \to \mathbf{R}^n$ by $T(t)(x_0) = x(t; x_0)$, we obtain from [4] that $T(t)$ is nonexpansive and that $\lim_{t \to \infty} T(t)(x_0) := r(x_0) \in S$ exists. The map r is a nonexpansive retraction of $\mathbf{R}^n$ onto S. If f is order-preserving, standard results about monotonic flows imply that $T(t)$ is order-preserving, so r is order-preserving. If f satisfies (2.2), one easily shows that $T(t)$ satisfies (2.2) and hence r satisfies (2.2).

Our next theorem is a special case of Theorem 4.7 on p. 128 of [26].

Theorem 4.2 *Let* K *be a cone with nonempty interior* $\overset{\circ}{K}$ *in a finite dimensional Banach space* X*. Assume that* $f : \overset{\circ}{K} \to \overset{\circ}{K}$ *is nonexpansive with respect to the part metric* $\bar{d}$ *and that* $S = \{x \in \overset{\circ}{K} \mid f(x) = x\}$ *is nonempty. Then there exists a retraction* r *of* $\overset{\circ}{K}$ *onto* S *such that* r *is nonexpansive with respect to* $\bar{d}$*. If* f *is order-preserving and homogeneous of degree one, the nonexpansive retraction* r *can also be chosen to be order-preserving and homogeneous of degree one.*

One reason that results like Theorem 4.1 and 4.2 are of interest is that the fixed point set S can be quite complex and very far from convex.

Example 4.1. Define $f : \mathbf{R}^n \to \mathbf{R}^n$ by

$$f_i(x) = \min\{\max(x_i, x_j) \mid 1 \le j \le n, j \ne i\}.$$

One can see that f is order-preserving and satisfies (2.2) and 0 is in the fixed point set S of f. Thus f is nonexpansive in the sup norm and there is an order-preserving, nonexpansive retraction r of $\mathbf{R}^n$ onto S and r satisfies (2.2). We leave to the reader the exercise of proving that

$$S = \{x \in \mathbf{R}^n \mid f(x) = x\} = \{x \in \mathbf{R}^n \mid \text{there exist integers } i \text{ and } l$$
$$\text{with } 1 \le i < l \le n \text{ and } x_i = x_l = \min\{x_j \mid 1 \le j \le n\}\}.$$

The reader can also verify that $f(x) \in S$ for all $x \in \mathbf{R}^n$, so f itself is the desired retraction.

The aforementioned explicit example (see Section 3) of a map of $\mathbf{R}^n$ to $\mathbf{R}^n$ which is nonexpansive in the sup norm and has a periodic point of period 2^n involves the construction of a nonexpansive retraction onto a fixed point set which is more complicated than that in Example 4.1.

References

[1] M.A. Akcoglu and U. Krengel (1987), 'Nonlinear models of diffusion on a finite space', *Prob. Theor. Rel. Fields*, **76** 411–420.

[2] N. Aronszajn and P. Panitchpakdi (1956), 'Extension of uniformly continuous transformations and hyperconvex metric spaces', *Pacific J. Math.*, **6** 405–439.

[3] F.L. Baccelli, G. Cohen, G.J. Olsder and J.-P. Quadrat (1992), *Synchronization and Linearity, an Algebra for Discrete Event Systems*, John Wiley and Sons, New York.

[4] J.B. Baillon, R.E. Bruck and S. Reich (1978), 'On the asymptotic behaviour of nonexpansive mappings and semigroups in Banach spaces', *Houston J. Math.*, **4** 1–9.

[5] A. Blokhuis and H.A. Willbrink (1992), 'Alternative proof of Sine's theorem on the size of a regular polygon in $\mathbf{R}^n$ with the ℓ_∞-metric', *Discrete and Comp. Geometry*, **7** 433–434.

[6] J.M. Borwein, A.S. Lewis and R.D. Nussbaum (1994), 'Entropy minimization, DAD problems, and doubly stochastic kernels', *J. Functional Analysis*, **123** 264–307.

[7] R. Bruck (1973), 'Properties of fixed point sets of nonexpansive mappings in Banach spaces', *Trans. Amer. Math. Soc.*, **179** 251–262.

[8] P. Bushell (1973), 'Hilbert's metric and positive contraction mappings in Banach space', *Arch. Rat. Mech. Anal.*, **52** 330–338.

[9] W. Chou and R.J. Duffin (1987), 'An additive eigenvalue problem of physics related to linear programming', *Adv. Appl. Math.*, **8** 486–498.

[10] W. Chou and R.B. Griffiths (1986), 'Ground states of one-dimensional systems using effective potentials', *Phys. Rev. B*, **34** 6219–6234.

[11] M.G. Crandall and L. Tartar (1980), 'Some relations between non-expansive and order-preserving mappings', *Proc. Amer. Math. Soc.*, **78** 385–390.

[12] R. Cuninghame-Green (1962), 'Describing industrial processes with interference and approximating their steady-state behaviour', *Op. Res. Quart.*, **13** 95–100.

[13] R. Cuninghame-Green (1979), *Minimax Algebra*, Lecture Notes in Economics and Math. Systems, vol. 166, Springer Verlag, Berlin.

[14] L.M. Floria and R.B. Griffiths (1989), 'Numerical procedure for solving a minimization eigenvalue problem', *Numerische Math.*, **55** 565–574.

[15] R.B. Griffiths and W. Chou (1986), 'Effective potentials, a new approach and new results for one-dimensional systems with competing length scales', *Phys. Rev. Lett.*, **56** 1929–1931.

[16] R.B. Griffiths (1990), *Frenkel–Kontorova models of commensurate-incommensurate phase transitions*, in Fundamental Problems in Statistical Mechanics VII, H. van Beijeren (editor), Elsevier Science Publishers B.V., 69–110.

[17] Jeremy Gunawardena, 'Cycle times and fixed points of min-max functions', preprint.

[18] Jeremy Gunawardena, 'Min-max functions', to appear in *Discrete Event Dynamic Systems*.

[19] Jeremy Gunawardena, 'Timing analysis of digital circuits and the theory of min-max functions', Hewlett Packard Laboratories Technical Report, April 1994.

[20] S. Ishikawa (1976), 'Fixed points and iteration of a nonexpansive mapping in a Banach space', *Proc. Amer. Math. Soc.*, **59** 65–71.

[21] S. Ishikawa and R.D. Nussbaum (1991), 'Some remarks on differential equations of quadratic type', *J. Dynamics and Differential Equations*, **3** 457–490.

[22] M.G. Krein and M.A. Rutman, 'Linear operators leaving invariant a cone in a Banach space', *Uspekhi Mat Nauk 3(1)*, **23**; English translation in *Amer. Math. Soc. Transl.*, No. 26.

[23] Shih-Kung Lo (1989), 'Estimates for rigid sets in $\mathbf{R}^m$ with ℓ_∞ or polyhedral norm', Diplomarbeit, Univ. Göttingen (in German).

[24] R.N. Lyons and R.D. Nussbaum (1992), 'On transitive and commutative finite groups of isometries', *Fixed Point Theory and Applications*, World Scientific, 189-228.

[25] P. Martus (1989) 'Asymptotic properties of nonstationary operator sequences in the nonlinear case', Ph.D. dissertation, Erlangen University, Germany, November 1989 (in German).

[26] R.D. Nussbaum (1988), 'Hilbert's projective metric and iterated nonlinear maps', *Memoirs of the Amer. Math. Soc.*, No. 391.

[27] R.D. Nussbaum (1989), 'Hilbert's projective metric and iterated nonlinear maps, II', *Memoirs of the Amer. Math. Soc.*, No. 401.

[28] R.D. Nussbaum (1990), 'Omega limit sets of nonexpansive maps: finiteness and cardinality estimates', *Differential and Integral Equations*, **3** 523-540.

[29] R.D. Nussbaum (1991), 'Estimates of the periods of periodic points for nonexpansive operators', *Israel J. Math.*, **76** 345-380.

[30] R.D. Nussbaum (1991), 'Convergence of iterates of a nonlinear operator arising in statistical mechanics', *Nonlinearity*, **4** 1223-1240.

[31] R.D. Nussbaum (1994), 'Lattice isomorphisms and iterates of nonexpansive maps', *Nonlinear Analysis, T.M. and A.*, **22** 945-970.

[32] R.D. Nussbaum (1993), 'Entropy minimization, Hilbert's projective metric and scaling integral kernels', *J. Functional Analysis*, **115** 45-99.

[33] R.D. Nussbaum (1994), 'Finsler structures for the part metric and Hilbert's projective metric and applications to ordinary differential equations', *Differential and Integral Equations*, **7** 1649-1707.

[34] R.D. Nussbaum and M. Scheutzow, 'Admissible arrays and periodic points of nonexpansive maps', submitted for publication.

[35] R.D. Nussbaum and S. Verduyn Lunel, in preparation.

[36] R.D. Nussbaum, M. Scheutzow and S. Verduyn Lunel, in preparation.

[37] M. Scheutzow (1988), 'Periods of nonexpansive operators on finite ℓ_1-spaces', *European J. Combinatorics,* **9** 73-81.

[38] M. Scheutzow (1991), 'Corrections to periods of nonexpansive operators on finite ℓ_1-spaces', *European J. Combinatorics,* **12** 183.

[39] R. Sine (1990), 'A nonlinear Perron–Frobenius theorem', *Proc. Amer. Math. Soc.,* **109** 331-336.

[40] D. Weller, 'Hilbert's metric, part metric and self-mappings of a cone', Ph.D. dissertation, Univ. of Bremen, Germany, December 1987.

A System-Theoretic Approach for Discrete-Event Control of Manufacturing Systems

Ayla Gürel, Octavian C. Pastravanu and Frank L. Lewis

1 Introduction

With advances in computer technology came the emergence of a new class of dynamic systems, examples of which include – but are not limited to – automated manufacturing systems, computer networks, communication systems, air traffic control systems, and office information systems. These, often manmade, complex systems are formed by interconnecting several subsystems with well understood dynamics. The interconnected system is then driven according to the *asynchronous* occurrences of discrete events resulting from the interactions between the subsystems. These systems are now commonly known as *discrete event dynamic systems* (DEDS) [3].

Over the past decade, a great deal of work has been done on the many aspects of the efficient functioning of these systems. There are now a variety of approaches based on Markov chains [14], queueing networks [15], Petri nets [20], formal language theory [22], sample path methods [13], perturbation analysis [24], object-oriented techniques [10], and $(max, +)$ or dioid algebra [1].

Within the field of study of DEDS, issues related to modeling, control, and performance analysis of a particular class, namely the automated manufacturing systems, have recently attracted a lot of interest from researchers in engineering, computer science, and mathematics. In this paper, the term *discrete event manufacturing systems* (DEMS) is used to refer to these systems, in order to stress the event-driven nature of their dynamics. Special cases of this broad classification are *flexible manufacturing systems (FMS)* [11], and *computer integrated manufacturing (CIM) systems* [23].

Apart from being event-driven, two other characterizing features of DEMS are concurrency (many operations are causally independent [20], and/or occur simultaneously [8]), and asynchronous operations (certain tasks are not always completed in the same amount of time [8]). An effective operation of such a system, while meeting its production objectives, requires a controller, the task of which is to establish and maintain a desired sequence of events.

Despite the proliferation of theory on the control of DEDS [22], [10], [16], [2], the lack of a unified modeling framework continues to complicate the mechanism of transferring this knowledge to practice, especially in the area of DEMS. Moreover, the relation of several tools available for manufacturing scheduling (e.g., partial assembly tree, bill of materials, resource requirements matrix, design scheduling matrix) to discrete event systems is obscure. Another result of the absence of solid bridges between different trends is that one usually ends up requiring different models to study different issues of operation of the same system.

In previous contributions [17]–[18], a modern system theoretic point of view has been developed for the design of *rule-based controllers* for DEMS, providing a matrix paradigm which yields a rigorous, repeatable design algorithm. Within this framework, it is obvious how to use the standard manufacturing tools mentioned above for designing DEMS sequencing controllers. The key in this formulation is to separate the workcell functions from the controller, in contrast to the Petri net model, where the workcell and the controller functions are undifferentiated. The controller is represented as a set of matrix equations over a nonstandard algebra.

Based on these concepts and results, and once again emphasizing the rigor induced by the system theoretic perspective, the paper presents a DEMS control strategy which incorporates the classic principles of control: the workcell functions constitute the 'plant', and the state transitions occur according to the decisions issued by a DEMS controller after processing the state information from the workcell. The matrix equations are written in $(min, +)$ algebra. This not only gets rid of the nonlinear parts, which were present in [17]–[18], but also allows the accommodation of pools of machining centres, pools of material handling systems and buffers more successfully into the model. The controller is capable of handling asynchronous events and, in the case of concurrent events, of maintaining synchronization while excluding the conflicting operations. The relation between this description and the models based on Petri nets and $(max, +)$ algebra points to the unifying potential of the proposed paradigm and is worth further exploitation.

In the rest of the paper, Section 2 describes the workcell activity as a set of operations changing state due to the occurences of a set of events, and gives a mathematical model of this dynamics. The rule-based controller is presented in Section 3. Section 4 explains the connection between the proposed formulation and two other, in effect closed-loop representations, the Petri net and $(max, +)$ models. The potential of the proposed framework to address various real-time control issues is mentioned in the conclusion.

2 Workcell Activity of DEMS

In this section the workcell activity is considered and its dynamics modeled, as separate from the sequencing controller.

2.1 An example

A simple manufacturing system to help one understand the nature of the workcell activity is presented in Figure 1. The resources of the workcell consist of three machines M1, M2, M3, one robot R, one buffer B with two slots for storing intermediate products, and four pallets P on which raw parts are fixtured before being processed. Only one type of product is produced as a result of two machine operations: a raw part from input storage S1 is automatically loaded into M1, where it is fixtured on a pallet and processed (operation $M1P$). When M1 finishes its operation, R unloads it and takes the intermediate product to B (operation $RU1$). After the intermediate product is stored in B (operation BS) it can be loaded automatically into M2 or M3 for further processing to yield the final product (operations $M2P$ or $M3P$, respectively). It is assumed that the buffer operates serially, i.e. only one part can go in or come out of it at a time. R removes any finished product on M2 or M3, defixtures it from the pallet, and places it into output storage S2, returning the empty pallet to M1 (operation $RU2$).

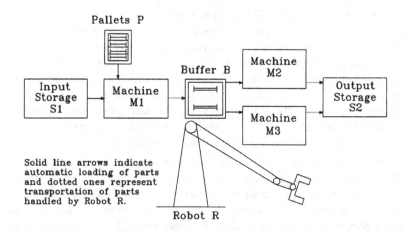

Figure 1: Sketch of the workcell example

From the setting of the production process, one observes that during a normal run of the system: $M1P$ occurs when there is a part in S1 (input operation PI complete), a pallet is available (pallet release operation PA complete), and M1 is idle (M1 release operation $M1A$ complete). For $RU1$

to start, R must be available (R release operation *RA* complete), and *M*1*P* must be complete. Concurrently with *RU*1 starts *M*1*A*. *BS* takes place when a buffer slot is available (buffer slot release operation *BA* complete), and *RU*1 is complete. This also initiates *RA*. *M*2*P* can occur when *BS* is complete, and M2 is available (M2 release operation *M*2*A* complete). Similarly, *M*3*P* requires that *BS* and M3 release operation *M*3*A* are complete. Starting either one of *M*2*P* and *M*3*P* causes the release of a buffer slot (start *BA*). When *RA* and *M*2*P* (or *M*3*P*) are complete, *RU*2 can begin, also resulting in the start of *M*2*A* (or respectively *M*3*A*). The completion of *RU*2 starts *PA* and *RA*, as the finished product goes into S2 (start output operation *PO*).

2.2   Operations and events

As the example illustrates, the workcell activity is a set of operations performed by the resources. Each operation is one of two types: it is either a *task operation* which involves a resource action on a part or an intermediate product, or a *resource release operation* which is the process of resetting a resource after the completion of a task operation using it. Now, one can define the states of an operation as *not started* (*ns*), *on-going* (*og*), and *complete and waiting for acknowledge* (*cwa*). The transitions from one state to another are determined by the discrete events of *start operation* (*so*), *complete operation* (*co*), and *acknowledge complete operation* (*aco*). Figure 2 shows the state transition diagram for an operation.

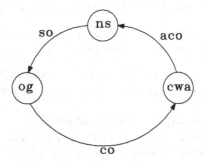

Figure 2: State transition diagram for an operation

It is important to note here that *so* and *aco* events correspond to decisions taken by the DEMS controller whereas *co* events occur within the cell. Obviously, starting an operation requires that all its antecedent operations are complete. Thus, an *so* event for an operation and the *aco* events associated with its antecedent operations occur simultaneously.

2.3   A model for the workcell dynamics

Let $O = V \cup R = \{o_i\}_{i=1}^{n}$ represent the set of distinct operations to be performed in the workcell, with $V = \{v_i\}_{i=1}^{n_v}$, and $R = \{r_i\}_{i=1}^{n_r}$, $n_v + n_r = n$, denoting the sets of distinct task and resource release operations, respectively. Identical operations performed by similar resources in a pool are not considered as distinct operations and will be referred to as *multiple operations*. Associate with the events *so*, *co*, and *aco* the n-dimensional vectors $o_{\delta s}(t)$, $o_{\delta c}(t)$, and $o_{\delta a}(t)$, respectively, where $o_{\delta s_i}(t)$, $o_{\delta c_i}(t)$, and $o_{\delta a_i}(t)$ are sums of shifted δ-pulses representing the counters, until time t, of the occurences of events *so*, *co*, and *aco*, respectively, for operation o_i. $o_{\delta c_i}(t)$ is a signal from the workcell, with each δ-pulse corresponding to another complete o_i operation. $o_{\delta s_i}(t)$ and $o_{\delta a_i}(t)$ are constructed by the DEMS controller as discussed in Section 3.3. Consider two piecewise constant vectors $o_c(t) \in \mathcal{N}_0^n$ and $o_o(t) \in \mathcal{N}_0^n$ associated with the operation states *cwa* and *og*, respectively ($\mathcal{N}_0 = \mathcal{N} \cup \{0\}$, where $\mathcal{N}$ denotes the set of natural numbers). The values of $o_{c_i}(t)$ and $o_{o_i}(t)$ represent the number of o_i operations in states *cwa* and *og* at time t, respectively. The operation state dynamics can be described as stated in the following theorem [19].

Theorem 2.1 *For any initial time t_0, the solutions of the differential equations*

$$Do_{c_i}(t) \;=\; o_{\delta c_i}(t) - o_{\delta a_i}(t), \quad o_{c_i}(t_0) \in \mathcal{N}_0 \qquad (2.1)$$
$$Do_{o_i}(t) \;=\; o_{\delta s_i}(t) - o_{\delta c_i}(t), \quad o_{o_i}(t_0) \in \mathcal{N}_0 \qquad (2.2)$$

are equivalent to the evolution given by the state diagram for each operation o_i, $1 \le i \le n$, in the workcell, including multiple operations.

The vectors $o_c(t)$ and $o_o(t)$ are the *workcell state vectors*, and the *workcell state equations* can be compactly written as

$$Do_c(t) \;=\; o_{\delta c}(t) - o_{\delta a}(t), \quad o_c(t_0) \in \mathcal{N}_0^n \qquad (2.3)$$
$$Do_o(t) \;=\; o_{\delta s}(t) - o_{\delta c}(t), \quad o_o(t_0) \in \mathcal{N}_0^n. \qquad (2.4)$$

Remark 2.1 *It will be useful to describe activities, such as parts going into the workcell and products going out of it, as input and output operations, respectively. Then, using the notations u_i, $1 \le i \le k$, for the input, and y_i, $1 \le i \le p$, for the output operations, and preserving the same meaning for the subscripts used above, one can write the workcell input and output equations as*

$$Du_c(t) \;=\; u_{\delta c}(t) - u_{\delta a}(t), \quad u_c(t_0) \in \mathcal{N}_0^k \qquad (2.5)$$
$$Dy_o(t) \;=\; y_{\delta s}(t) - y_{\delta c}(t), \quad y_o(t_0) \in \mathcal{N}_0^p. \qquad (2.6)$$

Note that the equations for $u_o(t)$ and $y_c(t)$ are omitted as these are irrelevant from the point of view of the workcell dynamics.

3   Rule-Based Controller

This section introduces a DEMS control strategy whereby the sequencing controller is designed in such a way that certain high level scheduling/dispatching decisions can be incorporated into the closed-loop system.

3.1   Causal labeling of task operations

Consider the set $V \subset O$ of distinct task operations in the workcell, and define a task labeling mapping $\mathcal{T}$ as a one-to-one function

$$\mathcal{T} : V \to \mathcal{N}_v = \{1, 2, \ldots, n_v\} \tag{3.1}$$

which labels the elements of V as $v_1, v_2, \ldots, v_{n_v}$. Assume that the *precedence matrix* (which is referred to as the sequencing matrix in [18], and is a structure matrix representing the temporal sequencing or precedence relations among the workcell operations required to produce a finished product) is available and is denoted by M. Note that M is nothing but Steward's design structure matrix which has been used extensively in representing task sequencing for complex manufacturing processes [9]. In this matrix the columns and rows correspond to tasks, and the (i, j) entry is '1' if $v_j \mathcal{S} v_i$ (i.e. v_j is an antecedent of v_i), and '0' otherwise, where $\mathcal{S}$ is used to denote the precedence relation. The precedence relation obviously is asymmetric, i.e. if $v_i \mathcal{S} v_j$ then $v_j \bar{\mathcal{S}} v_i$ (bar stands for negation), and it induces on V a partial ordering with $v_i \bar{\mathcal{S}} v_i$ [25].

For a given set V, one can obtain different precedence matrices for different task labeling mappings.

Definition 3.1 *A task labeling mapping is said to be causal if the corresponding precedence matrix is strictly lower triangular.*

Note that it is always possible to define a causal labeling on V, since the elements of V are intrinsically causally related, i.e. the precedence relation $\mathcal{S}$ on V does not involve feedback. The theorem stated below establishes the existence of a causal task labeling mapping, and follows from the results presented in [25].

Theorem 3.1 *For any set of task operations V, partially ordered by $\mathcal{S}$, there exists a causal task labeling mapping.*

A strictly lower triangular precedence matrix, reflecting a causal labeling of tasks, yields the hierarchical structure of the sequencing among the tasks and/or groups of tasks, thus providing a convenient method for determining the *task sequencing rules*.

3.2   The controller states

Defining the states of the controller is a crucial step in the design of the DEMS controller. Suppose that the elements of the task operation set V are causally labeled. Consider the set $V \overset{\triangle}{=} U \cup V \cup Y$, where $U = \{u_i\}_{i=1}^{k}$ and $Y = \{y_i\}_{i=1}^{p}$ are the input and output operations, respectively, with $u_i \bar{S} u_j$, and $y_i \bar{S} y_j$, for any i, j.

The controller is to ensure the sequencing of the operations within the workcell. This can be done via a set of rules involving synchronization of events (as meant in [1]) in accordance with the precedence and concurrency relations between the operations.

In the following, a method for defining the controller states is proposed. An immediate by-product is the task sequencing rules as a set of synchronization rules which can then be transformed into the controller rules [17], [18]. One assumption is made here: each task operation commences according to a unique synchronization rule (a reasonable assumption for a DEMS).

For each $v_j \in V$, obtain the set $V_j \subset V$ as the set of operations which have v_j as a predecessor, i.e.

$$V_j = \{v \in V|\ v_j S v\}, \quad 1 \leq j \leq n_v + k. \tag{3.2}$$

Using the concurrency relation between its elements, V_j can be partitioned into subsets V_{ji_j}, $i_j = 1, 2, \ldots, z_j$, $z_j \leq |V_j|$, such that each V_{ji_j} contains all those $v \in V_j$ which are concurrent. The elements of the set V_{ji_j} correspond to operations which start synchronously, since they have the same antecedent and are concurrent, i.e. cannot be conflicting. Note that, due to the assumption of a unique synchronization rule for each distinct task operation, V_{ji_j}, $i_j = 1, 2, \ldots, z_j$, are disjoint sets. V_{ji_j} will be referred to as a *maximal set of concurrent successors* (MSCS) for v_j. Denote by m the number of distinct MSCS. Associating a unique state x_i with each distinct MSCS gives the set of controller state variables obtained as $X = \{x_1, x_2, \ldots, x_m\}$.

Now rename, for convenience, the MSCS associated with controller state x_l as V_l^s, $l = 1, 2, \ldots, m$. It is obvious from the construction of these sets that V_l^s can be an MSCS for more than one $v_j \in V$ (e.g. when there is an assembly operation). So define the *predecessor sets*

$$V_l^p = \{v \in V|\ v \text{ is an antecedent of } V_l^s\}, \quad l = 1, 2, \ldots, m. \tag{3.3}$$

The task sequencing rules can now be easily determined in terms of the sets V_l^s and V_l^p, $l = 1, 2, \ldots, m$.

Remark 3.1 *Each $x_l \in X$ corresponds to a set of (task or output) operations which start synchronously – almost – immediately after the completion of the predecessor (task or input) operations. Thus each state monitors the completion of one set of operations and subsequently initiates another set of operations.*

Remark 3.2 *Given the precedence matrix corresponding to $\mathcal{V}$, each $\mathcal{V}_j$ can be obtained by scanning the j-th column; on the other hand, the $\mathcal{V}_{ji_j}$'s can be constructed according to the concurrency relation between the elements of $\mathcal{V}_j$.*

3.3 Rules of the controller

Given the workcell and controller states, the rules of the controller can be written using the task sequencing rules and the resource requirements of the tasks. The rules will depend on the workcell states, and are to ensure that the workcell operations occur in the desired sequence and according to the scheduling policy.

The example of Section 2.1 can be used to illustrate the basic idea. The workcell operations are task operations $V = \{M1P, RU1, BS, M2P, M3P, RU2\}$, input operations $U = \{PI\}$, output operations $Y = \{PO\}$, and resource release operations $R = \{PA, M1A, M2A, M3A, BA, RA\}$. The precedence matrix, augmented to correspond to the set $\mathcal{V} = U \cup V \cup Y$, with elements labeled in the same order as they appear above, is

$$M = \begin{bmatrix} 0 & 0 & 0 & 0 & 0 & 0 & 0 & 0 \\ 1 & 0 & 0 & 0 & 0 & 0 & 0 & 0 \\ 0 & 1 & 0 & 0 & 0 & 0 & 0 & 0 \\ 0 & 0 & 1 & 0 & 0 & 0 & 0 & 0 \\ 0 & 0 & 0 & 1 & 0 & 0 & 0 & 0 \\ 0 & 0 & 0 & 1 & 0 & 0 & 0 & 0 \\ 0 & 0 & 0 & 0 & 1 & 1 & 0 & 0 \\ 0 & 0 & 0 & 0 & 0 & 0 & 1 & 0 \end{bmatrix}, \tag{3.4}$$

strictly lower triangular due to the obviously causal labeling of the operations.

The sets $\mathcal{V}_j$ of equation 3.2 can be obtained as $\mathcal{V}_1 = \{M1P\}$, $\mathcal{V}_2 = \{RU1\}$, $\mathcal{V}_3 = \{BS\}$, $\mathcal{V}_4 = \{M2P, M3P\}$, $\mathcal{V}_5 = \{RU2\}$, $\mathcal{V}_6 = \{RU2\}$, $\mathcal{V}_7 = \{PO\}$. Clearly, only $\mathcal{V}_4$ has to be checked for concurrency between its elements: operations $M2P$ and $M3P$ are in conflict, and can never start synchronously, i.e. they are not concurrent. This yields subsets $\mathcal{V}_{41} = \{M2P\}$, and $\mathcal{V}_{42} = \{M3P\}$. The MSCS can then be written as $\mathcal{V}_1^s = \{M1P\}$, $\mathcal{V}_2^s = \{RU1\}$, $\mathcal{V}_3^s = \{BS\}$, $\mathcal{V}_4^s = \{M2P\}$, $\mathcal{V}_5^s = \{M3P\}$, $\mathcal{V}_6^s = \{RU2\}$, $\mathcal{V}_7^s = \{RU2\}$, $\mathcal{V}_8^s = \{PO\}$, and the set of controller states becomes $X = \{x_i\}$, $i = 1, \ldots, 8$.

The predecessor sets, which are needed for determining the rule base, are $\mathcal{V}_1^p = \{PI\}$, $\mathcal{V}_2^p = \{M1P\}$, $\mathcal{V}_3^p = \{RU1\}$, $\mathcal{V}_4^p = \{BS\}$, $\mathcal{V}_5^p = \{BS\}$, $\mathcal{V}_6^p = \{M2P\}$, $\mathcal{V}_7^p = \{M3P\}$, $\mathcal{V}_8^p = \{RU2\}$. With the resource requirements included, the rules can now be written in logical terms as

$$\begin{aligned} x_1 &= PI \wedge PA \wedge M1A, & x_2 &= M1P \wedge RA, & x_3 &= RU1 \wedge BA, \\ x_4 &= BS \wedge M2A, & x_5 &= BS \wedge M3A, & x_6 &= M2P \wedge RA, \\ x_7 &= M3P \wedge RA, & x_8 &= RU2. \end{aligned} \tag{3.5}$$

In these equations, the variables are considered as logical to help the argument. This form of the rules obviously is not of much use as regards the purpose of evaluating the quantitative information about the complete workcell operations, and accordingly supplying signals for the commencement of new ones. To achieve this end, associate with the controller states a piecewise constant vector denoted by $x_M(t) \in \mathcal{N}_0^m$. The value of $x_{M_i}(t)$ should reflect how many of each operation belonging to set $\mathcal{V}_i^s$ are to be started, depending on how many of each preceding operation are complete. Clearly, this reflects a synchronization phenomenon which can best be captured by minimization operating on the workcell states of complete operation. With this in mind, and defining the time dependent workcell state variables as conceived in Section 2.3, the rules above are transformed into

$$
\begin{aligned}
&x_{M_1}(t) = min\{PI_c(t), PA_c(t), M1A_c(t)\}, \quad x_{M_2}(t) = min\{M1P_c(t), RA_c(t)\}, \\
&x_{M_3}(t) = min\{RU1_c(t), BA_c(t)\}, \quad x_{M_4}(t) = min\{BS_c(t), M2A_c(t)\}, \\
&x_{M_5}(t) = min\{BS_c(t), M3A_c(t)\}, \quad x_{M_6}(t) = min\{M2P_c(t), RA_c(t)\}, \\
&x_{M_7}(t) = min\{M3P_c(t), RA_c(t)\}, \quad x_{M_8}(t) = RU2_c(t)
\end{aligned}
\tag{3.6}
$$

or more compactly

$$
x_M(t) =
\begin{bmatrix}
\varepsilon & \varepsilon & \varepsilon & \varepsilon & \varepsilon & \varepsilon \\
0 & \varepsilon & \varepsilon & \varepsilon & \varepsilon & \varepsilon \\
\varepsilon & 0 & \varepsilon & \varepsilon & \varepsilon & \varepsilon \\
\varepsilon & \varepsilon & 0 & \varepsilon & \varepsilon & \varepsilon \\
\varepsilon & \varepsilon & 0 & \varepsilon & \varepsilon & \varepsilon \\
\varepsilon & \varepsilon & \varepsilon & 0 & \varepsilon & \varepsilon \\
\varepsilon & \varepsilon & \varepsilon & \varepsilon & 0 & \varepsilon \\
\varepsilon & \varepsilon & \varepsilon & \varepsilon & \varepsilon & 0
\end{bmatrix}
\odot v_c(t) \hat{\oplus}
\begin{bmatrix}
0 & 0 & \varepsilon & \varepsilon & \varepsilon & \varepsilon \\
\varepsilon & \varepsilon & \varepsilon & \varepsilon & \varepsilon & 0 \\
\varepsilon & \varepsilon & \varepsilon & \varepsilon & 0 & \varepsilon \\
\varepsilon & \varepsilon & 0 & \varepsilon & \varepsilon & \varepsilon \\
\varepsilon & \varepsilon & \varepsilon & 0 & \varepsilon & \varepsilon \\
\varepsilon & \varepsilon & \varepsilon & \varepsilon & \varepsilon & 0 \\
\varepsilon & \varepsilon & \varepsilon & \varepsilon & \varepsilon & 0 \\
\varepsilon & \varepsilon & \varepsilon & \varepsilon & \varepsilon & \varepsilon
\end{bmatrix}
\odot r_c(t) \hat{\oplus}
\begin{bmatrix}
0 \\ \varepsilon \\ \varepsilon \\ \varepsilon \\ \varepsilon \\ \varepsilon \\ \varepsilon \\ \varepsilon
\end{bmatrix}
\odot u_c(t)
\tag{3.7}
$$

where

$$
\begin{aligned}
v_c(t) &= [M1P_c(t), RU1_c(t), BS_c(t), M2P_c(t), M3P_c(t), RU2_c(t)]^T \\
r_c(t) &= [PA_c(t), M1A_c(t), M2A_c(t), M3A_c(t), BA_c(t), RA_c(t)]^T \\
u_c(t) &= [PI_c(t)].
\end{aligned}
$$

In this matrix equation, $\hat{\oplus}$ and $\odot$ refer to minimization and addition, replacing addition and multiplication of the standard algebra, respectively. 'ε', which denotes numerical value ∞, and '0' are neutral elements with respect to minimization and addition, respectively.

There are two problems yet to be handled here: the shared usage of the robot, and the choice at the buffer in the sequencing of tasks. In the first case, there are three controller rules, namely those associated with $x_{M_2}(t)$, $x_{M_6}(t)$ and $x_{M_7}(t)$, all of which depend on the completion of RA, and so at some point more than one of $x_{M_2}(t)$, $x_{M_6}(t)$, and $x_{M_7}(t)$ might simultaneously become '1' (note that, in this particular example, $x_{M_i}(t) \in \{0,1\}$, $1 \le m$, as there are no pools of identical machines). Thus a conflict (also termed 'confusion' in the framework of Petri nets [20]) arises between these three rules, as they all depend on the availability of the same resource. In the second case, $x_{M_4}(t)$ and $x_{M_5}(t)$ are potentially in conflict, as both depend on the completion of BS. To deal with these situations, a conflict resolution input is employed [19], [18]. This input is selected according to a high level scheduling/dispatching function in order to meet the production objectives, as well as avoid undesirable behaviour (e.g. deadlock). For this example, an appropriate conflict resolution input is

$$\boldsymbol{u}_R(t) = [u_{R_1}(t), u_{R_2}(t), u_{R_3}(t), u_{R_4}(t), u_{R_5}(t)]^T \tag{3.8}$$

where

$$
\begin{aligned}
u_{R_1} &= f_{R_1}(x_{M_2}, BS_c, BS_o, M2P_c, M2P_o, M3P_c, M3P_o) \\
&= \begin{cases} 0 & \text{if } x_{M_2} = 1 \text{ and } BS_c + BS_o + M2P_c + M2P_o + \\ & \quad M3P_c + M3P_o = 4 \\ 1 & \text{otherwise} \end{cases} \\
u_{R_2} &= f_{R_2}(x_{M_2}, BS_c, BS_o, M2P_c, M2P_o, M3P_c, M3P_o) \\
&= 1 - f_{R_1} \\
u_{R_3} &= f_{R_3}(x_{M_2}) \\
&= \begin{cases} 1 & \text{if } x_{M_2} = 0 \\ 0 & \text{otherwise} \end{cases} \\
u_{R_4} &= f_{R_4}(x_{M_4}, M3A_c) \\
&= \begin{cases} 1 & \text{if } x_{M_4} = 1 \text{ and } M3A_c = 1 \\ 0 & \text{otherwise} \end{cases} \\
u_{R_5} &= f_{R_5}(x_{M_4}, M3A_c) \\
&= 1 - f_{R_4},
\end{aligned}
$$

all variables evaluated at time t. The complete state equation of the rule-based controller is given by

$$\boldsymbol{x}(t) = \boldsymbol{x}_M(t) \hat{\oplus} \begin{bmatrix} \varepsilon & \varepsilon & \varepsilon & \varepsilon & \varepsilon \\ 0 & \varepsilon & \varepsilon & \varepsilon & \varepsilon \\ \varepsilon & \varepsilon & \varepsilon & \varepsilon & \varepsilon \\ \varepsilon & \varepsilon & \varepsilon & 0 & \varepsilon \\ \varepsilon & \varepsilon & \varepsilon & \varepsilon & 0 \\ \varepsilon & 0 & \varepsilon & \varepsilon & \varepsilon \\ \varepsilon & \varepsilon & 0 & \varepsilon & \varepsilon \\ \varepsilon & \varepsilon & \varepsilon & \varepsilon & \varepsilon \end{bmatrix} \odot \boldsymbol{u}_R(t) \tag{3.9}$$

where $\boldsymbol{x}(t) \in \mathcal{N}_0^m$ is again a piecewise constant vector which depicts the controller state variables incorporating the high level decision. It is worth noting that the condition $BS_c + BS_o + M2P_c + M2P_o + M3P_c + M3P_o = 4$, which is checked when evaluating f_{R_1}, is crucial as regards ensuring a

deadlock-free workcell; it is due to a deadlock avoidance scheme, the details
of which can be found in [12]. The scheduling/dispatching function resolves
the conflict in favor of x_2, unless doing so results in deadlock.

Remark 3.3 *In this formulation, one can represent the decision-making controller as a set of linear (i.e. matrix) equations defined over $(min, +)$ algebra,
where min plays the role of 'addition' (hence the notation $\hat{\oplus}$), and $+$ plays the
role of 'multiplication'(hence the notation $\odot$); $\varepsilon = \infty$, and 0, are the neutral
elements with respect to min, and $+$, respectively. It works even when the
system involves conflict (i.e. shared resources and/or choice in the sequencing of tasks), which is resolved by 'adding' a high level decision input to the
appropriate variables.*

To generalize, the rules of the controller can be written in $(min, +)$ algebra
as

$$x(t) = x_M(t)\hat{\oplus}F_R \odot u_R(t) \tag{3.10}$$

where $x(t), x_M(t) \in \mathcal{N}_0^m$, and $u_R \in \mathcal{N}_0^{m_R}$, $m_R \leq m$, are the *controller
state, workcell monitoring*, and *conflict resolution input vectors*, respectively.
Equation (3.10) is called the *controller state equation*. The vectors $x_M(t)$ and
$u_R(t)$ are precalculated according to the *workcell monitoring equation*

$$x_M(t) = F \odot o_c(t)\hat{\oplus}F_u \odot u_c(t) \tag{3.11}$$

and the *conflict resolution equation*

$$u_R(t) = f_R(o_c(t),\ o_o(t),\ u_c(t),\ x_M(t)) \tag{3.12}$$

where $o_c = [v_c^T\ r_c^T]^T$, and $o_o = [v_o^T\ r_o^T]^T$. The $m \times n$ matrix F in equation (3.11) can be partitioned as $F = [F_v\ F_r]$.

Columns of F_v correspond to the task operations, and those of F_r correspond to the resource release operations. The matrices $F_v \in \mathcal{B}^{m \times n_v}$ and
$F_r \in \mathcal{B}^{m \times n_r}$, with $\mathcal{B} = \{\varepsilon, 0\}$, are the task sequencing and resource requirements matrices, respectively. Basically, they are the same as those in [17], [19],
or [18]; the only difference here is that matrix entries '0' and '1' are replaced
by '$\varepsilon = \infty$' and '0', respectively. Multiple '0' entries in a column of F_v indicate that choice is involved in the sequencing of tasks succeeding the task
associated with that column. Such a structure in F_r, on the other hand,
means that the associated resource is a shared one. Conflict may arise during
the running of the workcell when either one of these structures exists.

It is worth noting that the rule-based controller action fully depends on
processing the workcell feedback $o_c(t)$ and $o_o(t)$, with equations (3.11) and
(3.12) as the linear and nonlinear parts, respectively. In order to close the
loop, the controller should produce the actuating signals. This is done by
means of the *operation start* and the *operation acknowledge* vectors $o_s(t)$,

$o_a(t) \in \mathcal{N}^n$ obtained according to the *operation start* and *operation acknowledge equations*

$$o_s(t) = S\,x(t) \qquad (3.13)$$
$$o_a(t) = \bar{F}^T x(t). \qquad (3.14)$$

The $n \times m$ matrix S in equation (3.13) can be partitioned as

$$S = \begin{bmatrix} S_v \\ S_r \end{bmatrix} \qquad (3.15)$$

where the rows of the *task start matrix* $S_v \in \mathcal{N}_0^{n_v \times m}$ correspond to the task operations and those of the *resource release start matrix* $S_r \in \mathcal{N}_0^{n_r \times m}$ correspond to the resource release operations, as in the earlier contributions. Construction of these matrices is a straightforward matter once the sets $\mathcal{V}_l^s$ and $\mathcal{V}_l^p$, $1 \le l \le m$, of Section 3.2 are obtained. Matrix $\bar{F}$ is structurally equivalent to F of equation (3.11), obtained by replacing the matrix entries 'ε' and '0' (neutral elements with respect to *min* and +) of the latter, by '0' and '1' (neutral elements with respect to addition and multiplication in conventional algebra), respectively. Note that equations (3.13) and (3.14) are written in the conventional algebra as they involve no decisions but only structures and values.

Now the δ-pulse vectors $o_{\delta s}(t)$ and $o_{\delta a}(t)$, which change the states of the operations in the workcell when the control loop is closed (equations (2.3) and (2.4)), can be obtained by simply δ-sampling the values of $o_s(t)$ and $o_a(t)$, respectively [19].

Going back to the example to complete the design of the rule-based controller, one obtains the operation start equations for the task, resource release, and output operations, respectively, as

$$v_s(t) = S_v\,x(t) = \begin{bmatrix} 1 & 0 & 0 & 0 & 0 & 0 & 0 & 0 \\ 0 & 1 & 0 & 0 & 0 & 0 & 0 & 0 \\ 0 & 0 & 1 & 0 & 0 & 0 & 0 & 0 \\ 0 & 0 & 0 & 1 & 0 & 0 & 0 & 0 \\ 0 & 0 & 0 & 0 & 1 & 0 & 0 & 0 \\ 0 & 0 & 0 & 0 & 0 & 1 & 1 & 0 \end{bmatrix} x(t) \qquad (3.16)$$

$$r_s(t) = S_r\,x(t) = \begin{bmatrix} 0 & 0 & 0 & 0 & 0 & 0 & 0 & 1 \\ 0 & 1 & 0 & 0 & 0 & 0 & 0 & 0 \\ 0 & 0 & 0 & 0 & 0 & 1 & 0 & 0 \\ 0 & 0 & 0 & 0 & 0 & 0 & 1 & 0 \\ 0 & 0 & 0 & 1 & 1 & 0 & 0 & 0 \\ 0 & 0 & 1 & 0 & 0 & 0 & 0 & 1 \end{bmatrix} x(t) \qquad (3.17)$$

$$y_s(t) = S_y\,x(t) = \begin{bmatrix} 0 & 0 & 0 & 0 & 0 & 0 & 0 & 1 \end{bmatrix} x(t) \qquad (3.18)$$

where S_y is constructed in the same way as S_v and S_r of equation (3.15), and

$$
\begin{aligned}
v_s(t) &= [M1P_s(t), \; RU1_s(t), \; BS_s(t), \; M2P_s(t), \; M3P_s(t), \; RU2_s(t)]^T \\
r_s(t) &= [PA_s(t), \; M1A_s(t), \; M2A_s(t), \; M2A_s(t), \; BA_s(t), \; RA_s(t)]^T \\
y_s(t) &= [PO_s(t)].
\end{aligned}
$$

The operation acknowledge equations for the task, resource release, and input operations, respectively, are

$$
v_a(t) = \bar{F}_v^T \, x(t) =
\begin{bmatrix}
0 & 1 & 0 & 0 & 0 & 0 & 0 & 0 \\
0 & 0 & 1 & 0 & 0 & 0 & 0 & 0 \\
0 & 0 & 0 & 1 & 1 & 0 & 0 & 0 \\
0 & 0 & 0 & 0 & 0 & 1 & 0 & 0 \\
0 & 0 & 0 & 0 & 0 & 0 & 1 & 0 \\
0 & 0 & 0 & 0 & 0 & 0 & 0 & 1
\end{bmatrix}
x(t) \qquad (3.19)
$$

$$
r_a(t) = \bar{F}_r^T \, x(t) =
\begin{bmatrix}
1 & 0 & 0 & 0 & 0 & 0 & 0 & 0 \\
1 & 0 & 0 & 0 & 0 & 0 & 0 & 0 \\
0 & 0 & 0 & 1 & 0 & 0 & 0 & 0 \\
0 & 0 & 0 & 0 & 1 & 0 & 0 & 0 \\
0 & 0 & 1 & 0 & 0 & 0 & 0 & 0 \\
0 & 1 & 0 & 0 & 0 & 1 & 1 & 0
\end{bmatrix}
x(t) \qquad (3.20)
$$

$$
u_a(t) = \bar{F}_u^T \, x(t) = \begin{bmatrix} 1 & 0 & 0 & 0 & 0 & 0 & 0 & 0 \end{bmatrix} x(t) \qquad (3.21)
$$

where

$$
\begin{aligned}
v_a(t) &= [M1P_a(t), \; RU1_a(t), \; BS_a(t), \; M2P_a(t), \; RU2_a(t)]^T \\
r_a(t) &= [PA_a(t), \; M1A_a(t), \; M2A_a(t), \; BA_a(t), \; RA_a(t)]^T \\
u_a(t) &= [PI_a(t)]
\end{aligned}
$$

4   The Closed-Loop System

The scenario, in this matrix formulation, consists of a plant, i.e. a DEMS workcell modeled by equations (2.3) and (2.4), and a sequencing controller modeled by equations (3.10)–(3.14). This is just a closed-loop system in the classic control sense. On the other hand, in the case of the Petri net- and $(max, +)$-based models of DEMS, controller design and system representation are inseparable, and in both cases the model is in effect a closed-loop system description. In the following, the matrix formulation is bridged with these two approaches.

4.1   Connection with Petri net models

Consider O, the set of distinct operations to be performed in the workcell, and X_M, the set of workcell monitoring variables. Stated below are results characterizing the closed-loop behavior [19].

Theorem 4.1 *The controller equations (3.11)–(3.14) organize the sets O and X_M as a timed Petri net $PN(O, X_M, w^-, w^+)$ which is externally synchronized according to equation (3.10). Mappings $w^- : X_M \to O$ and $w^+ : X_M \to O$ are defined by the input and the output incidence matrices $W^- = \bar{F}^T$, and $W^+ = S$, respectively.*

Corollary 4.1 *If random durations are assumed for the workcell operations, then $PN(O, X_M, w^-, w^+)$ is an externally synchronized stochastic Petri net.*

Remark 4.1 *Deadlock avoidance and conflict resolution, which are important real-time control issues in DEMS, can be addressed in the presented matrix framework, using the insight provided by Theorem 4.1 and some results in [21] and [12].*

Remark 4.2 *Corollary 4.1 allows one to model machine failures as disturbance inputs and to incorporate in this framework the machine (resource) states down/broken defined in [22].*

4.2   Connection with $(max, +)$ models

It is well known that for decision-free DEDS $(max, +)$ based models offer a very efficient framework for addressing performance evaluation and optimization issues [6], [7]. A decision-free DEMS is one where neither choices nor shared resources are involved in the sequencing of tasks. Needless to say, practical manufacturing systems, e.g. flexible manufacturing systems, do not in general satisfy this constraint. The current $(max, +)$ approaches to modeling and analysis of such systems are mostly limited to providing closed-loop system descriptions, where all the necessary decisions concerning the scheduling and sequencing of tasks have already been made by some controller. An extension of the $(max, +)$ approach to a broader class of systems involving decisions, proposed in [4] and [5], can be applied to DEMS which require decisions in the routing of parts (i.e. downstream choices in relation to task operations) but it is not suitable when the system contains shared resources and/or upstream choices in relation to task operations. Another limitation of the $(max, +)$ models lies in their inability to accommodate finite intermediary storage areas (buffers) and material handling systems (robots, carts, etc.).

This section focuses on the relation between the approach proposed so far and the $(max, +)$ representation of DEMS which contain no pools of identical machines/machine centres. They may contain buffers and/or pools of material handling systems, which are assumed to operate according to a FIFO discipline (i.e. a part waiting upstream of a machine is dealt with as soon as possible). The simpler no-conflict case is employed in presenting the theory underlying the connection between the two representations. Once this is established for the decision-free system, then the extension to the more general case comes straightforwardly.

4.2.1 No-conflict case

Consider a manufacturing system which contains no shared resources, and which produces a single type of part according to a predetermined processing sequence, i.e. there are no choices in the sequencing of tasks either. Parts are fixtured on pallets; when a part completes its processing sequence it is sent to a load/unload station, and another part is fixtured on the pallet. The pallet is sent back to repeat the same processing sequence. Now let this system be controlled by the rule-base of equations (3.10)–(3.14). Due to the assumption of single machines, all the workcell operations, excluding those associated with buffers and automated material handling devices (robots, shuttle carts, etc.), are of multiplicity one.

Clearly, this is a decision-free system. This means that the conflict resolution vector can be omitted from equations (3.10)–(3.14). One can then attach, to each of the operation (task or resource release) and controller state variables, a superscript denoting the number of the current activation. For the k-th activation of the controller rule base, the equations are

$$x^k(t) = x_M^k(t) \tag{4.1}$$
$$x_M^k(t) = F_v \odot v_c^k(t)\hat{\oplus}F_r \odot r_c^k(t)\hat{\oplus}F_u \odot u_c^k(t) \tag{4.2}$$
$$v_s^k(t) = S_v\, x^k(t) \tag{4.3}$$
$$r_s^k(t) = S_r\, x^k(t) \tag{4.4}$$
$$y_s^k(t) = S_y\, x^k(t) \tag{4.5}$$

It is obvious that equations (4.1)–(4.5) have an equivalent in terms of time, that is to say, one can associate to each variable the moment when its current value is attained. For this, let $\xi(t)$ denote a generic variable, and define the *time projection operator* g which provides the time value $\tilde{\xi}^k$ corresponding to the moment when the variable attains its value of the k-th activation. The functional dependence established by g can be given as $g(\xi^k(t)) = \tilde{\xi}^k$ where $\xi^k(t)$ should be meant as a pulse function of strength α_ξ^k, and g is defined by

$$g(\xi^k(t)) = \frac{1}{\alpha_\xi^k} \int_0^\infty t\, D(\xi^k(t)), \tag{4.6}$$

with D denoting differentiation in the distribution space [26].

Define the binary time mappings $h : \{0,1\} \rightarrow \{-\varepsilon, 0\}$ by $h(0) = -\varepsilon$, $h(1) = 0$ and $\hat{h} : \{\varepsilon, 0\} \rightarrow \{-\varepsilon, 0\}$ by $\hat{h}(\varepsilon) = -\varepsilon$, $\hat{h}(0) = 0$.

Now, denote by G the multidimensional time projection operator that applies g to a vector, i.e. $G(\gamma^k) = \tilde{\gamma}^k$, where $\gamma^k(t)$ is a time value vector, with $\tilde{\gamma}_i^k = g(\gamma_i^k(t))$. Let H and $\hat{H}$ denote the multidimensional binary time mappings that apply h and $\hat{h}$ to binary matrices (i.e. if $N \in \{0,1\}^{m \times n}$, then $\tilde{N} \overset{\triangle}{=} H(N) \in \{-\varepsilon, 0\}^{m \times n}$, with $\tilde{N}_{ij} = h(N_{ij})$; and similarly, if $B \in \{\varepsilon, 0\}^{m \times n}$, then $\tilde{B} \overset{\triangle}{=} \hat{H}(B) \in \{-\varepsilon, 0\}^{m \times n}$, with $\tilde{B}_{ij} = \hat{h}(B_{ij})$). Note here that, for the sake of simplicity, G and H are defined using generic dimensions for the vectors and binary matrices, respectively.

In terms of these constructions, and using $\oplus$ and $\otimes$ to mean 'addition' and 'multiplication', respectively, in $(max, +)$ (i.e. $\oplus$ means maximum and $\otimes$ means sum), one can state the following result, which is proved in [21].

Lemma 4.1 *The $(max, +)$ representation of the rule-based controller behavior can be obtained directly from equations (4.1)–(4.5) according to*

$$\tilde{x}^k = \tilde{x}_M^k \tag{4.7}$$

$$x_M^k = (\tilde{F}_v \otimes \tilde{v}_c^k) \oplus (\tilde{F}_r \otimes \tilde{r}_c^k) \oplus (\tilde{F}_u \otimes \tilde{u}_c^k) \tag{4.8}$$

$$\tilde{v}_s^k = \tilde{S}_v \otimes \tilde{x}^k \tag{4.9}$$

$$\tilde{r}_s^k = \tilde{S}_r \otimes \tilde{x}^k \tag{4.10}$$

$$\tilde{y}_s^k = \tilde{S}_y \otimes \tilde{x}^k. \tag{4.11}$$

The aim is to write a $(max, +)$ representation for the closed-loop system. Towards this end, define the *task duration matrix* $T_v = diag\{t_{vi}\}_{i=1}^{n_v}$ and the *resource release duration matrix* $T_r = diag\{t_{rj}\}_{j=1}^{n_r}$, where t_{vi} and t_{rj} denote the durations of task operation v_i and resource release operation r_j, respectively.

A multiple resource release operation r_j (initially characterized by $r_{cj}(t_0)$ resources available) introduces, in k, a delay $d_j = r_{cj}(t_0)$ between the k-th release of any of these resources and its k-th allocation (note that the delay is constant because of the FIFO assumption). To account for this, define the *resource release matrix* $D = diag\{z^{-d_j}\}_{i=1}^{n_r}$ with $z^0 = 0$ and the *resource reset time matrix* $T_{rd} = T_r \otimes D = diag\{t_{rj} \otimes z^{-d_j}\}_{i=1}^{n_r}$ where z^{-1} represents the operator with the property $z^{-d_j}\tilde{\gamma}^k = \tilde{\gamma}^{k-d_j}$. Taking into consideration the times required for preparing the input parts, and delivering the output products, one can define the respective matrices $T_u = diag\{t_{uj}\}_{j=1}^{n_u}$, and $T_y = diag\{t_{yj}\}_{j=1}^{n_y}$, and write

$$\tilde{u}_c^k = T_u \otimes \tilde{p}_{in}^k$$

$$\tilde{p}_{out}^k = T_y \otimes \tilde{y}_s^k,$$

where $\tilde{p}_{in}^k$ and $\tilde{p}_{out}^k$ are the *parts in* (corresponding to the k-th input of parts), and *products out* (corresponding to the k-th delievery of products) vectors, respectively. Lemma 4.1 and these definitions together lead to the following result [21].

Theorem 4.2 *Consider a non-conflict DEMS, with no pools of identical machines, controlled by the rule-base of equations (3.10)–(3.14). The $(max, +)$ representation of the closed-loop system behaviour is given by*

$$
\begin{aligned}
\tilde{x}^k &= [(\tilde{F}_v \otimes T_v \otimes \tilde{S}_v) \oplus (\tilde{F}_r \otimes T_{rd} \otimes \tilde{S}_r)] \otimes \tilde{x}^k \oplus \\
&\quad (\tilde{F}_u \otimes T_u) \otimes \tilde{p}_{in}^k && (4.12) \\
\tilde{p}_{out}^k &= (T_y \otimes \tilde{S}_y) \otimes \tilde{x}^k. && (4.13)
\end{aligned}
$$

4.2.2 More general case

Consider now a more realistic DEMS where there are shared resources and/or choice in the sequencing of tasks. It is assumed once again that there are no pools of identical machines. The closed-loop system description given by equations (3.10)–(3.14) is then in its most general form. The equations for the k-th activation of the controller rule base are the same as before except that equation (4.1), the controller state equation, becomes

$$
x^k(t) = x_M^k(t) \hat{\oplus} F_R \odot u_R^k(t) \tag{4.14}
$$

where $u_{R_i}^k(t)$ denotes the value of the i-th conflict resolution input when it is 'on' for the k-th time, meant as a pulse function of strength $\alpha_{u_{R_i}}^k$ (which is unity in this case since pools of identical machines are not allowed). Then the result of Theorem 4.2 can be extended as follows:

Theorem 4.3 *Consider a DEMS, with no pools of identical machines, controlled by the rule-base of equations (3.10)–(3.14). The $(max, +)$ representation of the closed-loop system behavior is given by*

$$
\begin{aligned}
\tilde{x}^k &= [(\tilde{F}_v \otimes T_v \otimes \tilde{S}_v) \oplus (\tilde{F}_r \otimes T_{rd} \otimes \tilde{S}_r)] \otimes \tilde{x}^k \oplus && (4.15) \\
&\quad (\tilde{F}_u \otimes T_u) \otimes \tilde{p}_{in}^k \oplus \tilde{F}_R \otimes \tilde{u}_R^k && (4.16) \\
\tilde{p}_{out}^k &= (T_y \otimes \tilde{S}_y) \otimes \tilde{x}^k. && (4.17)
\end{aligned}
$$

It is clear from Theorems 4.2 and 4.3 that the rule-based approach allows one to obtain a model which, together with the description of the workcell, can then be readily used for implementing $(max, +)$ analysis procedures. The $(max, +)$ representation obtained from the closed-loop description of the rule-base controlled DEMS is more general than that proposed by [6] as regards accommodating finite storage areas (buffers) and/or pools of material handling systems. Moreover, the formulation applies to more general situations which involve choice and shared resources in the sequencing of tasks.

5 Conclusion

A system theoretic approach is employed to devise a DEMS control strategy, wherein the 'plant' (workcell functions) and the controller are separated. The workcell is modeled as a set of differential equations, providing the complete quantitative information as to the number and durations of the operations that occur in the workcell. The controller, represented as a set of matrix equations over $(min, +)$ algebra, processes the 'feedback' from the workcell. Based on this, the controller sequences the tasks according to a set of rules, resource requirements, and high level decision making scheduling/dispatching funtions. The formulation not only captures the characteristic DEMS features, such as concurrency and asynchronous operations, but is also capable of resolving conflicts, while avoiding deadlocks on a real-time basis, via schemes which can be incorporated into the high level decision making funtions. Thus, the controller equations allow a formal approach to design for operation with desired behavioral properties. Further research should address the issue of conflict resolution as a suitably defined optimization problem. As Section 4 shows, the structure provided by the formulation can be used to construct Petri net or $(max, +)$ descriptions for DEMS. Finally, the matrix framework, underlying the systematic design of the controller presented in Section 3, reveals a potential to develop new software tools for DEMS simulation, analysis, and design.

References

[1] F. Bacelli, G. Cohen, G.J. Olsder, J.-P. Quadrat (1992), *Synchronization and Linearity*, Wiley, New York.

[2] J.-L. Boimond (1993), 'Internal model control of discrete-event processes in the max-algebra', *Proc. European Control Conf.*, 150–157.

[3] C.G. Cassandras (1993), *Discrete Event Systems: Modeling and Performance Analysis*, Irwin, Boston.

[4] D.D. Cofer and V.K. Garg (1992), 'A timed model for the control of discrete event systems involving decisions in the max/plus algebra', *Proc. 31st Decision and Control Conf.*, 3363–3368.

[5] D.D. Cofer and V.K. Garg (1993), 'A generalized max-algebra model for performance analysis of timed and untimed discrete event systems', *Proc. American Control Conf.*, 2288–2292.

[6] G. Cohen, D. Dubois, J.P. Quadrat, and M. Viot (1985), 'A linear system-theoretic view of discrete-event processes and its use for performance evaluation in manufacturing', *IEEE Trans. Automatic Control*, **AC-30**, 210–220.

[7] G. Cohen, P. Moller, J.P. Quadrat, and M. Viot (1989), 'Algebraic tools for the performance evaluation of discrete event systems', *Proc. IEEE*, **77**, 39–58.

[8] A.A. Desrochers (1992), 'Performance analysis using Petri nets', *Journal of Intelligent and Robotic Systems*, **6**, 65–79.

[9] S.D. Eppinger, D.E. Whitney, and R.P. Smith (1990), 'Organizing the tasks in complex design projects', *Proc. ASME Int. Conf. Design Theory and Methodology*, 39–46.

[10] C.R. Glassey and S. Adiga (1990), 'Berkeley library of objects for control and simulation of manufacturing (BLOCS/M)', *Applications of Object-Oriented Programming* (ed. L.J. Pinson and R.S. Wiener), Addison-Wesley, Reading, MA.

[11] M.P. Groover (1980), *Automation Production Systems and Computer-Aided Manufacturing*, Prentice-Hall, New Jersey.

[12] A. Gürel, O.C. Pastravanu, F.L. Lewis (1994) 'A robust approach in deadlock-free and live FMS design', *Proc. IEEE Mediterranean Symp. New Directions in Control and Automation*, 40–47.

[13] Y.C. Ho (1989), 'Dynamics of discrete event systems', *Proc. IEEE*, **77**, 3–6.

[14] R.A. Howard (1971), *Dynamic Probabilistic Systems, vol.1, (Markov Models)*, Wiley, New York.

[15] L. Kleinrock (1975), *Queueing Systems, vol.1: Theory*, Wiley, New York.

[16] L.E. Lawrence and B.H. Krogh (1990), 'Synthesis of feedback control logic for a class of controlled Petri nets', *IEEE Transactions on Automatic Control*, **35**, 514–523.

[17] F.L. Lewis, O.C. Pastravanu, and H.-H. Huang (1993), 'Controller design and conflict resolution for discrete event manufacturing systems', *Proc. 32nd Control and Decision Conf.*, 3288–3293.

[18] F.L. Lewis, H.-H. Huang, O.C. Pastravanu, and A. Gürel (1994), 'Control system design for manufacturing systems', *Flexible Manufacturing Systems: Recent Developments*, (ed. A. Raouf and M. Ben Daya), Elsevier, New York.

[19] F.L. Lewis, O.C. Pastravanu, A. Gürel, and H.-H. Huang, (1994) 'Digital control of discrete event manufacturing systems', *Proc. 2nd IFAC/IFIP/IFORS Workshop on Intelligent Manufacturing Systems*, 311–315.

[20] T. Murata (1989), 'Petri nets: Properties, analysis, and applications', *Proc. IEEE*, **77**, 541–580.

[21] O.C. Pastravanu, A. Gürel, F.L. Lewis, and H.-H. Huang (1994), 'Rule-based controller design for discrete event manufacturing systems', *Proc. American Control Conf.*, 299–305.

[22] P.J. Ramadge and W.M. Wonham (1989), 'The control of discrete event systems', *Proc. IEEE*, **77**, 81–98.

[23] P. Ranky (1985), *Computer Integrated Manufacturing: An Introduction with Case Studies*, Prentice-Hall, New Jersey.

[24] R. Suri (1989), 'Perturbation analysis: The state of the art and research issues explained via G/G/G1 queue', *Proc. IEEE*, **77**, 114–137.

[25] J.N. Warfield (1973), 'Binary matrices in system modeling', *IEEE Trans. Systems Man. and Cybernetics*, **SMC-3** 441–449.

[26] L.A. Zadeh and C.A. Desoer (1963), *Linear System Theory: The State Space Approach*, McGraw-Hill, New York.

Idempotent Structures in the Supervisory Control of Discrete Event Systems

Darren D. Cofer and Vijay K. Garg

1 Introduction

Discrete event systems (DES) are characterized by a collection of events, such as the completion of a job in a manufacturing process or the arrival of a message in a communication network. The system state changes only at time instants corresponding to the occurrence of one of the defined events. At the logical level of abstraction, the behavior of a DES is described by the sequences of events that it performs. However, if time constraints are of explicit concern in the system dynamics and its performance specification, its behavior can be characterized by sequences of occurrence times for each event. The objective of this paper is to demonstrate how underlying algebraic similarities between certain logical and timed DES can be exploited to study the control of timed DES.

Logical DES are often modelled by automata known as *finite state machines* (FSM). A FSM consists of a set of states Q, a collection of events Σ, and a state transition function δ. The occurrence of an event causes the system to move from one state to another as defined by the transition function.

Timed DES which are subject to synchronization constraints can be modelled by automata known as *timed event graphs* (TEG). A TEG is a timed place Petri net in which forks and joins are permitted only at transitions. A delay or processing time is associated with each place connecting pairs of transitions. Each transition in the graph corresponds to an event in the system. When an event occurs it initiates the processes connecting it to successor events. These events will then occur when all the processes connecting them to their predecessor events are completed.

It turns out that TEG and FSM can both be described by idempotent algebraic structures known as *semirings* or *dioids*. Work on the performance analysis of TEG modelled in the "max–plus" semiring is described in [8] and [3]. Discussion of FSM modelled with a semiring based on regular languages can be found in [9] and [1]. In Section 2 we will briefly review these algebraic structures and elaborate on their similarities.

A well-developed theoretical framework has been established by Ramadge and Wonham for studying the control of logical DES behavior [13, 16]. In this framework, event sequences are normally assumed to be generated by a FSM which represents the system. Certain events in the system are designated as being controllable and may be disabled by a *supervisor* which is separate from the system. The goal of the supervisor is to restrict the system to some specified desirable behavior. In Section 3 we will show how the algebraic similarity between FSM and TEG can be used to define an analogous control framework for certain timed DES.

The idempotency property of the semirings we consider implies the existence of a lattice structure. Consequently, fixed-point results for lattices which have been used to study the existence and computation of optimal supervisors for logical DES [10] can also be applied to timed DES modelled by TEG. This is the subject of Section 4. These results are described in greater detail in [7].

In [11] control of logical DES is studied by the *synchronous composition* of two FSM, one representing the system and one representing the supervisor. In their synchronous composition an event which is common to both the system and the supervisor may only occur if it is enabled in both. In Section 5, we use an analogous definition of synchronous composition for TEG to study the controllability of behaviors specified in terms of minimum separation times between events. Additional details of this approach are discussed in [6].

2 Idempotent Structures

In [3] and [8] the dynamic behavior of TEG is studied using algebraic techniques based on idempotent semirings. Recall that a semiring differs from a ring most notably in that its elements form a monoid rather than a group under the sum operation (denoted $\oplus$).

Consider a TEG $G \equiv (T, A)$ where T is a set of N transitions and A is an $N \times N$ matrix of delays at places connecting the transitions. Let $x_i(k)$ denote the kth firing time of transition t_i. For a TEG where each place contains a single token, firing times satisfy a set of equations of the form

$$x_i(k) = \max\{ \max_{1 \le j \le N}\{A_{ij} + x_j(k-1)\}, v_i(k)\} \qquad (2.1)$$

where $v_i(k)$ specifies the earliest firing time for t_i. In the semiring structure of $(\mathcal{R} \cup \{\pm\infty\}, \max, +)$, where maximization is the idempotent sum and addition is the product, these equations can be written in the matrix form

$$x(k) = Ax(k-1) \oplus v(k).$$

In order to pursue the analogy with logical DES we employ a somewhat more general algebraic structure based on sequences, as in [4]. When studying

logical DES we are interested in sequences of event labels produced by a
system, while in the timed case we are concerned with sequences of event
occurrence times. Let $x_i \equiv \{x_i(k) \mid k \in \mathcal{N}\}$ denote the sequence of firing
times of transition t_i. Then x_i belongs to the set of sequences over $\{\mathcal{R} \cup \pm\infty\}$,
which we will call S. Let $\oplus$ denote pointwise maximization of sequences in S.
Let the delay matrix A now consist of functions which map sequences in S to
S. The absence of a place connecting t_j to t_i is indicated by setting $A_{ij} = \varepsilon$.
The function ε maps all sequences to the constant sequence $\{-\infty, -\infty, \ldots\}$,
and thus does not impose any delay or synchronization constraints.

The delay functions in A are required to belong to the set $\mathcal{F}$ of *lower-
semicontinuous* functions over S, meaning that they distribute over the $\oplus$
operator in S. Functions in $\mathcal{F}$ may be combined by a derived $\oplus$ operation
defined by

$$(f \oplus g)(x) = f(x) \oplus g(x).$$

Functions in $\mathcal{F}$ may also be combined by composition. The function ε is the
identity for $\oplus$. The function 0 which adds zero to every term in a sequence is
the identity for composition. With these operations $\mathcal{F}$ and S form a structure
called a *moduloid* or *semimodule* [3, 4, 15].

The behavior of all the transitions in a TEG is a vector sequence in S^N
which satisfies (2.1). Using the semimodule representation, this equation can
be written

$$x = Ax \oplus v. \tag{2.2}$$

The actual behavior of the system is given by the least solution to (2.2), which
is $x = A^*v$, where

$$A^* = \bigoplus_{i \geq 0} A^i$$

and $A^0 = I$ is the identity matrix having the function 0 on the diagonal
and ε elsewhere. Borrowing notation from untimed automata we denote the
behavior of a TEG G by $L(G)$ and so for the system in (2.2) we have $L(G) = A^*v$.

To model the effect of tokens in the initial marking of a TEG, we use the
index backshift function γ defined by

$$\gamma x(k) = \begin{cases} x(k-1) & : k > 0 \\ \varepsilon & : k = 0. \end{cases}$$

Each of the m tokens in the initial marking of a place causes a backshift since
the $(k - m)$th token to enter the place will be the kth to depart and enable
its successors. Therefore a place which adds a delay of 5 and contains two
initial tokens is described by composing the unary addition function 5 with
two γ functions, yielding $5\gamma^2$.

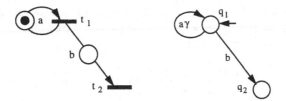

Figure 1: (a) Timed event graph and (b) finite state machine with same structure.

Example 2.1 The TEG of Figure 1(a) can be described by the equation

$$\begin{bmatrix} x_1 \\ x_2 \end{bmatrix} = \begin{bmatrix} a\gamma & \varepsilon \\ b & \varepsilon \end{bmatrix} \begin{bmatrix} x_1 \\ x_2 \end{bmatrix} \oplus \begin{bmatrix} \bar{0} \\ -\bar{\infty} \end{bmatrix},$$

where the overbar notation denotes a constant sequence. Thus, no action occurs before time 0. The behavior of the system is then given by

$$L(G) = A^*v = \begin{bmatrix} (a\gamma)^* & \varepsilon \\ b(a\gamma)^* & 0 \end{bmatrix} \begin{bmatrix} \bar{0} \\ -\bar{\infty} \end{bmatrix}.$$

For the case where $a = 2$ and $b = 3$ this results in the sequence

$$\begin{bmatrix} x_1 \\ x_2 \end{bmatrix} = \left\{ \begin{bmatrix} 0 \\ 3 \end{bmatrix}, \begin{bmatrix} 2 \\ 5 \end{bmatrix}, \begin{bmatrix} 4 \\ 7 \end{bmatrix}, \dots \right\}.$$

□

Timed event graphs are structurally very similar to finite state machines. Both are directed graphs with labelled edges. While in a TEG the nodes correspond to events and the edges to process delays, the nodes in a FSM represent the system states and the edges correspond to events. This is illustrated by comparing the automata in Figure 1. Both TEG and FSM are classes of Petri nets. While a TEG has its forks and joins only at transitions, a FSM has forks and joins only at places. The structures are dual in the sense that a FSM models nondeterministic choice but not synchronization while a TEG models synchronization but not choice.

Because of the structural similarities between these automata there is an algebraic similarity as well. If we consider sets of event sequences together with the operation of set union, the sequences generated by a FSM can be described in a semimodule framework. The functions which are assigned to the edges in this case are concatenations of event labels. The system is therefore governed by an equation of the form $x = Ax \oplus v$ as for the TEG.

Example 2.2 Consider the FSM of Figure 1(b) with initial state q_1 and event set $\Sigma = \{a, b, \gamma\}$. We are interested in determining all of the event sequences that can be generated by the system when it terminates in either state q_1 or q_2. In the semimodule framework, the corresponding sets of event sequences x_1 and x_2 must satisfy

$$\begin{bmatrix} x_1 \\ x_2 \end{bmatrix} = \begin{bmatrix} a\gamma & \emptyset \\ b & \emptyset \end{bmatrix} \begin{bmatrix} x_1 \\ x_2 \end{bmatrix} \oplus \begin{bmatrix} 1 \\ \emptyset \end{bmatrix},$$

where 1 denotes the set consisting of a zero-length sequence and indicates the initial state of the FSM. The behavior of the system is then given by

$$L(G) = A^*v = \begin{bmatrix} (a\gamma)^* & \emptyset \\ b(a\gamma)^* & 0 \end{bmatrix} \begin{bmatrix} 1 \\ \emptyset \end{bmatrix}.$$

Due to the convention of composing functions from the left, the sequences are generated in reverse order. Accounting for this we have

$$x_1 = (a\gamma)^*, \quad x_2 = (a\gamma)^*b.$$

Note that in this semimodule $(a\gamma)^* = \bigcup_{i \geq 0}\{(a\gamma)^i\}$.   □

It is this algebraic similarity which suggests that control of timed event graphs may be studied using techniques developed for untimed DES.

3   Supervisory Control

There is a natural control technology to use with TEG which extends the similarities with logical DES modelled by FSM. In a logical DES model with event set Σ, the *language L* of the system is the set of all event sequences it can generate. Events are classified as either *controllable*, meaning that their occurrence may be prevented, or *uncontrollable*. The control objective is to restrict the system to some desired language $K \subseteq L$. Control is accomplished by dynamically disabling certain of the controllable events to avoid undesirable behaviors. Observe that such a controller or *supervisor* may restrict system behavior, but not introduce any new behaviors. Since uncontrollable events may not be disabled, not all behaviors are realizable.

For a supervisor to be able to restrict the system to a desired language, uncontrollable actions which are executable by the system must not result in sequences which lie outside of that language. This invariance property yields the definition of controllability for a desired behavior.

Definition 3.1   ([13]) *Language K is said to be* controllable *with respect to L and the uncontrollable events Σ_u if*

$$pr(K).\Sigma_u \cap L \subseteq pr(K) \tag{3.1}$$

where $pr(k)$ is the prefix closure *of K.*

Now consider a timed event graph governed by (2.2). Suppose that some events $T_c \subseteq T$ are designated as controllable, meaning that their transitions may be delayed from firing until some arbitrary later time. The delayed enabling times $u_i(k)$ for the controllable events are to be provided by a supervisor. Let u represent the sequence of transition enabling times provided by the supervisor, with $u_i(k) = -\infty$ for t_i uncontrollable. Then the supervised system is described by $x = Ax \oplus v \oplus u$.

To compute the effect of uncontrollable events, let I_c denote the matrix having the identity function 0 on diagonal elements i for which $t_i \in T_c$ and ε elsewhere. Suppose Y is a set of desired event occurrence time sequences for the system. Then for any desired sequence $y \in Y$ the supervisor may provide firing times $u = I_c y$, which results in $L(G) = A^*(I_c y \oplus v)$. In other words we want the system to execute events at times given by y, but because of the uncontrollable events they actually occur at $A^*(I_c y \oplus v)$.

For an analogous definition for controllability, the desired behavior must be invariant under uncontrollable actions. This results in the following definition of controllability.

Definition 3.2 *A set of sequences $Y \subseteq \mathcal{S}^N$ is* controllable *with respect to A, v, and T_c if*

$$A^*(I_c Y \oplus v) \subseteq Y. \tag{3.2}$$

Intuitively, this means that enabling controllable events at any time allowed by the specification set Y must result in behavior within Y for all events. Notice that, as in the untimed model, no new behavior is introduced by the supervisor. System operation can never be accelerated – events can only be delayed.

As an example suppose that we wish to slow the system down as much as possible without causing any event to occur later than some sequence of execution times y. Such a specification could be used to maintain a production schedule in a factory. This type of specification is described by

$$Y = \{x \in \mathcal{S}^N \mid x \le y\} \tag{3.3}$$

where $y \in \mathcal{S}^N$ is a fixed sequence. For the remainder of the section we will consider acceptable behaviors of this form.

Theorem 3.1 ([7]) *If Y specifies sequences no later than y as in (3.3) then Y is controllable if and only if*

$$A^*(I_c y \oplus v) \le y. \tag{3.4}$$

Example 3.1 For the system in Figure 1(a) with t_1 controllable, suppose we are given as a specification the set Y of sequences less than or equal to

$$y = \left\{ \begin{bmatrix} 0 \\ 3 \end{bmatrix}, \begin{bmatrix} 3 \\ 5 \end{bmatrix}, \begin{bmatrix} 6 \\ 7 \end{bmatrix}, \ldots \right\}.$$

To determine whether Y is controllable, we compute

$$A^* I_c y \oplus A^* v = \begin{bmatrix} (2\gamma)^* & \varepsilon \\ 3(2\gamma)^* & 0 \end{bmatrix} \begin{bmatrix} y_1 \\ \varepsilon \end{bmatrix} \oplus A^* v$$

$$= \left\{ \begin{bmatrix} 0 \\ 3 \end{bmatrix}, \begin{bmatrix} 3 \\ 6 \end{bmatrix}, \begin{bmatrix} 6 \\ 9 \end{bmatrix}, \dots \right\} > y$$

so Y is not controllable.   □

When a specification of type (3.3) is found to be uncontrollable, a less restrictive controllable specification can always be generated from it.

Proposition 3.1 ([7]) *If $y \geq v$ then $Y = \{x \in \mathcal{S}^N \mid x \leq A^* y\}$ is controllable.*

Example 3.2 Consider y from Example 3.1. Then

$$A^* y = \left\{ \begin{bmatrix} 0 \\ 3 \end{bmatrix}, \begin{bmatrix} 3 \\ 6 \end{bmatrix}, \begin{bmatrix} 6 \\ 9 \end{bmatrix}, \dots \right\}$$

and

$$A^* I_c (A^* y) \oplus A^* v = \left\{ \begin{bmatrix} 0 \\ 3 \end{bmatrix}, \begin{bmatrix} 3 \\ 6 \end{bmatrix}, \begin{bmatrix} 6 \\ 9 \end{bmatrix}, \dots \right\} \leq A^* y$$

so by Theorem 3.1 $A^* Y$ is controllable.   □

It is more likely that a given specification cannot be relaxed. In this case we must find an optimal set of controllable behaviors which meets the specification. Depending on the situation this may mean finding the largest set which is contained in the specified behavior or the smallest set which contains the specified behavior. Determining these extremal behaviors is the subject of the next section.

4   Control of Extremal Behaviors

Determining the optimal controllable behavior of a DES requires that we define an order on the set of possible behaviors. The idempotent semirings we have considered each implicitly define a partial order according to the relation

$$x \leq y \Leftrightarrow x \oplus y = y.$$

In fact, they are also *complete lattices* with respect to this ordering [8], meaning that every arbitrary subset X has both a greatest lower bound (inf X) and a least upper bound (sup X).

We next recall some useful properties of functions over a general lattice $(\mathcal{X}, \leq)$. A function $f : \mathcal{X} \to \mathcal{X}$ is *monotone* if

$$\forall x, y \in \mathcal{X} : x \leq y \Rightarrow f(x) \leq f(y).$$

For $\mathcal{X}$ a complete lattice, a function is *disjunctive* if

$$\forall X \subseteq \mathcal{X} : f(\sup X) = \sup_{x \in X}\{f(x)\}.$$

A function is *conjunctive* if

$$\forall X \subseteq \mathcal{X} : f(\inf X) = \inf_{x \in X}\{f(x)\}.$$

Disjunctive and conjunctive functions can both be shown to be monotone as well. Note that lower-semicontinuous functions are disjunctive and therefore monotone.

If f is disjunctive then its *dual*, denoted $f^{\perp}$, is defined by

$$f^{\perp}(y) = \sup\{x \in \mathcal{X} | f(x) \leq y\}.$$

If f is conjunctive then its *co-dual*, denoted $f^{\top}$, is defined by

$$f^{\top}(y) = \inf\{x \in \mathcal{X} | y \leq f(x)\}.$$

A lattice *inequation* is an expression of the form

$$f(x) \leq g(x)$$

over a lattice $(\mathcal{X}, \leq)$, where we wish to find $x \in \mathcal{X}$ to satisfy the given expression. In the previous section we saw that the controllability of logical and timed DES is characterized by inequations (3.1) and (3.2) on their respective lattices.

The problem of finding optimal controllable behaviors is equivalent to finding extremal solutions to one or more of these inequations. It is shown in [10] that finding such extremal solutions can be reduced to extremal fixed-point computations for certain induced functions. The relevant results are summarized in the following two theorems.

Theorem 4.1 ([10]) *Given the system of inequations*

$$\{f_i(x) \leq g_i(x)\}_{i \leq n} \tag{4.1}$$

over a complete lattice $(\mathcal{X}, \leq)$, *let*

$$Y = \{y \in \mathcal{X} | \forall i \leq n : f_i(y) \leq g_i(y)\}$$

be the set of all solutions of the system of inequations. Consider the sets of all fixed points of functions h_1 *and* h_2 *defined by*

$$h_1(y) = \inf\{f_i^{\perp}(g_i(y))\}, \quad Y_1 = \{y \in \mathcal{X} | h_1(y) = y\}$$
$$h_2(y) = \sup\{g_i^{\top}(f_i(y))\}, \quad Y_2 = \{y \in \mathcal{X} | h_2(y) = y\}.$$

1. If f_i is disjunctive and g_i is monotone $\forall i \leq n$, then $\sup Y \in Y$, $\sup Y_1 \in Y_1$, and $\sup Y = \sup Y_1$.

2. If f_i is monotone and g_i is conjunctive $\forall i \leq n$, then $\inf Y \in Y$, $\inf Y_2 \in Y_2$, and $\inf Y = \inf Y_2$.

Under the stated conditions, the induced functions h_1 and h_2 are monotone. On a complete lattice, monotonicity guarantees the existence of supremal and infimal fixed points [14]. When the f's and g's satisfy these conditions, extremal solutions of (4.1) exist and correspond to the extremal fixed points of h_1 and h_2. The next theorem uses these functions to compute extremal solutions to (4.1).

Theorem 4.2 ([10]) *Consider the system of inequations (4.1) over a complete lattice $(\mathcal{X}, \leq)$ and the set Y of all solutions of the system.*

1. *Let f_i be disjunctive and g_i be monotone. Consider the iterative computation*

$$y_0 := \sup \mathcal{X}, \quad y_{k+1} := h_1(y_k).$$

If $y_{m+1} = y_m$ for some $m \in \mathcal{N}$ then $y_m = \sup Y$.

2. *Let f_i be monotone and g_i be conjunctive. Consider the iterative computation*

$$y_0 := \inf \mathcal{X}, \quad y_{k+1} := h_2(y_k).$$

If $y_{m+1} = y_m$ for some $m \in \mathcal{N}$ then $y_m = \inf Y$.

In the controllability condition for logical DES (3.1) on the lattice of languages, the prefix closure function is disjunctive and monotone and the concatenation function is disjunctive. With some additional modifications Theorems 4.1 and 4.2 demonstrate the existence and computation of the supremal controllable sublanguage of any specified language. Next we show that these results can be applied analogously to timed DES using the controllability condition (3.2).

Suppose we are given a set $Y \subseteq \mathcal{S}^N$ of acceptable sequences. We wish to find the least subset or the greatest superset of Y such that enabling the controllable events at times given by a sequence in the extremal set results in an actual behavior which lies in the extremal set. That is, we seek extremal sets of sequences which are invariant under uncontrollable actions.

To find the supremal controllable subset of Y we must find the supremal solution to the pair of inequations

$$\begin{aligned} A^*(I_c X \oplus v) &\subseteq X \\ X &\subseteq Y. \end{aligned} \tag{4.2}$$

It is easy to show that $A^*(I_c(\cdot) \oplus v)$ and the identity function are disjunctive and that the constant function Y is monotone. Therefore, we meet the conditions of the first part of Theorem 4.1 with

$$h_1(X) = Y \cap (A^*(I_c X \oplus v))^\perp.$$

Similarly, to find the infimal solution greater than Y let the second inequation in (4.2) be $Y \subseteq X$. Since the constant function Y is monotone and the identity function is conjunctive we can use the second part of Theorem 4.1 with

$$h_2(X) = Y \cup A^*(I_c X \oplus v)$$

where we use the fact that the co-dual of the identity is itself.

Summarizing, we have the following result.

Theorem 4.3 *Given $G = (T, A)$ with controllable events T_c and a set of acceptable behaviors $Y \in \mathcal{S}^N$, the supremal controllable subset of Y and the infimal controllable superset of Y both exist.*

Both of these solutions are computable by the iterative method of Theorem 4.2. For the special case where the function $f(\cdot) = A^*(I_c(\cdot) \oplus v)$ is disjunctive and also idempotent the iterative computation reduces to a single step.

Corollary 4.1 *Consider a complete lattice $(\mathcal{X}, \leq)$, a disjunctive and idempotent function f on $\mathcal{X}$, and a fixed $\hat{x} \in \mathcal{X}$.*

1. *The supremal solution less than $\hat{x}$ of $f(x) \leq x$ is $\inf\{\hat{x}, f^\perp(\hat{x})\}$.*

2. *The infimal solution greater than $\hat{x}$ of $f(x) \leq x$ is $\sup\{\hat{x}, f(\hat{x})\}$.*

Suppose that there are multiple constraints on the desired system behavior that must be satisfied. The desired behavior is then specified as the intersection of two or more sets. In computing the supremal controllable behavior it is possible to take a modular approach and consider the constraint sets individually without sacrificing overall performance.

Theorem 4.4 *Let Z_1 and Z_2 be the supremal controllable subsets of Y_1 and Y_2, respectively. Then the supremal controllable subset of $Y_1 \cap Y_2$ is given by $Z_1 \cap Z_2$.*

This modular approach is illustrated in the following example.

Example 4.1 Suppose a cat chases a mouse through the three-room house in Figure 2. One of the doors can be held shut to prevent the cat from entering the next room. The mouse's movement may not be restricted. Initially the

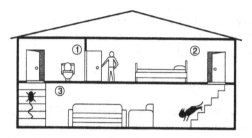

Figure 2: The cat and mouse problem.

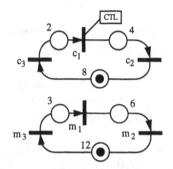

Figure 3: TEG for cat and mouse.

cat and mouse are at opposite ends of room 3. Our (admittedly questionable) objective is to shut the controllable door at the proper times to keep the cat from catching the mouse.

A timed event graph for the cat and mouse is shown in Figure 3. Event m_1 is the mouse leaving room 1, and so on. The controllable event is c_1. The initial condition gives the mouse a head start since it can exit room 3 immediately but the cat will take 8 seconds to cross the room and exit. The cat and mouse are governed by

$$
x_m = \begin{bmatrix} \varepsilon & \varepsilon & 3 \\ 6 & \varepsilon & \varepsilon \\ \varepsilon & 12\gamma & \varepsilon \end{bmatrix} x_m \oplus \begin{bmatrix} \bar{0} \\ \bar{0} \\ \bar{0} \end{bmatrix}
$$

$$
x_c = \begin{bmatrix} \varepsilon & \varepsilon & 2 \\ 4 & \varepsilon & \varepsilon \\ \varepsilon & 8\gamma & \varepsilon \end{bmatrix} x_c \oplus \begin{bmatrix} \bar{0} \\ \bar{0} \\ \bar{8} \end{bmatrix}.
$$

Our objective of sparing the mouse is equivalent to the following two specifications:

1. The cat must leave each room after the mouse does, so that the mouse stays ahead of the cat for all time. This is expressed in terms of the event occurrence times by $x_c \geq x_m$.

2. The mouse must not enter a room until the cat has left. This means that the mouse must not "lap" the cat; otherwise, the cat could turn around (within the same room) and catch it. Then we must have

$$
x_c \leq S x_m = \begin{bmatrix} \varepsilon & \varepsilon & \gamma^{-1} \\ \gamma^{-1} & \varepsilon & \varepsilon \\ \varepsilon & 0 & \varepsilon \end{bmatrix} x_m.
$$

Specification (1) can be met if $Y_1 = \{x \in \mathcal{S}^N \mid x \geq y_1 = x_m\}$ is controllable with respect to the cat system. However, we find that $C^*(I_c y_1 \oplus v_c) \notin Y_1$, showing that Y_1 is not controllable. Using Corollary 4.1, we compute the supremal controllable subset of Y_1, which is $Z_1 = \{x \in \mathcal{S}^N \mid x \geq z_1\}$, where

$$
z_1 = \left\{ \begin{bmatrix} 10 \\ 14 \\ 8 \end{bmatrix}, \begin{bmatrix} 30 \\ 34 \\ 22 \end{bmatrix}, \begin{bmatrix} 51 \\ 55 \\ 42 \end{bmatrix}, \ldots \right\}.
$$

Similarly, we find that for specification (2) $Y_2 = \{x \in \mathcal{S}^N \mid x \leq y_2 = S x_m\}$ is not controllable. Its supremal controllable subset is $Z_2 = \{x \in \mathcal{S}^N \mid x \leq z_2\}$, where

$$
z_2 = \left\{ \begin{bmatrix} 18 \\ 24 \\ 9 \end{bmatrix}, \begin{bmatrix} 39 \\ 45 \\ 30 \end{bmatrix}, \begin{bmatrix} 60 \\ 66 \\ 51 \end{bmatrix}, \ldots \right\}.
$$

Finally, by Theorem 4.4 we know that the supremal controllable subset of the overall desired behavior given by specifications (1) and (2) is $Z_1 \cap Z_2 = \{x \in \mathcal{S}^N \mid z_1 \leq x \leq z_2\}$. □

5 Control of Separation Times

In the previous section desired behavior was specified as a set or range of acceptable schedules (the sequences of event occurrence times). In this section we consider the situation where minimum separation times are to be enforced between certain event occurrences. For example, we may want to ensure that there is sufficient time between the successive outputs of a machine to perform an inspection, or we may want to delay the departure of an airplane until all of its connecting flights have been on the ground for some minimum time. Such a specification will take the form

$$
x_i \geq S_{ij} x_j
$$

where S_{ij} is a function giving the required delay between events t_i and t_j. The set of acceptable behaviors is then

$$
Y = \{x \in \mathcal{S}^N \mid x \geq S x\}. \tag{5.1}
$$

The problem now is to find the earliest controllable schedule (i.e. a single sequence y such that $A^*(I_c y \oplus v) = y$) which belongs to this set.

Note that the minimum required separation times S essentially form another TEG in the same way as the system delay times A. Thus a supervisor which provides the delayed enabling times to the controllable events in the plant can be defined by $x = Sx \oplus v$. We must now determine whether the system and supervisor automata can be combined in a way consistent with the controllable event set to yield an actual behavior that meets the specification in (5.1). To do so we will again pursue an analogy with logical DES.

Often the desired behavior for a logical DES will be specified by another FSM. The specification FSM is made to serve as a supervisor for the plant FSM by operating the two in synchrony [13]. The *synchronous composition* of the plant and the supervisor will execute an event common to both machines only if it is enabled in both machines. Thus, the supervisor may prevent or disable events which otherwise would have been permitted in the plant. A supervisor is said to be *complete* if it does not disable any uncontrollable events.

Synchronization of timed event graphs can be defined in a similar manner. Assume that we are given two TEG defined by delay matrices A_1 and A_2. Transitions which are common to both graphs fire only when they are enabled in both graphs. Transitions which appear in only one of the graphs fire when enabled by their own local predecessors, as before. An important observation is that synchronization may be used to realize a control policy for a TEG since some of its events may be delayed if they are synchronized to events in another TEG.

In the semimodule representation, the construction of the synchronous composition of two TEG is straightforward. If their respective event sets T_1 and T_2 are identical, then their synchronous composition, denoted $G_1 \| G_2$, is governed by

$$x = (A_1 \oplus A_2)x \oplus v$$

where $v \equiv v_1 \oplus v_2$. The sequence of firing times is then given by

$$L(G_1 \| G_2) = (A_1 \oplus A_2)^* v.$$

If the event sets are not identical, events are mapped into $T_1 \cup T_2$. For simplicity we assume hereafter that all TEG are defined over a common event set unless stated otherwise.

It is easy to see that synchronization results in event times which are greater than or equal to those of the individual systems operating independently.

Proposition 5.1 ([6]) $L(G_1 \| G_2) \geq L(G_1) \oplus L(G_2)$.

In fact, this inequality may be strict as illustrated in the next example.

Example 5.1 Consider the systems G_1 and G_2 with delays

$$A_1 = \begin{bmatrix} 1\gamma & \varepsilon \\ 2 & \varepsilon \end{bmatrix} x \oplus \begin{bmatrix} \bar{0} \\ \bar{0} \end{bmatrix}, \quad A_2 = \begin{bmatrix} \varepsilon & 1\gamma \\ \varepsilon & 1\gamma \end{bmatrix} x \oplus \begin{bmatrix} \bar{0} \\ \bar{0} \end{bmatrix}.$$

While the individual behavior of these graphs is given by

$$L(G_1) = \left\{ \begin{bmatrix} 0 \\ 2 \end{bmatrix}, \begin{bmatrix} 1 \\ 3 \end{bmatrix}, \begin{bmatrix} 2 \\ 4 \end{bmatrix}, \ldots \right\}, \quad L(G_2) = \left\{ \begin{bmatrix} 0 \\ 0 \end{bmatrix}, \begin{bmatrix} 1 \\ 1 \end{bmatrix}, \begin{bmatrix} 2 \\ 2 \end{bmatrix}, \ldots \right\}$$

their synchronous composition yields

$$\begin{aligned} L(G_1 \| G_2) &= \left\{ \begin{bmatrix} 0 \\ 2 \end{bmatrix}, \begin{bmatrix} 3 \\ 5 \end{bmatrix}, \begin{bmatrix} 6 \\ 8 \end{bmatrix}, \ldots \right\} \\ &> L(G_1) \oplus L(G_2). \end{aligned}$$

$\square$

Synchronous composition of TEG may introduce new circuits. Thus there is the potential to introduce a deadlock (a circuit containing no tokens) that was not present before. This condition is checked by examining $A_0 = (A_1)_0 \oplus (A_2)_0$, the subgraph of places in the synchronous composition with no initial tokens. If there exists a permutation of the rows and columns of this matrix that makes it strictly upper triangular then it is circuit-free. Equivalently, if there exists $n \leq N$ such that $(A_0)^n = \varepsilon$ then there is no deadlocked circuit.

Suppose that two TEG to be synchronized represent a plant A and a supervisor S. In defining synchronization we have implicitly assumed that all events in the plant are controllable. If this is actually the case, the problem becomes trivial since synchronization yields the sequence $y = (A \oplus S)^* v$ which is controllable and satisfies the minimum separation time specification.

Proposition 5.2 *If all events are controllable then $y \equiv L(G_p \| G_s)$ is controllable and $Sy \leq y$.*

Proof Since all events are controllable, $I_c = I$, the identity matrix in $\mathcal{F}$. The sequence $y = (A \oplus S)^* v$ is controllable since

$$A^*((A \oplus S)^* v \oplus v) = A^*(A \oplus S)^* v = (A \oplus S)^* v.$$

Since $S \leq A \oplus S$ we also have

$$S(A \oplus S)^* v \leq (A \oplus S)^* v.$$

$\square$

We must now account for the effect of the uncontrollable events. The specification is a legitimate supervisor for the plant if it does not *directly* delay the occurrence of uncontrollable events. That is, the specification may require uncontrollable event t_i to occur d seconds after some other event t_j, but this delay cannot be imposed directly; only through the action of controllable events and delays within the plant. This is analogous to the requirement that a FSM supervisor should not disable any uncontrollable events.

To verify this we must check whether the synchronous composition of the plant and supervisor is changed by deleting from the supervisor all incoming edges to uncontrollable events. Let I_c be defined as before. Then $\hat{G}_s = (T, I_c S)$ removes from the supervisor graph all incoming edges to events which are uncontrollable in the plant. We may sum up this condition with the following definition.

Definition 5.1 *A supervisor S is* complete *with respect to A and T_c if the synchronous operation of the supervisor and plant does not depend upon any delay of uncontrollable events imposed directly by the supervisor; that is,*

$$L(G_p\|\hat{G}_s) = L(G_p\|G_s)$$
$$\text{or } (A \oplus I_c S)^* v = (A \oplus S)^* v.$$

Note that completeness may or may not depend on the initial condition v. If $(A \oplus I_c S)^* = (A \oplus S)^*$ then S is complete for any initial condition.

Example 5.2 Consider the following plant and supervisor

$$A = \begin{bmatrix} \varepsilon & \varepsilon & 1\gamma \\ \varepsilon & \varepsilon & 1\gamma \\ 2 & 1 & \varepsilon \end{bmatrix}, \quad S = \begin{bmatrix} \varepsilon & 1 & \varepsilon \\ \varepsilon & \varepsilon & \varepsilon \\ \varepsilon & 3 & \varepsilon \end{bmatrix}$$

with t_1 controllable in the plant (see Figure 4). Letting $D = A \oplus S$ and $\hat{D} = A \oplus I_c S$ we have

$$D = \begin{bmatrix} \varepsilon & 1 & 1\gamma \\ \varepsilon & \varepsilon & 1\gamma \\ 2 & 3 & \varepsilon \end{bmatrix}, \quad \hat{D} = \begin{bmatrix} \varepsilon & 1 & 1\gamma \\ \varepsilon & \varepsilon & 1\gamma \\ 2 & 1 & \varepsilon \end{bmatrix}.$$

We find that $D^* = \hat{D}^*$ and therefore the supervisor is complete.   □

If the completeness condition holds, then the specification is implemented simply by means of the synchronous composition of the plant and the specification event graphs. The resulting behavior is controllable and meets the separation time requirements. To show this we first note that synchronization with a supervisor having no incoming edges to uncontrollable events always results in controllable behavior.

Lemma 5.1 *The sequence $L(G_p\|\hat{G}_s)$ is controllable.*

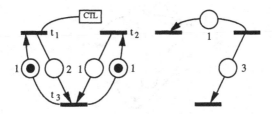

Figure 4: Plant A and complete supervisor S.

Theorem 5.1 *If supervisor S is complete with respect to A and T_c then $y \equiv L(G_p\|G_s)$ is a controllable sequence and $y \geq Sy$. Furthermore, y is the least sequence satisfying both of these properties.*

Proof Since $L(G_p\|G_s) = L(G_p\|\hat{G}_s)$, y is controllable by Lemma 5.1. We have already showed $Sy \leq y$ in Proposition 5.2. Now let z be any other controllable sequence such that $Sz \leq z$. Then z controllable implies $A^*v \leq z$ and $A^*z = z$, and so $Az \leq z$. Therefore $(A \oplus S)z \leq z$ and $(A \oplus S)^*z \leq z$. Finally,

$$y = (A \oplus S)^*v = (A \oplus S)^*A^*v \leq (A \oplus S)^*z \leq z$$

so y is the least sequence satisfying both properties. □

Suppose that the specification may be decomposed as a collection of k TEG whose separation times must all be satisfied. Equivalently, the desired behavior may be specified in a modular fashion. In either case we have an overall specification of the form

$$S = S_1 \oplus \cdots \oplus S_k.$$

It turns out that the component specifications may be evaluated individually to make a valid judgement regarding the completeness and, therefore, the controllability of the whole specification.

Theorem 5.2 *If each of the supervisors in the collection $\{S_i\}$ is complete then $\bigoplus_i S_i$ is complete.*

The next theorem gives us a check for completeness which is both necessary and sufficient.

Theorem 5.3 *A supervisor S is complete for any initial condition if and only if*

$$S \leq (A \oplus I_c S)^*. \tag{5.2}$$

Proof

$$S \leq (A \oplus I_c S)^* \;\Leftrightarrow\; A \oplus S \leq (A \oplus I_c S)^*$$
$$\Leftrightarrow\; (A \oplus S)^* \leq (A \oplus I_c S)^*$$

which is equivalent to S being complete for any initial condition. □

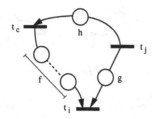

Figure 5: Delay h added to achieve required separation g.

Suppose that the given specification S is not complete. Assuming that it is important to achieve at least the event separation times given by S we would like to find the smallest $\hat{S} \geq S$ which is complete. Unfortunately, such a specification does not always exist [6]. Consider a transition t_i that fails the completeness test in (5.2). For the supervisor to influence the behavior of t_i so as to meet the specification, there must be a path from some controllable transition to t_i. We characterize the existence of such a path as follows.

Definition 5.2 *A transition t_i is* structurally controllable *if it is reachable from some controllable transition; that is, there exists $n < N$ and $t_k \in T_c$ such that $[(A)^n]_{ik} > \varepsilon$. In this case, $[(A)^n]_{ik}$ may be decomposed as $\bigoplus_p f_p$ with $f_p = a_p \gamma^{m_p}$. Each f_p is called a* controllable path.

We may now state a sufficient condition for the existence of a complete specification greater than S.

Theorem 5.4 *Suppose S is not complete. Let U denote the set of transitions for which the completeness condition (5.2) fails to hold. Then a complete specification greater than S exists if the following two conditions are satisfied:*

1. *Every transition in U is structurally controllable.*

2. *For each $t_i \in U$ with $[S]_{ij} \equiv g > \varepsilon$ there exists a controllable path f containing fewer tokens in its initial marking than g.*

Consider a typical situation shown in Figure 5. If every event $t_i \in U$ which requires a delay g from some t_j is reachable from a controllable event t_c via controllable path f (condition 1) then the result follows by adding delay h from t_j to t_c such that $fh \geq g$. Condition 2 is necessary since we must have a non-negative initial marking for h.

Example 5.3 To illustrate, return to the cat and mouse example. This time we specify the desired behavior in terms of minimum separation times between the cat and the mouse.

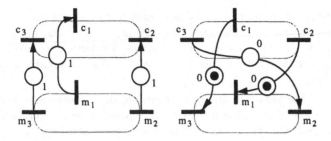

Figure 6: TEG for specifications S_1 and S_2.

1. The cat must leave each room at least one second after the mouse does. This is expressed in terms of the event occurrence times by $x_c \geq S_1 x_m$.

2. The mouse must not enter a room until the cat has left, so we must have $x_m \geq S_2 x_c$.

$$S_1 = \begin{bmatrix} 1 & \varepsilon & \varepsilon \\ \varepsilon & 1 & \varepsilon \\ \varepsilon & \varepsilon & 1 \end{bmatrix}, \quad S_2 = \begin{bmatrix} \varepsilon & \gamma & \varepsilon \\ \varepsilon & \varepsilon & 0 \\ \gamma & \varepsilon & \varepsilon \end{bmatrix}.$$

These specifications are equivalent to the TEG in Figure 6. Now the goal is to find a complete supervisor for both specifications using Theorem 5.4.

We first consider specification (1), and find that S_1 fails completeness with $U = \{c_2, c_3\}$. These transitions are found to be structurally controllable from c_1. Adding a delay of 1 from m_2 to c_1 achieves the desired delay for both transitions. This results in

$$\hat{S}_1 = \begin{bmatrix} 1 & 1 & \varepsilon \\ \varepsilon & 1 & \varepsilon \\ \varepsilon & \varepsilon & 1 \end{bmatrix} \geq S_1$$

which is complete. Next, we can verify that S_2 is complete because of the initial condition. By Theorem 5.2, this shows that the supervisor $S = \begin{bmatrix} \varepsilon & S_2 \\ \hat{S}_1 & \varepsilon \end{bmatrix}$ is complete. □

6   Conclusion

We have shown that timed DES which can be modelled by timed event graphs are structurally related to finite state machines. Both can be described by linear equations over an idempotent semiring. Using this structural similarity, we have extended supervisory control techniques developed for untimed DES to the timed case. When behavioral constraints are given as a range of acceptable schedules, it is possible to compute an extremal controllable subset or superset of the desired behavior. When constraints are expressed in terms of minimum separation times between events, it is possible to determine whether there is a controllable schedule which realizes the desired behavior. If none exists, we provide conditions for the existence of an acceptable alternative specification.

Due to of the wealth of literature regarding supervisory control of logical DES, there are many areas in which corresponding results for timed DES can be studied. One area of current interest is the case where the supervisor receives only an incomplete observation of the system state. For example, observations could be delayed or only certain events in the system may be observable. This problem has been studied in [12] and [11], and by others. Another area of interest concerns uncertain delays in the system. Suppose that the delay functions are not known exactly but rather may take values in a given range. Bounds on system performance are studied for this case in [2]. The control question is to examine the existence and synthesis of a supervisor to achieve some desired performance in this situation.

References

[1] A. V. Aho, J. E. Hopcroft, J. D. Ullman (1974), *The Design and Analysis of Computer Algorithms*, Addison-Wesley.

[2] T. Amon, H. Hulgaard, S. M. Burns, G. Borriello (1993), 'An algorithm for exact bounds on the time separation of events in concurrent systems,' in *Proc. IEEE Int'l Conf. on Computer Design*, 166–173.

[3] F. Baccelli, G. Cohen, G. J. Olsder, J. P. Quadrat (1992), *Synchronization and Linearity*, Wiley, New York.

[4] D. D. Cofer (1995), *Control and Analysis of Real-time Discrete Event Systems*, Ph.D. Thesis, Dept. of Elec. & Comp. Eng., The University of Texas at Austin.

[5] D. D. Cofer, V. K. Garg (1994), 'A Max-algebra Solution to the Supervisory Control Problem for Real-Time DES,' in *Lecture Notes in Control and Information Sciences 199*, G. Cohen, J. P. Quadrat, eds, Springer, New York, 283–289.

[6] D. D. Cofer, V. K. Garg (1994), 'Supervisory control of timed event graphs,' in *Proc. 1994 IEEE Conf. on Sys., Man. & Cyb.*, San Antonio, 994–999.

[7] D. D. Cofer, V. K. Garg (1995), 'Supervisory control of real-time discrete event systems using lattice theory,' *Proc. 33rd IEEE Conf. Dec. & Ctl.*, Orlando, 978–983 (also to appear in *IEEE Trans. Auto. Ctl.*).

[8] G. Cohen, P. Moller, J. P. Quadrat, M. Viot (1989), 'Algebraic tools for the performance evaluation of DES,' *Proc. IEEE*, **77** 39–58.

[9] M. Gondran, M. Minoux (1984), *Graphs and Algorithms*, Wiley, New York.

[10] R. Kumar, V. K. Garg (1995), *Modeling and Control of Logical Discrete Event Systems*, Kluwer.

[11] R. Kumar, V. K. Garg, S. I. Marcus (1991), 'On controllability and normality of discrete event dynamical systems,' *Sys. & Ctl. Ltrs*, **17** 157–168.

[12] F. Lin, W. Wonham (1988), 'On observability of discrete event systems,' *Info. Sci.*, **44** 173–198.

[13] P. J. Ramadge, W. M. Wonham (1989), 'The control of discrete event systems,' *Proc. IEEE*, **77** 81–98.

[14] A. Tarski (1955), 'A lattice-theoretical fixpoint theorem and its applications,' *Pac. J. Math.*, **5**, 285–309.

[15] E. Wagneur (1991), 'Moduloids and pseudomodules: 1. Dimension theory,' *Discrete Mathematics*, **98** 57–73.

[16] W. M. Wonham, P. J. Ramadge (1987), 'On the supremal controllable sublanguage of a given language,' *SIAM J. Ctl. & Opt.*, **25** 637–659.

Maxpolynomials and Discrete-Event Dynamic Systems

Raymond A. Cuninghame-Green

Abstract

If the output $\{g_j\}$ of a discrete-event dynamic system has the property that the maxpolynomial $\phi_{k,r}(t) = \sum_{j=0}^{2r} {}^{\oplus} g_{k+j} \otimes t^{(j)}$ has for any k a corner of order r arising from $r+1$ consecutive strictly inessential terms, then there is no realisation over $(I\!R, \max, +)$ of the form $g_j = c \otimes A^j \otimes b$ with A of dimension r or less.

Given the sequence $\{g_j\}$ of Markov parameters, we seek an (A, b, c) realisation over $(\max, +)$ algebra, i.e. a suitably-dimensioned matrix A and tuples b, c such that

$$g_j = c \otimes A^{(j)} \otimes b \qquad (j = 0, 1, \ldots). \tag{1}$$

The **dimension** of the realisation is the order of the matrix A; in general, we seek a realisation of least possible dimension.

We introduce the Hankel matrix $H_{k,r}$, where

$$H_{k,r} = \begin{bmatrix} g_k & g_{k+1} & \cdots & g_{k+r} \\ g_{k+1} & g_{k+2} & \cdots & g_{k+r+1} \\ & & \cdots & \\ g_{k+r} & g_{k+r+1} & \cdots & g_{k+2r} \end{bmatrix}. \tag{2}$$

The following lemma is easily proved.

Lemma 1 *Let $\Gamma = \{g_u\}$ be a set of $r+1$ elements chosen from $H_{k,r}$, with no two from the same row or column. If there is an index p $(k \le p \le k+2r)$ and a constant α such that*

$$g_p \ge g_j + (p - j)\alpha \qquad (j = k, \ldots, k+2r), \tag{3}$$

then summing (3) over all $g_j \in \Gamma$ gives

$$(r+1)g_p \ge \sum_\Gamma g_u + (r+1)(p - k - r)\alpha \tag{4}$$

Interpretation via Maxpolynomials

For the definition and basic properties of maxpolynomials, the reader is referred to [1]. Let a term $\beta_n \otimes t^{(n)}$ be called **essential** in the maxpolynomial $\Sigma^{\oplus}\beta_j \otimes t^{(j)}$ (taken over some set J of indices j) if there is a value of t for which

$$\beta_n \otimes t^{(n)} > \sum_{J\backslash n}^{\oplus} \beta_j \otimes t^{(j)}. \tag{5}$$

Otherwise, the term is called **inessential**. If the inequality (5) is reversed ($<$) for all t, then the term is called **strictly inessential**.

From [1] we know the following.

Lemma 2 *A necessary and sufficient condition that $\beta_n \otimes t^{(n)}$ be essential is that (n, β_n) be an extreme point of the concave hypograph of the point-set $\{(j, \beta_j) : j \in J\}$.*

Given the sequence $\{g_j\}$ of Markov parameters, let us define the following maxpolynomial:

$$\phi_{k,r}(t) = \sum_{j=0}^{2r}{}^{\oplus} g_{k+j} \otimes t^{(j)}.$$

From [1] we now recall the concept of a *corner* of a maxpolynomial to establish our main result.

Theorem 1 *If, for some k, r, the maxpolynomial $\phi_{k,r}$ has a block of exactly r consecutive strictly inessential terms, then $\phi_{k,r}$ has an exactly $(r + 1)$-fold corner, and there is no realisation of dimension r or less.*

Proof From the theory in [1], we know that an exactly $(r + 1)$-fold corner is associated with a block of r consecutive terms which are strictly inessential. Specifically, let us assume, for some p in the range $k, \ldots, k + r - 1$, that the terms involving g_p and g_{p+r+1} are essential and that those involving $g_{p+1}, \ldots, g_{p+r}$ are strictly inessential. To simplify the notation, write q for $p + r + 1$.

Since, by Lemma 2, (p, g_p) and (q, g_q) are consecutive extreme points of the concave hypograph, the join of these two points lies strictly above the join of either of them to any third member of the point-set $\{(j, g_j) : j \in \{k, \ldots, k + 2r\}\}$. Hence, if we define the slope α by $\alpha = (g_q - g_p)/(r+1)$, it easily follows that for all $u \in \{k, \ldots, k + 2r\}$ we have

$$g_p \geq g_u + (p - u)\alpha \text{ with equality only if } u = p \text{ or } q \tag{6}$$

and

$$g_q \geq g_u + (q - u)\alpha \text{ with equality only if } u = p \text{ or } q. \tag{7}$$

Now let $\Gamma = \{g_u\}$ be a set of $r + 1$ elements chosen from $H_{k,r}$, with no two from the same row or column. Summing each of (6), (7) over Γ, we obtain following Lemma 1:

$$(r+1)g_p \geq \sum_\Gamma g_u + (r+1)(p-k-r)\alpha \text{ and } (r+1)g_q \geq \sum_\Gamma g_u + (r+1)(q-k-r)\alpha.$$

If we add $(p - k + 1)$ times the first of these inequalities to $(r - p + k)$ times the second, we obtain after some simplification

$$(p - k + 1)g_p + (r - p + k)g_q \geq \sum_\Gamma g_u. \tag{8}$$

It follows from (6), (7) that the inequality in (8) is strict unless every index u in Γ is either p or q. For this to happen with no two entries on the same row or column of $H_{k,r}$, there is but one possibility, namely that Γ consists of the two 'broken antidiagonals' consisting respectively of $(p - k + 1)$ copies of g_p and $(r - p + k)$ copies of g_q.

It now follows that the Classical Assignment Problem based on the matrix $H_{k,r}$ has a unique solution, so by the theory given in [3], $H_{k,r}$ has linearly independent columns over $(\mathbb{R}, \oplus, \otimes)$. Hence no realisation of dimension r or less can exist.   □

An example

Consider the sequence $\{g_j\} = 0, 2, 0, 1, 0, 1, \ldots$. Resolving the sixth-degree maxpolynomial $0 \oplus 2 \otimes t \oplus t^{(2)} \oplus 1 \otimes t^{(3)} \oplus t^{(4)} \oplus 1 \otimes t^{(5)} \oplus t^{(6)}$ into linear factors as discussed in [1], we find $(t \oplus 1) \otimes (t \oplus 0.25)^{(4)} \otimes (t \oplus -2)$. Since there is a 4th-order corner, we check and find the terms in $t^{(2)}, t^{(3)}, t^{(4)}$ to be strictly inessential, so we conclude that there is no realisation of dimension 3 or less. A realisation of dimension 4 is easily constructed using a companion matrix, since the sequence clearly satisfies $g_{j+4} = g_{j+2}$; this is then evidently a minimal-dimensional realisation.

References

[1] Cuninghame-Green, R.A. and Meijer, P.F.J. An algebra for piecewise-linear minimax problems, *Discr. Appl. Math.* **2** (1980) 267–294.

[2] Cuninghame-Green, R.A. Convexity/concavity in discrete-event dynamic systems, Preprint 93/19, School of Mathematics and Statistics, The University of Birmingham.

[3] Gondran, M. and Minoux, M. L'indépendence linéaire dans les dioïdes, *Bulletin de la Direction des Etudes et Recherches EdF, Série C - Mathématiques, Informatique* **1** (1978) 67–90.

The Stochastic HJB Equation and WKB Method

Vassili N. Kolokoltsov

Abstract *The paper is devoted to the study of the Cauchy problem for a stochastic version of the Hamilton–Jacobi–Bellman equation, its applications to stochastic optimal control theory, and to a new WKB-type method of constructing asymptotics for stochastic pseudodifferential equations.*

1 Introduction. Stochastic Optimal Control

This paper is devoted to the study of the equation

$$dS + H\left(t, x, \frac{\partial S}{\partial x}\right) dt + \left(c(t, x) + g(t, x)\frac{\partial S}{\partial x}\right) \circ dW = 0, \qquad (1.1)$$

where $x \in \mathcal{R}^n$, $t \geq 0$, $W = (W^1, \ldots, W^n)$ is the standard n-dimensional Brownian motion ($\circ$, as usual, denotes the Stratonovich stochastic differential), $S(t, x, [W])$ is an unknown function, and the Hamiltonian $H(t, x, p)$ is convex with respect to p. This equation can be naturally called the stochastic Hamilton–Jacobi–Bellman equation. First of all we explain how this equation appears in the theory of stochastic optimization. In Section 2 we develop the stochastic version of the method of characteristics to construct classical solutions of this equation and then, on the basis of the methods of idempotent analysis (and similarly to the deterministic case [KM1, KM2]), we construct the theory of the generalized solutions of the Cauchy problem for this equation. In Section 3 we apply these results to the construction of the WKB-type asymptotics for a certain stochastic pseudodifferential equation, namely, for the Belavkin quantum filtering equation describing a quantum particle under the continuous nondemolition observation of diffusion type. In Section 4 we give examples of explicitly solvable problems and also prove the well-posedness theorem for the Cauchy problem of an important class of these quantum filtering equations.

We shall now show the connections between equation (1.1) and optimal control theory. Let the controlled stochastic dynamics be defined by the equation

$$dx = f(t, x, u) dt + g(t, x) \circ dW, \qquad (1.2)$$

where the controlled parameter u belongs to some metric space U and the functions f, g are continuous in t, u and Lipschitz continuous in x. Let the income along the trajectory $x(\tau)$, $\tau \in [t, T]$, defined by the starting point

$x = x(0)$ and the control $[u] = u(\tau), \tau \in [t, T]$, be given by the integral

$$I_t^T(x, [u], [W]) = \int_t^T b(\tau, x(\tau), u(\tau)) \, d\tau + \int_t^T c(\tau, x(\tau)) \circ dW. \qquad (1.3)$$

We are looking for the equation on the cost (or Bellman) function

$$S(t, T, x, [W]) = \sup_{[u]} I_t^T(x, [u], [W]) + S_0(x(T)), \qquad (1.4)$$

where sup is taken over all piecewise smooth (or equivalently, piecewise constant) controls $[u]$ and S_0 is some given function (terminal income). One can use two alternative ways to deduce the equation for S: by means of the Ito formalism or by means of the Stratonovich integral. We have chosen here the second approach, because it seems to be simpler and intuitively clearer. The arguments are based on the following well known fact [SV, Su]: if we approximate the noise W in some stochastic Stratonovich equation by means of smooth functions

$$W(\tau) = \int_0^\tau q(s) \, ds, \qquad (1.5)$$

then the solutions of the corresponding classical (deterministic) equations will tend to the solution of the given stochastic equation. So, considering W to be of the form (1.5) we have dynamics in the form

$$\dot{x} = f(\tau, x, u) + g(\tau, x)q(\tau)$$

and the integral income in the form

$$\int_t^T [b(\tau, x(\tau), u(\tau)) + c(\tau, x(\tau))q(\tau)] \, d\tau.$$

Writing down the Bellman equation for the corresponding deterministic (nonhomogeneous) optimization problem we have

$$\frac{\partial S}{\partial t} + \sup_u \left(b(t, x, u) + f(t, x, u)\frac{\partial S}{\partial x} \right) + \left(c(t, x) + g(t, x)\frac{\partial S}{\partial x} \right) q(t) = 0.$$

Rewriting this equation in stochastic Stratonovich form we obtain (1.1) with

$$H(t, x, p) = \sup_u (b(t, x, u) + pf(t, x, u)).$$

Let us consider, in particular, two special cases.

(i) $c = 0$ and $g = g(t)$ does not depend on x. Then differentiating (1.1) we obtain

$$d\frac{\partial S}{\partial x} + \left(\frac{\partial H}{\partial x} + \frac{\partial H}{\partial p}\frac{\partial^2 S}{\partial x^2}\right) dt + g\frac{\partial^2 S}{\partial x^2} \circ dW = 0,$$

and using the connection between the Ito and Stratonovich differentials

$$v \circ dW = v\, dW + \frac{1}{2}dv\, dW \tag{1.6}$$

we obtain the equation for S in the Ito form:

$$dS + H\left(t, x, \frac{\partial S}{\partial x}\right) + \frac{1}{2}g^2\frac{\partial^2 S}{\partial x^2}\, dt + g\frac{\partial S}{\partial x}\, dW = 0. \tag{1.7}$$

For the mean optimal cost function $\tilde{S}$ this implies the standard second order Bellman equation of stochastic control theory

$$\frac{\partial \tilde{S}}{\partial t} + H\left(t, x, \frac{\partial \tilde{S}}{\partial x}\right) + \frac{1}{2}g^2\frac{\partial^2 \tilde{S}}{\partial x^2} = 0.$$

(ii) $g = 0$. Then equation (1.1) takes the form

$$dS + H\left(t, x, \frac{\partial S}{\partial x}\right) dt + c(t, x)\, dW = 0, \tag{1.8}$$

because in that case the Ito and Stratonovich differential forms coincide.

2 The Cauchy Problem for the Stochastic HJB Equation

For simplicity we shall study here in detail a case of equation (1.1) that is very important for applications, namely, equation (1.8) with H and c not depending explicitly on t (for remarks on the general equation (1.1) see the end of this section). The main tool of the investigation is the stochastic Hamiltonian system

$$\begin{cases} dx = \dfrac{\partial H}{\partial p}\, dt \\[2mm] dp = -\dfrac{\partial H}{\partial x}\, dt - c'(x)\, dW. \end{cases} \tag{2.1}$$

The general theory of such systems (with Hamiltonians $H(x, p)$ quadratic in p), in particular, existence and uniqueness theorems for their solutions, can be found in [AHZ].

Theorem 1 *For fixed $x_0 \in \mathcal{R}^n$ and $t > 0$ let us consider the map $P : p_0 \mapsto x(t, p_0)$, where $x(\tau, p_0)$, $p(\tau, p_0)$ is the solution of (2.1) with initial values*

(x_0, p_0). *Let all the second derivatives of the functions H and c be uniformly bounded, the matrix $Hess_p H$ of the second derivatives of H with respect to p be uniformly positive (i.e. $Hess_p H \geq \lambda E$ for some constant λ), and for any fixed x_0 all matrices $Hess_p H(x_0, p)$ commute. Then the map P is a diffeomorphism for small $t \leq t_0$ and all x_0.*

We postpone the technical proof of this theorem to the end of the section in order to explain first the main construction of the classical and generalized solutions of equation (1.8). Let us define the two-point stochastic action

$$S_W(t, x, \xi) = \inf \int_o^t L(q, \dot{q}) \, d\tau - c(q) \, dW, \qquad (2.2)$$

where inf is taken over all piecewise smooth curves $q(\tau)$ such that $q(0) = \xi$, $q(t) = x$, and the Lagrangian L is, as usual, the Legendre transform of the Hamiltonian H with respect to its last argument.

Theorem 2 *Let the suppositions of Theorem 1 be fulfilled and let $t \leq t_0$. Then*

$$(i) \quad S_W(t, x, \xi) = \int_0^t (p \, dq - H(x, p) \, dt - c(x) \, dW), \qquad (2.3)$$

where the integral is taken along the trajectory $x(\tau), p(\tau)$ that joins the points ξ and x (and which exists due to Theorem 1),

$$(ii) \quad p(t) = \frac{\partial S}{\partial x}, \qquad p_0 = -\frac{\partial S}{\partial \xi},$$

(iii) S satisfies equation (1.8), as a function of x,
(iv) $S(t, x, \xi)$ is convex in x and ξ.

Proof The proof can be carried out by rather long and tedious direct differentiations with the use of the Ito formula. But fortunately, we can avoid this by using as in Section 1 the approximation of the Wiener trajectories W by the sequence of smooth functions W_n of form (1.5). For these functions equation (1.8) and system (2.1) become classical and the results of the theorem become well known (see, for instance, [A1, MF, KM2]). By the approximation theorem mentioned above the sequence of corresponding diffeomorphisms P_n of Theorem 1 converges to the diffeomorphism P. Moreover, due to the uniform estimates on their derivatives (see formulas (2.12) and (2.13) below), the convergence of $P_n(t, p_0)$ to $P(t, p_0)$ is locally uniform, and so is the convergence of the inverse diffeomorphisms $P_n^{-1}(t, x) \to P^{-1}(t, x)$. This implies the convergence of the corresponding solutions S_n to the function (2.3) together with their derivatives in x. Again by the approximation arguments we conclude that the limit function satisfies equation (1.8). Let us note also that the convex property of S is due to the equations

$$\frac{\partial^2 S}{\partial x^2} = \frac{\partial p}{\partial x} = \frac{\partial p}{\partial p_0} \left(\frac{\partial x}{\partial p_0} \right)^{-1} = \frac{1}{t}(1 + O(t^{1/2})),$$

$$\frac{\partial^2 S}{\partial \xi^2} = -\frac{\partial p_0}{\partial \xi} = \left(\frac{\partial x}{\partial p_0}\right)^{-1} \frac{\partial x}{\partial \xi} = \frac{1}{t}(1 + O(t^{1/2}))$$

that follow from (ii), formulas (2.12), (2.13) (see below) and again by the approximation arguments.

By similar arguments one proves the following:

Theorem 3 *Suppose $S_0(x)$ is a smooth function and for all $t \leq t_0$ and $x \in \mathcal{R}^n$ there exists a unique $\xi = \xi(t, x)$ such that $x(t, \xi) = x$ for the solution $x(\tau, \xi), p(\tau, \xi)$ of system (2.1) with initial data $x_0 = \xi$, $p_0 = (\partial S_0/\partial x)(\xi)$. Then*

$$S(t, x) = S_0(\xi) + \int_0^t (p \, dq - H(x, p) \, dt - c(x) \, dW) \qquad (2.4)$$

(where the integral is taken along the trajectory $x(\tau, \xi), p(\tau, \xi)$) is a unique classical solution of the Cauchy problem for equation (1.8) with initial function $S_0(x)$.

Remark 1 This theorem can be generalized to the complex case. Namely, the statement of the theorem is still true when H is a complex function holomorphic in x, p, and S_0 is holomorphic in x.

Theorems 1, 2 serve, in fact, the purpose to give simple sufficient conditions, when the suppositions of Theorem 3 are fulfilled. The following result is a direct corollary of Theorem 2.

Theorem 4 *Let the suppositions of Theorem 1 hold and let the function $S_0(x)$ be smooth and convex. Then for $t \leq t_0$ there exists a unique classical (i.e. smooth) solution of the Cauchy problem for equation (1.8) with initial function $S_0(x)$. This solution is given by equation (2.4) or equivalently by the formula*

$$R_t S_0(x) = S(t, x) = \min_{\xi}(S_0(\xi) + S_W(t, x, \xi)). \qquad (2.5)$$

Now one can directly apply the method of constructing the generalized solution of the deterministic Bellman equation from [M2, KM1, KM2] to the stochastic case. Let us give for completeness the main line of argument. Let A denote the metric semiring $\mathcal{R} \cup \{+\infty\}$ with the metric $\rho(a, b) = |e^{-a} - e^{-b}|$ and with the commutative binary operations $\oplus = \min$, $\odot = +$. Clearly these operations are continuous and have neutral elements $+\infty$ and 0 respectively. From (2.5) it follows that for smooth convex initial functions (and small t) the classical resolvent operator R_t is linear with respect to the binary operations $\oplus = \min$ and $\odot = +$, i.e.

$$R_t(a_1 \odot S_1 \oplus a_2 \odot S_2) = a_1 \odot R_t S_1 \oplus a_2 \odot R_t S_2$$

holds for any $a_1, a_2 \in A$. With the help of this "linearity" it is possible to apply here the classical concept of weak solutions developed for standard

linear partial differential equations. We introduce first an inner product for
A-valued functions that are bounded below:

$$(f, g)_A = \inf_x (f(x) \odot g(x)). \tag{2.6}$$

This product is linear (in the sense of the operations $\oplus, \odot$) and is the idempotent analogue of the ordinary L^2-product. We say that the Cauchy problem for the equation

$$dS + H\left(t, x, -\frac{\partial S}{\partial x}\right) dt + c(x) \, dW = 0$$

is adjoint to the Cauchy problem for equation (1.8). This terminology is due
to the fact that its resolvent operator

$$(R_t^* S_0)(x) = \inf_\xi (S_0(\xi) + S_W(t, \xi, x))$$

(note that in comparison with (2.5) the arguments in S_W have changed
their places) obviously satisfies the identity

$$(R_t f, g)_A = (f, R_t^* g)_A \tag{2.7}$$

and is therefore adjoint to the resolvent operator R_t of the initial equation (1.8). Furthermore, let us define generalized functions as linear (in the
sense of operations $\oplus, \odot$) continuous A-valued functionals on the semimodule (with uniform topology corresponding to the metric ρ) of continuous
functions $g : \mathcal{R}^n \to A$ that approach $+\infty$ at infinity. Obviously, any function $f : \mathcal{R}^n \to A$ that is bounded below defines such a functional by means
of formula (2.6). The theorem proved in [KM2] states that two such functionals coincide if and only if the functions which determine them have the
same lower semicontinuous closure, and that the functionals thus obtained
exhaust the entire set of generalized functions. Now, for any initial generalized function f let us define the generalized solution $R_t f$ of equation
(1.8) with initial data f as a linear functional that acts on convex smooth
functions g by formula (2.7). It remains to note that the smooth convex
functions form a "complete" set in the semimodule of continuous A-valued
functions which approach $+\infty$ at infinity, because they approximate the
idempotent "δ"-function [M2, KM2]:

$$\delta_A(x - \xi) = \lim_{n \to \infty} n(x - \xi)^2,$$

where the limit is understood in the weak sense. Therefore each functional
is uniquely determined by its values on the set of smooth convex functions.
Note also that if the resolving operator R_t is defined for small times $t \leq t_0$,

then for any times $t > 0$ it can be naturally defined by iterations. Consequently, we obtain the following result.

Theorem 5 *For any initial function $S_0(x)$ that is bounded below there exists a unique generalized solution of the Cauchy problem for equation (1.8) that is given by formula (2.5) for all $t \geq 0$.*

Remark 2 Approximating nonsmooth Hamiltonians by means of smooth ones and defining the generalized solutions as the limits of the solutions corresponding to smooth Hamiltonians, we deduce (as in the deterministic case [KM2]) that formula (2.5) for generalized solutions will be still valid for nonsmooth (but convex in p) Hamiltonians.

Proof of Theorem 1 Clearly the solution of the linear matrix equation

$$dG = B_1 G \, dt + B_2 G \, dW, \quad G|_{t=0} = G_0, \tag{2.8}$$

where $B_j = B_j(t, [W])$ are given uniformly bounded and nonanticipating functionals on the Wiener space, can be represented by the convergent series

$$G = G_0 + G_1 + G_2 + \cdots \tag{2.9}$$

with

$$G_k = \int_0^t B_1(\tau) G_{k-1}(\tau) \, d\tau + \int_0^t B_2(\tau) G_{k-1}(\tau) \, dW(\tau). \tag{2.10}$$

Differentiating (2.1) with respect to the initial data (x_0, p_0) one gets that the matrix

$$G = \frac{\partial(x, p)}{\partial(x_0, p_0)} = \begin{pmatrix} \frac{\partial x}{\partial x_0} & \frac{\partial x}{\partial p_0} \\ \frac{\partial p}{\partial x_0} & \frac{\partial p}{\partial p_0} \end{pmatrix} (x(\tau, [W]), p(\tau, [W]))$$

satisfies a particular case of (2.8):

$$dG = \begin{pmatrix} \frac{\partial^2 H}{\partial p \partial x} & \frac{\partial^2 H}{\partial p^2} \\ -\frac{\partial^2 H}{\partial x^2} & -\frac{\partial^2 H}{\partial x \partial p} \end{pmatrix} (x, p) G \, dt - \begin{pmatrix} 0 & 0 \\ c''(x) & 0 \end{pmatrix} G \, dW,$$

$$G_0 = \begin{pmatrix} E & 0 \\ 0 & E \end{pmatrix}. \tag{2.11}$$

Let us denote by $\tilde{O}(t^\alpha)$ any function that is of order $O(t^{\alpha - \epsilon})$ for any $\epsilon > 0$, as $t \to 0$. Applying the log log law for stochastic integrals [A2] to the solutions of system (2.2) and then calculating G_1 by (2.10) we obtain

$$G_1 = \left(t \begin{pmatrix} \frac{\partial^2 H}{\partial p \partial x} & \frac{\partial^2 H}{\partial p^2} \\ -\frac{\partial^2 H}{\partial x^2} & -\frac{\partial^2 H}{\partial x \partial p} \end{pmatrix} (x_0, p_0) + \begin{pmatrix} 0 & 0 \\ c''(x_0) \tilde{O}(t^{1/2}) & 0 \end{pmatrix} \right) \begin{pmatrix} E & 0 \\ 0 & E \end{pmatrix}$$

up to a term of order $\tilde{O}(t^{3/2})$. Again applying the log log law for the next terms of series (2.9) we obtain easily that the remainder $G - G_0 - G_1$ is of order $\tilde{O}(t^{3/2})$. Thus, we have the convergence of series (2.9) for system (2.11) and the following approximate formula for its solutions:

$$\frac{\partial x}{\partial x_0} = E + t\frac{\partial^2 H}{\partial p \partial x}(x_0, p_0) + \tilde{O}(t^{3/2}), \quad \frac{\partial x}{\partial p_0} = t\frac{\partial^2 H}{\partial p^2}(x_0, p_0) + \tilde{O}(t^{3/2}),$$
$$(2.12)$$

$$\frac{\partial p}{\partial x_0} = \tilde{O}(t^{1/2}), \quad \frac{\partial p}{\partial p_0} = E + t\frac{\partial^2 H}{\partial x \partial p}(x_0, p_0) + \tilde{O}(t^{3/2}). \tag{2.13}$$

From these formulas, one concludes that the map $P : p_0 \mapsto x(t, p_0)$ is a local diffeomorphism. Let us prove that it is injective. In fact, since

$$x(t, p_1) - x(t, p_2) = \int_0^t \frac{\partial x}{\partial p_0}(p_1 + \tau(p_2 - p_1))(p_2 - p_1)\, d\tau,$$

we have

$$(x(t, p_1) - x(t, p_2))^2$$

$$= \int_0^1 \int_0^1 \left(\left(\frac{\partial x}{\partial p_0}\right)^t (p_1 + s(p_2 - p_1)) \left(\frac{\partial x}{\partial p_0}\right) (p_1 + \tau(p_2 - p_1))(p_2 - p_1) \right.$$

$$\left. (p_2 - p_1) \right) d\tau ds \geq C\|p_2 - p_1\|^2. \tag{2.14}$$

The last inequality is due to the second formula in (2.12) and to the properties of $Hess_p H$ pointed out in the conditions of Theorem 1. It follows from estimate (2.14) that the considered map is injective and moreover, that $x(t, p_0) \to \infty$, if $p_0 \to \infty$. From this one deduces that the image of this map is simultaneously closed and open and therefore coincides with the whole space. The proof of the Theorem 1 is thus complete.

To conclude this section let us note that the general equation (1.1) can be treated analogously. The difference is that the corresponding Hamiltonian system will take the form

$$\begin{cases} dx = \dfrac{\partial H}{\partial p}\, dt + g(t, x) \circ dW \\ dp = -\dfrac{\partial H}{\partial x}\, dt - c'(t, x) \circ dW \end{cases}$$

(so the Stratonovich form will no longer coincide with the Ito form, as it does for system (2.1)) and therefore estimates (2.12), (2.13) should be replaced by coarser ones.

3 The Stochastic WKB Method

We shall show here how equation (1.8) appears in the construction of WKB-type asymptotics for the stochastic PDE

$$d\psi + \frac{i}{h}H\left(t,x,\frac{h}{i}\frac{\partial}{\partial x}\right)\psi\,dt + \frac{|\alpha|^2}{2}c(x)^2\psi\,dt = \alpha c(x)\psi\,dW, \qquad (3.1)$$

where α is a complex constant, $H(t,x,\frac{h}{i}\frac{\partial}{\partial x})$ is a pseudodifferential operator in $L^2(\mathcal{R}^n)$ with the smooth real symbol $H(t,x,p)$, $c(x)$ is a smooth real vector-function, and W is a standard n-dimensional Brownian motion. Equations (3.1) are the most important particular cases of the Belavkin quantum filtering equation [B1]. They are intensively studied now in connection with the theory of quantum measurement and quantum filtering, see [B1, B2, BS, D, GP, K1, K2] and references there. Let us note also that for a purely imaginary constant α we obtain an example of the Hudson–Parthasarathy quantum stochastic unitary evolution [HP] (with unbounded operator-valued coefficients). The simplest example of (3.1)

$$d\psi + \frac{i}{h}\left(-\frac{1}{2}h^2\Delta + V(x)\right)\psi\,dt + \frac{|\alpha|^2}{2}c(x)^2\psi\,dt = \alpha c(x)\psi\,dW \qquad (3.2)$$

corresponds to the standard Schrödinger operator H with the symbol $\frac{1}{2}p^2 + V(x)$.

Let us look for the solution of equation (3.1) in the form

$$\psi = \varphi(t,x,[W])\exp\left\{\frac{i}{h}S(t,x,h,[W])\right\}. \qquad (3.3)$$

Note that this form differs from the standard WKB substitution by a more complicated dependence of the phase on h. This dependence will be made more explicit below. By the Ito formula we have

$$d\psi = \left(d\varphi + \varphi\left(\frac{i}{h}dS - \frac{1}{2h^2}(dS)^2\right) + \frac{i}{h}d\varphi dS\right)\exp\left\{\frac{i}{h}S\right\}.$$

Substituting (3.3) in (3.1) and using standard formulas of PDO (pseudo-differential operator) calculus [MF] we obtain

$$d\varphi + \left[\frac{i}{h}H\left(t,x,\frac{\partial S}{\partial x}\right)\varphi + \left(\frac{\partial H}{\partial p},\frac{\partial \varphi}{\partial x}\right) + \frac{1}{2}Sp\frac{\partial^2 H}{\partial p^2}\frac{\partial^2 S}{\partial x^2}\varphi + \frac{|\alpha|^2}{2}c(x)^2\varphi\right]dt$$

$$+O(h)\,dt + \varphi\left(\frac{i}{h}dS - \frac{1}{2h^2}(dS)^2\right) + \frac{i}{h}d\varphi dS = \alpha c(x)\varphi\,dW. \qquad (3.4)$$

The main idea of the approach proposed here is to add extra terms (linearly dependent on h) to the Hamilton–Jacobi equation in such a way that the corresponding transport equation takes the standard deterministic form. To this end, let us write the Hamilton–Jacobi equation in the form

$$dS + H\left(t, x, \frac{\partial S}{\partial x}\right) dt + \frac{h}{2i}(\alpha^2 + |\alpha|^2)c(x)^2 \, dt + ih\alpha c(x) \, dW = 0. \quad (3.5)$$

Therefore, equation (3.4) is satisfied up to terms of order $O(h)$, if (3.5) is fulfilled and the following transport equation holds:

$$d\varphi + \alpha c(x) \, d\varphi \, dW + \left(\frac{\partial H}{\partial p}, \frac{\partial \varphi}{\partial x}\right) dt + \frac{1}{2}Sp\frac{\partial^2 H}{\partial p^2}\frac{\partial^2 S}{\partial x^2}\varphi \, dt = 0.$$

It follows that the differential $d\varphi$ has no stochastic terms and therefore $d\varphi dW = 0$ and the transport equation takes, in fact, the standard (see [MF]) form:

$$d\varphi + \left(\frac{\partial H}{\partial p}, \frac{\partial \varphi}{\partial x}\right) dt + \frac{1}{2}Sp\frac{\partial^2 H}{\partial p^2}\frac{\partial^2 S}{\partial x^2}\varphi \, dt = 0. \quad (3.6)$$

Remark 3 In the proposed approach we thus obtain a stochastic Hamilton–Jacobi equation and a deterministic transport equation. Another approach to the stochastic generalization of the WKB method was proposed in [BK, K1], where the Hamilton–Jacobi equation was deterministic and the transport equation was stochastic. It seems that the asymptotics constructed by means of the stochastic Hamilton–Jacobi equation are more suitable for some problems. For instance, these asymptotics give exact solutions for quadratic Hamiltonians and linear functions $c(x)$ (just as in the deterministic case, the WKB asymptotics for the quantum oscillator or a free particle give the exact solution).

Remark 4 For equation (3.2) the term $O(h)$ in (3.4) is equal to $h\Delta\varphi/2$. It follows that if the solution of the transport equation (3.6) does not depend on x, then the constructed asymptotic solution (3.3) is an exact solution of equation (3.1).

To solve equation (3.4) by the method of the previous section, we need to consider the corresponding system (2.1), which takes now the following form:

$$\begin{cases} dx = \dfrac{\partial H}{\partial p} \, dt, \\[2mm] dp = -\left(\dfrac{\partial H}{\partial x} - ih(\alpha^2 + |\alpha|^2)c'(x)c(x)\right) dt - ih\alpha c'(x) \, dW. \end{cases} \quad (3.7)$$

Along the trajectories of this system equation (3.6) can be written in the form

$$\frac{d\varphi}{dt} + \frac{1}{2}Sp\frac{\partial^2 H}{\partial p^2}\frac{\partial^2 S}{\partial x^2} = 0, \quad (3.8)$$

which is again the same as in the deterministic case and thus can be solved explicitly. In general, the solution of the complex system (3.7) is a difficult problem and a special investigation of the complex case will be published elsewhere (see, however, an important complex example in the next section). Let us confine ourselves now to the real case, when $\alpha = -i$. Then we have just a real Hamilton–Jacobi equation that can be solved by the method developed in Section 2. Thus we obtain

Theorem 6 *Let $\alpha = -i$ and*

$$\psi_0(x) = \varphi_0(x) \exp\left\{ \frac{i}{h} S_0(x) \right\},$$

where φ_0 and S_0 are smooth, φ_0 is finite, and for S_0 the suppositions of Theorem 3 hold. Then for $t \leq t_0$ the solution of the Cauchy problem for equation (3.1) with initial function ψ_0 exists and is equal to (we drop the dependence on h and $[W]$)

$$\psi(t, x) = J^{-1/2}(t, x)\varphi_0(\xi) \exp\left\{ \frac{1}{2} \int_0^t Sp \frac{\partial^2 H}{\partial p \partial x} \right\} \exp\left\{ \frac{i}{h} S(t, x) \right\} + O(h), \tag{3.9}$$

where S is given by (2.4) or (2.5), $J = \det \frac{\partial x(t, \xi)}{\partial \xi}$, and the point ξ is defined in Theorem 3.

Proof By construction, the WKB approximation (3.3) satisfies equation (3.1) up to a term of order $O(h)$. The proof of the fact that this asymptotic solution differs from the exact one by a term $O(h)$ (the justification of formal asymptotics) can be carried out by the precisely the same procedure as in deterministic case [MF].

Remark 5 For large t the solution can be constructed by means of (a stochastic generalization of) the Maslov canonical operator.

Let us give a representation for the Green function of equation (3.2) with $\alpha = -i$. When the stochastic Hamilton–Jacobi equation is written and solved, the remaining part is quite similar to deterministic case [MF]. Thus we obtain

Theorem 7 *Let the functions $V(x)$ and $c(x)$ have second derivatives uniformly bounded in $\mathcal{R}^n$. Then for small $t \leq t_0$ the Green function of the Cauchy problem for equation (3.2) (with $\alpha = -i$) is equal to*

$$G_W(t, x, \xi, h) = (2\pi h)^{-n/2} J_W^{-1/2}(t, x, \xi, h) \exp\left\{ \frac{i}{h} S_W(t, x, \xi, h) \right\} (1 + O(h)), \tag{3.10}$$

where $J_W = \det \frac{\partial x(t, p_0)}{\partial p_0}$ and p_0 is defined as the pre-image of the point x under the diffeomorphism P, which was described in Theorem 2.

As mentioned above, the construction of the corresponding asymptotics for all finite time can be carried out by means of (a stochastic version of) the

Maslov canonical operator. In particular, to find the asymptotic solution of equation (3.1) at some point (t, x) we need to know all branches of the (multivalued) solution of the corresponding equation (3.5). Let us now point out partial differential equations, where the generalized (hence, nonsmooth, but single-valued) solution (2.5) must be used to construct WKB-type asymptotics for finite times. It is the case of stochastic generalizations of the so called tunnel equations [M1, DKM]. The simplest example of such an equation is the stochastic version of the heat equation

$$du = \frac{1}{h}\left(\frac{h^2}{2}\Delta - V(x) + \frac{1}{2}c(x)^2\right)u\,dt + c(x)u\,dW. \qquad (3.11)$$

Looking for the solution of this equation in the form

$$u(t, x) = \varphi(t, x)\exp\{-S(t, x, h)/h\},$$

we construct functions φ and S for small times as previously. But for finite times we shall have an essential difference. Namely, since for $S_1 < S_2$ one has

$$\varphi_1 e^{-S_1/h} + \varphi_2 e^{-S_2/h} \approx \varphi_1 e^{-S_1/h}$$

up to an exponentially small term, one concludes that when a point (t, x) can be reached by several characteristics, one should take into consideration only the characteristic with the minimal value of the action. This implies that the logarithmic asymptotics $\lim_{h\to 0} h\log u$ of the solution of (3.11) is given by the generalized solution (2.5) of the corresponding stochastic Hamilton–Jacobi–Bellman equation. The rigorous arguments here are rather long (and we shall not give them here in detail) but they are quite similar to the deterministic case [M1, DKM]: one should first justify the constructed WKB-type asymptotics for small times (see Section 2 in [DKM]) and then develop (a stochastic version of) the theory of the tunnel canonical operator (see Section 3 in [DKM]), so that one can give global formulas for the solutions and thus take into consideration the times when the so called focal points exist and are essential.

4   Examples and a Well-Posedness Theorem

We find here the two-point action $S(t, x, \xi, h)$ for the equation

$$dS + \left(\frac{1}{2}\left(\frac{\partial S}{\partial x}\right)^2 + \frac{h}{2i}(\alpha^2 + |\alpha|^2)x^2\right)dt + ih\alpha x\,dW = 0. \qquad (4.1)$$

This corresponds to the equation

$$d\psi + \left(\frac{h}{2i}\Delta + \frac{|\alpha|^2}{2}x^2\right)\psi dt = \alpha x\psi\,dW, \qquad (4.2)$$

which is a particular case of (3.2) with $V(x) = 0$ and $c(x) = x$ (and to equation (3.11) also with $V(x) = 0$, $c(x) = x$). Then we apply this result to obtain the existence and uniqueness theorem for a more general Belavkin quantum filtering equation.

First, let us consider the simplest case, when $\alpha = -i$. Then the evolution (4.2) is unitary and equation (4.1) is real. The corresponding Hamiltonian system (2.1) that stands for the Hamilton–Jacobi equation (4.1) has the form

$$\begin{cases} dx = p\,dt \\ dp = -h\,dW \end{cases}$$

and can be trivially solved:

$$\begin{cases} x = x_0 + p_0 t - h \int_0^t W(s)\,ds \\ p = p_0 - hW. \end{cases} \tag{4.4}$$

Obviously for any $t > 0$ and any x, x_0 there exists a unique $p_0(t, x, x_0)$ such that $x = x(t, p_0, x_0)$, namely

$$p_0 = \frac{1}{t}\left(x - x_0 + h \int_0^t W(s)\,ds \right).$$

Now we can calculate the function S by formula (2.3). After some algebraic manipulations (including integration by parts) we obtain

$$S_W(t, x, \xi, h) = -hxW(t) - \frac{1}{2}h^2 \int_0^t W^2(s)\,ds + \frac{1}{2t}\left(x - \xi + h \int_0^t W(s)\,ds \right)^2. \tag{4.5}$$

Clearly S is smooth and strongly convex in x and ξ for all $t > 0$. Therefore, the generalized solution (2.5) of the Cauchy problem for this example will be smooth for all $t > 0$, if the initial function is smooth and convex. Now we can write down the asymptotic Green function (3.10) for equation (4.2) with $\alpha = -i$, which due to Remark 3 turns out to be exact:

$$G = \frac{1}{(2\pi h t)^{n/2}} \times$$

$$\exp\left\{ \frac{i(x - x_0 + h \int_0^t W(s)\,ds)^2}{2th} - \frac{ih}{2} \int_0^t W^2(s)\,ds - ixW(t) \right\} \tag{4.6}$$

If α is not purely imaginary, the calculations are more complicated but in principle quite similar. Let us outline the main steps. Let us now consider α to have a nonzero real part. Then the complex number β is uniquely defined by the conditions

$$\beta^2 = ih(\alpha^2 + |\alpha|^2), \quad 0 < \arg \beta < \pi/2.$$

Equation (4.1) takes the form

$$dS + \frac{1}{2}\left(\frac{\partial S}{\partial x}\right)^2 dt - \frac{\beta^2}{2}x^2 dt + ih\alpha x\, dW = 0, \qquad (4.7)$$

and the corresponding Hamiltonian system is

$$\begin{cases} dx = p\, dt \\ dp = \beta^2 x\, dt - ih\alpha\, dW. \end{cases} \qquad (4.8)$$

Obviously, one can write down its solution explicitly:

$$\begin{cases} x = x_0 \cosh \beta t + p_0 \beta^{-1} \sinh \beta t - ih\alpha \beta^{-1} \int_0^t \sinh \beta(t-\tau)\, dW(\tau), \\ p = x_0 \beta \sinh \beta t + p_0 \cosh \beta t - ih\alpha \int_0^t \cosh \beta(t-\tau)\, dW(\tau). \end{cases} \qquad (4.9)$$

Therefore, for all x, ξ and each $t > 0$ there exists a unique

$$p_0 = \frac{\beta}{\sinh \beta t}\left(x - \xi \cosh \beta t + \frac{ih\alpha}{\beta}\int_0^t \sinh \beta(t-\tau)\, dW(\tau)\right)$$

such that the solution of (4.8) with initial values ξ, p_0 joins the points ξ and x in time t. And consequently, the two-point function (2.4) can be calculated explicitly by the formula

$$S_W(t, x, \xi, h) = \frac{1}{2}\int_0^t (p^2(\tau) + \beta^2 x^2(\tau))\, d\tau - ih\alpha \int_0^t x(\tau)\, dW(\tau), \quad (4.10)$$

where the integral is taken along this solution. Simple but rather long calculations (which we omit) give the result

$$S_W(t, x, \xi, h) = \frac{\beta}{2\sinh \beta t}((x^2 + \xi^2)\cosh \beta t - 2x\xi) + ax + b\xi + \gamma, \quad (4.11)$$

where

$$a = -\frac{ih\alpha}{\sinh \beta t}\int \sinh \beta\tau\, dW(\tau),$$

$$b = \beta \int_0^t \frac{a(\tau)}{\sinh \beta\tau}\, d\tau, \quad \gamma = -\frac{1}{2}\int a^2(\tau)\, d\tau. \qquad (4.12)$$

It follows from (4.9) that the Jacobian $J = \det(\partial x/\partial p_0)$ equals $(\sinh \beta t/\beta)^n$ (where n is the dimension) and by Remark 4 the asymptotic Green function is exact and is equal to

$$G_W(t, x, \xi, h) = (2\pi h)^{-n/2}\left(\frac{\beta}{\sinh \beta t}\right)^{n/2}\exp\left\{\frac{i}{h}S_W(t, x, \xi, h)\right\} \qquad (4.13)$$

with S_W given by (4.11), (4.12).

For the most important case $\alpha = 1$ formula (4.13) was obtained already by quite another method in [BK, K2]. Now we shall use this example to give a correctness theorem for the Cauchy problem of a more general equation, namely for the equation

$$d\psi + \left(\frac{h}{2i}\Delta - \frac{1}{ih}V(x) + \frac{|\alpha|^2}{2}x^2 \right) \psi \, dt = \alpha x \psi \, dW, \qquad (4.14)$$

where $V(x)$ is a measurable uniformly bounded complex function (potential) and the constant α has nonvanishing real part (for instance, $\alpha = 1$). A general existence and uniqueness theorem for a wide class of abstract linear stochastic equations, including (4.14) as a particular case, was obtained recently by abstract methods in [Ho]. But there only weak operator solutions were discussed. Here we establish the existence of the strong solution for equation (4.14) and prove also the martingale property of its squared norms. Note also that a sketch of another proof of this result was given by the author in [K1].

Let us rewrite equation (4.14) in the integral form

$$\psi(t) = U_0^t \psi_0 + \int_0^t U_{t-\tau}^t V(x)\psi(\tau) \, d\tau, \qquad (4.15)$$

where U_s^t is the (random) resolving operator for equation (4.2) that gives the solution at the time $t > s$, the initial data being given at the time $s > 0$. It is natural to solve this integral equation by iterations. Let us note the following simple fact. If in the abstract equation (4.15) (where V is some bounded linear operator in some Banach space) each random operator U_s^t is bounded and the integral $\int_0^t \|U_{t-\tau}^t\| \, d\tau$ exists (for almost all W), then the iteration series for (4.15) converges (for a.a. W). Using now the explicit formulas (4.11)–(4.13) together with the log log law for the Wiener process we find that $\|U_s^t\| \sim -\log(t-s)$ which satisfies the above condition. Hence, the first part of the next theorem is proved.

Theorem 8 *The solution of equation (4.15) exists and is unique for all $\psi_0 \in L^2(\mathcal{R}^n)$ and a.a. W. Moreover, the square of its norm $\|\psi\|^2$ is a positive martingale; in particular, its mean (with respect to the Wiener measure) does not depend on t.*

Proof. It remains to prove the martingale property. Standard manipulations with Ito's formula [K2, BS, Ho] show that $\|\psi\|^2$ is a positive supermartingale and a local martingale. Now, by construction it is bounded by $C\|U_0^t \psi_0\|^2$, where C is some constant. But the last random process is a martingale, which was proved in [K2]. Therefore $\|\psi\|^2$ is also a martingale. This completes the proof. $\square$

Remark 6 The results of this paper were reported at the conference on Idempotency, which was organized by the Hewlett-Packard scientific laboratory in Bristol, 3–7 October 1994.

Remark 7 When the paper was already almost finished, the author became aware of the very interesting new preprint of A. Truman and H. Zhao [TZ1] devoted to similar problems (the results of this work were presented at the International Conference on Stochastic PDE at the ICMS in Edinburgh in March 1994). In particular, Theorem 3 was obtained there for the Hamiltonians of the form $H = p^2 + V(x)$ (the proof was based on the direct calculations using Ito's formula). Further, A. Truman and H. Zhao also apply stochastic first order partial differential equations to the investigation of certain stochastic differential equations of quantum mechanics. They develop the stochastic version of the path integral representation and thus obtain other formulas for solutions of unitary evolution (3.2) (with $\alpha = -i$) and for the stochastic heat equation (3.11). For the specific unitary example (4.2) (with $\alpha = -i$) they obtain the same formula (4.6) as was obtained here by means of our stochastic WKB method. Some developments of the results of [TZ1] can be found in [TZ2, TZ3].

Acknowledgements

I am grateful to Prof. J. Gunawardena for organizing a very interesting conference on Idempotency in Bristol and thus stimulating my work on the Hamilton-Jacobi equation, and to S. Albeverio, Z. Brzesniak, V. Belavkin, and A. Ponosov for useful discussions. The work was carried out with the financial support of the Russian Fund of Basic Investigation, Project no. 93-012-1075. It was finished while the author was in the Ecole des Mines de Nantes with a visiting position, on leave from the Institute of New Technology, Nizhnaya Radischevska 9, 109004 Moscow.

References

[A1] I.V. Arnold. Mathematical methods in classical mechanics. Springer-Verlag, 1978.

[A2] L. Arnold. The log log law for multidimensional stochastic integrals and diffusion processes. Bull. Austral. Math. Soc. **5** (1971), 351–356.

[AHZ] S. Albeverio, A. Hilbert, E. Zehnder. Hamiltonian systems with a stochastic force: Nonlinear versus linear and a Girsanov formula. Stochastics and Stochastics Reports **39** (1992), 159–188.

[B1] V.P. Belavkin. Nondemolition measurement, nonlinear filtering and dynamic programming of quantum stochastic processes. In: Modelling and Control of Systems, Proc. Bellman Continuous Workshop, Sophia-Antipolis 1988, Lect. Notes in Contr. and Inform. Sci., **121** (1988), 245–265.

[B2] V.P. Belavkin. A new wave equation for a continuous nondemolition measurement. Phys. Let. A **140** (1989), 355–358.

[BK] V.P. Belavkin, V.N. Kolokoltsov. Quasi-classical asymptotics of quantum stochastic equations. Teor. i Mat. Fiz. **89** (1991), 163–178. Engl. transl. in Theor. Math. Phys.

[BS] V.P. Belavkin, P.Staszewski. A stochastic solution of Zeno's paradox for quantum Brownian motion. Centro Mat. V. Volterra, Univ. degli studi di Roma 2. Preprint **28** (1990). Published in Phys. Rev. A **45:3** (1992), 1347–1356.

[D] L. Diosi. Continuous quantum measurement and Ito formalism. Phys. Lett. A **129** (1988), 419–423.

[DKM] S.Yu. Dobrokhotov, V.N. Kolokoltsov, V.P. Maslov. Quantization of the Bellman equation, exponential asymptotics and tunneling. Adv. Sov. Math. 13 (1992), Idempotent Analysis, ed. V. Maslov, S. Samborski, 1–46.

[GP] N. Gisin, I. Percival. The quantum state diffusion picture of physical processes. J. Phys. A: Math. Gen. **26** (1993), 2233–2260.

[Ho] A.S. Holevo. On dissipative stochastic equations in a Hilbert space. Preprint. Steklov Math. Inst., (1994).

[HP] R.L. Hudson, K.R. Parthasarathy. Quantum Ito's formula and stochastic evolution. Comm. Math. Phys. **93** (1984), 301–323.

[K1] V.N. Kolokoltsov. Application of quasi-classical method to the investigation of the Belavkin quantum filtering equation. Mat. Zametki **50** (1993), 153–156. Engl. transl. in Math. Notes.

[K2] V.N. Kolokoltsov. Scattering theory for the Belavkin equation describing a quantum particle with continuously observed coordinate. Ruhr University Bochum, SFB 237, Preprint **228** (1994). Published in Journ. Math. Phys. **36:6** (1995).

[KM1] V.N. Kolokoltsov, V.P. Maslov. The Cauchy problem for the homogeneous Bellman equation. Dokl. Akad. Nauk **296:4** (1987). Engl. transl. in Sov. Math. Dokl. **36:2** (1988), 326–330.

[KM2] V.N. Kolokoltsov, V.P. Maslov. Idempotent calculus as the apparatus of the control theory, part 1. Funkt. Anal. i Prilozh. **23:1** (1989), 1–14; part 2, **23:4** (1989), 53-62. Engl. transl. in Funct. Anal. Applic.

[KM3] V.N. Kolokoltsov, V.P. Maslov. Idempotent analysis and applications. A monograph, tKluver, in press.

[M1] V.P. Maslov. Global exponential asymptotics of the solutions of the tunnel equations and the large deviation problems. Trudy Steklov Math. Inst. **163** (1984), 150–180. Engl. transl. in Proc. Steklov Inst. Math.

[M2] V.P. Maslov. On a new superposition principle for optimization problems. Uspekhi Mat. Nauk **42:3** (1987), 39–48. Engl. transl. in Russian Math. Surveys **42** (1987).

[MF] V.P. Maslov, M.V. Fedoriuk. Semi-classical approximation in quantum mechanics. Reidel Publishing Company, Dordrecht, 1981.

[SV] D.W. Stroock, S.R. Varadhan. On the support of diffusion processes with applications to the strong maximum principle. Proc. Sixth Berkley Symp. **3** (1972), 333–359.

[Su] H. Sussman. On the gap between deterministic and stochastic ordinary differential equations. Ann. of Prob. **6:1** (1978), 19–41.

[TZ1] A. Truman, H.Z. Zhao. The stochastic Hamilton–Jacobi equation, stochastic heat equations and Schrödinger equations. University of Wales Swansea, Preprint, 1994.

[TZ2] A. Truman, H.Z. Zhao. Stochastic Hamilton–Jacobi equations and related topics. University of Wales Swansea, Preprint, 1994. To appear in: Stochastic Partial Differential Equations, ed. A.M. Etheridge, London Mathematical Society Lecture Notes Series, Cambridge University Press.

[TZ3] A.Truman, H.Z. Zhao. On stochastic diffusion equations and stochastic Burgers equations. University of Wales Swansea, Preprint, 1994.

[WZ] E. Wong, H. Zakai. Riemann–Stieltjes approximation of stochastic integrals. Z. Wahrscheinlichkeitstheorie und Verw. Gebiete **12** (1969), 87–97.

The Lagrange Problem from the Point of View of Idempotent Analysis

Serguei Samborskiĭ

1 Introduction

This paper is devoted to the Lagrange problem, which is a classical one in variational calculus. It deals with Inf of the integral $\int_0^t L(x(\tau), \dot{x}(\tau)) d\tau$ over the set of absolutely continuous functions (trajectories) $x(\cdot)$. Since the time t is available, we can consider this problem as an evolutionary process for which the Hamilton–Jacobi differential equation is the evolutionary equation. But the functions describing this evolution are far from always being differentiable in a classical sense. The approach proposed in this paper is an extension of a differential operator in the Hamilton–Jacobi equation. This extension makes it possible to include nondifferentiable (in the classical sense) functions in the domain of definition of this operator. Usually such an extension requires a detailed definition of the spaces in which it is constructed.

The results of this paper can be explained more easily if one compares them with the classical results for elliptic equations and with the corresponding construction of Sobolev spaces. As usual in applications of mathematics, an evolutionary process is given by the semigroup of transformations of the state space of a physical or mechanical system, the state space being a space of functions. For example, the diffusion process is described by the semigroup of linear operators of a linear function space, and the corresponding stationary equation is Laplace. If the initial or boundary value conditions are not regular, e.g. if they contain δ-functions, then also the solutions of the Laplace equation are not classical twice differentiable functions as well. It is more convenient to pass to the nonhomogeneous equation $\Delta y = U$ with nonregular right-hand part U. One can present the analysis of solutions of this equation by means of the following scheme.

First we consider a space Φ (of test functions) such that U is identified with an element from Φ^*, i.e. with a linear continuous functional on Φ. This space Φ^* with the weak topology is the space containing the image of the desired extension of the differential operator $\Delta = \sum \frac{\partial^2}{\partial x_i^2}$. And now we can say that the corresponding Sobolev space W is the subset in Φ^* uniquely defined by the following requirements:

(1) W is a normed complete space (i.e. a space in which the standard operations of a vector space – addition and multiplication by a scalar – are continuous).

(2) The set of twice differentiable (in the classical sense) functions is dense in W.

(3) The differential operator Δ can be extended to a continuous mapping from W in Φ^*.

(4) The topology in W is the weakest one in which (1)–(3) hold.

After the extension of Δ the equation $\Delta y = U$ is examined for the functions from W that may no longer be twice differentiable in the classical sense. The well known existence and uniqueness theorems justify the extension procedure indicated above.

And now return to the Lagrange problem. Let $\mathcal{F}$ be the set of functions on $\mathbb{R}^n$ that are bounded below and possibly take the value $+\infty$. This set is assumed to be endowed with the operations $(f \oplus g) = \min(f+g)$ and $(\lambda \odot f) = \lambda + f$, where λ belongs to $\mathbb{R} \cup \{\infty\}$. (Thus this is a so-called semimodule over the semiring $\mathcal{R} = (\mathbb{R} \cup \{\infty\}, \min, +)$.) The considered evolutionary process is given in the following explicit form (similarly to the functional integral for diffusion processes):

$$(U_t f)(x) = \mathrm{Inf}(\int_0^t L(x(\tau), \dot{x}(\tau))d\tau + f(y)), \qquad (1.1)$$

where Inf is taken over the set of absolutely continuous trajectories $x(\cdot)$: $[0, t] \rightarrow \mathbb{R}^n$, $x(0) = y$ and $x(t) = x$. If L takes values in $\mathcal{R}$, then the properties of this evolution are fundamental in solving problems of optimal control theory.

From expression (1.1) it follows that the mappings U_t are endomorphisms of the semimodule $\mathcal{F}$ over the semiring $\mathcal{R}$ (that is, $U_t(f \oplus g) = U_t(f) \oplus U_t(g)$ and $U_t(\lambda \odot f) = \lambda \odot U_t(f)$).

The corresponding stationary equation, called a stationary Hamilton–Jacobi equation, has the form

$$\mathcal{H}(y)(x) = H(x, Dy(x)) = \lambda, \qquad (1.2)$$

where H is the Legendre transform of the Lagrangian L by the second argument. Note that $\mathcal{H}$ is not an endomorphism of the semimodule $\mathcal{F}$ although the semigroup consists of endomorphisms. It is well known that the functions $U_t f$ and, specifically, the stationary regimes of the form $(U_t f)(x) = \lambda t + \phi(x)$ are not, generally speaking, differentiable functions even for a differentiable f. Hence there arises the problem: to give a real rather than formal meaning to the Hamilton–Jacobi equation (1.2). An extensive bibliography is devoted to solving this problem. We claim that the scheme described above for the construction of Sobolev spaces, but with the structure of the semimodule $(\mathcal{F}, \oplus, \odot)$ replacing the linear one, gives a solution of the problem. This solution is admissible for variational problems to the same extent as is the approach of distribution theory for linear evolutionary problems.

The first step is connected with the choice of the space Φ (of test functions). We choose it so that its dual space Φ^*, i.e. the space of continuous homomorphisms from Φ into $\mathcal{R}$, contains images of the mapping $\mathcal{H}$. From Kolokoltsov's and Maslov's results about the general form of homomorphisms of idempotent semimodules it follows that homomorphisms from Φ into $\mathcal{R}$ have the form $\phi \mapsto (f, \phi)$, where the scalar product (f, g) is $\oplus \int f \odot g = \mathrm{Inf}(f(x) + g(x))$. Thus the elements of such a space can be defined as equivalence classes of mappings with the following equivalence relation: f is equivalent to g, whenever $(f, \phi) = (g, \phi)$ for any continuous function ϕ from Φ. This space endowed with the weak convergence ($f_i \to f$, whenever $(f_i, \phi) \to (f, \phi)$ for any continuous ϕ) plays the same part for equation (1.2) as a space of distributions for the Laplace equation considered above.

Next define the space $\mathcal{P}_H$ as the subset of Φ^* with the following properties, similar to the properties characterizing Sobolev spaces:

(1) $\mathcal{P}_H$ is a complete metric semimodule over $(\mathcal{R}, \oplus, \odot)$ (i.e. the operations of addition $\oplus$ and multiplication $\odot$ by a $\lambda \in \mathcal{R}$ are continuous in $\mathcal{P}_H$).

(2) The set of differentiable (in the classical sense) functions is dense in $\mathcal{P}_H$.

(3) The differential expression $\mathcal{H}$ can be extended to a continuous mapping from $\mathcal{P}_H$ to Φ^*.

(4) The topology in $\mathcal{P}_H$ is the weakest one in which (1)–(3) hold.

It is shown in this paper that such spaces exist (Theorem 1). We also obtain the expression for their metrics in an explicit form. This explicit expression is an analog of the graph-norm of a closed linear operator in its domain of definition and turns into it when we substitute $(+, \cdot)$ for $(\oplus, \odot)$. But undoubtedly the main fact is: the functions $(t, x) \to U_t f$ of the form $\lambda t + f(x)$ (the stationary regimes) belong to these spaces for every t (Theorem 2).

The solutions of boundary value problems for the Hamilton–Jacobi equation (1.2) belong to the space $\mathcal{P}_H$, and uniqueness holds for these solutions (Theorem 3).

Let us remark finally that the use of the classical extension of an operator, adapted from functional analysis to the algebraic structure which is conserved by an evolution, allows us to work in the functional spaces that are at the same level of manipulation as, for example, the space L^2. But in our case the Lebesgue integral is replaced by its idempotent analog, i.e. by Inf. In particular this allows us to avoid using many-valued mappings as is done in the approach of convex analysis, or objects of a considerably more complicated character such as, say, singularly perturbed equations of second order in the limit of disappearing viscosity.

Once more, we want to emphasize that just the algebraic structure which is conserved by the evolution connected with variational problems defines the analogs of classical constructions of linear functional analysis. This idea is due to V. Maslov and was first purposefully developed in his papers and papers of his school. This idea was got by the author from V. Maslov as a result of discussions for which the author expresses his sincere gratitude.

2   The Maslov Spaces

We shall denote by $\mathcal{R}$ the set $\mathbb{R}\cup\{\infty\}$ endowed with the operations $a \oplus b = \min(a, b)$ and $a \odot b = a + b$ under the usual agreement that $\min(\infty, a) = a$ and $\infty + a = \infty$ for any $a \in \mathbb{R}\cup\{\infty\}$. This set has the structure of an associative commutative semiring with usual distributivity $(a \oplus b) \odot c = a \odot c \oplus b \odot c$ and the neutral elements $\mathbf{0}$ $(= \infty)$ and $\mathbf{1}(= 0)$ with respect to the operations $\oplus$ and $\odot$. It is convenient to provide $\mathcal{R}$ with a metric coordinated with the operations. Without discussing the total possibilities, we turn our attention to one such metric

$$\rho(a, b) = |\exp(-a) - \exp(-b)|$$

supposing naturally that $\exp(-\infty) = 0$.

Let X be an open domain in $\mathbb{R}^n$ whose closure $cl\, X$ is a compact set, and let $\Gamma = \partial X = cl\, X \setminus X$ be its boundary. For every pair of bounded mappings from X into $\mathcal{R}$ (i.e. the mappings from X into $\mathbb{R}\cup\{\infty\}$ that are semibounded below: $\underset{x\in X}{\mathrm{Inf}}\, f(x) \geq c(f) \in \mathbb{R}$) introduce a scalar product with values in $\mathcal{R}$:

$$(f, g) = \underset{x\in X}{\mathrm{Inf}}\, (f(x) + g(x))$$

which is written in terms of the operations of $\mathcal{R}$ in the form

$$(f, g) = \oplus \int_X f \odot g,$$

where the idempotent integral $\oplus \int = \mathrm{Inf}$ fulfills the same function in the structure of the idempotent semiring $(\mathcal{R}, \oplus, \odot)$ as the usual integral in the structure of $(\mathbb{R}, +, \cdot)$, i.e. it expresses continual addition.

We shall introduce two sets of test functions: Φ is the set of continuous mappings from X into $\mathcal{R}$ that are extended continuously $cl X$, and Φ_0 is the set of continuous mappings from X into $\mathcal{R}$ whose supports (Supp) belong to

X, where as usual Supp $\phi = cl\{x \in X | \phi(x) \neq 0\}$. We define the following equivalence relation on the set of bounded mappings from X into $\mathcal{R}$:

$$f \sim g \quad \text{whenever} \quad (f, \phi) = (g, \phi),$$

for any $\phi \in \Phi$ and any $\phi \in \Phi_0$, respectively, and weak convergence on the sets of equivalence classes:

$$f_i \rightarrow g \quad \text{whenever} \quad (f_i, \phi) \rightarrow (f, \phi),$$

for any $\phi \in \Phi$ and any $\phi \in \Phi_0$, respectively (in the expression on the right the convergence is understood in $\mathcal{R}$, i.e. in the metric ρ of $\mathcal{R}$ introduced above). Denote by $\mathcal{P}$ and $\mathcal{P}^0$, respectively, the spaces obtained.

Here are some examples in which $X =]-1, 1[$, and elements from $\mathcal{P}$ (or from $\mathcal{P}^0$), i.e. equivalence classes, are given by representatives.

Example 1 $f_n = \sin nx$. The weak limit of f_n in $\mathcal{P}$ (or in $\mathcal{P}^0$) is given by an equivalence class of the function $x \mapsto -1$.

Example 2 $f_n = e^{-n|x|}$. The weak limit of f_n in $\mathcal{P}$ (or in $\mathcal{P}^0$) equals the equivalence class of the mapping $x \mapsto 0$.

Example 3 $f_n = -e^{-n|x|}$. The weak limit of f_n in $\mathcal{P}$ equals the class defined by the representative g, where $g(x) = 0$ for $x \neq 0$ and $g(0) = -1$.

Example 4 The sequences $f_n = e^{-n|x-1|}$ and $f_n = -e^{-n|x-1|}$ have different weak limits in $\mathcal{P}(]-1, 1[)$, but one and the same in $\mathcal{P}^0(]-1, 1[)$, since the mappings f, where $f(x) = 0$ for $x \neq 1$, $f(1) = 1$ and g, where $g(x) = 0$ for $x \neq 1$, $g(1) = -1$ are contained in the same equivalence class in $\mathcal{P}^0(]-1, 1[)$.

Example 5 Let V be a nowhere dense closed subset in X, and V_i a sequence of closed subsets in X such that the Hausdorff distance between V and V_i tends to 0 as $i \rightarrow \infty$. Let $f_i : X \rightarrow \mathcal{R}$ be a sequence of mappings such that $f_i = 0$ outside of V_i and $f_i(x) \geq 0$ for $x \in V_i$. Then the weak limit of f_i exists and coincides with the equivalence class of the mapping $f \equiv 0$.

Proposition 1 (Kolokoltsov, Maslov [1]) *Every continuous homomorphism (i.e. functional) from the semimodule Φ endowed with the metric of the uniform convergence $m(\phi, \psi) = \underset{X}{Sup}(\rho(\phi(x), \psi(x)))$ in the semiring $\mathcal{R}$ can be represented in the form $\phi \rightarrow (f, g)$, where f is some bounded mapping from X into $\mathcal{R}$. Therefore the conjugate semimodule Φ^* of the semimodule Φ coincides with the semimodule $\mathcal{P}(X)$.*

Suppose that $\delta_{n,x}$, $n \geq 1$, is a δ-formed sequence of continuous functions, i.e. the sequence for which $\delta_{n,x}(x) = 1$, $\delta_{n,x}(z) \rightarrow 0$, if $z \neq x$, and $\delta_{n+1,x} \geq \delta_n(x)$ (for example, one can take $\delta_{n,x}(z) = n(z-x)^2$). Then the sequence $(f, \delta_{n,x})$ is monotone and bounded and hence has a limit in $\mathcal{R}$. It is evident that this limit is the same for equivalent mappings f; it does not depend on the choice of the δ-sequence.

Definition 1 The value of $\lim_{n\to\infty}(f,\delta_{n,x})$ will be called the value of a class $f \in \mathcal{P}(x)$ at the point $x \in clX$ and the value of a class $f \in \mathcal{P}^0(X)$ at the point $x \in X$, respectively.

If $\bar{f}$ is the semicontinuous representative of a class f, then $\bar{f}(x) = f(x)$. We also have the equality $f(x) = \underset{f}{\text{Inf}}\,\bar{f}(x)$, for $f \in \mathcal{P}(X)$, where Inf is taken over all possible representatives $\bar{f}$ of the class f.

Proposition 2 *There exist metrics r and r^0 on the sets $\mathcal{P}(X)$ and $\mathcal{P}^0(X)$, respectively, such that the convergences defined by these metrics coincide with the corresponding weak convergence introduced above. The spaces $\mathcal{P}(X)$ and $\mathcal{P}^0(X)$ endowed with these metrics are complete.*

Proof Description of the metrics. Let f be a mapping from X into $\mathcal{R}$ and $\tilde{f}$ be a closed subset in $clX \times \mathcal{R}$ that is the closure of the epigraph of f, i.e. $\tilde{f} = cl\{(x,a)|\ f(x) \le a\}$. It is easy to see that $f \sim g$ in $\mathcal{P}(X)$ if $\tilde{f} = \tilde{g}$, and $f \sim g$ in $\mathcal{P}^0(X)$ if $\tilde{f} \cup \{\mathcal{R} \times \partial X\} = \tilde{g} \cup \{\mathcal{R} \times \partial X\}$.

We now introduce the following metrics: $r(f,g)$ is the Hausdorff distance between $\tilde{f}$ and $\tilde{g}$, and $r^0(f,g)$is the Hausdorff distance between $\tilde{f} \cup \{\mathcal{R} \times \partial X\}$ and $\tilde{g} \cup \{\mathcal{R} \times \partial X\}$. The validity of Proposition 2 follows from the properties of Hausdorff distance.   □

By introducing in $\mathcal{P}(X)$ and $\mathcal{P}^0(X)$ the structure of a semimodule over the semiring R according to the rule

$$(f \oplus g, \phi) = (f,\phi) \oplus (g,\phi)$$
$$(\lambda \odot f, \phi) = \lambda \odot (f,\phi)$$

for any ϕ in Φ (respectively, in Φ^0), we obtain complete metric semimodules, i.e. the operations $\oplus : \mathcal{P} \times \mathcal{P} \to \mathcal{P}$ and $\odot : \mathcal{R} \times \mathcal{P} \to \mathcal{P}$ (respectively, for $\mathcal{P}^0$) are continuous.

If $f \in \mathcal{P}(X)$ possesses the property $(f,\phi) < (c,\phi)$ for any $\phi \in \Phi$, where c is the equivalence class of the constant mapping $x \mapsto c$ and $c \ne \mathbf{0}$,then we uniquely define the class which will be denoted by $-f$ and which is given by the representative $x \mapsto -f(x)$. Note that in general for values of f and $-f$ the equality $(-f)(x) = -f(x)$ is not valid.

3   The Lagrange Problem

Assume that there is a function $L : \mathbb{R}^n \times \mathbb{R}^n \to \mathcal{R}$. (The conditions upon L will be imposed below.)

For every fixed $t \ge 0$ let us consider the following transformation : $f \to U_t f$, where $f : X \to \mathcal{R}$ and

$$(U_t f)(x) = \text{Inf}(\int_0^t L(x(\tau), \dot{x}(\tau))d\tau + f(y)). \qquad (3.1)$$

Here Inf is taken over all absolutely continuous functions $x(\cdot) : [0, t] \to \mathbb{R}^n$ such that $x(0) = y \in X$, $x(t) = x \in X$.

We shall now make the assumptions concerning L that are usual for Lagrange problem [5],[6]. The Euclidean norm and scalar product will be denoted below by $\| \cdot \|$ and $\langle \cdot, \cdot \rangle$, respectively.

(1) For each $x \in \mathbb{R}^n$, $L(x, \cdot)$ is a convex lower semicontinuous function that does not identically equal $\mathbf{0}$.

(2) $\frac{1}{\|v\|} L(x, v) \to \mathbf{0}$ if $\|v\| \to \infty$ (the so-called coercivity).

(3) The mapping $x \mapsto L(x, \cdot)$ from $\mathbb{R}^n$ into $\mathcal{P}(\mathbb{R}^n)$ is Lipschitzian.

The last assumption needs to be explained, since the metric has been defined above only for spaces $\mathcal{P}(X)$ with $cl\,X$ compact. In fact, it means that for any ball B in $\mathbb{R}^n$ the restriction to B of the mapping indicated in (3) is Lipschitz-continuous from $\mathbb{R}^n$ into $\mathcal{P}(B)$, with the Lipschitz constant not depending on B. This assumption is equivalent to the following:

(3') For the Legendre transform $H(x, \cdot)$ of the function $L(x, \cdot)$ the so-called strong Lipschitz condition is true: there exists a $\lambda > 0$ such that for each x, y and $p \in \mathbb{R}^n$ the inequality $|H(x,p) - H(y,p)| \leq \lambda(1 + \| p \|) \| x - y \|$ holds.

For every $t \geq 0$, the mapping U_t is an integral one in the sense that

$$(U_t f)(x) = \oplus \int V^{(t)}(y, x) \odot f(y),$$

where

$$V^{(t)}(y, x) = \operatorname*{Inf}_{x(\cdot)} \int_0^t L(x(\tau), \dot{x}(\tau)) d\tau,$$

and Inf is taken over $x(\cdot)$ from the set of absolutely continuous functions $x(\cdot) : [0, t] \to \mathbb{R}^n$ such that $x(0) = y \in X$, $x(t) = x \in X$.

Recall that the set $\Phi(X)$ endowed with the uniform metric is a metric semimodule over $\mathcal{R}$.

Assigning to every continuous function on X its equivalence class in $\mathcal{P}(X)$, we obtain a natural mapping $\Phi \to \mathcal{P}(X)$, which is an embedding: if $f \neq g$, then $f(x) \neq g(x)$ for x from some open set U. Therefore $(f, \phi) \neq (g, \phi)$ for ϕ with $\mathrm{Supp}\phi \subset U$.

Proposition 3 *For every t, the mapping U_t is a continuous endomorphism of the semimodule $\Phi(X)$, i.e. it is continuous and $U_t(f \oplus g) = U_t(f) \oplus U_t(g)$, $U_t(\lambda \odot f) = \lambda \odot U_t(f)$. This endomorphism has a unique extension to a continuous endomorphism of the semimodule $\mathcal{P}(X)$.*

Proof From properties (1)–(3) of the Lagrangian L it follows that this endomorphism satisfies all the requirements of the Kolokoltsov–Maslov theorem on the general form of continuous endomorphisms of the semimodule Φ [1],[3]. To prove the second part consider an endomorphism U_t^* defined by the rule

$$(U_t^* f)(x) = \oplus \int V^t(x, y) \odot f(y).$$

Under conditions (1)–(3) on the Lagrangian L indicated above U_t^* is also a continuous endomorphism of the semimodule Φ. It is easy to see that for every $f, \phi \in \Phi$ the equality

$$(U_t f, \phi) = (f, U_t^* \phi) \tag{3.2}$$

is valid.

Since the semimodule $\mathcal{P}$ is dual to Φ (Proposition 1), the formula (3.2), where $f \in \mathcal{P}(X)$, defines a continuous endomorphism of the semimodule $\mathcal{P}(X)$. $\square$

The following obvious fact is known also as the "optimality principle".

Assertion 1 *The correspondence $t \to U_t$ is a semigroup of continuous endomorphisms of the semimodule $\mathcal{P}(X)$: $U_0 = Id$, $U_{t_1+t_2} = U_{t_1} + U_{t_2}$.*

Assertion 2 *In terms of the kernels $V^{(t)}(\cdot, \cdot)$ of the integral operators U_t the semigroup property can be written as follows:*

$$V^{(t_1+t_2)}(y, x) = \operatorname*{Inf}_{z \in X}(V^{t_1}(y, z) + V^{t_2}(z, x)) = \oplus \int_{z \in X} V^{t_1}(y, z) \odot V^{t_2}(z, x)$$

Let $\delta > 0$. Define one more endomorphism of the semimodule Φ as follows. For $f \in \Phi$ let $A_\delta f$ be a function defined by the rule

$$(A_\delta f)(x) = \oplus \int_{y \in Y} \delta L(y, \frac{x-y}{\delta}) \odot f(y) = \oplus \int_{\xi \in \mathbb{R}^n} \delta L(x - \delta\xi, \xi) \odot f(x - \delta\xi)$$

(we suppose that $f = \mathbf{0}$ outside of X in the last expression).

Proposition 4 *Under the assumptions imposed upon L the mapping A_δ is a continuous integral endomorphism of the semimodule Φ that has a unique extension to a continuous integral endomorphism of the semimodule $\mathcal{P}(X)$.*

Concerning the proof one can repeat the same remarks as those in the proof of Proposition 3.

It is evident that the endomorphism A_δ is connected with the quantization of the Lagrange problem in terms of t, i.e. with the replacement of the trajectories $x(\cdot)$ in (3.1) by segments of straight lines.

The infinitesimal connection between the opertors U_δ and A_δ has been studied in terms of their kernels in the paper of P. Wolenski [5]. The main result of that paper can be formulated as follows.

Proposition 5 (P. Wolenski) *Let $V^{(t)}(\cdot,\cdot)$ be the kernel of the endomorphism U_t and $W^{(\delta)}(y,v) = \frac{1}{\delta}V^{(\delta)}(y, y+\delta v)$. Then $r(W^{(\delta)}(y,\cdot), L(y,\cdot)) \to 0$ in $\mathcal{P}(B_R)$ on balls B_R with radius R uniformly with respect to all $R > 0$ and $y \in X$ as $\delta \to 0$.*

Make a change of variable. Let $y + \delta v = x$. From the property of the metric r in $\mathcal{P}(B_R)$ we have

$$r(\frac{V^{(\delta)}(y,\cdot)}{\delta}, L(y, \frac{\cdot - y}{\delta})) \to 0, \tag{3.3}$$

For every $y \in X$ and $\phi \in \Phi$ we have from (3.3)

$$\rho((\frac{V^{(\delta)}(y,\cdot)}{\delta}, \phi(\cdot)), (L(y, \frac{\cdot - y}{\delta}), \phi(\cdot))) \to 0.$$

Let $f \in \mathcal{P}$ and $f \geq 0$. Then $\forall y \in X$

$$\rho((\frac{V^{(\delta)}(y,\cdot)}{\delta}, \phi(\cdot)) \odot \frac{f}{\delta}(y), (L(y, \frac{\cdot - y}{\delta}), \phi(\cdot)) \odot \frac{f}{\delta}(y)) \to 0.$$

Hence

$$\rho((\frac{\oplus \int_{y \in Y} V^{(\delta)}(y,\cdot) \odot f(y)}{\delta}, \phi(\cdot)), (\frac{\oplus \int_{y \in Y} \delta L(y, \frac{\cdot-y}{\delta}) \odot f(y)}{\delta}, \phi(\cdot))) \to 0.$$

Hence

Proposition 6 *Let $\phi \in \Phi$ and $f \in \mathcal{P}(X)$, $f \geq 0$. Then*

$$\lim_{\delta \to 0}(\rho((\frac{A_\delta f}{\delta}, \phi), (\frac{U_\delta f}{\delta}, \phi))) = 0. \tag{3.4}$$

We give now more convenient form to the last equality.

Assertion 3 *Let $\delta \in [0, \delta_0]$, let $\{f_\delta\}$ and $\{g_\delta\}$ be two families of elements from $\mathcal{P}^0(X)$ and suppose:*

(1) *for any δ there exists a constant $c_\delta \in R$ such that $f_\delta(x) \leq c_\delta$, $x \in X$;*

(2) *for every $x \in X$ $\delta \in [0, \delta_0]$ the inequality $g_\delta(x) \geq f_\delta(x)$ holds;*

(3) *for every $\phi \in \Phi^0$ we have $\rho((g_\delta, \phi), (f_\delta, \phi)) \overset{\delta \to 0}{\to} 0$.*

Then the sequence of the classes $(g_\delta - f_\delta)$ defined by the representatives $x \to g_\delta(x) - f_\delta(x)$ tends to the class $x \equiv 0$ in $\mathcal{P}^0(X)$ as $\delta \to 0$.

Proof Suppose, by contradiction, that $g_i - f_i$ does not tend in $\mathcal{P}^0$ to the class defined by the mapping $x \to 0$, $x \in X$. Since $g_i - f_i \geq 0$, there exist a neighborhood $\mathcal{O}$ in X and a number $\alpha > 0$ such that for $x \in \mathcal{O}$ the inequality $g_i(x) \geq f_i(x) + \alpha$ holds. But this contradicts condition (3). □

Let $f \in \mathcal{P}$ and $C \in \mathcal{R}$, $C \neq 0$ so that $f_1 = f \odot C > 0$. Then $A_\delta f_1 - U_\delta f_1 = A_\delta f - U_\delta f$. We apply Assertion 3 to the families $f_\delta = \frac{1}{\delta} U_\delta f_1$ and $g_\delta = \frac{1}{\delta} A_\delta f_1$. The trivial inequality for kernels of the endomorphisms A_δ and U_δ yields $(A_\delta f_1)(x) \geq (U_\delta f_1)(x)$. Hence by using formula (3.4) we obtain

Proposition 7 *Let a class* $f \in \mathcal{P}(X)$ *be such that for all* $\delta \in [0, \delta_0]$ *there exists a constant* c_δ *for which the inequality* $U_\delta f(x) < c_\delta$ *holds for every* $x \in X$. *Then for any* $\phi \in \Phi$

$$\lim_{\delta \to 0} (\frac{A_\delta f - U_\delta f}{\delta}, \phi) = 0.$$

4   Spaces Defined by Hamiltonians

Let $H : X \times \mathbb{R}^n \to \mathbb{R}$ be a mapping such that for every x the mapping $H(x, \cdot) : \mathbb{R}^n \to \mathbb{R}$ is concave (generally speaking, not strictly) and the strong Lipschitz condition $|H(x, p) - H(y, p)| \leq \lambda(1 + \| p \|) \| x - y \|$ holds for some $\lambda > 0$. The function H gives rise to a differential operator defined on the set of differentiable functions in clX by the rule

$$(\mathcal{H}\phi)(x) = H(x, D\phi(x)). \tag{4.1}$$

We shall consider the values of this mapping in $\mathcal{P}^0(X)$. First, extend this operator over the algebraic semimodule $\mathcal{P}_{C^1}(X)$ with the operations $\oplus$ and $\odot$ generated by bounded differentiable functions, i.e. over finite functions of the form $f_1 \oplus f_2 \oplus ... \oplus f_k$, where f_i are differentiable in clX. Now define $\mathcal{H}(f_1 \oplus f_2)$ for $f_1, f_2 \in C^1(clX, \mathbb{R})$. Suppose that $\mathcal{H}(f_1 \oplus f_2)$ is an equivalence class in $\mathcal{P}^0(X)$ containing the following representative:

$$x \mapsto \begin{cases} \mathcal{H}(f_1)(x), & \text{if } f_1(x) < f_2(x) \\ \mathcal{H}(f_2)(x), & \text{if } f_2(x) < f_1(x) \\ \mathcal{H}(f_1)(x) \oplus \mathcal{H}(f_2)(x), & \text{if } f_1(x) = f_2(x). \end{cases} \tag{4.2}$$

It is easy to see that if $(f_1 \oplus f_2)(x) = (g_1 \oplus g_2)(x)$, then $\mathcal{H}(f_1 \oplus f_2) = \mathcal{H}(g_1 \oplus g_2)$ as elements from $\mathcal{P}^0(X)$. By the same formula (4.2) by induction on k define also $\mathcal{H}(f_1 \oplus f_2 \oplus \cdots \oplus f_k)$. Using the concavity of $H(x, \cdot)$, it is not difficult to show that the class $\mathcal{H}(f)$ does not depend on the representation of f in the form $f_1 \oplus f_2 \oplus \cdots \oplus f_k$. Introduce a metric in the set of differentiable functions from clX to $\mathcal{R}$ according to the following rule. Let $A(X)$ be the semimodule generated by the set of affine mappings from X in $\mathbb{R}$, i.e., if $h \in A$, then $h = \phi_1 \oplus \cdots \oplus \phi_k$, where $\phi_i(x) = \langle a_i, x \rangle + b_i$, $a_i \in \mathbb{R}^n$, $b_i \in \mathbb{R}$, $i = 1, \ldots, k$. Assume that

$$r_H(f, g) = \text{Max}\{r(f, g), \sup_{h \in A(X)} r^0(\mathcal{H}(f \oplus h), \mathcal{H}(g \oplus h))\}.$$

For any differentiable f and g the expression on the right is finite, since $r^0(a,b)$ never exceeds the diameter of the domain X. Obviously this formula introduces a metric.

Definition 2 The completion of the set of differentiable functions by this metric r_H will be called the $\mathcal{P}_H(X)$ space.

It is evident that assigning to every sequence f_i fundamental in the metric r_H its limit f in the space $\mathcal{P}(X)$, we obtain a canonical embedding $i : \mathcal{P}_H(X) \to \mathcal{P}(X)$ which allows us to consider the elements from $\mathcal{P}_H(X)$ as classes in $\mathcal{P}(X)$. In particular, one can speak about continuous, differentiable and so on functions (classes) from $\mathcal{P}_H(X)$ and the values of $f(x)$ at every point $x \in X$ of a class $f \in \mathcal{P}_H(X)$.

Assertion 4 $\mathcal{P}_H(X)$ *is a complete metric semimodule over* $\mathcal{R}$, *i.e., the mappings* $\oplus : \mathcal{P}_H \times \mathcal{P}_H \to \mathcal{P}_H$ *and* $\odot : \mathcal{P}_H \times \mathcal{R} \to \mathcal{P}_H$ *are continuous in the metrics* r_H *in* $\mathcal{P}_H$ *and* ρ *in* $\mathcal{R}$, *respectively.*

Assertion 5 *The mapping* $\mathcal{H}$ *is uniquely extended to a continuous mapping from* $\mathcal{P}_H(X)$ *in* $\mathcal{P}^0(X)$ *that also will be denoted by* $\mathcal{H}$.

We shall now give some examples.

Example 6 Let $X =]-1,1[$, $H(x,p) = p$, $\mathcal{H}(f) = \frac{df}{dx}$. Let $f_1(x) = x, f_2(x) = -x$ so that $(f_1 \oplus f_2)(x) = -|x|$. Then $\mathcal{H}(f_1 \oplus f_2)$ is an equivalence class with the representative g such that $g(x) = 1$ for $x < 0$, $g(x) = -1$ for $x > 0$ and $g(0) = \alpha$, where α is an arbitrary number not smaller than -1. The class $\mathcal{H}(f_1 \oplus f_2)$ is the same one for each $\alpha \geq -1$.

Example 7 Show that the function $f : x \mapsto |x|$ does not belong to $\mathcal{P}_H$, with H from the previous example. Consider the sequence of the functions $h_i(x) = \frac{1}{i}$ tending to 0 as $i \to \infty$. Then $\lim_{i \to \infty} \mathcal{H}(f \oplus h_i) = g$, where g is the mapping such that $g(x) = 0$ for $x \neq 0$ and $g(0) = -1$, and this mapping does not equal $\mathcal{H}(f)$ which is equal to 0 everywhere. The last fact contradicts the continuity of the mapping $h \mapsto f \oplus h \mapsto \mathcal{H}(f \oplus h)$.

In Examples 8–12 let X be the set $]0,1[\times]0,1[$.

Example 8 Let $H(x,p) = p_1$. Every class in $\mathcal{P}(X)$ defined by a function of the type $(x_1,x_2) \mapsto f(x_2)$ for arbitrary f belongs to $\mathcal{P}_H(X)$, the class defined by $(x_1,x_2) \mapsto f(x_2) - |x_1|$ belongs to $\mathcal{P}_H(X)$, but the class defined by $(x_1,x_2) \mapsto f(x_2) + |x_1|$ does not belong to $\mathcal{P}_H(X)$.

Example 9 Let $H(x,p) = p_1^2 + p_2^2$. The function $(x_1,x_2) \mapsto \min(x_1,x_2)$ belongs to $\mathcal{P}_H$ but $(x_1,x_2) \mapsto \max(x_1,x_2)$ does not belong to $\mathcal{P}_H$.

Example 10 Let $H = |p_1| + |p_2|$. Then the functions $(x_1,x_2) \mapsto \max(x_1,x_2)$ and $(x_1,x_2) \mapsto \min(x_1,x_2)$ both belong to $\mathcal{P}_H$.

Example 11 Let $H = |p_1 + p_2|$. The discontinuous function $f(x_1,x_2)$ which is equal to x_1 for $x_1 \geq x_2$ and $x_2 + 1$ for $x_1 < x_2$ determines a class which belongs to $\mathcal{P}_H$.

Example 12 Let $H(x,p) = p_1 + p_2^2$. The discontinuous function $f(x_1, x_2)$ which is equal to ∞ for $|x_2| > 1 + x_1$ and 0 for $|x_2| \le 1 + x_1$ belongs to $\mathcal{P}_H$.

Theorem 1 *The set $\mathcal{P}_H(X)$ as a subset of $\mathcal{P}(X)$ is uniquely characterized by the following properties:*

(1) *$\mathcal{P}_H$ is a complete metric semimodule over $\mathcal{R}$ (i.e. the operations $\oplus$ and $\odot$ are continuous in $\mathcal{P}_H$);*

(2) *The set of differentiable bounded functions in X is dense in $\mathcal{P}_H$;*

(3) *The differential operator $\mathcal{H}$ generated according to the rule (4.1) may be extended to a continuous mapping from $\mathcal{P}_H(X)$ to $\mathcal{P}^0(X)$;*

(4) *The topology in $\mathcal{P}_H(X)$ is the weakest one in which properties (1)–(3) hold.*

Proof Suppose that there is a some metric satisfying properties (1)–(3), $\mathcal{P}_H$ is the completion by this metric and the sequence f_i of differentiable functions converges to $f \in \mathcal{P}(X)$ in this metric. Then by virtue of properties (1)–(3) for any $h \in A(X)$ the limit $\mathrm{Max}(r(f_i, f), r^0(\mathcal{H}(f_i \oplus h), \mathcal{H}(f \oplus h)))$ is equal to 0 as $i \to \infty$. Let $R \in \mathbb{R}$ and $K(R)$ be a precompact set in $\mathcal{P}_H(X)$ which consists of mappings $h \in A(X)$ such that $h \ge R$ and $\mathcal{H}(h)(x) \ge R$ for any $x \in X$. Due to continuity of the operations in $\mathcal{P}(H)$ and $\mathcal{H}$ we see that for any R the limit of the expression $\mathrm{Max}(r(f_i, f), \underset{h \in K(R)}{\mathrm{Sup}}\, r^0(\mathcal{H}(f_i \oplus h), \mathcal{H}(f \oplus h)))$ equals 0 as $i \to \infty$. We shall fix $R < 0$ with $|R|$ sufficiently large. Let now $h = h_1 \oplus \cdots \oplus h_k$, $g = f \oplus h_1 \oplus \cdots \oplus h_{k-1}$, $g_i = f_i \oplus h_1 \oplus \ldots \oplus h_{k-1}$, $h_k(x) = \langle a, x - b \rangle$ and suppose that $h_\lambda(x) = \lambda \langle a, x - b \rangle$, $\lambda \in \mathbb{R}$.

The sets $\Lambda_{i,\lambda} = \{x | h_\lambda(x) < g_i(x)\}$ and $\Lambda_\lambda = \{x | h_\lambda(x) < g(x)\}$ are open because of the lower semicontinuity of the mappings $x \mapsto g_i(x)$ and $x \mapsto g(x)$. The Hausdorff distance d of their closures $cl\Lambda_{i,\lambda}$ and $cl\Lambda_\lambda$ tends to 0 as $i \to \infty, \lambda \to \infty$. Moreover, since $r(f_i, f) \to 0$ as $i \to \infty$, $\underset{\lambda \ge 1}{\mathrm{Sup}}\, d(cl\Lambda_{i,\lambda}, cl\Lambda_\lambda) \to 0$ as $i \to \infty$. On the set $\Lambda_{i,\lambda}$ the values of $\mathcal{H}(g_i \oplus h_\lambda)$ coincide with the values of $\mathcal{H}(h_\lambda) = H(\cdot, \lambda a)$, and on the set Λ_λ the values of $\mathcal{H}(g \oplus h_\lambda)$ also coincide with those of $\mathcal{H}(h_\lambda)$. Outside of $cl\Lambda_{i,\lambda}$ the values of $\mathcal{H}(g_i \oplus h_\lambda)$ coincide with $\mathcal{H}(g_i)$, and outside of $cl\Lambda_\lambda$ the values of $\mathcal{H}(g \oplus h_\lambda)$ coincide with $\mathcal{H}(g)$. We have that $r^0(\mathcal{H}(g_i), \mathcal{H}(g)) \to 0$ as $i \to \infty$. Therefore to show that $\underset{\lambda \ge 1}{\mathrm{Sup}}\, r^0(\mathcal{H}(g_i \oplus h_\lambda), \mathcal{H}(g \oplus h_\lambda)) \to 0$ as $i \to \infty$ it is sufficient to show that at the points $x \in cl\Lambda_\lambda \setminus \Lambda_\lambda$ such that $\mathcal{H}(h_\lambda)(x) < \mathrm{Inf}\, \mathcal{H}(g_i)(x)$ the inequality $\mathcal{H}(g)(x) \ge \mathcal{H}(h_\lambda)(x) = H(x, \lambda a)$ hold for all $\lambda \ge 1$. For large $|R|$ and $\lambda \ge 1$ the function $\lambda \mapsto H(x, \lambda a)$ is monotone for every x via the concavity of H. Hence it remains to show that if at some point $x \in cl\Lambda_1 \setminus \Lambda_1$ the inequality $H(x, a) \le \mathcal{H}(g_i)(x)$ is true, then $H(x, a) \le \mathcal{H}(g)(x)$ is true as well. The last implication arises when one compares the derivatives $\langle Df_i(x), \xi \rangle$ in the directions ξ "toward the interior" of $\Lambda_{i,1}$ with $\langle a, \xi \rangle$. This comparison follows from the inequalities $g_i \geq h_k$ that are valid inside of $\Lambda_{i,1}$ in some neighborhood of the sets $cl\Lambda_{i,1} \setminus \Lambda_{i,1}$, whereas $g_i(x) = h_k(x)$ for $x \in cl\Lambda_{i,1} \setminus \Lambda_{i,1}$.

Thus for sufficiently large $|R|$ depending only on H

$$\text{Max}(r(f_i, f), \underset{h = h^1 \oplus h^2}{\text{Sup}} r^0(\mathcal{H}(f_i \oplus h), \mathcal{H}(f \oplus h))),$$

where $h^1 \in A(R)$ and h^2 are arbitrary affine mappings, tends to 0 as $i \to \infty$. Dealing with the remaining h_i in the sum $h = h_1 \oplus \cdots \oplus h_k$ in an analogous way, we obtain that $r_H(f_i, f) \to 0$ as $i \to \infty$. The last fact proves property (4), and the theorem follows from Assertions 4, 5. $\qquad\square$

Proposition 8 *Let $f, g \in \mathcal{P}_H(X)$ and suppose that at $x_0 \in X$ the inequality $f(x_0) < g(x_0)$ is true. Then $\mathcal{H}(f \oplus g)(x_0) = \mathcal{H}(f)(x_0)$.*

Proof If f and g are differentiable at the point x_0 this proposition is obvious due to equality (4.2). For arbitrary f and g we now choose sequences f_i and g_i of differentiable functions which tend to f and g, respectively, in the metric of $\mathcal{P}_H$ and such that $f_i(x_0) = f(x_0)$ and $g_i(x_0) = g(x_0)$. Then for every i the equality $\mathcal{H}(f_i \oplus g_i)(x) = \mathcal{H}(f_i)(x)$ is satisfied at all points x from some neighborhood U_i of the point x_0. The result follows from the continuity of the mapping $\mathcal{H}$.

5   Smoothness of Eigenfunctions

Definition 3 An element $f \in \mathcal{P}(X)$ that is not identically equal to $\mathbf{0}$ is called an *eigenelement* (eigenfunction) of an endomorphism A of the semimodule $\mathcal{P}(X)$ corresponding to an eigenvalue $\lambda \in \mathcal{R}$, whenever

$$Af = \lambda \odot f.$$

In the case when A is a compact integral endomorphism of $C(X, \mathcal{R})$, there are theorems asserting that this eigenvalue exists and under certain conditions on the kernel of A is unique [4]. In the general Lagrange problem the integral endomorphisms U_t need not be compact. But when investigating the behavior asymptotic in t of the mappings $U_t\phi$, the eigenelements of the endomorphisms U_t play a very important part [4].

Let f be an eigenelement of the endomorphism U_t for any t. The corresponding eigenvalue $\alpha(t)$ has the form $\lambda \cdot t$ because the semigroup property of U_t yields

$$\alpha(t_1 + t_2) = \alpha(t_1) \odot \alpha(t_2) = \alpha(t_1) + \alpha(t_2) \qquad \text{for all } t_1, \ t_2.$$

Theorem 2 *For any $t \geq 0$ let the class $f \in \mathcal{P}(X)$ be the eigenelement of the endomorphism U_t from (3.1) corresponding to an eigenvalue λt, and let $f \leq C$ for some $C \in \mathcal{R}$, $C \neq 0$. Let*

$$H(x, p) = \underset{\xi \in \mathbb{R}^n}{Inf} (L(x, \xi) - \langle \xi, p \rangle).$$

Then $f \in \mathcal{P}_H$ and $\mathcal{H}(f) = \lambda$, where $\mathcal{H}$ is the extension of the differential expression $y \to H(x, Dy(x))$ to the space $\mathcal{P}_H$.

Proof For any $\delta > 0$ and $\delta < \delta_0$ let the mapping $\mathcal{H}_\delta$ be defined on the set of upper bounded functions from X in $\mathbb{R}$ by the rule

$$\mathcal{H}_\delta(\phi) = \frac{1}{\delta}(A_\delta \phi - \phi).$$

Passing to classes in $\mathcal{P}(H)$, we obtain the mapping $\mathcal{H}_\delta$ which assigns to every class $\phi \in \mathcal{P}(X)$ such that $\phi(x) < c(\phi)$, $c(\phi) \in \mathbb{R}$ the class in $\mathcal{P}^0(X)$ with the representative $x \to \frac{1}{\delta}((A_\delta \phi)(x) - \phi(x))$.

Proposition 9 *Let $\phi \in \mathcal{P}(X)$. Suppose that*

(1) *For any $h \in A(X)$ in the metric of $\mathcal{P}^0(X)$ there exists $\lim_{\delta \to 0} \mathcal{H}_\delta(\phi \oplus h) = \mathcal{M}_{\phi,h} \in \mathcal{P}^0(X)$;*

(2) *$\lim \mathcal{M}_{\phi,h_i} = \mathcal{M}_{\phi,h}$ in $\mathcal{P}^0(X)$ whenever $h_i \in A(X)$ and $h_i \to h$ (in $\mathcal{P}_H$).*

Then $\mathcal{H}(\phi) = \lim_{i \to \infty} \mathcal{M}_{\phi,c_i}$ where c_i is a sequence of constant mappings tending to $\mathbf{0}$.

Let f be an eigenelement of the endomorphisms U_t and λ be the element from $\mathcal{R}$ associated with it, so that $U_t f = \lambda t + f$ in $\mathcal{P}(X)$, $h \in A(X)$.

For any $\delta > 0$ divide the set X into the following four disjoint sets $\mathcal{K}_\delta^i$ ($i = 1, \ldots, 4$) :

$$\mathcal{K}_\delta^1 = \{x \mid f(x) \leq h(x) \text{ and } (A_\delta f)(x) \leq (A_\delta h)(x)\},$$
$$\mathcal{K}_\delta^2 = \{x \mid f(x) > h(x) \text{ and } (A_\delta f)(x) > (A_\delta h)(x)\},$$
$$\mathcal{K}_\delta^3 = \{x \mid f(x) > h(x) \text{ and } (A_\delta f)(x) \leq (A_\delta h)(x)\},$$
$$\mathcal{K}_\delta^4 = \{x \mid f(x) \leq h(x) \text{ and } (A_\delta f)(x) > (A_\delta h)(x)\}.$$

Evaluate $\mathcal{H}_\delta(f \oplus h) = \frac{1}{\delta}(A_\delta(f \oplus h) - (f \oplus h))$ on each of these sets.

1. At the point x from $Int\mathcal{K}_\delta^1$ we have

$$\mathcal{H}_\delta(f \oplus h)(x) = \frac{1}{\delta}(A_\delta f(x) - f(x)).$$

But $f(x) = U_\delta f(x) - \lambda\delta$. Therefore we see that the equality

$$\mathcal{H}_\delta(f \oplus h)(x) = \frac{1}{\delta}(A_\delta f(x) - U_\delta f(x)) + \lambda \qquad (5.1)$$

holds in $Int\mathcal{K}_\delta^1$.

2. In $\mathcal{K}_\delta^2$ we have

$$\mathcal{H}_\delta(f \oplus h)(x) = \frac{1}{\delta}(A_\delta h(x) - h(x)). \qquad (5.2)$$

3. In $\mathcal{K}_\delta^3$ by virtue of the inequality $(A_\delta f)(x) \geq (U_\delta f)(x)$ we obtain

$$\mathcal{H}_\delta(f \oplus h)(x) = \frac{1}{\delta}(A_\delta f(x) - h(x)) \geq \frac{1}{\delta}(U_\delta f(x) - h(x))$$

$$= \lambda + \frac{1}{\delta}(f(x) - h(x)) \geq \lambda. \qquad (5.3)$$

4. In $\mathcal{K}_\delta^4$

$$\mathcal{H}_\delta(f \oplus h)(x) = \frac{1}{\delta}(A_\delta h(x) - f(x)) \geq \frac{1}{\delta}(A_\delta h(x) - h(x)). \qquad (5.4)$$

Now we shall give a property of the operators A_δ that will be important in what follows, and arises from their definition on taking account of property (2) of the Lagrangian L.

Assertion 6 *The operators A_δ possess the following localization property: if $f(x) \leq g(x)$ in some neighborhood of a point x_0, then for δ sufficiently small the inequality $(A_\delta f)(x_0) \leq (A_\delta g)(x_0)$ is valid.*

The mappings $x \mapsto f(x)$ and $x \mapsto (A_\delta f)(x)$ are lower semicontinuous and the mappings $x \mapsto h(x)$ and $x \mapsto (A_\delta h)(x)$ are continuous. Hence the sets $\{x \mid f(x) \leq h(x)\}$ and $\{x \mid (A_\delta f)(x) \leq (A_\delta h)(x)\}$ are closed, but the sets $\{x \mid f(x) > h(x)\}$ and $\{x \mid (A_\delta f)(x) > (A_\delta h)(x)\}$ are open.

This implies that the set $\mathcal{K}_\delta^2$ is open and for each $\mathcal{K}_\delta^i$ ($i = 1, 2, 3$) the complement of its interior in $\mathcal{K}_\delta^i$ (i.e. its boundary) is nowhere dense.

Thus by Assertion 6 the set $\mathcal{K}_\delta^3$ (respectively, $\mathcal{K}_\delta^4$) shrinks to a nowhere dense set belonging to the closure of the limit of $\mathcal{K}_\delta^1$ (respectively, $\mathcal{K}_\delta^2$) as $\delta \to 0$.

If a neighborhood $U \subset \text{Int}\mathcal{K}_\delta^1$ for all $\delta \geq 0$ sufficiently small, then by virtue of Proposition 7 and (5.1) we have $(\mathcal{H}_\delta(f \oplus h), \phi) \to (\lambda, \phi)$ as $\delta \to 0$ for every $\phi \in \Phi$, Supp $\phi \in U$.

If a neighborhood $U \subset \mathcal{K}_\delta^2$ for all $\delta \geq 0$ is sufficiently small, then by virtue of (5.2) we have $(\mathcal{H}_\delta(f \oplus h), \phi) \to (\mathcal{H}(h), \phi)$ as $\delta \to 0$ for $\phi \in \Phi$, Supp $\phi \in U$.

Since the sets $\mathcal{K}_\delta^3$ and $\mathcal{K}_\delta^4$ shrink to nowhere dense sets and the values of $\mathcal{H}_\delta(f \oplus g)$ satisfy inequalities (5.3) and (5.4) on these sets, respectively, the limit of the family $\mathcal{H}_\delta(f \oplus h)$ in $\mathcal{P}^0$ as $\delta \to 0$ (see Example 5) exists and equals the class from $\mathcal{P}_0(X)$ with a representative which takes values λ on some open subset U_h^1 of the set X and $\mathcal{H}(h)$ on another open subset U_h^2 of X, where U_h^1 and U_h^2 are such that the complement V of $U_h^1 \cup U_h^1$ in X is nowhere dense. Since $V = cl\, U_h^1 \cap cl\, U_h^2$, the values of this representative on V can be assumed to be equal to $\mathcal{H}(h) \oplus \lambda$.

This establishes that for every $h \in A(X)$ the limit of $\mathcal{M}_{f,h}$ from Proposition 9 exists. We now investigate the dependence of $\mathcal{M}_{f,h}$ on h. Let $h_i \to h$ and let $d(\cdot, \cdot)$ be the Hausdorff distance for compact sets in $cl\, X$. Hence, if the closures of the sets U_h^1 and U_h^2 satisfy the relations $d(cl\, U_{h_i}^1, cl\, U_h^1) \to 0$ and $d(cl\, U_{h_i}^2, cl\, U_h^2) \to 0$ as $i \to \infty$, then $\mathcal{M}_{f,h_i} \to \mathcal{M}_{f,h}$ in the metric of $\mathcal{P}^0$.

These relations hold unless there exists a neighborhood W in X such that $W \cap \Gamma$ is the empty set and one of the two sets $cl\, U_h^1$ or $cl\, U_h^2$ is empty. Remark that at the same time the set $cl\, U_{h_i}^1$ or $cl\, U_{h_i}^2$, respectively, is not empty for any i. Suppose that one of these cases occurs.

First let $cl\, U_h^2 \cap W$ be empty and $cl\, U_{h_i}^2 \cap W$ be nonempty for all i. This means that $f(x) \le h(x)$ for all x from W. But for any i the set $\{x \in N \mid f(x) > h_i(x)\}$ is nonempty.

Let $x \in W$ be a limiting point for a sequence $\{x_i\}$, where $x_i \in U_{h_i}^2$ such that $f(x_i) \ge h_i(x_i)$. By virtue of Assertion 6 the inequality $f(x) \le h(x)$ in W yields for δ sufficiently small the inequality $(A_\delta f)(x) \le (A_\delta h)(x)$ which must be valid in some neighborhood W' smaller than W.

From the convergence of the family $\mathcal{H}_\delta f$ to λ in $\mathcal{P}^0(X)$ as $\delta \to 0$ it follows that for $\delta > 0$ sufficiently small there exists a positive function $\delta \to \varepsilon(\delta)$ such that $\varepsilon(\delta) \to 0$ as $\delta \to 0$ and $(\mathcal{H}_\delta f)(x) \ge \lambda - \varepsilon(\delta)$ for all $x \in W$.

Let $\delta > 0$ be fixed. Then

$$\lambda - \varepsilon(\delta) \le (\mathcal{H}_\delta f)(x_i) = \frac{1}{\delta}(A_\delta f(x_i) - f(x_i)) \le$$

$$\le \frac{1}{\delta}(A_\delta h(x_i) - h(x_i) + h(x_i) - h_i(x_i)) = \frac{1}{\delta}(A_\delta h(x_i) - h(x_i)) + \frac{1}{\delta}(h(x_i) - h_i(x_i)).$$

Since the function $x \mapsto (A_\delta h)(x)$ is continuous, we obtain on passing to the limit as $i \to \infty$ that $\lambda - \varepsilon(\delta) \le \mathcal{H}_\delta(h)(x)$, whence $\lambda \le \mathcal{H}(h)(x)$.

And so (Example 5), when the sets $cl\, U_{h_i}^2$ vanish as $i \to \infty$, we have the convergence $\mathcal{M}_{f,h_i} \to \mathcal{M}_{f,h} = \mathcal{M}_f = \lambda$ in W.

The other case is that $cl\, U_h^1 \cap W$ is empty, whereas for any i the sets $cl\, U_{h_i}^1 \cap W$ are nonempty. This means that $h(x) \le f(x)$ for all x from W and the set $\{x \in W \mid h_i(x) > f(x)\}$ is nonempty for any i. Once again, let $x \in W$ be a limiting point for a sequence $\{x_i\}$, where $x_i \in U_{h_i}^1$. For every $\varepsilon > 0$ choose a number $\delta > 0$ and a sequence x_i such that the inequality

$$(\mathcal{H}_\delta f)(x_i) \le \lambda + \varepsilon$$

holds. From $h(x) < f(x)$ in W, for δ sufficiently small it follows that the inequality $(A_\delta h)(x) \le (A_\delta f)(x)$ holds in some neighborhood $W' \subset W$.

We have

$$(\mathcal{H}_\delta h)(x_i) = \frac{1}{\delta}((A_\delta h)(x_i) - h_i(x_i) + h_i(x_i) - h(x_i)) \le$$

$$\frac{1}{\delta}((A_\delta f)(x_i) - f(x_i)) + \frac{1}{\delta}(h_i(x_i) - h(x_i)) \le$$

$$(\mathcal{H}_\delta f)(x_i) + \frac{1}{\delta}(h_i(x_i) - h(x_i)) \le \lambda + \varepsilon + \frac{1}{\delta}(h_i(x_i) - h(x_i)).$$

Passing to the limit first as $i \to \infty$ and then as $\varepsilon \to 0$, we obtain the inequality

$$(\mathcal{H}h)(x) \le \lambda.$$

And so when the sets $cl\,U_{h_i}^1$ vanish as $i \to \infty$ we have the convergence $\mathcal{M}_{f,h_i} \to \mathcal{M}_{f,h} = \mathcal{M}_h = \mathcal{H}(h)$ in W. The property (2) from Proposition 9 is proved, and thus Theorem 2 is proved. $\qquad\square$

6 On the Uniqueness of Solutions of Stationary Hamilton–Jacobi Equations

Theorem 3 *Suppose that there exist $p_0 \in \mathbb{R}^n$ and $\lambda_0 > \lambda$ such that for all $x \in X$ the inequality*

$$H(x, p_0) \ge \lambda_0$$

is valid. Let y_1 and y_2 be two solutions from $\mathcal{P}_H(X)$ of the equation $\mathcal{H}(y) = \lambda$ for which there exist limiting values on the boundary Γ of the domain X (i.e. $\forall x_0 \in \Gamma$, $\lim y_i(x)$ exists in $\mathcal{R}$ as $x \to x_0$, $x \in X$). Then $y_1 \equiv y_2$ in X, whenever these limiting values coincide.

Proof Suppose, by contradiction, that there are two classes y_1 and y_2 in $\mathcal{P}_H$ such that $\mathcal{H}(y_i) = \lambda$ $(i = 1, 2)$ and $y_1(x) = y_2(x)$ for $x \in \Gamma = \partial X$, but for some x_0 belonging to X $y_1(x_0) \ne y_2(x_0)$. Adding to y_1 and y_2 one and the same constant, one can consider that $\underset{x \in X}{\text{Inf }} y_1(x) \ge \gamma$, where γ is a preassigned positive number. Let $y_1(x_0) - y_2(x_0) = \beta > 0$, and let $y_{(\alpha)}(\cdot)$ be an element from $\mathcal{P}_H(X)$ defined for $0 < \alpha < 1$ by the rule

$$y_{(\alpha)} = \alpha \cdot y_1 + (1 - \alpha)\langle \cdot, p_0 \rangle.$$

We shall show that one can choose α such that the following requirements are satisfied:

(1) There exists a neighborhood U of the boundary Γ such that for $x \in U$ the inequality $y_{(\alpha)}(x) < y_2(x)$ holds;

(2) $y_2(x_0) < y_{(\alpha)}(x_0) < y_1(x_0)$;

(3) There exists $\delta > 0$ such that $\mathcal{H}(y(\alpha)) \ge \lambda + \delta$ in $\mathcal{P}^0(X)$.

To guarantee the validity of (1) for any α it is sufficient to choose γ such that $\|p_0\| \cdot M < \gamma$, where M is the radius of a ball in $\mathbb{R}^n$ containing X.

Property (2) can be rewritten in the following way:

$$y_1(x_0) - \frac{\beta}{1-\alpha} < \langle x_0, p_0 \rangle < y_1(x_0)$$

and to guarantee its validity it is enough to choose γ sufficiently large and α sufficiently close to 1.

Note first the following: suppose that $\varepsilon > 0$ and $H(x,p) \geq \lambda - \varepsilon$ for some $p \in \mathbb{R}^n$. Then due to the concavity of the mapping $p \mapsto H(x,p)$

$$H(x, \alpha p + (1-\alpha)p_0) \geq \alpha H(x,p) + (1-\alpha)H(x,p_0) \geq$$
$$\geq \alpha \lambda + (1-\alpha)\lambda_0 - \alpha \varepsilon = \lambda + \delta(\alpha) - \alpha \varepsilon$$

with $\delta(\alpha) > 0$ for $\alpha < 1$.

Let $y_{1,i}$ be a sequence of functions differentiable in X that converges to y_1 in the metric of $\mathcal{P}_H$. This means, in particular, that there exists a sequence ε_i tending to 0, $\varepsilon_i > 0$ such that $\mathcal{H}(y_{1,i}) \geq \lambda - \varepsilon_i$ in $\mathcal{P}^0(X)$.

So it follows that

$$\mathcal{H}(y_{1,i})(x) = H(x, Dy_{1,i}(x)) = H(x, \alpha Dy(x) + (1-\alpha)p_0) \geq \lambda + \delta(\alpha) - \alpha \varepsilon_i.$$

The continuity of $\mathcal{H}$ yields $\mathcal{H}(y_{(\alpha)}) \geq \lambda + \delta(\alpha)$. Hence, property (3) is proved.

Now fix α so that properties (1)–(3) hold. Consider the family of elements from $\mathcal{P}_H$

$$h_\mu = y_{(\alpha)} \oplus \mu \odot y_2 = \min(y_{(\alpha)}, y_2 + \mu)$$

for $\mu \geq 0$.

Let $\bar{\mu}$ be the smallest value of μ for which $y_{(\alpha)} \oplus \mu \odot y_2 = y_{(\alpha)}$. Observe the behavior of $\mathcal{H}(h_\mu)$ as $\mu \to \bar{\mu}$ with $\mu < \bar{\mu}$. The continuity of the operations $\oplus$ and $\odot$ in $\mathcal{P}_H$ and of $\mathcal{H} : \mathcal{P}_H \to \mathcal{P}^0$ must yield in $\mathcal{P}^0$

$$\lim_{\mu \to \bar{\mu}} \mathcal{H}(h_\mu) = \mathcal{H}(h_{\bar{\mu}}) = \mathcal{H}(y_{(\alpha)}) \geq \lambda + \delta$$

via property (3). But now we shall show that this is impossible. For every $\mu < \bar{\mu}$ the set $U_\mu = \{x \mid y_2(x) + \mu < y_{(\alpha)}(x)\}$ is nonempty due to property (2) and by Proposition 8 we have $\mathcal{H}(h_\mu)(x) = \lambda$ on this set. The sets U_μ for $\mu < \bar{\mu}$ contain a common point that is in the interior of X as shown by property (1). Hence $\lim_{\mu \to \bar{\mu}} \mathcal{H}(h_\mu)$ is the class in $\mathcal{P}^0(X)$ taking the value λ at some point of X. This contradiction proves the theorem. □

The existence of a p_0 common for all $x \in X$ and such that $H(x,p_0) \geq \lambda_0 > \lambda$ is a purely technical condition which is used here for simplicity. But if for some x_0 $\mathrm{Sup}_{p \in \mathbb{R}^n} H(x_0, p) = \lambda$, then the uniquiness of solutions from $\mathcal{P}_H$ cannot hold. For example: $H(x,p) = -|p| + x^2$, $x \in]-1,1[$. The functions $y = \frac{1}{3}|x|^3 - \frac{1}{3}$ and $y = -\frac{1}{3}|x|^3 + \frac{1}{3}$ are both differentiable solutions of $\mathcal{H}(y) = 0$ with the same boundary values.

References

[1] V. Kolokoltsov and V. Maslov "The general form of endomorphisms in the space of continuous functions with values in a numerical semiring with idempotent addition" Dokl. Akad. Nauk SSSR, 295 no. 2 283–287 (1987); English transl. Soviet Math. Dokl., 36 no. 1 55–59 (1988)

[2] V. Kolokoltsov and V. Maslov "Idempotent calculus as the apparatus of optimization theory" Funktsional. Anal. i Prilozhen., 23 no. 1 1–14 (1989), 23 no. 4 53–62 (1989); English transl. in Functional Anal. Appl. 23

[3] V. Kolokoltsov and V. Maslov "Idempotent analysis" Nauka, Moscow (in Russian) (1993)

[4] V. Maslov and S. Samborskiĭ (Editors) Idempotent Analysis Advances in Soviet Mathematics, vol. 13, Amer. Math. Soc. (1992)

[5] P. Wolenski "The semigroup property of value functions in Lagrange problems" Transactions of the Amer. Math. Soc., vol. 335 131–154 (1993)

[6] F.H. Clarke "Optimization and nonsmooth analysis" Wiley Interscience, New York (1983)

New Differential Equation for the Dynamics of the Pareto Sets

Vassili N. Kolokoltsov and Victor P. Maslov

Abstract *Idempotent structures suitable for multicriteria optimization are various semirings of sets or (in another representation) semirings of functions with an idempotent analog of convolution playing the role of multiplication. Here we use these structures to derive the differential equation describing the continuous dynamics of Pareto sets in multicriteria optimization problems and to define and construct its generalized solutions.*

1 Semiring of Pareto Sets

Let $\leq$ denote the Pareto partial order on $\mathbf{R}^k$. For any subset $M \subset \mathbf{R}^k$ we denote by $Min(M)$ the set of minimal elements of the closure of M in $\mathbf{R}^k$. Let us introduce the class $P(\mathbf{R}^k)$ of subsets $M \subset \mathbf{R}^k$ whose elements are pairwise incomparable,

$$P(\mathbf{R}^k) = \{M \subset \mathbf{R}^k \mid Min(M) = M\}.$$

Obviously, $P(\mathbf{R}^k)$ is a semiring with respect to the operations $M_1 \oplus M_2 = Min(M_1 \cup M_2)$ and $M_1 \odot M_2 = Min(M_1 + M_2)$; the empty set is the neutral element $\mathbf{0}$ with respect to addition in this semiring, and the neutral element $\mathbf{1}$ with respect to multiplication is the zero vector in $\mathbf{R}^k$. It is also clear that $P(\mathbf{R}^k)$ is isomorphic to the semiring of so-called normal sets, that is, closed subsets $N \subset \mathbf{R}^k$ such that $b \in N$ implies $a \in N$ for any $a \geq b$; the sum and the product of normal sets are defined as their usual union and sum, respectively. Indeed, if N is normal, then $Min(N) \in P(\mathbf{R}^k)$, and conversely with each $M \in P(\mathbf{R}^k)$ we can associate its normalization $Norm(M) = \{a \in \mathbf{R}^k \mid \exists b \in M : a \geq b\}$.

For any set X and any (partially) ordered set A, we shall denote by $B(X, A)$ the set of mappings $X \to A$ bounded below. Let us recall that in each idempotent semiring a partial order is naturally defined: $a \leq b$ if and only if $a \oplus b = a$. Now, if A is an idempotent semiring, then $B(X, A)$ is a semimodule with respect to pointwise addition $\oplus$ and multiplication $\odot$ by elements of A.

The semirings and semimodules arise naturally in multicriteria dynamic programming problems. Let a mapping $f : X \times U \to X$ specify a controlled dynamical system on X, so that any choice of $x_0 \in X$ and of a sequence of controls $\{u_1, \ldots, u_k\}$, $u_j \in U$, determines an admissible trajectory $\{x_j\}_{j=0}^{k}$ in X, where $x_j = f(x_{j-1}, u_j)$. Let $\varphi \in B(X \times U, \mathbf{R}^k)$, and let $\Phi(\{x_j\}_{j=0}^{k}) = \sum_{j=1}^{k} \varphi(x_{j-1}, u_j)$ be the corresponding vector criterion on the set of admissible trajectories. The element $Min(\bigcup_{\{x_j\}} \Phi(\{x_j\}_{j=0}^{k}))$, where $\{x_j\}_{j=0}^{k}$ are all possible k-step trajectories issuing from x_0, is denoted by $\omega_k(x_0)$ and is called the *Pareto set* for the criterion Φ and initial point x_0. Let us define a *Bellman operator* $\mathcal{B}$ on the semimodule $B(X, P(\mathbf{R}^k))$ by setting

$$(\mathcal{B}\omega)(x) = Min\left(\bigcup_{u \in U} (\varphi(x, u) \odot \omega(f(x, u))) \right).$$

Obviously, $\mathcal{B}$ is linear in the semimodule $B(X, P(\mathbf{R}^k))$, and it follows from Bellman's optimality principle that the Pareto sets in k-step optimization problems satisfy the recursion relation

$$\omega_k(x) = \mathcal{B}\omega_{k-1}(x). \tag{1}$$

Sometimes it is convenient to use another representation of the set $P(\mathbf{R}^k)$. Proposition 1 below, which describes this representation, is a specialization of a more general result stated in [ST]. Let L denote the hyperplane in $\mathbf{R}^k$ determined by the equation

$$L = \left\{ (a^j) \in \mathbf{R}^k \mid \sum a^j = 0 \right\},$$

and let $CS(L)$ denote the semiring of functions $L \to \mathbf{R} \cup \{+\infty\}$ with the pointwise minimum as the addition and the idempotent convolution

$$(g \star h)(a) = \inf_{b \in L} (g(a - b) + h(b))$$

as the multiplication. Let us define a function $n \in CS(L)$ by setting $n(a) = \max_j(-a^j)$. Obviously, $n \star n = n$; that is, n is a multiplicatively idempotent element of $CS(L)$. Let $CS_n(L) \subset CS(L)$ be the subsemiring of functions h such that $n \star h = h \star n = h$. It is easy to see that $CS_n(L)$ contains the function identically equal to $\mathbf{0} = \infty$, and the other elements of $CS_n(L)$ are just the functions that take the value $\mathbf{0}$ nowhere and satisfy the inequality $h(a) - h(b) \leq n(a - b)$ for all $a, b \in L$. In particular, for each $h \in CS_n(L)$ we have

$$|h(a) - h(b)| \leq \max_j |a^j - b^j| = \|a - b\|,$$

which implies that h is differentiable almost everywhere.

Proposition 1 *The semirings $CS_n(L)$ and $P(\mathbf{R}^k)$ are isomorphic.*

Proof The main idea is that the boundary of each normal set in $\mathbf{R}^k$ is the graph of some real function on L, and vice versa. More precisely, consider the vector $e = (1, \ldots, 1) \in \mathbf{R}^k$ normal to L and define the function $h_M : L \to \mathbf{R}$ corresponding to a set $M \in P(\mathbf{R}^k)$ as follows:

$$h_M(a) = \inf\{\lambda \in \mathbf{R} \mid a + \Lambda e \in Norm(M)\}.$$

Then the functions corresponding to singletons $\{\varphi\} \in \mathbf{R}^k$ have the form

$$h_\varphi(a) = \max_j(\varphi^j - a^j) = \overline{\varphi} + n(a - \varphi_L), \tag{2}$$

where $\overline{\varphi} = k^{-1} \sum_j \varphi^j$ is the mean of the coordinates of φ and $\varphi_L = \varphi - \overline{\varphi} e$ is the projection of φ on L. Since idempotent sums $\oplus$ of singletons in $P(\mathbf{R}^k)$ and of functions (2) in $CS_n(L)$ generate $P(\mathbf{R}^k)$ and $CS_n(L)$, respectively, to prove the proposition it suffices to verify that the $\odot$-multiplication of vectors in $\mathbf{R}^k$ corresponds to the convolution of functions (2), i.e. that

$$h_\varphi \star h_\psi = h_{\varphi \oplus \psi}.$$

By (2) it suffices to show that

$$n_\varphi \star n_\psi = n_{\varphi \oplus \psi},$$

where $n_\varphi(a) = n(a - \varphi_L)$, and the latter identity is valid since

$$n_\varphi \star n_\psi = n_0 \star n_{\varphi + \psi} = n \star n_{\varphi + \psi} = n_{\varphi + \psi}.$$

$\square$

2   The Evolutionary Differential Equation

Let us consider the controlled process in $\mathbf{R}^n$ with continuous time $t \in \mathbf{R}_+$ specified by the controlled differential equation $\dot{x} = f(x, u)$ (where u belongs to a metric control space U and the function f is Lipschitz continuous with respect to both arguments) and by a continuous function $\varphi \in B(\mathbf{R}^n \times U, \mathbf{R}^k)$, which determines a vector-valued integral criterion

$$\Phi(x(\cdot)) = \int_0^t \varphi(x(\tau), u(\tau)) \, d\tau$$

on the trajectories. Let us pose the problem of finding the Pareto set $\omega_t(x)$ for a process of duration t issuing from x and with the terminal set determined by some function $\omega_0 \in B(\mathbf{R}^n, \mathbf{R}^k)$. Therefore,

$$\omega_t(x) = B_t\omega_0(x) = Min \bigcup_{x(\cdot)} (\Phi(x(\cdot)) \odot \omega_0(x(t))), \tag{3}$$

where $x(\,\cdot\,)$ are all possible admissible trajectories issuing from x, i.e. $x(\,\cdot\,)$ are all solutions of the equation $\dot{x} = f(x, u)$ with initial condition $x(0) = x$ corresponding to piecewise constant functions $u : [0, t] \mapsto U$.

Due to (1) (or, in other words, due to the semigroup property of the family of the Bellman operator $\mathcal{B}_t$), we have

$$\omega_t(x) = \mathcal{B}_\tau \omega_{t-\tau}(x). \tag{4}$$

For small τ, one can rewrite this recursive equation up to terms of order $O(\tau^2)$ in a simpler form:

$$\omega_t(x) = \tilde{\mathcal{B}}_\tau \omega_{t-\tau}(x) = Min \bigcup_{u \in U} \tau\varphi(x, u) \odot \omega_{t-\tau}(x + \tau f(x, u)). \tag{5}$$

By Proposition 1 and using the natural imbedding

$$in : B(\mathbf{R}^n, CS_n(L)) \to B(\mathbf{R}^n \times L, \mathbf{R} \cup \{\infty\}), \tag{6}$$

we can encode functions $\omega_t \in B(\mathbf{R}^n, PR^k)$ by functions $S(t, x, a) : \mathbf{R}_+ \times \mathbf{R}^n \times L \mapsto \mathbf{R} \cup +\infty$ and thus rewrite (5) in the following form

$$S(t, x, a) = \tilde{\mathcal{B}}_\tau S(t - \tau, x, a) = \min_u (h_{\tau\varphi(x,u)} \star S(t - \tau, x + \tau f(x, u)))(a). \tag{7}$$

Using (2) and the fact that n is the multiplicative unit in $CS_n(L)$ we have

$$S(t, x, a) = \min_u (\tau\overline{\varphi}(x, u) + S(t - \tau, x + \Delta x(u), a - \tau\varphi_L(x, u))).$$

Now we expand S into a series modulo $O(\tau^2)$, and collect like terms, which gives the equation

$$\frac{\partial S}{\partial t} + \max_u \left(\varphi_L(x, u)\frac{\partial S}{\partial a} - f(x, u)\frac{\partial S}{\partial x} - \overline{\varphi}(x, u) \right) = 0. \tag{8}$$

3   Main Results

In our deduction of equation (8) we have followed (in our setting) the standard heuristic procedure of the Bellman dynamic programming method. In order to give to equation (8) a real sense, one should give a rigorous definition of its generalized solution. But although the presence of a vector criterion has resulted in a larger dimension, this equation coincides in its form with the usual Hamilton–Jacobi–Bellman differential equation and we can use the standard definitions of the resolving operator for the generalized solutions to the Cauchy problem for this equation, the most natural of them being obtained by the method of introducing the vanishing viscosity (see, for instance, [Kr] or [CL]) or by the method of idempotent analysis (see

[KM1, KM2]). All these theories give the following formula for the resolving operator:

$$R_t : B(\mathbf{R}^n \times L, \mathbf{R} \cup \{+\infty\}) \to B(\mathbf{R}^n \times L, \mathbf{R} \cup \{+\infty\})$$

to the Cauchy problem for equation (8):

$$R_t S_0(x, a) = \inf_{x(\cdot)} \int_0^t \overline{\varphi}(x(\tau), u(\tau)) \, d\tau + S_0(x(t), a(t)), \qquad (9)$$

where inf is taken among all admissible solutions of the system

$$\dot{x} = f(x, u), \quad \dot{a} = -\varphi_L(x, u) \qquad (10)$$

in $\mathbf{R}^n \times L$ (i.e. with piecewise constant functions $u : [0, t] \mapsto U$) starting in (x, a) with a free right end and fixed time t. Equivalently,

$$R_t S_0(x, a) = \inf_{\xi, \eta}(S_0(\xi, \eta) + S(t, x, a, \xi, \eta)), \qquad (11)$$

where

$$S(t, x, a, \xi, \eta) = \inf_{x(\cdot)} \int_0^t \overline{\varphi}(x(\tau), u(\tau)) \, d\tau$$

and inf is taken among all admissible solutions of the system

$$\dot{x} = -f(x, u), \quad \dot{a} = \varphi_L(x, u)$$

such that

$$x(0) = \xi, \quad a(0) = \eta, \quad \text{and} \quad x(t) = x, \quad a(t) = a.$$

In terms of the idempotent integral [KM1], formula (10) can be written as

$$R_t S_0(x, a) = \int^{\oplus} S(t, x, a, \xi, \eta) \odot S_0(\xi, \eta) \odot d\xi \odot d\eta. \qquad (11')$$

Proposition 2 *The image of imbedding (6) is invariant with respect to R_t.*

The proof follows directly from the well known formula (see, for instance, [KM1])

$$R_t = \lim_{m \to \infty} (\tilde{\mathcal{B}}_{t/m})^m, \qquad (12)$$

where the operator $\tilde{\mathcal{B}}_\tau$ is defined in (5), and the observation that the image of imbedding (6) is evidently invariant with respect to $\tilde{\mathcal{B}}_\tau$.

Comparing formulas (3) and (9) one sees that (9) can be in fact obtained simply by rewriting formula (3) (which defines Pareto sets $\omega_t(x)$) in the functional representation of the semiring $P(\mathbf{R}^k)$. Thus we obtain our main result:

Theorem 1 *The Pareto set $\omega_t(x)$ defined in (3) is determined by a function*

$$S_t = R_t S_0 \in B(\mathbf{R}^n \times L, \mathbf{R} \cup +\infty)$$

that is the generalized solution (9),(11) to equation (8) with the initial condition

$$S_0(x) = h_{\omega_0(x)}.$$

More precisely, let the linear mapping R_t^{CS} in $B(\mathbf{R}^n, CS_n(L))$ be defined by the equation

$$R_t \circ in = R_t^{CS} \circ in.$$

Then $\omega_t(x)$ corresponds to $R_t^{CS} S_0$ by the isomorphism of Proposition 1.

Let us state a similar result for the case in which the time is not fixed. Namely, the problem is to find the Pareto set

$$\omega(x) = Min \bigcup_{x(\cdot)} \Phi(x(\cdot)), \tag{13}$$

where Min is taken over the set of all admissible trajectories of the equation $\dot{x} = f(x, u)$ joining a point $x \in \mathbf{R}^n$ with a given point ξ. For the corresponding function $S(x, a)$ we now obtain (using the same line of arguments as in Section 2) the stationary equation

$$\max_u \left(\varphi_L \frac{\partial S}{\partial a} - f(x, u) \frac{\partial S}{\partial x} - \overline{\varphi}(x, u) \right) = 0 \tag{14}$$

with the additional condition

$$S(x, a)|_{x=\xi} = 0 = 1, \tag{15}$$

where $0 = 1$ is the neutral element for multiplication in the semiring $\mathbf{R} \cup +\infty$ with operations min and +.

For defining and constructing generalized solutions to the problem (14, 15) one can again use (as for the evolutionary case) the method of vanishing viscosity and the method of idempotent analysis. In the latter one can further distinguish two approaches, one of which (presented in [KM1, Ko, KM2]) defines the generalized solution for stationary Bellman equation as the eigenfunctions of the resolving operator to the Cauchy problem, while the other (presented in [SM, S]) defines idempotent analogues of Sobolev spaces in which reasonable (from the point of view of the optimization theory) solutions exist. Both of these approaches give the following formula for the solution of (14, 15):

$$S(x, a) = \inf_{x(\cdot)} \int_0^t \overline{\varphi}(x, u) \, d\tau + n(a(t)), \tag{16}$$

where inf is taken among all admissible solutions of system (10) issuing from (x, a) and satisfying the boundary condition $x(t) = \xi$. In other words

$$S(x, a) = \inf_{t \geq 0} R_t S_0(x, a),$$

where R_t is given by (9) with

$$S_0(x, a) = \delta_\xi(x, a) = \begin{cases} 0 = 1, & x = \xi \\ +\infty = 0, & x \neq \xi. \end{cases}$$

In view of the representations (11), (11′) this can be also rewritten in the form

$$S(x, a) = \int^{\oplus} S(t, \xi, \eta, x, a) \odot d\eta \odot dt. \tag{17}$$

As in the evolutionary case, we see that (16) is simply the functional representation of (13) and therefore we obtain the following theorem.

Theorem 2 *The Pareto set (8) is determined (by virtue of the isomorphism in Proposition 1) by the function* $(in)^{-1}S \in B(\mathbf{R}^n, CS_n(L))$, *where* $S(x, a)$: $\mathbf{R}^n \times L \to \mathbf{R}$ *is the generalized solution of the problem (14, 15) given by formulas (16, 17).*

It is worth noting here that the fundamental role in the calculation of the Pareto sets belongs not to equation (8) or (14), but to formulas (10, 11) and (16, 17) which give an (idempotent) integral representation for the solution in a convenient (in fact quite standard for optimization theory with only one criterion of quality) functional semimodule. For specific calculations one can use either recursive approximation (12), or the Pontryagin maximum principle. The simplest example of such a calculation is given below. Let us note also that the limit (12) should be understood in the pointwise sense for continuous initial functions and in a weak sense in general. Different equivalent formulations of this (idempotent) weak convergence are given in [KM1].

4   Example

Let us find the curves $x(\tau)$ in $\mathbf{R}$ joining the points $0 \in \mathbf{R}$ and $\xi \in \mathbf{R}$ and minimizing (in Pareto's sense) the integral functionals

$$\Phi_j = \int_0^t L_j(x(\tau), \dot{x}(\tau)) \, d\tau, \qquad j = 1, 2,$$

with quadratic Lagrangians $L_j(x, v) = A_j x + xv + \frac{1}{2}v^2$, where A_j, $j = 1, 2$, are constants (for definiteness, we assume $A_1 > A_2$).

For the case of two criteria the hyperplane L is a line whose points are parametrized by a single number, namely, by the first coordinate on $\mathbf{R}^2$; we have $n(a) = |a|$ for $a \in L$. The auxiliary system (7) in our case has the form

$$\dot{x} = u, \quad u \in \mathbf{R}, \qquad \dot{a} = -\frac{1}{2}(L_1 - L_2) = \frac{1}{2}(A_2 - A_1)x.$$

The corresponding Hamilton (Pontryagin) function has the form

$$H(x, a, p, \psi) = \max_u \left(pu + \psi \frac{L_2 - L_1}{2} - \frac{L_1 + L_2}{2} \right),$$

where ψ is the dual variable associated with a. The transversality condition on the right edge means that $\psi = 1$ if $a(t) < 0$, $\psi = -1$ if $a(t) > 0$, and $\psi \in [-1, 1]$ if $a(t) = 0$ (recall that ψ is constant on the trajectory, since a does not occur explicitly in the right-hand side of system (7)). Thus, we must solve a two-point variational problem with Lagrangian L_1 in the first case, L_2 in the second case, and $(L_1 + L_2)/2 + \psi(L_1 - L_2)/2$ in the third case, where ψ is determined from the condition $a(t) = 0$. These problems are uniquely solvable, and the solutions are given by quadratic functions of time. By finding these trajectories explicitly and by calculating

$$S(a, t) = \int_0^t \frac{L_1 + L_2}{2} (x(\tau), \dot{x}(\tau)) \, d\tau + |a(t)|$$

along these trajectories, we obtain

(1) $S(a, t) = \dfrac{1}{2} A_1 \xi t - \dfrac{A_1^2 t^3}{24} + \dfrac{\xi^2(t + 1)}{2t} - a,$

if $\dfrac{24a}{(A_1 - A_2)t} \leq 6\xi - A_1 t^2;$

(2) $S(a, t) = \dfrac{1}{2} A_2 \xi t - \dfrac{A_2^2 t^3}{24} + \dfrac{\xi^2(t + 1)}{2t} + a,$

if $\dfrac{24a}{(A_1 - A_2)t} \geq 6\xi - A_2 t^2;$

(3) $S(a, t) = \dfrac{A_1 + A_2}{A_1 - A_2} a - \dfrac{12\xi a}{t^2(A_1 - A_2)} + \dfrac{24a^2}{t^3(A_1 - A_2)^2} + \dfrac{\xi^2(t + 4)}{2t},$

if $6\xi - A_1 t^2 \leq \dfrac{24a}{(A_1 - A_2)t} \leq 6\xi - A_2 t^2.$

The Pareto set is given by (3) (parts (1) and (2) give the boundary of its normalization). It is a piece of a parabola. For example, for $A_1 = 6$, $A_2 = 0$, and $\xi = t = 1$ we obtain $S(a, t) = \frac{2}{3}a^2 - a + \frac{5}{2}$, where $a \in [0, \frac{3}{2}]$. Carrying out the rotation according to Proposition 1, we see that the Pareto set is a piece of a parabola with endpoints $(2.5, 2.5)$ and $(4, 1)$.

The results of this paper were first announced in [KM3]. Here we give complete (and at the same time simplified) arguments.

5   Acknowledgements

The work was carried out with the financial support of the Russian Fund of Basic Investigation, Project no. 93-012-1075. The first author is grateful

to Prof. E. Wagneur, who helped him to obtain a short visiting position in
the Ecole de Mines de Nantes, where this work was finished.

6 References

[CL] M.G. Crandall, P.L. Lions. *Viscosity solutions of Hamilton–Jacobi equations.* Trans. Amer. Math. Soc. **277** (1983), 1–42.

[Ko] V.N. Kolokoltsov. *On linear, additive, and homogeneous operators in idempotent analysis.* Adv. Sov. Math. **13** (1992), 87–101.

[KM1] V.N. Kolokoltsov, V.P. Maslov. *Idempotent calculus as the apparatus of optimization theory, part II.* Funkz. Anal. i pril. **23:4** (1989), 53–62. Engl. transl. in Function. Anal. and Appl.

[KM2] V.N. Kolokoltsov, V.P. Maslov. *Idempotent analysis and its application to optimal control theory.* Moscow, Nauka, 1994 (in Russian).

[KM3] V.N. Kolokoltsov, V.P. Maslov. *Differential Bellman equation and the Pontryagin maximum principle for multicriterial optimization problems.* Dokl. Akad. Nauk **324:1** (1992), 29–34. Engl. transl. in Sov. Math. Dokl.

[Kr] S.N. Kruzhkov. *Generalized solutions of nonlinear equations of first order with several variables.* Matem. Sbornik **70:3** (1966), 394–415.

[MS] V.P. Maslov, S.N. Samborski. *Stationary Hamilton–Jacobi and Bellman equations (Existence and uniqueness of solutions).* Adv. Sov. Math. **13** (1992), 119–133.

[S] S.N. Samborski. *The Lagrange problem from the point of view of idempotent analysis.* In this volume.

[ST] S.N. Samborski, A.A. Taraschan. *On semirings appearing in multicriteria optimization problems and in the problems of the analysis of computer media.* Dokl. Akad. Nauk **308:6** (1989), 1309–1312. Engl. transl. in Sov. Math. Dokl.

Duality between Probability and Optimization

Marianne Akian, Jean-Pierre Quadrat and Michel Viot

1 Introduction

Following the theory of idempotent Maslov measures, a formalism analogous
to probability calculus is obtained for optimization by replacing the classical
structure of real numbers $(\mathbb{R}, +, \times)$ by the idempotent semifield obtained by
endowing the set $\mathbb{R} \cup \{+\infty\}$ with the "min" and "+" operations. To the
probability of an event corresponds the cost of a set of decisions. To random
variables correspond decision variables.

Weak convergence, tightness and limit theorems of probability have an
optimization counterpart which is useful for approximating the Hamilton–
Jacobi–Bellman (HJB) equation and obtaining asymptotics for this equation.
The introduction of tightness for cost measures and its consequences is the
main contribution of this paper. A link is established between weak convergence and the epigraph convergence used in convex analysis.

The Cramér transform used in the large deviation literature is defined as
the composition of the Laplace transform by the logarithm by the Fenchel
transform. It transforms convolution into inf-convolution. Probabilistic results about processes with independent increments are then transformed into
similar results on dynamic programming equations. The Cramér transform
gives new insight into the Hopf method used to compute explicit solutions of
some HJB equations. It also explains the limit theorems obtained directly as
the image of the classic limit theorems of probability.

2 Cost Measures and Decision Variables

Let us denote by $\mathbb{R}_{\min}$ the idempotent semifield $(\mathbb{R} \cup \{+\infty\}, \min, +)$ and by
extension the metric space $\mathbb{R} \cup \{+\infty\}$ endowed with the exponential distance
$d(x, y) = |\exp(-x) - \exp(-y)|$. We start by defining cost measures which
can be seen as normalized idempotent Maslov measures in $\mathbb{R}_{\min}$ [24].

Definition 2.1 We call a *decision space* the triplet $(U, \mathcal{U}, \mathbb{K})$ where U is a
topological space, $\mathcal{U}$ the set of open sets of U and $\mathbb{K}$ a mapping from $\mathcal{U}$ to
$\mathbb{R}_{\min}$ such that

1. $\mathbb{K}(U) = 0$,

2. $\mathbb{K}(\emptyset) = +\infty$,

3. $\mathbb{K}\left(\bigcup_n A_n\right) = \inf_n \mathbb{K}(A_n)$ for any $A_n \in \mathcal{U}$.

The mapping $\mathbb{K}$ is called a *cost measure*.

A function $c : U \to \mathbb{R}_{\min}$ such that $\mathbb{K}(A) = \inf_{u \in A} c(u) \; \forall A \in \mathcal{U}$ is called a *cost density* of the cost measure $\mathbb{K}$.

The set $D_c \overset{\text{def}}{=} \{u \in U \mid c(u) \neq +\infty\}$ is called the *domain* of c.

Theorem 2.2 *Given a l.s.c. function c with values in $\mathbb{R}_{\min}$ such that $\inf_u c(u) = 0$, the mapping $A \in \mathcal{U} \mapsto \mathbb{K}(A) = \inf_{u \in A} c(u)$ defines a cost measure on $(U, \mathcal{U})$. Conversely any cost measure defined on open sets of a second countable topological space[1] admits a unique minimal extension $\mathbb{K}_*$ to $\mathcal{P}(U)$ (the set of subsets of U) having a density c which is a l.s.c. function on U satisfying $\inf_u c(u) = 0$.*

Proof This precise result is proved in Akian [1]. See also Maslov [24] and Del Moral [15] for the first part and Maslov and Kolokoltsov [23, 25] for the second part.   □

Remark 2.3 This theorem shows that on second countable spaces there is a bijection between l.s.c. functions and cost measures. In this paper, we will consider cost measures on $\mathbb{R}^n$, $\mathbb{R}^{\mathbb{N}}$, separable Banach spaces and separable reflexive Banach spaces with the weak topology, which are all second countable topological spaces.

Example 2.4 We will very often use the following two cost densities defined on $\mathbb{R}^n$ with $\|.\|$ the euclidean norm.

1. $\chi_m(x) \overset{\text{def}}{=} \begin{cases} +\infty & \text{for } x \neq m. \\ 0 & \text{for } x = m, \end{cases}$

2. $\mathcal{M}^p_{m,\sigma}(x) \overset{\text{def}}{=} \frac{1}{p} \|\sigma^{-1}(x - m)\|^p$ for $p \geq 1$ with $\mathcal{M}^p_{m,0} \overset{\text{def}}{=} \chi_m$.

By analogy with conditional probability we define the conditional cost excess.

Definition 2.5 The *conditional cost excess* to take the best decision in A knowing that it must be taken in B is

$$\mathbb{K}(A|B) \overset{\text{def}}{=} \mathbb{K}(A \cap B) - \mathbb{K}(B).$$

By analogy with random variables we define decision variables and related notions.

---

[1] i.e. a topological space with a countable basis of open sets.

Definition 2.6

1. A *decision variable* X on $(U, \mathcal{U}, \mathbb{K})$ is a mapping from U to E (a second countable topological space). It induces a cost measure $\mathbb{K}_X$ on $(E, \mathcal{B})$ ($\mathcal{B}$ denotes the set of open sets of E) defined by $\mathbb{K}_X(A) = \mathbb{K}_*(X^{-1}(A))$ for all $A \in \mathcal{B}$. The cost measure $\mathbb{K}_X$ has a l.s.c. density denoted by c_X. When $E = \mathbb{R}$, we call X a real decision variable; when $E = \mathbb{R}_{\min}$, we call it a *cost variable*.

2. Two decision variables X and Y are called *independent* when

$$c_{X,Y}(x, y) = c_X(x) + c_Y(y) .$$

3. The *conditional cost excess* of X knowing Y is defined by

$$c_{X|Y}(x, y) \overset{\text{def}}{=} \mathbb{K}_*(X = x \mid Y = y) = c_{X,Y}(x, y) - c_Y(y) .$$

4. The *optimum* of a decision variable is defined by

$$\mathbb{O}(X) \overset{\text{def}}{=} \arg\min_{x \in E} \operatorname{conv}(c_X)(x)$$

when the minimum exists. Here conv denotes the l.s.c. convex hull and arg min the point where the minimum is reached. When a decision variable X with values in a linear space satisfies $\mathbb{O}(X) = 0$ we say that it is *centered*.

5. When the optimum of a decision variable X with values in $\mathbb{R}^n$ is unique and when near the optimum, we have

$$\operatorname{conv}(c_X)(x) = \frac{1}{p}\|\sigma^{-1}(x - \mathbb{O}(X))\|^p + o(\|x - \mathbb{O}(X)\|^p) ,$$

we say that X is of *order p* and we define its *sensitivity of order p* by $\mathbb{S}^p(X) \overset{\text{def}}{=} \sigma$. When $\mathbb{S}^p(X) = I$ (the identity matrix) we say that X is of *order p and normalized*.

6. The *value* of a cost variable X is $\mathbb{V}(X) \overset{\text{def}}{=} \inf_x(x + c_X(x))$; *the conditional value* is $\mathbb{V}(X \mid Y = y) \overset{\text{def}}{=} \inf_x(x + c_{X|Y}(x, y))$.

Example 2.7 For a real decision variable X of cost $\mathcal{M}_{m,\sigma}^p$ with $p > 1$ and $1/p + 1/p' = 1$, we have

$$\mathbb{O}(X) = m, \ \mathbb{S}^p(X) = \sigma, \ \mathbb{V}(X) = m - \frac{1}{p'}\sigma^{p'} .$$

3   Vector Spaces of Decision Variables

Theorem 3.1 *For $p > 0$, the numbers*

$$|X|_p \overset{\text{def}}{=} \inf \left\{ \sigma \mid c_X(x) \geq \frac{1}{p}|(x - \mathbb{O}(X))/\sigma|^p \right\} \text{ and } \|X\|_p \overset{\text{def}}{=} |X|_p + |\mathbb{O}(X)|$$

define respectively a seminorm and a norm on the vector space $\mathbb{L}^p$ of classes[2] of real decision variables having a unique optimum and such that $\|X\|_p$ is finite.

Proof Let us denote $X' = X - \mathbb{O}(X)$ and $Y' = Y - \mathbb{O}(Y)$. We first remark that $\sigma > |X|_p$ implies

$$c_X(x) \geq \frac{1}{p}(|x - \mathbb{O}(X)|/\sigma)^p \quad \forall x \in \mathbb{R} \Leftrightarrow \mathbb{V}(-\frac{1}{p}|X'/\sigma|^p) \geq 0 . \tag{3.1}$$

If there exists $\sigma > 0$ and $\mathbb{O}(X)$ such that (3.1) holds, then $c_X(x) < 0$ for any $x \neq \mathbb{O}(X)$, and $c_X(x)$ tends to 0 implies x tends to $\mathbb{O}(X)$; therefore $\mathbb{O}(X)$ is the unique optimum of X. Moreover $|X|_p$ is the smallest σ such that (3.1) holds.

If $X \in \mathbb{L}^p$, $\lambda \in \mathbb{R}$ and $\sigma > |X|_p$ we have

$$\mathbb{V}(-\frac{1}{p}|\lambda X'/\lambda \sigma|^p) = \mathbb{V}(-\frac{1}{p}|X'/\sigma|^p) \geq 0 ,$$

whence $\lambda X \in \mathbb{L}^p$, $\mathbb{O}(\lambda X) = \lambda \mathbb{O}(X)$ and $|\lambda X|_p = |\lambda||X|_p$.

If X and $Y \in \mathbb{L}^p$, $\sigma > |X|_p$ and $\sigma' > |Y|_p$,

$$\mathbb{V}(-\frac{1}{p}(\max(|X'/\sigma|^p, |Y'/\sigma'|^p))) = \min(\mathbb{V}(-\frac{1}{p}|X'/\sigma|^p), \mathbb{V}(-\frac{1}{p}|Y'/\sigma'|^p)) \geq 0$$

and

$$\frac{|X' + Y'|}{\sigma + \sigma'} \leq \frac{\sigma}{\sigma + \sigma'}\frac{|X'|}{\sigma} + \frac{\sigma'}{\sigma + \sigma'}\frac{|Y'|}{\sigma'} \leq \max(\frac{|X'|}{\sigma}, \frac{|Y'|}{\sigma'}) ,$$

whence

$$\mathbb{V}(-\frac{1}{p}(|X' + Y'|/(\sigma + \sigma'))^p) \geq 0 .$$

Therefore we have proved that $X + Y \in \mathbb{L}^p$ with $\mathbb{O}(X + Y) = \mathbb{O}(X) + \mathbb{O}(Y)$ and $|X + Y|_p \leq |X|_p + |Y|_p$.

Then $\mathbb{L}^p$ is a vector space, $|.|_p$ and $\|.\|_p$ are seminorms and $\mathbb{O}$ is a linear continuous operator from $\mathbb{L}^p$ to $\mathbb{R}$. Moreover, $\|X\|_p = 0$ implies $c_X = \chi$. Hence $X = 0$ up to a set of infinite cost.   □

---

[2]for the almost sure equivalence relation: $X \overset{\text{a.s.}}{=} Y \Leftrightarrow \mathbb{K}_*(X \neq Y) = +\infty$.

Theorem 3.2 *For two independent real decision variables X and Y and $k \in \mathbb{R}$ we have (as long as the right and left hand sides exist)*

$$\mathbb{O}(X+Y) = \mathbb{O}(X) + \mathbb{O}(Y), \quad \mathbb{O}(kX) = k\mathbb{O}(X), \quad \mathbb{S}^p(kX) = |k|\mathbb{S}^p(X),$$

$$[\mathbb{S}^p(X+Y)]^{p'} = [\mathbb{S}^p(X)]^{p'} + [\mathbb{S}^p(Y)]^{p'}, \quad (|X+Y|_p)^{p'} \leq (|X|_p)^{p'} + (|Y|_p)^{p'},$$

where $1/p + 1/p' = 1$.

Proof Let us prove only the last inequality. Consider X and Y in $\mathbb{L}^p$ and $\sigma > |X|_p$ and $\sigma' > |Y|_p$. Let us denote $\sigma'' = (\sigma^{p'} + \sigma'^{p'})^{1/p'}$, $X' = X - \mathbb{O}(X)$ and $Y' = Y - \mathbb{O}(Y)$. The Hölder inequality $a\alpha + b\beta \leq (a^p + b^p)^{1/p}(\alpha^{p'} + \beta^{p'})^{1/p'}$ implies

$$(|X'+Y'|/\sigma'')^p \leq |X'/\sigma|^p + |Y'/\sigma'|^p,$$

then by the independence of X and Y we get

$$\mathbb{V}(-\frac{1}{p}(|X'+Y'|/\sigma'')^p) \geq 0,$$

and the inequality is proved. □

Theorem 3.3 *(Chebyshev) For a decision variable belonging to $\mathbb{L}^p$ we have*

$$\mathbb{K}(|X - \mathbb{O}(X)| \geq a) \geq \frac{1}{p}(a/|X|_p)^p,$$

$$\mathbb{K}(|X| \geq a) \geq \frac{1}{p}((a - \|X\|_p)^+/\|X\|_p)^p.$$

Proof The first inequality is a straightforward consequence of the inequality $c_Y(y) \geq (|y|/|Y|_p)^p/p$ applied to the centered decision variable $Y = X - \mathbb{O}(X)$. The second inequality comes from the nonincreasing property of the function $x \in \mathbb{R}^+ \mapsto (a-x)^+/x$. □

4 Convergence of Decision Variables and Law of Large Numbers

Definition 4.1 A *sequence of independent and identically costed (i.i.c.) real decision variables of cost c* on $(U, \mathcal{U}, \mathbb{K})$ is an application X from U to $\mathbb{R}^{\mathbb{N}}$ which induces the density cost

$$c_X(x) = \sum_{i=0}^{\infty} c(x_i), \quad \forall x = (x_0, x_1, \ldots) \in \mathbb{R}^{\mathbb{N}}.$$

Remark 4.2 The cost density is finite only on minimizing sequences of c; elsewhere it is equal to $+\infty$.

Remark 4.3 We have defined a decision sequence by its density and not by its value on the open sets of $\mathbb{R}^N$ because the density always exists and can be defined easily.

In order to state limit theorems, we define several type of convergence of sequences of decision variables.

Definition 4.4 For the sequence of real decision variables $\{X_n, n \in \mathbb{N}\}$ we say that

1. $X_n \in \mathbb{L}^p$ *converges in p-norm towards* $X \in \mathbb{L}^p$, *denoted* $X_n \xrightarrow{\text{L}^p} X$, *if* $\lim_n \|X_n - X\|_p = 0$;

2. X_n *converges in cost towards* X, *denoted* $X_n \xrightarrow{\text{K}} X$, *if for all* $\epsilon > 0$ *we have* $\lim_n \mathbb{K}\{u \mid |X_n(u) - X(u)| \geq \epsilon\} = +\infty$;

3. X_n *converges almost surely towards* X, *denoted* $X_n \xrightarrow{\text{a.s.}} X$, *if we have* $\mathbb{K}\{u \mid \lim_n X_n(u) \neq X(u)\} = +\infty$.

Some relations between these different kinds of convergence are given in the following theorem.

Theorem 4.5

1. *Convergence in p-norm implies convergence in cost but the converse is false.*

2. *Convergence in cost implies almost sure convergence but the converse is false.*

Proof See Akian [2] for points 1 and 2 and Del Moral [15] for point 2.   □

We have the analogue of the law of large numbers.

Theorem 4.6 *Given a sequence* $\{X_n, n \in \mathbb{N}\}$ *of i.i.c. decision variables belonging to* $\mathbb{L}^p$, $p \geq 1$, *we have*

$$\lim_{N \to \infty} Y_N \stackrel{\text{def}}{=} \frac{1}{N} \sum_{n=0}^{N-1} X_n = \mathbb{O}(X_0),$$

where the limit can be taken in the sense of almost sure, cost and p-norm convergence.

Proof We have only to estimate the convergence in p-norm. The result follows from simple computation of the p-seminorm of Y_N. By Theorem 3.2 we have $(|Y_N|_p)^{p'} \leq N(|X_0|_p)^{p'}/N^{p'}$ which tends to 0 as N tends to infinity.   □

5 Weak Convergence and Tightness of Decision Variables

In this section we introduce the notions of weak convergence and tightness of cost measures and show the relations between the weak convergence and epigraph convergence of functions introduced in convex analysis [5, 4, 22]. Weak convergence and tightness of decision variables will mean weak convergence of their cost measures.

Definition 5.1

1. Let $\mathbb{K}_n$ and $\mathbb{K}$ be cost measures on $(U, \mathcal{U})$. We say that $\mathbb{K}_n$ *converges weakly* towards $\mathbb{K}$, denoted $\mathbb{K}_n \xrightarrow{\text{w}} \mathbb{K}$, if for all f in[3] $C_b(U)$ we have[4] $\lim_n \mathbb{K}_n(f) = \mathbb{K}(f)$.

2. Let c_n and c be functions from U (a first countable topological space[5]) to $\mathbb{R}_{\min}$. We say that c_n converges in the epigraph sense (*epi-converges*) towards c, denoted $c_n \xrightarrow{\text{epi}} c$, if

$$\forall u, \quad \forall u_n \to u, \quad \liminf_n c_n(u_n) \geq c(u), \tag{5.1}$$

$$\forall u, \quad \exists u_n \to u : \limsup_n c_n(u_n) \leq c(u). \tag{5.2}$$

3. If U is a reflexive Banach space, we say that c_n *Mosco-epi-converges* towards c, denoted $c_n \xrightarrow{\text{M-epi}} c$, if the convergence of u_n holds for the weak topology in (5.1) and for the strong topology in (5.2).

Theorem 5.2 *Let $\mathbb{K}_n$, $\mathbb{K}$ be cost measures on a metric space U. Then the following three conditions are equivalent:*

1. $\mathbb{K}_n \xrightarrow{\text{w}} \mathbb{K}$;

2.

$$\liminf_n \mathbb{K}_n(F) \geq \mathbb{K}(F) \quad \forall F \, closed, \tag{5.3}$$

$$\limsup_n \mathbb{K}_n(G) \leq \mathbb{K}(G) \quad \forall G \, open; \tag{5.4}$$

3. $\lim_n \mathbb{K}_n(A) = \mathbb{K}(A)$ *for any set A such that $\mathbb{K}(\overset{\circ}{A}) = \mathbb{K}(\bar{A})$.*

---

[3] $C_b(U)$ denotes the set of continuous and lower bounded functions from U to $\mathbb{R}_{\min}$.
[4] $\mathbb{K}(f) \overset{\text{def}}{=} \inf_u(f(u) + c(u))$ where c is the density of $\mathbb{K}$.
[5] Each point admits a countable basis of neighborhoods.

Proof The proof is similar to those of classical probability theory. The main ingredients in both theories are: (a) U is normal, (b) a probability on the Borel sets of U or a cost measure on the open sets of U is "regular", (c) any bounded continuous function from U to $\mathbb{R}$ or $\mathbb{R}_{\min}$ may be approximated above and below by an $\mathbb{R}$ or $\mathbb{R}_{\min}$-linear combination of characteristic functions of "measurable" sets. Properties (a) and (b) are used in showing 1. $\Rightarrow$ 2. and the equivalence 2. $\leftrightarrow$ 3. and property (c) in showing 2. $\Rightarrow$ 1. The main difficulty in optimization theory compared to classical probability is that cost measures are not continuous for nonincreasing convergence of sets.

Let us make properties (a)–(c) precise in the case of optimization theory. Firstly, since U is a metric space, U is normal in the classical sense. Equivalently, by using the bi-continuous application $t \mapsto -\log(t)$ from $[0,1]$ to the subset $[0,+\infty]$ of $\mathbb{R}_{\min}$, U is normal with respect to $\mathbb{R}_{\min}$ (see Maslov [24] for this notion), that is for any open set G and closed set F such that $F \subset G$, there exists a continuous function f from U to $\mathbb{R}_{\min}$ such that $f \geq 0$, $f = 0$ on F and $f = +\infty$ on G^c and then $\chi_G \leq f \leq \chi_F$. A typical function f is

$$f(u) = -\log\left(\frac{d(u, G^c)}{d(u, F) + d(u, G^c)}\right).$$

Secondly, the regularity property of classical probabilities may be translated here into the following two conditions:

$$\mathbb{K}(F) = \sup_{G \supset F,\, G \in \mathcal{U}} \mathbb{K}(G) \quad \forall F \text{ closed} \tag{5.5}$$

and

$$\mathbb{K}(G) = \inf_{F \subset G,\, F \text{ closed}} \mathbb{K}(F) \quad \forall G \in \mathcal{U}. \tag{5.6}$$

The first condition is a consequence of the definition of the minimal extension; the second of the fact that in a metric space, any open set is a countable union of closed sets and of the continuity of cost measures for the nondecreasing convergence of sets. Let us note that in classical probability conditions (5.5) and (5.6) are equivalent, which is not the case here.

Finally, any lower bounded continuous function may be approximated by simple functions: above by an $\mathbb{R}_{\min}$-linear combination of characteristic functions of open sets, below by an $\mathbb{R}_{\min}$-linear combination of characteristic functions of closed sets. The first approximation follows easily from upper semi-continuity of continuous functions. The second one uses the relative compactness of lower bounded sets in $\mathbb{R}_{\min}$. □

Definition 5.3 A set of cost measures $\mathcal{K}$ is called *tight* if

$$\sup_{C \text{ compact } \subset U} \inf_{\mathbb{K} \in \mathcal{K}} \mathbb{K}(C^c) = +\infty.$$

A sequence $\mathbb{K}_n$ of cost measures is called *asymptotically tight* if

$$\sup_{C \text{ compact } \subset U} \liminf_{n} \mathbb{K}_n(C^c) = +\infty.$$

Theorem 5.4 *On asymptotically tight sequences $\mathbb{K}_n$ over a metric space U, the weak convergence of $\mathbb{K}_n$ towards $\mathbb{K}$ is equivalent to (5.4) and*

$$\liminf_n \mathbb{K}_n(C) \geq \mathbb{K}(C) \quad \forall C \, compact. \tag{5.7}$$

Remark 5.5 In a locally compact space conditions (5.7) and (5.4) are equivalent to the condition $\lim_n \mathbb{K}_n(f) = \mathbb{K}(f)$ for any continuous function with compact support. This is the definition of weak convergence used by Maslov and Samborski in [27]. These conditions are also equivalent to the epigraph convergence of densities (see Theorem 5.7 below). This type of convergence does not ensure that a weak-limit of cost measures is a cost measure (the infimum of the limit is not necessarily equal to zero).

Theorem 5.6 *Let us denote by $\mathcal{K}(U)$ the set of cost measures on U (a metric space) endowed with the topology of the weak convergence. Any tight set K of $\mathcal{K}(U)$ is relatively sequentially compact*[6].

Proof It is sufficient to prove that from any asymptotically tight sequence $\{\mathbb{K}_n\}$, we can extract a weakly convergent subsequence.

Let C_k be a compact set such that $\liminf_n \mathbb{K}_n(C_k^c) \geq k$ and $V = \bigcup_k C_k$; then $\liminf_n \mathbb{K}_n(V^c) = +\infty$. The convergence of $\mathbb{K}_n$ is then equivalent to the convergence of $\mathbb{K}_n$ on V, which is a separable metric space. Since $\mathbb{K}_n$ is still asymptotically tight on V, we suppose now $U = V$.

Let $\mathcal{B}$ be a countable basis of open sets of U. Since $\mathbb{K}_n$ takes its values in $[0, +\infty]$ (which is a compact set of $\mathbb{R}_{\min}$) and $\mathcal{B}$ is countable, we may extract a subsequence of $\mathbb{K}_n$, denoted also $\mathbb{K}_n$, such that $\lim_n \mathbb{K}_n(B) = \widetilde{\mathbb{K}}(B) \; \forall B \in \mathcal{B}$.

Since any open set is a countable union of elements of $\mathcal{B}$, we define $\mathbb{K}$ on $\mathcal{U}$ by:

$$\mathbb{K}(A) = \sup_A \inf_{B \in \mathcal{A}} \widetilde{\mathbb{K}}(B),$$

where the supremum is taken over subsets $\mathcal{A}$ of $\mathcal{B}$ such that $\bigcup_{B \in \mathcal{A}} B = A$. $\mathbb{K}$ is the minimal cost measure on $\mathcal{U}$ greater than $\widetilde{\mathbb{K}}$ on $\mathcal{B}$.

Its minimal extension to $\mathcal{P}(U)$ is

$$\mathbb{K}(A) = \sup_A \inf_{B \in \mathcal{A}} \widetilde{\mathbb{K}}(B),$$

where this time $\mathcal{A}$ satisfies $\bigcup_{B \in \mathcal{A}} B \supset A$.

Let us show that $\mathbb{K}_n \xrightarrow{w} \mathbb{K}$. By Theorem 5.4 it is enough to prove (5.4) and (5.7). If G is an open set, then for any $B \in \mathcal{B}$ such that $B \subset G$ we have

$$\limsup_n \mathbb{K}_n(G) \leq \limsup_n \mathbb{K}_n(B) = \widetilde{\mathbb{K}}(B).$$

---

[6]that is any sequence of K contains a weakly convergent subsequence.

Therefore if $G = \bigcup_{B \in \mathcal{A}} B$ with $\mathcal{A} \subset \mathcal{B}$, we have

$$\limsup_n \mathbb{K}_n(G) \leq \inf_{B \in \mathcal{A}} \widetilde{\mathbb{K}}(B) \leq \mathbb{K}(G).$$

If F is a compact set, and if $\bigcup_{B \in \mathcal{A}} B \supset F$, we may restrict $\mathcal{A}$ to be finite, and then we have

$$\liminf_n \mathbb{K}_n(F) \geq \liminf_n \inf_{B \in \mathcal{A}} \mathbb{K}_n(B) = \inf_{B \in \mathcal{A}} \liminf_n \mathbb{K}_n(B) = \inf_{B \in \mathcal{A}} \widetilde{\mathbb{K}}(B) \, .$$

By taking the supremum over all sets $\mathcal{A}$, we obtain condition (5.7). $\qquad\square$

Theorem 5.7 *On a first countable topological space, the epi-convergence of l.s.c. densities c_n of $\mathbb{K}_n$ towards the density c of $\mathbb{K}$ is equivalent to conditions (5.4) and (5.7).*

Proof We prove (5.1) $\Leftrightarrow$ (5.7) and (5.2) $\Leftrightarrow$ (5.4).

1. (5.1) $\Rightarrow$ (5.7).

   Let u_n be a point where c_n reaches its optimum in compact set C. For any converging subsequence $\{u_{n_k}\}$, the limit u belongs to C and we have $\liminf_k c_{n_k}(u_{n_k}) \geq c(u) \geq \mathbb{K}(C)$, therefore $\liminf_n \mathbb{K}_n(C) \geq \mathbb{K}(C)$.

2. (5.7) $\Rightarrow$ (5.1).

   The sets $C_N = \{u_n, n \geq N\} \cup \{u\}$ are compact and we have

   $$\liminf_n c_n(u_n) \geq \liminf_n \mathbb{K}_n(C_N) \geq \mathbb{K}(C_N) \, .$$

   As the l.s.c. of c implies $\sup_N \mathbb{K}(C_N) \geq c(u)$, the result follows.

3. (5.2) $\Rightarrow$ (5.4).

   Let us prove this assertion by contraposition. Let us suppose there exists an open set G and $\epsilon > 0$ such that $\limsup_n \mathbb{K}_n(G) > \mathbb{K}(G) + \epsilon$. By definition of the infimum there exists $u \in G$ such that $c(u) \leq \mathbb{K}(G) + \epsilon$. Therefore for any sequence u_n converging towards u, $u_n \in G$ for n large enough and we have $\limsup_n c_n(u_n) \geq \limsup_n \mathbb{K}_n(G) > \mathbb{K}(G) + \epsilon \geq c(u)$. This contradicts the hypothesis.

4. (5.4) $\Rightarrow$ (5.2).

   For all u there exists a decreasing family of open sets $\{G_k, k \in \mathbb{N}\}$ such that $\bigcap_k G_k = \{u\}$. By definition of the infimum, there exists u_n^k such that $\mathbb{K}_n(G_k) \geq c_n(u_n^k) - 1/n$. Then we have $\limsup_n c_n(u_n^k) \leq \limsup_n \mathbb{K}_n(G_k) \leq \mathbb{K}(G_k) \leq c(u)$. By diagonal extraction we obtain a sequence $u_n^{k(n)}$ which satisfies (5.2). $\qquad\square$

Remark 5.8 In Attouch [4] another definition of epigraph convergence is given in a general topological space and is mostly related to conditions (5.4) and (5.7).

Proposition 5.9 *If $\mathbb{K}_n$ and $\mathbb{K}'_n$ are (asymptotically) tight sequences of cost measures on U and U' and $\mathbb{K}_n \xrightarrow{w} \mathbb{K}$ and $\mathbb{K}'_n \xrightarrow{w} \mathbb{K}'$ then $\mathbb{K}_n \times \mathbb{K}'_n$ is (asymptotically) tight and $\mathbb{K}_n \times \mathbb{K}'_n \xrightarrow{w} \mathbb{K} \times \mathbb{K}'$.*

Proof The product of two measures $\mathbb{K}$ and $\mathbb{K}'$ is defined as in probability; then if $\mathbb{K}$ and $\mathbb{K}'$ have densities $c(u)$ and $c'(u')$, $\mathbb{K} \times \mathbb{K}'$ has density $c(u)+c'(u')$. In probability theory, tightness is not necessary, but the technique of proof does not work here. We need to impose the tightness condition, but in this case weak convergence is equivalent to epigraph convergence, for which the result is clear. □

Theorem 5.10 *If $X_n \xrightarrow{\mathbb{K}} X$ and X is tight then $X_n \xrightarrow{w} X$. More generally if $X_n \xrightarrow{w} X$, $X_n - Y_n \xrightarrow{\mathbb{K}} 0$ and X is tight, then $Y_n \xrightarrow{w} X$.*

Proof See [2]. □

6 Characteristic Functions

The role of the Laplace or Fourier transforms in probability calculus is played by the Fenchel transform in decision calculus.

Definition 6.1

1. Let $c \in \mathcal{C}_x$, where $\mathcal{C}_x$ denotes the set of l.s.c. and proper[7] convex functions from E (a reflexive Banach space with dual E') to $\mathbb{R}_{\min}$. The *Fenchel transform* of c is the function from E' to $\mathbb{R}_{\min}$ defined by $\hat{c}(\theta) \overset{\text{def}}{=} [\mathcal{F}(c)](\theta) \overset{\text{def}}{=} \sup_x [\langle \theta, x \rangle - c(x)]$.

2. The *characteristic function* of a decision variable is $\mathbb{F}(X) \overset{\text{def}}{=} \mathcal{F}(c_X)$.

3. Given two functions f and g from E to $\mathbb{R}_{\min}$, the *inf-convolution of f and g*, denoted $f \,\square\, g$, is the function $z \in E \mapsto \inf_{x,y} \{f(x)+g(y) \mid x+y = z\}$.

Theorem 6.2

1. *For $f, g \in \mathcal{C}_x$ we have*

   (a) *$\mathcal{F}(f) \in \mathcal{C}_x$,*
   (b) *$\mathcal{F}$ is an involution, that is $\mathcal{F}(\mathcal{F}(f)) = f$,*
   (c) *$\mathcal{F}(f \,\square\, g) = \mathcal{F}(f) + \mathcal{F}(g)$,*
   (d) *$\mathcal{F}(f + g) = \mathcal{F}(f) \,\square\, \mathcal{F}(g)$.*

---

[7]not always equal to $+\infty$

2. *For two independent decision variables X and Y and $k \in \mathbb{R}$, we have*

$$c_{X+Y} = c_X \,\square\, c_Y, \quad \mathbb{F}(X+Y) = \mathbb{F}(X) + \mathbb{F}(Y), \quad [\mathbb{F}(kX)](\theta) = [\mathbb{F}(X)](k\theta),$$

3. *A decision variable with values in $\mathbb{R}^n$ is of order p if we have*

$$\mathbb{F}(X)(\theta) = \langle \mathbb{O}(X), \theta \rangle + \frac{1}{p'} \|\mathbb{S}^p(X)\theta\|^{p'} + o(\|\theta\|^{p'}),$$

with $1/p + 1/p' = 1$.

Remark 6.3 The Fenchel transform (for l.s.c. proper convex functions) is bi-continuous for the Mosco-epi-convergence [22].

Theorem 6.4 *Let $\mathbb{K}_n$ and $\mathbb{K}$ be cost measures on a separable reflexive Banach space with (proper) l.s.c. convex densities c_n and c. Then $c_n \xrightarrow{\text{M-epi}} c$ iff the two conditions (5.4) and*

$$\liminf_n \mathbb{K}_n(C) \geq \mathbb{K}(C) \quad \forall C \text{ bounded closed and convex} \tag{6.1}$$

hold.

Proof In the proof of Theorem 5.7 we easily see that we can replace compact sets by bounded closed convex sets which are weakly compact on a reflexive Banach space and let open sets be those of the strong topology. □

Corollary 6.5 *For an asymptotically tight sequence X_n of decision variables with l.s.c. convex cost densities on a separable reflexive Banach space, X_n converges weakly towards X iff $\mathbb{F}(X_n)$ Mosco-epi-converges towards $\mathbb{F}(X)$.*

Proof By the tightness property and the previous result, weak convergence of X_n towards X is equivalent to Mosco-epi-convergence of X_n towards X and then to Mosco-epi-convergence of $\mathbb{F}(X_n)$ towards $\mathbb{F}(X)$. □

This may be used for proving the central limit theorem in a Banach space. For simplicity let us state it in the finite dimensional situation, where epigraph and Mosco-epigraph convergences are equivalent.

Theorem 6.6 *(Central limit theorem) Let $\{X_n, n \in \mathbb{N}\}$ be an i.i.c. sequence centered of order p with l.s.c. convex cost density and $1/p + 1/p' = 1$. Then*

$$Z_N \stackrel{\text{def}}{=} \frac{1}{N^{1/p'}} \sum_{n=0}^{N-1} X_n \xrightarrow{w} \mathcal{M}_{0,\mathbb{S}^p(X_0)}^p.$$

Proof We have $\lim_N [\mathbb{F}(Z_N)](\theta) = \frac{1}{p'} \|\mathbb{S}^p(X_0)\theta\|^{p'}$, where the convergence can be taken in the pointwise, uniform on any bounded set or epigraph sense. In order to obtain the weak convergence we have to prove the tightness of Z_N. But as the convergence is uniform on $B = \{\|\theta\| \leq 1\}$ we have for $N \geq N_0$, $\mathbb{F}(Z_N) \leq C$ on B where C is a constant. Therefore $c_{Z_N}(x) \geq \|x\| - C$ for $N \geq N_0$ and Z_N is asymptotically tight. □

The central limit theorem may be generalized to the case of nonconvex cost densities. This generalization essentially uses the strict convexity of the limiting cost density and was suggested by the Gärtner–Ellis theorem on large deviations of dependent random variables [18, 21]. Indeed, the large deviation principle for probabilities $\mathbb{P}_n$ with entropy I may be considered as the weak convergence of "measures" $\mathbb{K}_n = -h(n) \log \mathbb{P}_n$ (with $\lim_n h(n) = 0$) towards the cost measure with density I. We first need the following result, which is proved in [2].

Proposition 6.7 *If $X_n \xrightarrow{w} X$ in $\mathbb{R}^p$ and $(\mathbb{F}(X_n)(\theta))_n$ is bounded above for any $\theta \in \mathcal{O}$ where $\mathcal{O}$ is an open convex neighborhood of 0 in $\mathbb{R}^p$, then*

$$\mathbb{F}(X_n)(\theta) \xrightarrow[n \to +\infty]{} \mathbb{F}(X)(\theta) \quad \forall \theta \in \mathcal{O}.$$

In general, a l.s.c. function c on $\mathbb{R}^p$ is not characterized by its Fenchel transform, but when the Fenchel transform is essentially smooth, the convex hull of c is essentially strictly convex, thus c is necessarily convex (see Rockafellar [30] for definitions). A generalization of this remark leads to a result equivalent to the Gärtner–Ellis theorem.

Proposition 6.8 *If X_n is a sequence of decision variables with values in $\mathbb{R}^p$ such that*

$$\mathbb{F}(X_n)(\theta) \xrightarrow[n \to +\infty]{} \varphi(\theta) \quad \forall \theta \in \mathbb{R}^p,$$

where φ is an essentially smooth proper l.s.c. convex function such that $0 \in \overset{\circ}{D}_\varphi$, then $X_n \xrightarrow{w} \mathcal{F}(\varphi)$.

The particular case where $\varphi(\theta) = \frac{1}{p'} \|\sigma\theta\|^{p'}$, which has a strictly convex Fenchel transform, leads to the general central limit theorem.

7 Bellman Chains and Processes

We can generalize i.i.c. sequences to the analogue of Markov chains, which we call Bellman chains.

Definition 7.1 A finite valued Bellman chain (E, C, ϕ) with

1. E a finite set of $|E|$ elements called the state space,

2. $C : E \times E \to \mathbb{R}_{\min}$ satisfying $\inf_y C_{xy} = 0$ called the transition cost,

3. ϕ a cost measure on E called the initial cost,

is a decision sequence $X = \{X_n, \ n \in \mathbb{N}\}$ taking its values in $E^{\mathbb{N}}$, such that

$$c_X(x \overset{\text{def}}{=} (x_0, x_1, \ldots)) = \phi_{x_0} + \sum_{i=0}^{\infty} C_{x_i x_{i+1}} , \quad \forall x \in E^{\mathbb{N}}.$$

Theorem 7.2 *For any function f from E to $\mathbb{R}_{min}$, a Bellman chain satisfies the Markov property* $\mathbb{V}\{f(X_n) \mid X_0, \ldots, X_{n-1}\} = \mathbb{V}\{f(X_n) \mid X_{n-1}\}$.

The analogue of the forward Kolmogorov equation giving a way to compute recursively the marginal probability of being in a state at a given time is the following Bellman equation.

Theorem 7.3 *The marginal cost $v_x^n = \mathbb{K}(X_n = x)$ of a Bellman chain is given by the recursive forward equation:* $v^{n+1} = v^n \otimes C \stackrel{\text{def}}{=} \min_{x \in E}(v_x^n + C_{x.})$ *with $v^0 = \phi$.*

Remark 7.4 The cost measure of a Bellman chain is normalized, which means that its infimum on all the trajectories is 0. In some applications we would like to avoid this restriction. This can be done by introducing the analogue of the multiplicative functionals of the trajectories of a stochastic process.

We can easily define continuous time decision processes which correspond to deterministic controlled processes. We discuss here only decision processes with continuous trajectories.

Definition 7.5

1. A *continuous time Bellman process* X_t with continuous trajectories is a decision variable with values in[8] $C(\mathbb{R}^+)$ having the cost density

$$c_X(x(\cdot)) \stackrel{\text{def}}{=} \phi(x(0)) + \int_0^\infty c(t, x(t), x'(t))dt,$$

   with $c(t, \cdot, \cdot)$ a family of transition costs (that is a function c from $\mathbb{R}^3$ to $\mathbb{R}_{min}$ such that $\inf_y c(t, x, y) = 0$, $\forall t, x$) and ϕ a cost density on $\mathbb{R}$. When the integral is not defined the cost is by definition equal to $+\infty$.

2. The Bellman process is called *homogeneous* if c does not depend on time t.

3. The Bellman process is said to have *independent increments* if c does not depend on the state x. Moreover, if this process is homogeneous, c is reduced to the cost density of a decision variable.

4. The *p-Brownian decision process*, denoted by B_t^p, is the process with independent increments and transition cost density $c(t, x, y) = \frac{1}{p}|y|^p$.

As in the discrete time case, the marginal cost of being in state x at time t can be computed recursively using a forward Bellman equation.

---

[8] $C(\mathbb{R}^+)$ denotes the set of continuous functions from $\mathbb{R}^+$ to $\mathbb{R}$.

Theorem 7.6 *The marginal cost* $v(t,x) \stackrel{\text{def}}{=} \mathbb{K}(X_t = x)$ *is given by the Bellman equation*

$$\partial_t v + \hat{c}(\partial_x v) = 0, \quad v(0,x) = \phi(x), \tag{7.1}$$

where $\hat{c}$ *means here* $[\hat{c}(\partial_x v)](t,x) \stackrel{\text{def}}{=} \sup_y [y \partial_x v(t,x) - c(t,x,y)]$.

Let $p > 1$ and $1/p + 1/p' = 1$. For the Brownian decision process B_t^p starting from 0, the marginal cost of being in state x at time t satisfies the Bellman equation

$$\partial_t v + (1/p')|\partial_x v|^{p'} = 0, \quad v(0,\cdot) = \chi \, .$$

Its solution can be computed explicitly as $v(t,x) = \mathcal{M}_{0,t^{1/p'}}^p(x)$*, therefore*

$$\mathbb{V}[f(B_t^p)] = \inf_x \left[f(x) + \frac{x^p}{pt^{\frac{p}{p'}}} \right]. \tag{7.2}$$

8 Tightness in $\mathcal{C}([0,1])$ and Brownian Approximation

Theorem 8.1 *A sequence of decision variables* $\{X_n, n \in \mathbb{N}\}$ *with values in* $\mathcal{C}([0,1])$ *is tight if* $X_n(t) \in \mathbb{L}^p$ *for* $t \in [0,1]$, $\|X_n(0)\|_p$ *is bounded and*

$$\lim_{\delta \to 0^+} \sup_{t \in [0,1-\delta], n \in \mathbb{N}} \|X_n(t+\delta) - X_n(t)\|_p = 0 \, . \tag{8.1}$$

Proof By Ascoli's theorem, we know that relatively compact subsets of $\mathcal{C}([0,1])$ coincide with equi-continuous subsets taking bounded values in 0. Therefore, we can deduce a necessary and sufficient condition for the tightness of a sequence of decision variables $\{X_n\}$ in $\mathcal{C}([0,1])$. The sequence $\{X_n\}$ is tight iff (i) $X_n(0)$ is tight, that is for all η there exists a such that $\mathbb{K}(|X_n(0)| \geq a) \geq \eta$, and (ii) for all $\eta, \epsilon > 0$ there exists $\delta > 0$ such that

$$\inf_n \inf_{\substack{t,s \in [0,1] \\ |s-t| \leq \delta}} \mathbb{K}(|X_n(t) - X_n(s)| \geq \epsilon) \geq \eta.$$

Then condition (i) is a direct consequence of the fact that $\|X_n(0)\|_p$ is bounded and of the Chebyshev inequality applied to $X_n(0)$, and condition (ii) is a direct consequence of (8.1) and of the Chebyshev inequality applied to $X_n(t) - X_n(s)$. $\quad\square$

The following result shows that weak convergence in $\mathcal{C}([0,1])$ may be characterized by the convergence of finite dimensional marginal costs.

Theorem 8.2

1. *There may exist different cost measures* $\mathbb{K}$ *and* $\mathbb{K}'$ *on* $\mathcal{C}([0,1])$ *such that*

$$\mathbb{K}_\pi = \mathbb{K}'_\pi \quad \forall \pi : \mathcal{C}([0,1]) \to \mathbb{R}^k, x \mapsto (x(t_1), \ldots, x(t_k)). \tag{8.2}$$

2. *If $\mathbb{K}$ is tight and $\mathbb{K}$ and $\mathbb{K}'$ satisfy (8.2), then $\mathbb{K} = \mathbb{K}'$.*

3. *If the sequence $\mathbb{K}_n$ is asymptotically tight and if $(\mathbb{K}_n)_\pi \xrightarrow{w} \mathbb{K}_\pi$ for all $\pi : \mathcal{C}([0,1]) \to \mathbb{R}^k$, $x \mapsto (x(t_1), \ldots, x(t_k))$, then $\mathbb{K}_n \xrightarrow{w} \mathbb{K}$.*

Proof Condition (8.2) is equivalent to $\mathbb{K}(U) = \mathbb{K}'(U)$ for any open set U of the form $U = \{x \mid x(t_1) \in U_1, \ldots, x(t_k) \in U_k\}$ where the U_i are open subsets of $\mathbb{R}$, that is for any open set of the pointwise convergence topology. Since any ball of $\mathcal{C}([0,1])$ is a nonincreasing limit of such open sets, we could conclude $\mathbb{K} = \mathbb{K}'$ if cost measures were continuous for the nonincreasing convergence of sets as in classical probability. This is not the case in general, but it remains true for a sequence of closed sets if $\mathbb{K}$ is tight.

Let us prove the second assertion. Let $\overline{B}(x, \varepsilon)$ denotes the closed ball of center x and radius ε for the uniform convergence norm. There exists open sets U_n for the pointwise convergence topology such that $\overline{B}(x, \varepsilon) = \bigcap_n \overline{U}_n = \bigcap_n U_n$. Then, if $\mathbb{K}$ is tight

$$\mathbb{K}(\overline{B}(x, \varepsilon)) = \sup_n \mathbb{K}(\overline{U}_n) \leq \sup_n \mathbb{K}(U_n)$$

$$= \sup_n \mathbb{K}'(U_n) \leq \mathbb{K}'(\bigcap_n U_n) = \mathbb{K}'(\overline{B}(x, \varepsilon)).$$

As any open set of $\mathcal{C}([0,1])$ is a countable union of closed balls, we obtain $\mathbb{K}(U) \leq \mathbb{K}'(U)$. Then the tightness of $\mathbb{K}$ implies the tightness of $\mathbb{K}'$ which implies the reverse inequality.

For the first assertion, it is sufficient to exhibit a cost measure $\mathbb{K}$ such that $\mathbb{K}(G) \neq 0$ for some open set G and $\mathbb{K}(G) = 0$ for any open set G of the pointwise convergence topology. Since $\mathbb{K}$ necessarily has a density c, $\mathbb{K}(G) = \inf_{x \in G} c(x) = 0$ for any open set G of the pointwise convergence topology. This means that the l.s.c. envelope of c for this topology is equal to 0, whereas that for the uniform convergence topology is not equal to 0. The function $c(x) = \exp(-\|x\|_\infty)$ satisfies this property.

As $\mathbb{K}_n$ is asymptotically tight, there exists a weakly converging subsequence that we also denote by $\mathbb{K}_n$. Let $\mathbb{K}'$ be the limit. By the tightness of $\mathbb{K}_n$, $\mathbb{K}'$ is also tight and from $\mathbb{K}_n \xrightarrow{w} \mathbb{K}'$ we have $(\mathbb{K}_n)_\pi \xrightarrow{w} \mathbb{K}'_\pi$ and therefore $\mathbb{K}_\pi = \mathbb{K}'_\pi$ for any finite dimensional projection π. By the previous result and the tightness of $\mathbb{K}'$ we obtain $\mathbb{K} = \mathbb{K}'$. From the uniqueness of the limit we obtain $\mathbb{K}_n \xrightarrow{w} \mathbb{K}$. $\qquad\square$

The next result is the analogue of Donsker's theorem about time discretization of Brownian motion.

Theorem 8.3 *Given an i.i.c. sequence X_n of real decision variables centered with sensitivities of order p equal to $\mathbb{S}^p(X_1) = \sigma$, let $S_i = X_1 + \cdots + X_i$ be the* *partial sums and Z_n be the decision variable with values in $C([0,1])$ defined by*

$$Z_n(t) = \frac{1}{\sigma n^{1/p'}}(S_{[nt]} + (nt - [nt])X_{[nt+1]}),$$

with $1/p + 1/p' = 1$. Suppose in addition that $X_1 \in \mathbb{L}^p$. Then Z_n weakly converges towards the p-Brownian decision process:

$$Z_n \xrightarrow{\text{w}} B^p.$$

Proof From Theorem 8.2, we only have to prove tightness of Z_n on the one hand and convergence of finite dimensional distributions of Z_n towards those of B^p on the other hand. For second point, we follow the same technique as in Billingsley's proof of the probabilistic version [13], while tightness is proved by using the sufficient conditions of Theorem 8.1.

The tightness of $Z_n(0)$ is obvious since $Z_n(0) \equiv 0$. Using Theorem 3.2 we obtain for any $s, t \in [0, 1]$

$$\|Z_n(t) - Z_n(s)\|_p \leq (t - s)^{1/p'}\|X_1\|_p/\sigma,$$

whence $\sup_{n,t} \|Z_n(t + \delta) - Z_n(t)\|_p$ tends to 0 when δ tends to 0.

Let us now prove that the finite dimensional distributions of Z_n converge towards those of B^p, that is $\pi(Z_n) \xrightarrow{\text{w}} \pi(B^p)$ for any function of the form $\pi : x \mapsto (x(t_1), \dots x(t_k))$.

We first prove $Z_n(t) \xrightarrow{\text{w}} B^p(t)$ for $t > 0$ (it is clear for $t = 0$). By Theorem 7.6, $B^p(t)$ has cost density $\mathcal{M}^p_{0,t^{1/p'}}$. However, by the central limit theorem, we have $S_{[nt]}/[nt]^{1/p'} \xrightarrow{\text{w}} \mathcal{M}^p_{0,\sigma}$, whence

$$Y_n \stackrel{\text{def}}{=} \frac{t^{1/p'}S_{[nt]}}{\sigma[nt]^{1/p'}} \xrightarrow{\text{w}} B^p(t).$$

Since

$$Z_n(t) = \left(\frac{[nt]}{nt}\right)^{1/p'}Y_n + \left(\frac{nt - [nt]}{\sigma n^{1/p'}}\right)X_{[nt]+1},$$

$\lim_n \|Z_n(t) - Y_n\|_p = 0$, and by Chebyshev's inequality $Z_n(t) - Y_n \xrightarrow{\text{K}} 0$, and the convergence of $Z_n(t)$ towards $B^p(t)$ follows from Theorem 5.10.

Let us now prove $\pi(Z_n) \xrightarrow{\text{w}} \pi(B^p)$ for any function π. By using a bicontinuous transformation, we may replace π by $x \mapsto (x(t_1), x(t_2) - x(t_1), \dots x(t_k) - x(t_{k-1}))$, which we also denote by π. Now, by the same type of approximation as before, we may replace $Z_n(t_i) - Z_n(t_{i-1})$ (with $i = 1, \dots, k$ and $t_0 = 0$) by

$$Y_{n,i} = \frac{(t_i - t_{i-1})^{1/p'}(S_{[nt_i]} - S_{[nt_{i-1}]})}{\sigma([nt_i] - [nt_{i-1}])^{1/p'}}.$$

The decision variables $Y_{n,i}$ with $i = 1, \dots, k$ are independent for any n and separately tend to $B^p(t_i) - B^p(t_{i-1})$ which are also independent. Since the sequences $Y_{n,i}$ are tight for any i, the convergence of $\pi(Z_n)$ follows from Proposition 5.9. $\square$

See Maslov [24] for a weaker result when $p = 2$. See Dudnikov and Samborski [17] for an analogous result when the state space is also discretized.

9 The Inf-Convolution and the Cramér Transform

Definition 9.1 The Cramér transform $\mathcal{C}$ is a function from $\mathcal{M}$, the set of positive measures on $E = \mathbb{R}^n$, to $\mathcal{C}_x$ defined by $\mathcal{C} \overset{\text{def}}{=} \mathcal{F} \circ \log \circ \mathcal{L}$, where $\mathcal{L}$ denotes the Laplace transform[9].

From the properties of the Laplace and Fenchel transforms the following result is clear.

Theorem 9.2 *For $\mu, \nu \in \mathcal{M}$ we have $\mathcal{C}(\mu * \nu) = \mathcal{C}(\mu) \,\square\, \mathcal{C}(\nu)$.*

The Cramér transform transforms convolutions into inf-convolutions and consequently independent random variables into independent decision variables. In Table 1 we summarize the main properties and examples concerning the Cramér transform when $E = \mathbb{R}$. The difficult results of this table can be found in Azencott et al. [6]. In this table we have denoted

$$H(x) \overset{\text{def}}{=} \begin{cases} 0 & \text{for } x \geq 0, \\ +\infty & \text{elsewhere.} \end{cases}$$

Let us give an example of the use of these results in the domain of partial differential equations (PDEs). Processes with independent increments are transformed into decision processes with independent increments. This implies that a generator $\hat{c}(-\partial_x)$ of a stochastic process is transformed into the generator of the corresponding decision process $v \mapsto -\hat{c}(\partial_x v)$.

Theorem 9.3 *The Cramér transform v of the solution r of the PDE on $E = \mathbb{R}$*

$$-\partial_t r + [\hat{c}(-\partial_x)](r) = 0, \ r(0, .) = \delta \,,$$

(with $\hat{c} \in \mathcal{C}_x$) satisfies the HJB equation

$$\partial_t v + \hat{c}(\partial_x v) = 0, \ v(0, .) = \chi \,. \tag{9.1}$$

This last equation is the forward HJB equation of the control problem with dynamic $x' = u$, instantaneous cost $c(u)$ and initial cost χ.

Remark 9.4 First let us remark that $\hat{c}$ is convex l.s.c. and not necessarily polynomial, which means that fractional derivatives may appear in the PDE.

---

[9]$\mu \mapsto \int_E e^{\langle \theta, x \rangle} \mu(dx)$.

$\mathcal{M}$	$\log(\mathcal{L}(\mathcal{M})) = \mathcal{F}(\mathcal{C}(\mathcal{M}))$	$\mathcal{C}(\mathcal{M})$
μ	$\hat{c}_\mu(\theta) = \log \int e^{\theta x} d\mu(x)$	$c_\mu(x) = \sup_\theta(\theta x - \hat{c}(\theta))$
0	$-\infty$	$+\infty$
δ_a	θa	χ_a
$\lambda e^{-\lambda x - H(x)}$	$H(\lambda - \theta) + \log(\lambda/(\lambda - \theta))$	$H(x) + \lambda x - 1 - \log(\lambda x)$
$p\delta_0 + (1-p)\delta_1$	$\log(p + (1-p)e^\theta)$	$x\log(\frac{x}{1-p})$ $+(1-x)\log(\frac{1-x}{p})$ $+H(x) + H(1-x)$
stable distrib.	$m\theta + \frac{1}{p'}\lvert\sigma\theta\rvert^{p'} + H(\theta)$ $1 < p' < 2$	$c(x) = \mathcal{M}^p_{m,\sigma}\,, x \geq m$ $c(x) = 0,\ x < m,$ $1/p + 1/p' = 1$
Gauss distrib.	$m\theta + \frac{1}{2}\lvert\sigma\theta\rvert^2$	$\mathcal{M}^2_{m,\sigma}$
$\mu * \nu$	$\hat{c}_\mu + \hat{c}_\nu$	$c_\mu \,\square\, c_\nu$
$k\mu$	$\log(k) + \hat{c}$	$c - \log(k)$
$\mu \geq 0$	$\hat{c}$ convex l.s.c.	c convex l.s.c.
$m_0 \stackrel{\text{def}}{=} \int \mu$	$\hat{c}(0) = \log(m_0)$	$\inf_x c(x) = -\log(m_0)$
$m_0 = 1$	$\hat{c}(0) = 0$	$\inf_x c(x) = 0$
$S_\mu \stackrel{\text{def}}{=} \overline{\text{cvx}(\text{supp}(\mu))}$	$\hat{c}$ strictly convex in $D_{\hat{c}}$	$\overset{\circ}{D}_c = \overset{\circ}{S}_\mu$
$m_0 = 1$	$\hat{c}$ is C^∞ in $\overset{\circ}{D}_{\hat{c}}$	c is C^1 in $\overset{\circ}{D}_c$
$m_0 = 1,\ m \stackrel{\text{def}}{=} \int x\mu$	$\hat{c}'(0) = m$	$c(m) = 0$
$m_0 = 1,\ m_2 \stackrel{\text{def}}{=} \int x^2\mu$	$\hat{c}''(0) = \sigma^2 \stackrel{\text{def}}{=} m_2 - m^2$	$c''(m) = 1/\sigma^2$
$m_0 = 1,\ 1 < p' < 2$ $\hat{c} = \lvert\sigma\theta\rvert^{p'}/p' + o(\lvert\theta\rvert^{p'})$ $+H(\theta)$	$\hat{c}^{(p')}(0^+) = \Gamma(p')\sigma^{p'}$	$c^{(p)}(0^+) = \Gamma(p)/\sigma^p$

Table 1: Properties of the Cramér transform.

Proof The Laplace transform of r, denoted by q, satisfies:

$$-\partial_t q(t,\theta) + \hat{c}(\theta)q(t,\theta) = 0, \quad q(0,.) = 1 .$$

Therefore $w = \log(q)$ satisfies:

$$-\partial_t w(t,\theta) + \hat{c}(\theta) = 0, \quad w(0,.) = 0 , \qquad (9.2)$$

which can be easily integrated. As long as $\hat{c}$ is l.s.c. and convex, w is l.s.c. and convex and can be considered as the Fenchel transform of a function v. The function v satisfies a PDE which can be easily computed. Indeed we have

$$w(t,\theta) = \sup_x(\theta x - v(t,x)) \Rightarrow \begin{cases} \theta = \partial_x v , \\ \partial_t w = -\partial_t v . \end{cases}$$

Therefore v satisfies equation (9.1). This equation is the forward HJB equation of the control problem with dynamic $x' = u$, instantaneous cost $c(u)$ and initial cost χ because $\hat{c}$ is the Fenchel transform of c and the HJB equation of this control problem is

$$-\partial_t v + \min_u \{-u\partial_x v + c(u)\} = 0, \ v(0,.) = \chi.$$

$\square$

If $\hat{c}$ is independent of time the optimal trajectories are straight lines and $v(x) = tc(x/t)$. This can be obtained by using (9.2).

Solutions of linear PDEs with constant coefficients can be computed explicitly by means of the Fourier transform. The previous theorem shows that that nonlinear convex first order PDEs with constant coefficients are isomorphic to linear PDEs with constant coefficients and therefore can be computed explicitly. Such explicit solutions of the HJB equation are known as Hopf formulas [8]. Let us develop the computations on a nontrivial example.

Example 9.5 Let us consider the HJB equation

$$\partial_t v + \frac{1}{2}(\partial_x v)^2 + \frac{2}{3}(|\partial_x v|)^{3/2} = 0, \ v(0,.) = \chi.$$

From (9.2) we deduce that

$$w(t,\theta) = t(\frac{1}{2}\theta^2 + \frac{2}{3}|\theta|^{3/2}),$$

therefore using the fact that the Fenchel transform of a sum is an inf-convolution we obtain

$$v(t,x) = \frac{x^2}{2t} \ \square \ \frac{|x|^3}{3t^2}.$$

We can verify on this explicit formula a continuous time version of the central limit theorem. Using the scaling $x = yt^{2/3}$, we have

$$\lim_{t \to +\infty} v(t, yt^{2/3}) = y^3/3,$$

since the shape around zero of the corresponding instantaneous cost $c(u) = (u^2/2) \ \square \ (|u|^3/3)$ is $|u|^3/3$. Indeed a simple computation shows that $c(u)$ is obtained from

$$\begin{cases} c = y^4/2 + |y|^3/3, \\ u = |y|y + y, \end{cases}$$

by elimination of y. This system may be also considered as a parametric definition of $c(u)$.

Notes and comments Bellman [11] was aware of the interest of the Fenchel transform (which he calls max transform) for the analytic study of dynamic programming equations. The bicontinuity of the Fenchel transform has been well studied in convex analysis [22, 5, 4].

Maslov started the study of idempotent integration in [24]. He has been followed in particular by [23, 25, 26, 16, 15, 10, 3, 1, 2] and independently by [28]. In [27] idempotent Sobolev spaces have been introduced as a way to study the HJB equation as a linear object. In that paper the min-plus weak convergence has also been introduced, but for compact support test functions. This weak convergence is used in [17] for the approximation of HJB equations. In [29] and [7] the law of large numbers and the central limit theorem for decision variables have been given in the particular case $p = 2$. In two independent works [16, 15] and [10] the study of decision variables has been started. The second work has been continued in [3]. Many results announced in [3] are proved in [1] and [2].

The Cramér transform is an important tool in large deviations literature [6, 21, 31, 18]. In [16, 7, 3] the Cramér transform has been used in the min-plus context.

Some aspects of [32, 33, 9, 12], for instance the morphism between LQG and LEQG problems presented in [32, Section 6.1] and the separation principle developed in [12], provide other illustrations of the analogy between probability and decision calculus.

References

[1] Akian, M.: Densities of idempotent measures and large deviations. INRIA Report **2534** (1995).

[2] Akian, M.: Theory of cost measures: convergence of decision variables. INRIA Report **2611** (1995).

[3] Akian, M., Quadrat J.P., Viot M.: Bellman processes, in 11th Inter. Conf. on Analysis and Optimization of Systems. L.N. in Control and Inf. Sc. No. 199, Springer-Verlag (1994).

[4] Attouch, H: Variational convergence for functions and operators. Pitman (1984).

[5] Attouch, H., Wets, R.J.B.: Isometries for the Legendre–Fenchel transform. Transactions of the American Mathematical Society **296**, No. 1 (1986) 33–60.

[6] Azencott, R., Guivarc'h, Y., Gundy, R.F.: Ecole d'été de Saint Flour 8. Lect. Notes in Math., Springer-Verlag, Berlin (1978).

[7] Baccelli, F., Cohen, G., Olsder, G.J., Quadrat, J.P.: Synchronization and linearity: an algebra for discrete event systems. John Wiley & Sons, New York (1992).

[8] Bardi, M., Evans, L.C. On Hopf's formulas for solutions of Hamilton–Jacobi equations. Nonlinear Analysis, Theory, Methods & Applications, Vol.8., No. 11 (1984) 1373–1381.

[9] Basar, T., Bernhard, P.: H_∞ optimal control and relaxed minimax design problems. Birkhäuser (1991).

[10] Bellalouna, F.: Un point de vue linéaire sur la programmation dynamique.

Détection de ruptures dans le cadre des problèmes de fiabilité. Thesis dissertation, University of Paris-IX Dauphine (1992).

[11] Bellman, R., Karush, W.: Mathematical programming and the maximum transform. SIAM Journal of Applied Mathematics **10** (1962).

[12] Bernhard, P.: Discrete and continuous time partial information minimax control. Submitted to Annals of ISDG (1994).

[13] Billingsley, P.: Convergence of probability measures. John Wiley & Sons, New York (1968).

[14] Cuninghame-Green, R.: Minimax algebra. Lecture Notes in Economics and Mathematical Systems No. 166, Springer-Verlag (1979).

[15] Del Moral, P.: Résolution particulaire des problèmes d'estimation et d'optimisation non-linéaires. Thesis dissertation, Toulouse, France (1994).

[16] Del Moral, P., Thuillet, T., Rigal, G., Salut, G.: Optimal versus random processes: the non-linear case. LAAS Report, Toulouse, France (1990).

[17] Dudnikov, P.I., Samborski, S.: Networks methods for endomorphisms of semimodules over min-plus Algebras, in 11th Inter. Conf. on Analysis and Optimization of Systems. L.N. in Control and Inf. Sc. No.199, Springer-Verlag (1994).

[18] Ellis, R. S.: Entropy, large deviations, and statistical mechanics. Springer-Verlag, New York (1985).

[19] Feller, W.: An introduction to probability theory and its applications, John Wiley & Sons, New York (1966).

[20] Fenchel, W.: On the conjugate convex functions. Canadian Journal of Mathematics **1** (1949) 73–77.

[21] Freidlin, M.I., Wentzell, A.D.: Random perturbations of dynamical systems. Springer-Verlag, Berlin (1984).

[22] Joly, J.L.: Une famille de topologies sur l'ensemble des fonctions convexes pour lesquelles la polarité est bicontinue. J. Math. pures et appl. **52** (1973) 421–441.

[23] Kolokoltsov, V. N. and Maslov, V. P.: The general form of the endomorphisms in the space of continuous functions with values in a numerical commutative semiring (with the operation ⊕=max). Soviet Math. Dokl. Vol. 36, No. 1 (1988) 55–59.

[24] Maslov, V.: Méthodes opératorielles. Éditions MIR, Moscow (1987).

[25] Maslov, V. P. and Kolokoltsov, V. N.: Idempotent analysis and its applications to optimal control theory. Nauka, Moscow (1994) in Russian.

[26] Maslov, V., Samborski, S.N.: Idempotent analysis. Advances In Soviet Mathematics **13**, Amer. Math. Soc., Providence (1992).

[27] Maslov, V., Samborski, S.N.: Stationary Hamilton–Jacobi and Bellman equations (existence and uniqueness of solutions), in Idempotent analysis. Advances in Soviet Mathematics **13**, Amer. Math. Soc., Providence (1992).

[28] Pap, E.: Solution of nonlinear differential and difference equations. EUFIT'93, Aachen, Sept 7–10 (1993) 498–503.

[29] Quadrat, J.P.: Théorèmes asymptotiques en programmation dynamique. Note CRAS **311** Paris (1990) 745–748.

[30] Rockafellar, R.T.: Convex analysis. Princeton University Press Princeton, N.J. (1970).

[31] Varadhan, S.R.S.: Large deviations and applications. CBMS-NSF Regional Conference Series in Applied Mathematics No. 46, SIAM, Philadelphia (1984).

[32] Whittle, P.: Risk sensitive optimal control. John Wiley & Sons, New York (1990).

[33] Whittle, P.: A risk-sensitive maximum principle: the case of imperfect state observation. IEEE Trans. Auto. Control, AC-36 (1991) 793–801.

Maslov Optimization Theory: Topological Aspects

Pierre Del Moral

Abstract

Maslov integration theory allows a process optimization theory to be derived at the same level of generality as stochastic process theory. The approach followed here captures the main idea in forward time, and we therefore introduce Maslov optimization processes such as encountered in maximum likelihood problems. A reversal of time yields optimal control problems of regulation type. We briefly recall the common concepts and, in particular, that the Markov causality principle in this theory is the same as the Bellman optimality principle. We introduce several modes for the convergence of optimization variables and present some classical asymptotic theorems. We derive some optimization martingale properties, such as the inequalities and the (max,+)-version of the Doob up-crossing lemma which leads to new developments in the field of qualitative studies of optimization processes. Finally we show how some classical large deviation principles translate into this framework and focus on applications to nonlinear filtering.

Introduction

One purpose of this paper is to present a survey of an idempotent measure point of view on optimization theory. The author ([9], [7]) and, independently, Bellalouna ([3]) have introduced the appropriate methodology to derive an optimization theory at the same level of generality as probability and stochastic process theory. This work was clearly inspired by the idempotent measure and integration originally proposed by Maslov [23] as well as the pioneering works of Huillet–Salut [20], Huillet–Rigal–Salut [21] and Del Moral–Huillet–Salut [22].

Sections 1 and 2 are taken from Del Moral [9]. We introduce some basic concepts such as performance spaces, optimization variables, independence and conditioning. Similar developments can be found in Akian–Quadrat–Viot [1], and Del Moral [7], [13] or [14].

The second motivation of this text is to state new results and to introduce some of the relevant techniques. No detailed proofs are given, but some proofs are sketched. Each section is followed by bibliographical comments.

Section 3 is devoted to the introduction of new topological aspects such as Lebesgue L^p-spaces, the Ky-Fan metric, and the Bienaymé–Chebyshev, Hölder and Minkowski inequalities.

Section 4 proposes several modes for the convergence of optimization variables and how they can be used to generalize Quadrat's results on the law of large numbers [32] to the case where the optimization variables have not necessarily the same performance. We also state another type of large number law (proposition 7).

Section 5, we recall that the Markov causality principle corresponds to the Bellman optimality principle. The Bellman principle is, in fact, the (max,+)-version of the Chapman–Kolmogorov equation for stochastic processes. A general presentation can be founded in Del Moral–Rigal–Salut [9]. For a systematic development of the theory of optimization processes we refer to Del Moral [7] and [11]. Related results have been summarized independently by Akian–Quadrat–Viot [1]. We also introduce Dynkin's formula and some consequences of Doob's inequalities and up-crossing lemma. This work was first introduced by the author in [11], and futher developed in [7] and [12].

Section 6 focuses on large deviation principles for conditional measures and related asymptotics of filtering equations. We recall the Wentzell–Freidlin and Cramér transforms and we introduce the Log-Exp transform. All these transforms provide a way of computing the probability of rare events in terms of performance measures. We also focus on applications to nonlinear filtering problems. For applications to (max,+)-optimization problems, we refer to Del Moral–Salut [15] and the present volume.

I want to thank the Hewlett-Packard Laboratories of Bristol and in particular Jeremy Gunawardena for inviting me to give this lecture and for their warm hospitality.

1 Maslov integration theory

In this section, we recall Maslov integration theory and the semiring of reference ([23] and [7]). The elements of idempotent analysis (analysis of functions with values in a general idempotent semiring) are developed in [16], [23], [24], [25], [26], [28] and [29]. Let $\mathcal{A} \overset{def}{=} [-\infty, +\infty[$ be the semiring of real numbers endowed with the commutative semigroup laws $\oplus$, $\odot$, the neutral elements $\mathbb{0}$, $\mathbb{1}$ and the exponential metric ρ such that:

$$\mathbb{0} = -\infty \quad a \oplus b = \sup(a, b) \quad \rho(a, b) = \left| e^a - e^b \right|$$
$$\mathbb{1} = 0 \quad a \odot b = a + b \quad \frac{a}{b}\mathbb{0} = a - b \quad (b \neq \mathbb{0})$$

We also denote the exponential metric over $\mathcal{A}^n$ by $\rho(a, b) = \sup_{1 \leq i \leq n} |e^{a_i} - e^{b_i}|$. We specially mention a notion which will be used throughout the study of the (max,+)-version of Lebesgue spaces. We denote by $d(x, y) = \log \rho(x, y) \in \mathcal{A}$ the logarithm of the exponential metric ρ. This application is characterized by the following properties:

$$d(x, y) = \mathbb{0} \Longleftrightarrow x = y, \qquad d(x, y) = d(y, x)$$
$$d(x, y) \leq \log(\exp d(x, z) \odot \exp d(z, y)).$$

More generally, if $\mathbf{L}$ is a $(\oplus, \odot)$-semimodule, every application d from $\mathbf{L} \times \mathbf{L}$ into $\mathcal{A}$ satisfying these three conditions is called a $\oplus$-metric. Let us consider an auxiliary measurable space (Ω, σ). An $\mathcal{A}$-measure μ on (Ω, σ) is an application from σ into $\mathcal{A}$ such that $\mu(\emptyset) = \mathbb{0}$ and

$$\mu(A) = \sup_{\omega \in A} g(\omega)$$

for some $\mathcal{A}$-valued function g. The function g is called the density of the $\mathcal{A}$-measure μ. It is clear that, for all sets $A, B \in \sigma$,

$$\mu(A \cup B) = \mu(A) \oplus \mu(B).$$

μ is called bounded whenever $\mu(\Omega) < +\infty$.

Let $\mathcal{L}^o(\Omega, \sigma)$ be the semiring of $\mathcal{A}$-valued measurable functions, respectively step functions. We define the integral of $f \in \mathcal{L}^0(\Omega, \sigma)$ with respect to the $\mathcal{A}$-measure μ by the following formulas:

$$\int_\Omega^\oplus f \odot \mu \stackrel{def}{=} \sup_{\omega \in \Omega}(f(\omega) + g(\omega)) \stackrel{def}{=} \int_\Omega^\oplus f(\omega) \odot g(\omega) \odot d\omega.$$

It is obvious that this integration is linear over $\mathcal{A}$.

Finally, to complete this section we recall the notion of convolution of Maslov measures.

Let μ_1 and μ_2 be two $\mathcal{A}$-measures. We denote by $\mu_1 \circledast \mu_2$ the unique $\mathcal{A}$-measure defined, for every $f \in \mathcal{C}_\mathbb{0}(\Omega, \mathcal{A})$, by

$$\int_{\Omega \times \Omega}^\oplus f(x+y) \odot \mu_1(dx) \odot \mu_2(dy) = \int_\Omega^\oplus f(z) \odot (\mu_1 \circledast \mu_2)(dz) \qquad (1.1)$$

This law is commutative, associative, and distributive over the addition $\oplus$. There are also analogs for the well known theorems of Riesz, Hahn–Banach, Banach–Steinhaus, etc. See Litvinov–Maslov [27] for a complete bibliography on these subjects.

2   Performance Theory

All notations, assumptions and results of Section 1 are in force. The concepts of performance measure and optimization variable are basic notions. Their analysis in the semiring $\mathcal{A}$ offers an alternative to the classical representations for optimization problems. The purpose of this section is to recall certain of these axioms. The intuitive background of the concept of an optimization variable is as follows. Suppose that we are given an optimization problem described by a measurable space (Ω, σ), where Ω is the set of all possible controls and σ the sigma-algebra of controls that are possible or interesting in the framework of the optimization problem. Now we are given a reference value $Y(\omega)$ associated to the control event ω. This value depends on ω. The assumption of measurability means that for every reference value there is a meaningful control event in the original space. There is in general less information in the reference value $Y(\omega)$ than in the event ω, a fact expressed by the condition that $\sigma(Y) \subset \sigma$. Let $\Omega = \{(i,j) \ : \ 1 \leq i,j \leq n\}$, $n \in \mathbb{N}$, and let the set of subsets of Ω serve as σ. Then every function on Ω is measurable. On the other hand, if $\sigma(Y)$ is the sigma-algebra generated by the sets $S_k = \{(i,j) \ : \ i+j = k\}$ with $1 \leq k \leq 2n$ then $Y((i,j)) = i+j$ is $\sigma(Y)$-measurable but $Y((i,j)) = i - j$ is not $\sigma(Y)$-measurable. The general conditional optimization problem will be studied in Section 5.

Let (Ω, σ) and $(E, \mathcal{E})$ be two measurable spaces and E a semiring. An $\mathcal{A}$-measure $\mathbb{P} \in \mathbf{M}(\Omega, \sigma)$ such that $\mathbb{P}(\Omega) = \mathbb{1}$ is called a performance measure and $(\Omega, \sigma, \mathbb{P})$ a performance space. As in probability theory, we assign a performance to the events of Ω and we define optimization variables whose domain consists of the elements of Ω. A measurable function X from (Ω, σ) into $(E, \mathcal{E})$ is called an E-valued optimization variable, and we denote by $\mathbb{P}^X$ its performance measure, and by $\mathbb{p}^X$ its upper-semicontinuous density. Whenever the integral exists we write:

$$\mathbb{E}(X) = \int_\Omega^\oplus X \odot \mathbb{P} = \int_E^\oplus x \odot \mathbb{p}^X(x) \odot dx$$

$$\forall A \in \mathcal{E} \qquad \mathbb{P}(\{\omega \in \Omega \ : \ X(\omega) \in A\}) = \mathbb{P}^X(A) = \int_A^\oplus \mathbb{p}^X(x) \odot dx$$

$$\mathbb{p}^X(op(X)) = \mathbb{1}.$$

In our framework, the following implications hold:

$$\mathbb{P}(X \in \Omega - A) < \mathbb{1} \Rightarrow \mathbb{P}(X \in A) = \mathbb{1} \Rightarrow op(X) \in A.$$

These facts show that it is important to control the performance of an event. For this purpose we will state in Section 3 the (max,+)-version of the Bienaymé–Chebyshev, Minkowski and Hölder inequalities.

Definition 1 *Let X, Y be two $\mathcal{A}$-valued optimization variables. We say that X and Y are equal $\mathbb{P}$-almost everywhere ($\mathbb{P}$-a.e.) if*

$$\mathbb{P}(\{\omega \in \Omega : X(\omega) \neq Y(\omega)\}) = \mathbb{0}.$$

Several other characterizations of this equivalence relation are in order.

$$
\begin{aligned}
X = Y \quad \mathbb{P}\text{-}a.s. \quad &\Longleftrightarrow \quad \forall \epsilon > 0 \quad \mathbb{P}\{\omega \in \Omega : d(X(\omega), Y(\omega)) \geq \epsilon\} = 0 \\
&\Longleftrightarrow \quad \exists A \in \sigma \quad \mathbb{P}(\Omega - A) = 0 \text{ and } \forall \omega \in A \quad X(\omega) = Y(\omega) \\
&\Longleftrightarrow \quad \forall A \in \sigma \quad \mathbb{E}(1_A \odot X) = \mathbb{E}(1_A \odot Y)
\end{aligned}
$$

For example, the last condition is used to prove the uniqueness of the conditional expectation. Let $\mathbf{L}^0(\Omega, \sigma) = \mathcal{L}^0(\Omega, \sigma)/\mathbb{P}$-a.s. the induced quotient semiring. We now introduce the semiring of $\mathcal{A}$-valued and integrable optimization variables

$$
\mathcal{L}^1(\Omega, \sigma, \mathbb{P}) \overset{def}{=} \left\{ X \in \mathcal{L}^0(\Omega, \sigma) : d_1(X, 0) = \mathbb{E}(X) < +\infty \right\}
$$

$$
d_1(X, Y) \overset{def}{=} \mathbb{E}(d(X, Y)).
$$

In [8] we prove that $X = Y$ $\mathbb{P}$-a.s. if, and only if, $d_1(X, Y) = 0$ and d_1 is a $\oplus$-metric over $\mathbf{L}^1(\Omega, \sigma) = \mathcal{L}^1(\Omega, \sigma)/\mathbb{P}$-a.s. For instance, if X is a real-valued optimization variable whose performance law is given by $\mathbb{p}(x) = -\frac{1}{2}\left(\frac{x-m}{a}\right)^2$, then

$$
d_1(X, 0) = m \odot \left(\frac{a^2}{2}\right) \qquad d_1(X, m) = m \odot d_1\left(\frac{X}{m}\circ, 1\right).
$$

Moreover $d_1\left(\frac{X}{m}\circ, 1\right) = \sup_x(\log|e^x - 1| - \frac{1}{2}\left(\frac{x}{a}\right)^2) = \sup_{x \geq 0} \log \Theta_a(x)$ with, for $x \geq 0$

$$
\Theta_a'(x) = 0 \Longleftrightarrow e^x - 1 = \left(a^{-2}x - 1\right)^{-1}.
$$

By standard numerical approximations, $d_1\left(\frac{X}{m}\circ, 1\right) = \log \Theta_a(x(a))$ with $0 \leq x(a) \leq x'(a)$ and

$$
x'(a) = \left(a^{-2}x'(a) - 1\right)^{-1} \quad \left(\Longleftrightarrow a^{-2}x'(a)^2 - 1 = x'(a)\right).
$$

Similarly, if $a < 2$, $0 \leq x'(a) \leq x''(a)$ with $\frac{2}{a}(x''(a) - a) = x''(a)$ ($\Longleftrightarrow x''(a) = \frac{2a}{2-a}$), then, for $a < 2$, we have

$$
0 \leq d_1(X, m) \leq m \odot d(\frac{2a}{2-a}, 1).
$$

By the same line of argument, if $\mathbb{p}(x) = -\frac{1}{p}\left|\frac{x-m}{a}\right|^p$ for some $p \geq 2$, we obtain for $\frac{1}{q} + \frac{1}{p} = 1$

$$
d_1(X, 0) = m \odot \left(\frac{a^q}{q}\right) \quad \text{and} \quad a < p \Rightarrow 0 \leq d_1(X, m) \leq m \odot d(\frac{pa}{p-a}, 1).
$$

$$\tag{2.1}$$

Proposition 1 *Let $Q_m(a, p)$ be a real-valued optimization variable whose performance is*

$\mathbb{p}(x) = -\frac{1}{p}\left|\frac{x-m}{a}\right|^p$, *for some $p \geq 2, a > 0, m \in \mathbb{R}$. Then $\lim_{a\to 0} d_1(Q_m(a,p), m) = \mathbb{0}$, and for $\frac{1}{q} + \frac{1}{p} = 1$:*

$$d_1(Q_m(a,p), \mathbb{0}) = m \odot \left(\frac{a^q}{q}\right)$$

$$a < p \Rightarrow \mathbb{0} \leq d_1(Q_m(a,p), m) \leq m \odot d(\frac{p\,a}{p-a}, 1).$$

Our aim is to transpose probabilistic axioms to optimization theory. The **independence** concept in such a framework is defined by

Definition 2 *Let $(X_i)_{i\in I}$ be a (not necessarily finite) family of E-valued optimization variables on the same performance space $(\Omega, \sigma, \mathbb{P})$. We say they are $\mathbb{P}$-independent when, for every finite subset $J = \{t_1, \ldots, t_n\}$, $J \subset I$, $n \geq 1$, one of the following equivalent assertions is satisfied:*

1. *$\forall i \in \{1, \ldots, n\}$ $\forall A_i \in \sigma(X_{t_i})$*

   $$\mathbb{P}(A_1 \cap \ldots \cap A_n) = \mathbb{P}(A_1) \odot \ldots \odot \mathbb{P}(A_n)$$

2. *$\mathbb{p}_J^X(x_1, \ldots, x_n) = \bigodot_{j=1}^n \mathbb{p}_{t_j}^X(x_j)$ where $\mathbb{p}_J^X$, respectively $\mathbb{p}_{t_j}^X$, $1 \leq j \leq n$, is the performance density of $X_J = (X_{t_1}, \ldots, X_{t_n})$, respectively X_{t_i}, $j \in [1, n]$.*

At the opposite of the latter concept of independence, we introduce the extension of Bayes' formula to performance measures. The conditional performance of an event A, assuming an event B such that $\mathbb{P}(B) \neq \mathbb{0}$, and denoted by $\mathbb{P}(A/B)$, is by definition the ratio

$$\mathbb{P}(A/B) = \frac{\mathbb{P}(A \cap B)}{\mathbb{P}(B)}\circ. \tag{2.2}$$

The last conditions are the axioms of Maslov optimization theory. In the development of the theory all conclusions are based directly or indirectly on these axioms and only these axioms. We now examine how the notion of conditional expectation and optimization of random variable translates into such a framework.

Theorem–Definition 1 *Let X and Y be two E-valued optimization variables defined on the same performance space $(\Omega, \sigma, \mathbb{P})$.*

1. *Let $\sigma(X, Y)$ be the σ-algebra spanned by the optimization variables X and Y, $\sigma(Y)$ the σ-algebra spanned by Y.*

   For every function $\phi \in \mathbf{L}^1(\Omega, \sigma(X, Y), \mathbb{P})$, there exists a unique function in $\mathbf{L}^1(\Omega, \sigma(Y), \mathbb{P})$, called the conditional expectation of ϕ relative to $\sigma(Y)$ and denoted by $\mathbb{E}(\phi/\sigma(Y))$, such that

   $$\forall \psi \in \mathbf{L}^1(\Omega, \sigma(Y), \mathbb{P}) \quad \mathbb{E}(\phi \odot \psi) = \mathbb{E}(\mathbb{E}(\phi/\sigma(Y)) \odot \psi)$$

2. The measurable function $\mathbb{p}^{X/Y}(x/y)$ defined as $\frac{\mathbb{p}^{X,Y}(x,y)}{\mathbb{p}^Y(y)}\circ$ if $\mathbb{p}^Y(y) > 0$,
    $\mathbb{0}$ otherwise, is called the conditional performance density of X relative
    to Y. A conditional optimal state of X relative to Y is a measurable
    function $op(X/Y)$ such that $\mathbb{p}^{X/Y}(op(X/y)/y) = \mathbb{1}$, for all y.

Proposition 2 *Let X, Y, Z be three E-valued optimization variables on the
same performance space, $\varphi, \psi \in \mathbf{L}^1(\Omega, \sigma(X,Y), \mathbb{P})$ $\phi \in \mathbf{L}^1(\Omega, \sigma(X,Y,Z), \mathbb{P})$
and $a, b \in \mathcal{A}$ Then, $\mathbb{P}$-a.e.,*

1. $\mathbb{E}(a \odot \varphi \oplus b \odot \psi / \sigma(Y)) = a \odot \mathbb{E}(\varphi / \sigma(Y)) \oplus b \odot \mathbb{E}(\psi / \sigma(Y))$
    $\mathbb{E}(\phi) = \mathbb{E}(\phi / \{\emptyset, \Omega\})$

2. $\mathbb{E}(\varphi / \sigma(Y)) = \varphi \quad$ if $\varphi \in \mathbf{L}^1(\Omega, \sigma(Y))$
    $\mathbb{E}(\mathbb{E}(\phi / \sigma(X,Y)) / \sigma(Y)) = \mathbb{E}(\phi / \sigma(Y))$.

Proposition 3 *Let X, Y, Z be E-valued optimization variables defined on the
same performance space and let φ be a measurable function. Then*

$$op(op(X/Z, Y)/Y) = op(X/Y) \qquad op(\varphi(X)/Y) = \varphi(op(X/Y)).$$

For instance, let F be a measurable function from $\mathbb{R}^n$ into $\mathbb{R}^n$, C a $n \times m$-matrix, U and V two $\mathbb{P}$-independent optimization variables with $\mathbb{p}^U(u) \overset{def}{=}$ $-\frac{1}{2}u'Q^{-1}u$ and $\mathbb{p}^V(v) \overset{def}{=} -\frac{1}{2}v'R^{-1}v$, where Q^{-1} and R^{-1} are symmetric positive definite $n \times n$ and $m \times m$-matrices. Let $S^{-1} \overset{def}{=} Q^{-1} + C'R^{-1}C$ and $X_1 = F(X_0) + U$, $Y_1 = CX_1 + V$. Then

$$
\begin{aligned}
op(X_1/X_0) &= F(X_0),\ op(Y_1/X_0) = CF(X_0) \\
op(X_1/X_0, Y_1) &= F(X_0) + SC'R^{-1}(Y_1 - CF(X_0)).
\end{aligned}
$$

When X, Y are two $\mathbb{P}$-independent and E-valued optimization variables,

$$\mathbb{p}^{X,Y}(x,y) = \mathbb{p}^X(x) \odot \mathbb{p}^Y(y)$$

and the performance density of the sum $X + Y$ can be described by the (max,+)-convolution of the latter

$$\mathbb{p}^{X+Y}(z) = \int_E^{\oplus} \mathbb{p}^X(z-y) \odot \mathbb{p}^Y(y) \odot dy \overset{def}{=} \left(\mathbb{p}^X \circledast \mathbb{p}^Y\right)(z)$$

Similarly we claim that every (max,+)-linear and continuous operator S on the class of upper-semicontinuous functions (the notions of continuity and weak convergence were introduced by Maslov in [23] and futher developed by Kolokoltsov and Maslov in [25] and [26]) that commutes with the delays U_x (defined by $U_x(f)(y) = f(y-x)$ for every upper-semicontinuous function f) may be represented by a (max,+)-convolution. Indeed, let h_n be a weak approximation of the Dirac function $\delta_0(x) = \mathbb{1}_0(x) = \mathbb{1}$ if $x = 0$, $\mathbb{0}$ otherwise (if E is a normed space we may choose $h_n(x) = -\frac{n}{2}\|x\|^2$). Let $\mathcal{C}_{\mathbb{0}}$ be the semiring of continuous functions from E into $\mathcal{A}$ such that $f(x) = \mathbb{0}$ for every x outside some compact set $K \subset E$. Then for every function $f \in \mathcal{C}_{\mathbb{0}}$ we have

$$
\begin{aligned}
S(f \circledast h_n) &= S\left(\int_E^\oplus U_y(f) \odot h_n(y) \odot dy\right) = \int_E^\oplus S\left(U_y(f)\right) \odot h_n(y) \odot dy \\
&= \int_E^\oplus f(y) \odot U_y\left(S(h_n)\right) \odot dy = \int_E^\oplus U_y\left(S(f)\right) \odot h_n(y) \odot dy \\
&= S(f) \circledast h_n = f \circledast S(h_n)
\end{aligned}
$$

$$
\implies S(f) = f \circledast \lim_{n \to +\infty} S(h_n) = f \circledast S(\delta_0).
$$

For instance, if $S(\delta_0)(x) = g(x)$, then

$$
S(f)(x) = \sup_{y \in E}(g(x - y) + f(y)).
$$

We now recall the definition of **Fenchel transform**. Let $\mathcal{S}_+$ be the class of proper upper-semicontinuous and concave functions from $\mathbb{R}$ into $\mathbb{R}$. The Fenchel transform is the application

$$
\begin{aligned}
\mathcal{F} \; : \; \mathcal{S}_+ &\longrightarrow \mathcal{S}_+ \\
f &\longmapsto \mathcal{F}(f) \quad \text{such that} \quad -(\mathcal{F}f)(x^*) = \int_{\mathbb{R}}^\oplus x^*(x) \odot f(x) \odot dx.
\end{aligned}
$$

Let τ_a be the translation on $\mathbb{R}$ associated with $a \in \mathbb{R}$. Then, for every $x^* \in \mathbb{R}$,

$$
\left(-\mathcal{F}\left(f \circ \tau_a\right)\right)(x^*) = x^*(a) \odot \left(-\mathcal{F}(f)\right)(x^*). \tag{2.3}
$$

This property is characteristic of the Fourier transform (Maslov [23]). The Fenchel transform is then equal to the Fourier transform in our setting.

Proposition 4 *Let $f, g \in \mathcal{S}_+$ and X be an $\mathbb{R}$-valued optimization variable on $(\Omega, \sigma, \mathbb{P})$ whose optimal state $op(X) \in \mathbb{R}$ is unique. Assume $\mathcal{F}\mathbb{p}^X$ is twice continuously differentiable around 0 and*

$$
(f \circledast g)(z) \overset{def}{=} \int^\oplus f(z - x) \odot g(x) \odot dx = \sup_{x \in \mathbb{R}} \left(f(z - x) + g(x)\right).
$$

Then:

1. $\mathcal{F} \circ \mathcal{F} = \mathcal{I}d$, *and* $\mathcal{F}\left(f \circledast g\right) = \mathcal{F}(f) \odot \mathcal{F}(g)$

2. $\mathbb{E}(\omega\, X) = -\left(\mathcal{F}\mathbb{p}^X\right)(\omega)$

3. $-\left(\mathcal{F}\mathbb{p}^X\right)'(0) = op(X)$

4. $\left(\mathcal{F}\mathbb{p}^X\right)''(0) = \left((\mathbb{p}^X)''\right)^{-1}(op(X))$

5. *Let S be a continuous operator on the class of upper-semi continuous functions that commutes with the delays, $f \in \mathcal{C}_0$ and $H \overset{def}{=} \mathcal{F}(S(\delta_0))$. Then:*

$$\mathcal{F}(S(f)) = H \odot \mathcal{F}(f).$$

The following example is classical in convex analysis and it will be essential in the study of convergence modes. If $a \in \mathbb{R}^+$, $m \in \mathbb{R}$, $a \neq 0$, $p > 2$ and $\frac{1}{q} + \frac{1}{p} = 1$, then

$$-\mathcal{F}(f) = m + \frac{1}{p}|x\,a|^p \iff f = -\frac{1}{q}\left|\frac{x-m}{a}\right|^q. \tag{2.4}$$

The content of this section is based on the author's works [8] and [7]. Theorem–Definition 1 and Proposition 3 of this section have been proved independently by Bellalouna in [3].

3   Lebesgue semirings

In this section we introduce the analogues of Lebesgue spaces and the Markov, Minkowski and Hölder inequalities. Then we deal with several modes of convergence for optimization variables. Unfortunately we do not have enough room here for a thorough study in all details ([8] or [7]). To focus on the main idea, the treatment will leave aside technical issues. In the sequel all optimization variables are defined on the same optimization basis $(\Omega, \sigma, \mathbb{P})$. As for random variables we introduce the Ky-Fan metric of optimization variables. For every $\mathcal{A}$-valued optimization variables X, Y we write

$$\begin{aligned}
\mathcal{K}(X,Y) &= \left\{(\epsilon,\eta) \in \mathbb{R}^+ \times \mathbb{R}^+ : \mathbb{P}\{d(X,Y) > \log\eta\} \le \log\epsilon\right\} \\
\delta(X,Y) &= \inf\{\epsilon + \eta : (\epsilon,\eta) \in \mathcal{K}(X,Y)\} \\
e(X,Y) &= \log\delta(X,Y) \\
\tilde{\delta}(X,Y) &= 2\inf\{\epsilon : (\epsilon,\epsilon) \in \mathcal{K}(X,Y)\} \\
\tilde{e}(X,Y) &= \log\tilde{\delta}(X,Y).
\end{aligned}$$

Proposition 5 *δ and $\tilde{\delta}$ are metrics over $\mathbf{L}^0(\Omega,\sigma)$; e and $\tilde{e}$ are $\oplus$-metrics over $\mathbf{L}^0(\Omega,\sigma)$ with:*

$$e(X,Y) \le \tilde{e}(X,Y) \le c \odot e(X,Y) \qquad (c = \log 2) \tag{3.1}$$

One can also introduce the $\mathbf{L}^p$-semirings for $0 < p \le +\infty$. For $X, Y \in \mathcal{L}^0(\Omega,\sigma)$ we write

$$\begin{aligned}
d_p(X,Y) &= \mathbb{E}\left(d(X,Y)^{(p)}\right)^{\left(\frac{1}{p}\right)}, \\
\mathcal{L}^p(\Omega,\sigma,\mathbb{P}) &= \left\{X \in \mathcal{L}^0(\Omega,\sigma) : d_p(X,\mathbb{0}) < +\infty\right\}.
\end{aligned}$$

where $a^{(p)} \overset{def}{=} p\,a$. For $p = +\infty$ we write

$$d_\infty(X,Y) = \inf\{m \geq \mathbb{0} : \mathbb{P}(\{\omega \in \Omega : d(X(\omega),Y(\omega)) \geq m\}) = \mathbb{0}\}$$

$$\mathcal{L}^\infty(\Omega,\sigma,\mathbb{P}) = \{X \in \mathcal{L}^0(\Omega,\sigma) : d_\infty(X,\mathbb{0}) < +\infty\}.$$

For every $0 < p \leq +\infty$, we state ([7] or [8]) that:

$$X = Y \quad \mathbb{P}\text{-a.s.} \quad \Longleftrightarrow \quad d_p(f,g) = \mathbb{0}.$$

If $\mathbf{L}^p(\Omega,\sigma,\mathbb{P}) = \mathcal{L}^p(\Omega,\sigma,\mathbb{P})/\mathbb{P}$-a.e., then, for every $0 < p \leq +\infty$, d_p is a $\oplus$-metric over $\mathbf{L}^p$. Keeping in mind the notations of Proposition 1 we obtain

$$d_p(Q_m(a,r),m) = d_1(Q_m(a\,p^{\frac{1}{r}},r),m).$$

Moreover, if $X,Y \in L^\infty$, then $d_p(X,Y)$ is an increasing sequence in $\mathcal{A}$ that converges to $d_\infty(X,Y)$ when p goes to $+\infty$. The next theorem gives an exhaustive list of properties which lead to useful conclusions because they make explicit the relationship between the latter $\oplus$-metrics and the expectation

As usual, for every $a \in \mathcal{A}, p > 0$ we write $a^{(p)} = p\,a$, and $\mathbf{L}^p = \mathbf{L}^p(\Omega,\sigma,\mathbb{P})$.

Theorem 1

1. *For every $p > 0$, $X \in \mathbf{L}^p$*

$$\mathbb{E}\left(X^{(p)}\right)^{\left(\frac{1}{p}\right)} \leq d_\infty(X,\mathbb{0}).$$

2. **Bienaymé–Chebyshev inequality** $\forall X \in \mathbf{L}^p$, $p > 0$, $\epsilon \geq \mathbb{0}$

$$\mathbb{P}(X \geq \epsilon) \odot \epsilon^{(p)} \leq \mathbb{E}\left(X^{(p)}\right).$$

3. *Let g be an increasing function from $\mathcal{A}$ into $\mathcal{A}$. For every $a \geq \mathbb{0}$, $X \in \mathbf{L}^0$, we have*

$$g(a) \odot \mathbb{P}(\{\omega \in \Omega : X(\omega) \geq a\}) \leq \mathbb{E}(g(X)) \leq d_\infty(g(X),\mathbb{0}) \oplus g(a).$$

4. **Hölder inequality** *For every $0 < p \leq q < \infty$, $0 < n < +\infty$ such that $\frac{1}{p} + \frac{1}{q} = \frac{1}{n}$, $X \in \mathbf{L}^p$ and $Y \in \mathbf{L}^q$, we have $X \odot Y \in \mathbf{L}^n$ and*

$$\mathbb{E}\left((X \odot Y)^{(n)}\right)^{\left(\frac{1}{n}\right)} \leq \mathbb{E}\left(X^{(p)}\right)^{\left(\frac{1}{p}\right)} \odot \mathbb{E}\left(Y^{(q)}\right)^{\left(\frac{1}{q}\right)}.$$

5. **Minkowski inequality** *For every $0 < p < +\infty$, $X,Y \in \mathbf{L}^p$ we have $X \oplus Y \in \mathbf{L}^p$, and*

$$\mathbb{E}\left((X \oplus Y)^{(p)}\right)^{\left(\frac{1}{p}\right)} = \mathbb{E}\left(X^{(p)}\right)^{\left(\frac{1}{p}\right)} \oplus \mathbb{E}\left(Y^{(p)}\right)^{\left(\frac{1}{p}\right)}.$$

6. For every $X, Y \in L^1$, we have

$$d\left(\mathbb{E}(X), \mathbb{E}(Y)\right) \leq \mathbb{E}\left(d(X, Y)\right)$$
$$\tilde{e}(X, Y) \leq c \odot \mathbb{E}\left(d(X, Y)\right) \qquad \text{where } c = \log 2.$$

Consequently, for every $0 \leq p \leq +\infty$, $\mathbf{L}^p$ is a semiring, in other words $(\mathbf{L}^p\left(\Omega, \sigma, \mathbb{P}\right), \oplus, \mathbb{0})$ is a semi-group.

In view of the results stated in Proposition 1, we immediately obtain the following corollary.

Corollary 1 *All notations of Proposition 1 are in force. Let X^N be a sequence of real-valued optimization variables whose performance density satisfies $\mathbb{p}_N(x) \leq -\frac{1}{p}\left|\frac{x-m}{a_N}\right|^p$ for some real number m, $p > 0$ and $\lim_{N \to +\infty} a_N = 0$. Then, for N large enough, for every $\epsilon > \mathbb{0}$ and $c = \log 2$, we have*

$$\mathbb{P}\left(\left\{\omega \in \Omega : d\left(X^N(\omega), m\right) > \epsilon\right\}\right) \leq \frac{\mathbb{E}\left(d\left(Q_m(a_N, p), m\right)\right)}{\epsilon}$$

$$\leq \frac{m}{\epsilon} \odot d\left(\frac{p\, a_N}{p - a_N}, \mathbb{1}\right) \xrightarrow[N \to +\infty]{} \mathbb{0}$$

$$d\left(\mathbb{E}(X^N), m\right) \leq m \odot d\left(\frac{p\, a_N}{p - a_N}, \mathbb{1}\right) \quad \text{and} \quad \tilde{e}(X^N, m) \leq c \odot d\left(\frac{p\, a_N}{p - a_N}, \mathbb{1}\right).$$

Most of these results are taken from Del Moral [7], but they were published for the first time in Del Moral [8] and Del Moral–Rigal–Salut [9]. Other different and independent topological results can be found in Akian–Quadrat–Viot [1] or Bellalouna [3].

4   Convergence modes

One problem in the theory of performances is the determination of the asymptotic properties of optimization variables. In this section we concentrate on the clarification of the underlying concepts. We start with a simple problem. Suppose we wish to study the behaviour of a sequence of performance convolutions $\mathbb{p}_1 \circledast \mathbb{p}_2 \circledast \cdots \circledast \mathbb{p}_n$. This problem is also related to asymptotic studies of the solution of a certain Bellman equation (see [4], [20], [25], [32], [33], [34]). We claim that it is natural to analyse such equations by means of an appropriate optimization variable sequence. Indeed, let $(X^n)_{n \geq 1}$ be a sequence of independent optimization variables, and $\mathbb{p}_n$ the performance law of X^n.

Let S_n be the sequence defined by $S_n = \sum_{i=1}^{n} X^i$, and $\mathbb{p}^n$ its performance law. According to the latter results

$$\mathbb{p}^n = \mathbb{p}_1 \circledast \mathbb{p}_2 \circledast \cdots \circledast \mathbb{p}_n$$

For instance, let $X_k \stackrel{def}{=} \Delta X_0 + \sum_{l=1}^{k} \Delta X_l$, where ΔX_l denotes $k+1$ independent optimization variables whose performance laws are defined by:

$$\mathbb{p}^{\Delta X_l}(z) = -\frac{1}{q}\left|\frac{z - op(\Delta X_l)}{\sigma_l}\right|^q \qquad \sigma_l > 0, \ q \geq 2.$$

Then, using the properties of the Fenchel transform, for $\frac{1}{p} + \frac{1}{q} = 1$

$$\mathbb{p}^{X_k}(z) = \mathbb{p}^{X_0} \circledast \mathbb{p}^{\Delta X_1} \circledast \cdots \circledast \mathbb{p}^{\Delta X_k}$$
$$\Rightarrow$$
$$\mathbb{p}^{\frac{X_k}{k+1}}(z) = -\frac{(k+1)^{\frac{q}{p}}}{q}\left|\frac{z - \Delta \overline{X}_k}{\overline{\sigma}_k}\right|^q$$

and $\frac{X_k}{k+1}$ has the same performance law as $Q_{\Delta \overline{X}_k}\left(\frac{\overline{\sigma}_k}{(k+1)^{\frac{1}{p}}}, q\right)$ where

$$\Delta \overline{X}_k = \frac{1}{k+1}\sum_{l=0}^{k} op(\Delta X_l) = op\left(\frac{1}{k+1}\sum_{l=0}^{k}\Delta X_l\right) \qquad \overline{\sigma}_k = \left(\frac{1}{k+1}\sum_{l=0}^{k}\sigma_l^q\right)^{\frac{1}{q}}.$$

Consequently, if $\Delta \overline{X}_\infty \stackrel{def}{=} \lim_{k \to +\infty} \Delta \overline{X}_k < +\infty$ and $\lim_{k \to +\infty} \overline{\sigma}_k < +\infty$, then $\frac{1}{k+1}X_k$, $\mathcal{L}^1$-converges to $\Delta \overline{X}_\infty$. Indeed, on account of Proposition 1, if $c = \log 2$ and $a_k = \frac{\overline{\sigma}_k}{(k+1)^{\frac{1}{p}}}$ then

$$d_1\left(\frac{X_k}{k+1}, \Delta \overline{X}_\infty\right) \leq c \odot \left(d_1\left(\Delta \overline{X}_k, \Delta \overline{X}_\infty\right) \oplus d_1\left(\frac{X_k}{k+1}, \Delta \overline{X}_k\right)\right)$$
$$\leq c \odot \left(d_1\left(\Delta \overline{X}_k, \Delta \overline{X}_\infty\right) \oplus \Delta \overline{X}_k \odot d_1\left(\frac{q \, a_k}{q - a_k}, 1\right)\right)$$
$$\xrightarrow[k \to +\infty]{} \mathbb{0}$$

According to the Bienaymé–Chebyshev inequality, it is easy to bound the performance of the events associated with the nonconvergence. For instance, for every $\epsilon \geq \mathbb{0}$

$$\mathbb{P}\left(\left\{\omega \in \Omega : d\left(\frac{X_k(\omega)}{k+1}, \Delta \overline{X}_\infty\right) \geq \epsilon\right\}\right) \leq \frac{d_1\left(\frac{X_k}{k+1}, \Delta \overline{X}_\infty\right)}{\epsilon} \xrightarrow[k \to +\infty]{} \mathbb{0}$$

We then introduce various convergence modes involving sequences of optimization variables.

Definition 3 *Let $(X_n)_{n \geq 1}$ be a sequence of $\mathcal{A}$-valued optimization variables and X an $\mathcal{A}$-valued optimization variable.*

1. **Uniform convergence** $\lim_{n \to +\infty} \sup_{\omega \in \Omega} d\left(X_n(\omega), X(\omega)\right) = \mathbb{0}$

2. **P-convergence** *(e − P)*
   $\forall \epsilon > 0 \; \lim_{n \to +\infty} \mathbb{P}\left(\{\omega \in \Omega \; : \; d\left(X_n(\omega), X(\omega)\right) \geq \epsilon\}\right) = 0$

3. **L^∞-convergence** $\lim_{n \to +\infty} d_\infty\left(X_n, X\right) = 0$

4. **L^p-convergence** $(0 < p < +\infty)$ $\lim_{n \to +\infty} d_p\left(X_n, X\right) = 0$

5. **P-almost everywhere convergence** *(P-a.e.)*

$$\mathbb{P}\left(\left\{\omega \in \Omega \; : \; \limsup_{n \to +\infty} d\left(X_n(\omega), X(\omega)\right) > 0\right\}\right) = 0$$

6. **Weak convergence** $\forall \phi \in \mathcal{C}_0\left(\Omega\right)$

$$\lim_{n \to +\infty} \mathbb{E}\left(\phi(X_n)\right) = \mathbb{E}\left(\phi(X)\right).$$

Weak convergence was introduced by Maslov in [23]. Futher developments can be found in Kolokolsov–Maslov [24], [25], [26]. We introduce below the uniform integrability of a class of functions in $\mathbf{L}^1$.

Definition 4 *Let $\mathcal{H} \subset L^1\left(\Omega, \sigma, \mathbb{P}\right)$. $\mathcal{H}$ is called uniformly integrable whenever the integrals*

$$\int_{\{\omega \in \Omega \; : \; X(\omega) \geq c\}}^{\oplus} X \odot \mathbb{P} \quad \forall X \in \mathcal{H} \tag{4.1}$$

uniformly converge to 0 when $c \geq 0$ goes to $+\infty$.

These classes can be characterized as follows.

Proposition 6 *Let $\mathcal{H} \subset L^1\left(\Omega, \sigma, \mathbb{P}\right)$. $\mathcal{H}$ is uniformly integrable if, and only if, the Maslov expectations $\mathbb{E}(X)$, $X \in \mathcal{H}$, are uniformly bounded and for every $\epsilon > 0$, there exists $\delta > 0$ such that*

$$\forall A \in \sigma \quad \mathbb{P}(A) \leq \delta \implies \int_A^{\oplus} X \odot \mathbb{P} \leq \epsilon \quad \forall X \in \mathcal{H}.$$

Moreover, let G be a $\mathcal{A}$-valued function. If

$$\lim_{t \to +\infty} \frac{G(t)}{t}_\circ = +\infty \quad \text{and} \quad \sup_{X \in \mathcal{H}} \mathbb{E}\left(G(X)\right) < +\infty$$

then $\mathcal{H}$ is uniformly integrable. Other topological results such as the dominated convergence theorem may be found in references [7], [8] or [9]. The following theorem gives an exhaustive list of comparisons between the different convergence modes.

Theorem 2 *Let* X, X_n *be a sequence of optimization variables defined on the same performance space* $(\Omega, \sigma, \mathbb{P})$.

1. *For* $0 < p \leq q \leq +\infty$, $\mathbf{L}^q$*-convergence implies* $\mathbf{L}^p$*-convergence and the* $\mathbb{P}$*-convergence.* $\mathbb{P}$*-convergence implies* $\mathbb{P}$*-a.e.-convergence.*

2. *If, for every* $\nu > 0$, *the sequence* $d\left(X_{\nu+n}, X_n\right)$ $\mathbb{P}$*-converges to* $\mathbb{0}$, *then* X_n $\mathbb{P}$*-converges.*

3. X_n $\mathbb{P}$*-converges to* X *if, and only if,* $e(X_n, X)$ *or*
$$\mathbb{P}\left(\bigcup_{m=n}^{+\infty} \{d(X_m, X) \geq \epsilon\}\right) \text{ converges to } \mathbb{0}, \text{ for every } \epsilon > \mathbb{0}. \text{ Moreover, if}$$
$\{X_n\} = \mathcal{H}$ *is uniformly integrable then* X_n $\mathbf{L}^1$*-converges.*

In point 1, sentence 1 is an immediate consequence of Bienaymé–Chebyshev inequality. Sentence 2 is stated in Bellalouna [3] and proved in Del Moral [8]. The reverse implications are not true (see [8] or [7]). As already remarked, optimization problems involving independent optimization variables are useful in the time discrete case; the continuous case will be dealt with later on. In other words the useful case is typically a sequence $X^1, \ldots, X^N, \ldots$ of mutually independent optimization variables. Then, because of the independence property, the induced performance density is the convolution of the individual optimization variables. The key to studying the sums of these independent variables is that the partial sum performances $\mathbb{p}_N$ must be regular, in the sense that, there exists $p > 0$ and some sequences $a_N > 0, m_N$, such that $(\frac{1}{q} + \frac{1}{p} = 1)$

$$\mathbb{p}_N(x) \leq -\frac{1}{p}\left|\frac{x - m_N}{a_N}\right|^p \text{ and } \lim_{N \to +\infty} \frac{m_N}{N} = m \text{ and } \lim_{N \to +\infty} \frac{a_N}{N^{\frac{1}{q}}} < +\infty.$$

In that case, the performance $\mathbb{p}^N$ of the normalized sums $S_N = \frac{1}{N}\sum_{i=1}^N X^i$ satisfies (see Proposition 1 and Corollary 1):

$$\mathbb{p}^N(x) = \mathbb{p}_N(N\,x) \leq -\frac{1}{p}\left|\frac{x - m'_N}{a'_N}\right|^p \implies \lim_{N \to +\infty} d_1\left(S_N, m\right) = \mathbb{0}$$

with $m'_N = \frac{m_N}{N}$ and $a'_N = \frac{a_N}{N}$. These facts may be summarized as follows:

Theorem 3 *Let* $(X^i)_{i \geq 1}$ *be a sequence of real-valued optimization variables on* $(\Omega, \sigma, \mathbb{P})$ *whose optimal states are well defined. We assume they are* $\mathbb{P}$*-independent. Assume*

$$\lim_{N \to +\infty} \frac{1}{N}\sum_{i=1}^N op\left(X^i\right) = \lim_{N \to +\infty} op\left(\frac{1}{N}\sum_{i=1}^N X^i\right) \overset{def}{=} \overline{X} < +\infty$$

for every $i \geq 1$, $\mathbb{p}^{X^i} \in \mathcal{S}_+$, $\mathcal{F}\mathbb{p}^{X^i} \in C^2(\mathbb{R})$ and that there exist two reals $a > 0$, $r > 1$ such that for every $N \geq 1$, $\lambda \in [0,1]$, $x \in \mathbb{R}$

$$\frac{1}{2}x\left(\frac{1}{N}\sum_{i=1}^{N}\left|\left(\mathcal{F}\mathbb{p}^{X^i}\right)''\right|(\lambda x)\right)x \leq \frac{1}{r}(xax)^{\frac{r}{2}} \tag{4.2}$$

Then, for every $p > 0$, $\delta > 0$,

$$\frac{1}{N}\sum_{i=1}^{N}X^i\xrightarrow[N\to+\infty]{\mathbf{L}^p}\overline{X} \qquad \frac{1}{N^{\delta+1}}\sum_{i=1}^{N}X^i\xrightarrow[N\to+\infty]{\mathbf{L}^p}0 \tag{4.3}$$

Before proving the theorem we giv an equivalent formulation of these conditions in terms of the performances densities. If

$$\overline{X}^i = X^i - op(X^i) \qquad \text{and} \qquad \bigoplus_{i=1}^{+\infty}\mathbb{p}^{\overline{X}^i}(x) \leq -\frac{1}{r'}(xax)^{\frac{r'}{2}} \tag{4.4}$$

then for every $i \in \mathbb{N}$ we obtain (equation (2.4)):

$$\mathcal{F}\mathbb{p}^{\overline{X}^i}(x^*) \geq -\frac{1}{r}\left|x^*a^{-1}x^*\right|^{\frac{r}{2}} \qquad \text{with} \qquad \frac{1}{r}+\frac{1}{r'}=1$$

Because of the properties of the Fenchel transform (Proposition 4), for every $i \geq 1$, $x^* \in R$ there exists $\lambda^i(x^*) \in [0,1]$ such that

$$\frac{1}{N}\sum_{i=1}^{N}\left|\mathcal{F}\left(\mathbb{p}^{\overline{X}^i}\right)\right|(x^*) = \frac{1}{2}x^*\frac{1}{N}\sum_{i=1}^{N}\left|\mathcal{F}\mathbb{p}^{\overline{X}^i}\right|''\left(\lambda^i(x^*)x^*\right)x^* \leq \frac{1}{r}\left|x^*a^{-1}x^*\right|^{\frac{r}{2}}$$

Proof of Theorem 3 For every $N \geq 1$ we denote by $\mathbb{P}^N$ and $\mathbb{P}^{S_N}$ the performance measures of Σ_N and S_N. The optimization variables X^i being $\mathbb{P}$-independent:

$$\mathbb{p}^N = \mathbb{p}^{X^1}\circledast\cdots\circledast\mathbb{p}^{X^N} \Rightarrow \mathcal{F}\left(\mathbb{p}^N\right) = \bigodot_{i=1}^{N}\mathcal{F}\left(\mathbb{p}^{X^i}\right)$$

Because of the properties of the Fenchel transform, for every $i \geq 1$, $x^* \in R$ there exists $\lambda^i(x^*) \in [0,1]$ such that

$$\mathcal{F}\left(\mathbb{p}^{X^i}\right)(x^*) = -op\left(X^i\right) + \frac{1}{2}x^*\left(\mathcal{F}\mathbb{p}^{X^i}\right)''\left(\lambda^i(x^*)x^*\right)x^*.$$

Then

$$\mathbb{p}^N(x) = \mathcal{F}\left(\odot_{i=1}^{N}\mathcal{F}\left(\mathbb{p}^{X^i}\right)\right)(x) = -\sup_{x^*\in\mathbb{R}}\left(x^*(x)+\sum_{i=1}^{N}\mathcal{F}\left(\mathbb{p}^{X^i}\right)(x^*)\right)$$

$$\mathbb{p}^{S_N}(x) = N\Big(-\sup_{x^*\in\mathbb{R}}\Big(x^*\Big(x-\frac{1}{N}\sum_{i=1}^{N}op\left(X^i\right)\Big)$$

$$-\frac{1}{2}x^*\left(\frac{1}{N}\sum_{i=1}^{N}\left(-\mathcal{F}\mathbb{p}^{X^i}\right)''\left(\lambda^i(x^*)x^*\right)\right)x^*\Big)\Big)$$

By the second assumption, there exist two real numbers $r > 1$, $a \geq 0$ such that

$$\mathbb{p}^{S_N}(x) \leq N\left(-\sup_{x^* \in \mathbb{R}}\left(x^*\left(x - \frac{1}{N}\sum_{i=1}^{N} op\left(X^i\right)\right) - \frac{1}{r}(x^* a x^*)^{\frac{r}{2}}\right)\right)$$

We introduce for every $s > 1$, $m \in \mathbb{R}$, $b > 0$,

$$\mathcal{Q}_s\left(m,b\right)(x) \overset{def}{=} -\frac{1}{s}\left((x-m)b(x-m)\right)^{\frac{s}{2}}$$

According to the involution property of Fenchel's transform, for $\frac{1}{r} + \frac{1}{s} = 1$

$$\mathbb{0} \leq \mathbb{p}^{S_N} \leq N\,\mathcal{Q}_s\left(\frac{1}{N}\sum_{i=1}^{N} op\left(X^i\right), a^{-1}\right) = \mathcal{Q}_s\left(\frac{1}{N}\sum_{i=1}^{N} op\left(X^i\right), a_N^{-1}\right)$$

with
$$\lim_{N \to +\infty} a_N = \lim_{N \to +\infty} \frac{a}{N^{\frac{2}{s}}} = 0.$$

Then (4.3) is an immediate consequence of Corollary 1 □

We now state a result which concerns the estimation of the optimal state through the $\oplus$-sum of independent variables.

Proposition 7 *Let X be a real-valued optimization variable whose performance density equals $\mathbb{1}$ at only one point $op(X) \in \mathbb{R}$. Let $(X^i)_{1 \leq i \leq N}$, $n \geq 1$, be N independent optimization variables which have the same performance law as X. Then*

$$\forall \epsilon > \mathbb{0} \quad \lim_{N \to +\infty} \mathbb{P}\left(d\left(\oplus_{i=1}^{N} \mathbb{p}^X\left(X^i\right), \mathbb{1}\right) > \epsilon\right) = \mathbb{0}. \tag{4.5}$$

Assume for every $\epsilon > \mathbb{0}$ there exists $\eta > \mathbb{0}$ such that $d\left(x, op(X)\right) \leq \epsilon$ whenever $d\left(\mathbb{p}^X(x), \mathbb{1}\right) \leq \eta$. In that case

$$\forall \epsilon > \mathbb{0} \quad \lim_{N \to +\infty} \mathbb{P}\left(d\left(op_N(X)\right), op(X)\right) > \epsilon) = \mathbb{0}. \tag{4.6}$$

where $op_N(X) \overset{def}{=} Arg\sup_{x \in \Omega_N} \mathbb{p}^X(x)$ and $\Omega_N \overset{def}{=} \{X^1, \ldots, X^N\}$.

Proof It is sufficient to remark that $\mathbb{P}\left(d\left(\mathbb{p}^X(X), \mathbb{1}\right) > \epsilon\right) < \mathbb{1}$ and, because of the independence of the X^i,

$$\mathbb{P}\left(d\left(\oplus_{i=1}^{N} \mathbb{p}^X\left(X^i\right), \mathbb{1}\right) > \epsilon\right) = \mathbb{P}\left(\forall i \in \{1, \ldots, N\} \,:\, d\left(\mathbb{p}^X\left(X^i\right), \mathbb{1}\right) > \epsilon\right)$$
$$= N\,\mathbb{P}\left(d\left(\mathbb{p}^X(X), \mathbb{1}\right) > \epsilon\right)$$

□

Theorem 3 was proved for the first time in a very particular case by Quadrat [32], and improved in Bellalouna [3] and in Akian–Quadrat–Viot [1]. Our result is stronger than the results of those papers because the optimization variables need not have the same performance. The coercivity assumptions (4.2) and (4.4) stated in Theorem 3 were introduced for the first time in Del Moral [8] and further developed in [7]. It should be underlined that in [1] Akian–Quadrat–Viot introduced other $\mathbf{L}^p$-norms that correspond to a different treatment of the optimization variable $Q_m(a,p)$.

5 Maslov Optimization Processes

After the above introduction of Maslov optimization axiomatics, an important step consists in introducing sequences of optimization variables indexed by a real subset. This leads to a lot of additional structures and therefore to deeper results. A E-valued optimization process with space time I is a system $\left(\Omega, \sigma, \mathbb{P}, X = (X_t)_{t \in I}\right)$ defined by an optimization basis $(\Omega, \sigma, \mathbb{P})$ and a family of E-valued optimization variables defined on this basis. Let $\mathbf{F} = \{\mathcal{F}_t\}_{t \geq 0}$ be an increasing sequence of σ-algebras spanned by optimization variables. An E-valued optimization $\mathbf{F}$-martingale (resp. sub-martingale, super-martingale) M is an optimization process such that, for every $s \leq t$, $\mathbb{E}(M_t/\mathcal{F}_s) = M_s$ (resp. $\geq$, $\leq$). We denote by $\mathbb{L}og(\mathbf{F})$ the $\mathcal{A}$-semimodule of optimization martingales M such that $\mathbb{E}(M_0) = 1$. We stated in Del Moral [7], or [11], the (max,+)-version of Doob's inequalities and up-crossing lemma. These results lead to new developments in the field of qualitative studies of optimization processes ([10], [7]) mainly because they exhibit explicit bounds of the cost function over certain classes of optimization variables. Bellman–Hamilton–Jacobi theory is introduced here in forward time (with initial penalty) in order to show the central role played by the semi-group of optimal transition performances in the Bellman optimality principle. We claim here that this principle may be viewed as a basic definition of the optimization process rather than a deductive conclusion. A Maslov optimization process is an optimization process X that satisfies the (max,+)-version of Markov's causality principle. A reversal of time yields optimal control processes of regulation type. In other words X is a **Maslov process** when the past of X is $\mathbb{P}$-independent of the future when the present value is known. In that case its performance transitions satisfy the Bellman optimality principle:

$$\mathbb{p}_{t/r}^X(z/x) = \int_E^{\oplus} \mathbb{p}_{t/s}^X(z/y) \odot \mathbb{p}_{s/r}^X(y/x) \odot dy$$

with $\mathbb{p}_{\tau_1/\tau_2}^X \overset{def}{=} \mathbb{p}^{X_{\tau_1}/X_{\tau_2}}$, $\tau_1 \leq \tau_2$. The **Bellman optimality principle** is then the (max,+)-version of the **Chapman–Kolmogorov** equation for Markov stochastic processes. Let L be an upper-semicontinuous function from $\mathbb{R}^n \times \mathbb{R}^n$ into $\mathcal{A}$, L_0 an upper-semicontinuous function from $\mathbb{R}^n$ into $\mathcal{A}$.

For every $x \in \mathbb{R}^n$ we assume that $L(x, .)$ and L_0 are performance densities. Let $\Omega = \mathbb{R} \times (\mathbb{R}^n)^{[0,T]}$, $[0,T] \subset \mathbb{R}$, endowed with the uniform convergence, let σ be its Borel sigma-algebra and $\mathbb{P}$ a performance measure over Ω. Assume an optimization variable X_0 and an optimization process U are defined on this performance basis. Let $X = (X_t)_{t \in [0,T]}$ be the optimization process defined by $\dot{X}_t = F(X_t, U_t)$ (X_0 initial condition) where F satisfies the usual Lipschitz and boundeness conditions. Whenever X is a $\mathcal{C}([0,T], \mathbb{R}^n)$-valued optimization variable whose performance law $\mathbb{p}^X$ is the upper-semicontinuous function defined by

$$\mathbb{p}^X(x) \overset{def}{=} \sup \left\{ \mathbb{p}^{X_0, U}(x_0, u) \, / (x_0, u) \in \Omega \; : \; X(x_0, u) = x \right\}$$

$$= \sup \left\{ L_0(z_0) + \int_0^T L(x_\tau, u_\tau) \, d\tau \, /(z_0, u) \in \Omega, \; \dot{x} = F(x, u), \; x_0 = z_0 \right\}$$

if x is absolutely continuous, $\mathbb{0}$ otherwise, then one can easily check that X is a Maslov process. In that case we say that X is an $\mathbb{R}^n$-valued (F, L)-**Maslov process**.

We denote in the sequel $\mathbf{F}^X = \left\{ \mathcal{F}_t^X \right\}_{t \in [0,T}$ where $\mathcal{F}_t^X = \sigma(X_0, \ldots, X_t)$. We introduce next the Hamiltonian function associated with X. This function is essential to give the $(\max, +)$-version of the Kolmogorov operator and Dynkin formula of Markov stochastic processes. We denote for every $\xi, x, u \in \mathbb{R}^n$,

$$H^X(\xi, x, u) = \xi' F(x, u) + L(x, u) \qquad H^X(\xi, x) = \sup_{u \in \mathbb{R}} H^X(\xi, x, u).$$

Then for every differentiable function ϕ

1. $\phi(X_t) - \phi(X_0) - \int_0^t \mathcal{H}^X(\phi)(X_\tau) \, d\tau \in \mathbb{Log}\left(\mathbf{F}^X\right)$

2. $\mathbb{E}\left(\phi(X_t) - \int_0^t \mathcal{H}^X(\phi)(X_\tau) \, d\tau \, / \mathcal{F}_0^X\right) = \phi(X_0)$ (Dynkin's formula)

where $\mathcal{H}^X(\phi)(x) = H^X\left(x, \frac{d\phi}{dx}(x)\right)$ is the Hamilton–Jacobi operator associated with X. Whenever ϕ is time dependent, we obviously obtain the same equations with the operator ∂_t. In [29], Maslov and Samborskii introduce idempotent Sobolev spaces in order to study Hamilton–Jacobi equations as linear objects. These formulas allow the construction of optimization martingales as functions of the state X_t. The following theorem is an easy consequence of Doob's inequalities in our framework ([7] or [10]).

Theorem 4 *Let X be an $\mathbb{R}^n$-valued (F, L)-Maslov process and $\phi \in \mathcal{C}^1(\mathbb{R}^n)$.*

1. *If $\mathcal{H}^X(\phi) = 0$ then $\phi(X_t)$ is an optimization $\mathbf{F}^X$-martingale. For instance, for every $s \leq t$, if $\Omega_{s,x} = \{\omega \in \Omega \; : \; X_s(\omega) = x\}$*

$$\mathbb{E}\left(\phi(X_t) \, / \mathcal{F}_s^X\right) = \phi(X_s) = \sup_{(x_0, u) \in \Omega_{s, X_s}} \phi(x_t) + \int_s^t L(x_\tau, u_\tau) \, d\tau.$$

2. If $\mathcal{H}^X(\phi) \leq 0$ then $\phi(X_t)$ is an optimization $\mathbf{F}^X$-super-martingale and, for every $x \in \mathbb{R}$, $a \in \mathcal{A}$,

$$\sup_{(x_0,u) \in \Omega_{a,x}} \left(\int_0^T L\left(x_\tau, u_\tau\right) \right) \leq \frac{\phi(x)}{a} \circ .$$

where $\Omega_{a,x} = \left\{ \omega \in \Omega : X_0(\omega) = x, \ \sup_{0 \leq t \leq T} \phi\left(X_t(\omega)\right) \geq a \right\}$. Moreover, for every $p \geq 1$, $a \leq b$, if $M_{\phi(X)}\left([a,b],\omega\right)$ denotes the number of up-crossings in the interval $[a,b]$, then

$$\sup_{(x_0,u) \in \Omega_{[a,b],x}} \left(\int_0^T L\left(x_\tau, u_\tau\right) \right) \leq \left(\frac{a}{b} \circ\right)^{(p-1)} \odot \frac{\phi(x)}{b} \circ$$

where $\left(\frac{a}{b} \circ\right)^{(p-1)} = (p-1)(a-b)$, and

$$\Omega_{[a,b],x} = \left\{ \omega \in \Omega : X_0(\omega) = x, \ M_{\phi(X)}\left([a,b],\omega\right) \geq p \right\}.$$

Consequently, when $T = \infty$, there exists an optimization variable, denoted by $\phi(X_\infty)$, such that

$$\phi(X_t) \xrightarrow[t \to +\infty]{} \phi(X_\infty) \qquad \mathbb{P}\text{-}a.e.$$

3. Assume $\mathcal{H}^X(\phi) \leq -b < 0$ with $b > 0$, then, with the same notations as in 2,

$$\sup_{(x_0,u) \in \Omega_{a,x}} \left(\int_0^T L\left(x_\tau, u_\tau\right) \right) \leq \frac{\phi(x)}{a \odot b\, T} \circ \left(\xrightarrow[T \to +\infty]{} \mathbb{0} \right)$$

Consequently there exists some $T \gg 0$ such that

$$\sup_{0 \leq t \leq T} \phi\left(op\left(X_t/X_0 = x\right)\right) \leq a.$$

The material here comes from Del Moral–Salut [14]. The idea of introducing the (max,+)-version of the Markov causality principle was first introduced by Huillet–Salut in [20], [21]. The general presentation of such processes can be founded in Del Moral–Rigal–Salut [9], Del Moral [11] and [7]. The analogues of Dynkin's formula and Doob's inequalities and up-crossing lemma were published for the first time in Del Moral [11]. Some applications of these results are presented in Del Moral [12].

6   Maslov and Markov Processes

This section constitutes the final step on our way to the topological aspects of Maslov optimization theory. The main purpose of this section is to show that Maslov optimization processes and Markov stochastic processes can be mapped into each other via various transformations. We introduce some transformations between performance and probability measures which make clear the relationships between optimization and estimation problems.

6.1   The Wentzell–Freidlin transform

This subsection is devoted to two closely related kinds of result. One is the integral extension of the following equalities:

$$a \oplus b = \lim_{\epsilon \to +\infty} \epsilon \log \left(e^{\frac{a}{\epsilon}} + e^{\frac{b}{\epsilon}} \right) \quad \text{and} \quad a \odot b = \epsilon \log \left(e^{\frac{a}{\epsilon}} . e^{\frac{b}{\epsilon}} \right) \qquad a, b \in \mathbb{R}.$$

The first place where this study appears is Pontryagin–Andronov–Vitt [31], it is further developed in Friedlin–Wentzell [36] and Hijab [18]. It is well known that the study of various limit theorems for random processes is motivated by dynamical systems subject to the effect of random perturbations that are small compared with the deterministic constituents of the motion. In order to study the effect of perturbations on large time intervals we have to be able to estimate the probability of rare events. The so-called Wentzell-Freidlin transform provides a way of computing the probability of rare events in terms of performance measures. Roughly speaking, for some random variable sequence X^ϵ and some optimization variable X taking values in the same measurable space, we have

$$\forall A \in \mathcal{E} \qquad \epsilon \log P\left(X^\epsilon \in A \right) \overset{\epsilon \sim 0}{\approx} \mathbb{P}\left(X \in A \right). \tag{6.1}$$

This investigation also includes results analogous to the laws of large numbers and the central limit theorem. The second motivation is to introduce an asymptotic mapping between conditional Markov processes and conditional Maslov processes. In other words our interest is in large deviation results for conditional measures and related asymptotics of the filtering equations. The topology used in this mapping is the Prohorov topology of probabilities, which is equivalent to the weak topology. More precisely, we show that the conditional expectation weakly converges to the conditional optimal state. Maslov optimization theory allows a very tractable description of these results. We go by the shortest possible route, thus leaving aside a large number of properties. We assume the reader is familiar with the basics facts about nonlinear filtering and, in particular, with the fundamentals of the so-called change of measure approach. For simplicity we shall deal with the simplest case. First we consider a complete probability basis $(\Omega, \mathcal{F}_T, P)$, $T > 0$, and an increasing filtration $(\mathcal{F}_t)_{t \in [0,T]}$ on which are defined two independent and real-valued Wiener processes W and V and an independent real-valued random variable X_0^ϵ whose probability distribution μ_0^ϵ is given by

$$\mu_0^\epsilon(dx) = C_\epsilon \exp\left(\frac{1}{\epsilon} S_0(x) \right) dx$$

where C_ϵ is a positive normalization constant and S_0 a Lipschitz concave Maslov performance density such that $S_0\left(\overline{X}_0 \right) = 0$ and $S_0\left(x \right) < 0$ whenever $x \neq \overline{X}_0$.

We choose two real-valued functions f and g such that the following equations have a strong solution on $(\Omega, \mathcal{F}_T, P)$.

$$\begin{cases} dX_t^\epsilon = f(X_t^\epsilon)\, dt + \sqrt{\epsilon}\, dW_t \\ X^\epsilon(0) = X_0^\epsilon \end{cases} \quad \text{and} \quad \begin{cases} dY_t^\epsilon = h(X_t^\epsilon)\, dt + \sqrt{\epsilon}\, dV_t \\ Y^\epsilon(0) = 0. \end{cases}$$

We define $\Omega_t \overset{def}{=} C([0,T], \mathbb{R})$, $\Omega_{0,t} \overset{def}{=} \{\eta \in \Omega_t : \eta(0) = 0\}$, $\Omega_0 = \Omega_{0,T}$ and equip these spaces with the uniform topology. In the sequel $\mathbf{B}(\mathcal{X})$ denotes the Borel sigma-algebra of a topological space $\mathcal{X}$. The following theorem gives a functional integral representation for the conditional expectation, which is known as the Kallianpur–Striebel formula.

Theorem 5 *Let φ be bounded measurable function from Ω_T into $\mathbb{R}$. Then P-a.e.*

$$E\left(\varphi(X^\epsilon)\,/\mathcal{F}_T^{Y^\epsilon}\right) = \frac{\int \varphi(\theta)\, Z^\epsilon(\theta, Y^\epsilon)\, P^{X^\epsilon}(d\theta)}{\int Z^\epsilon(\theta, Y^\epsilon)\, P^{X^\epsilon}(d\theta)}$$

$$= \frac{E_0\left(\varphi(X^\epsilon)\, Z^\epsilon(X^\epsilon, Y^\epsilon)\,/\mathcal{F}_T^{Y^\epsilon}\right)}{E_0\left(Z^\epsilon(X^\epsilon, Y^\epsilon)\,/\mathcal{F}_T^{Y^\epsilon}\right)}$$

where $\quad \epsilon \log Z^\epsilon(\theta, \eta) \overset{def}{=} \int_0^T h(\theta_\tau)\, d\eta_\tau - \frac{1}{2}\int_0^T h(\theta_\tau)^2\, d\tau$

and $E_0(\cdot)$ is the expectation associated with the probability measure P_0 defined by

$$\frac{dP}{dP_0}\Big/_{\mathcal{F}_T} = Z^\epsilon(X^\epsilon, Y^\epsilon).$$

Using Ito's integration by parts formula,

$$h(X_t^\epsilon)\, dY_t^\epsilon = d\left(h(X_t^\epsilon)\, Y_t^\epsilon\right) - Y_t^\epsilon\,(\mathcal{L}^\epsilon h)(X_t^\epsilon)\, dt - Y_t^\epsilon\left(\frac{\delta h}{\delta x}\right)(X_t^\epsilon)\,\sqrt{\epsilon}\, dW_t$$

where $\mathcal{L}^\epsilon$ is the Kolmogorov operator associated with the diffusion X^ϵ. Then

$$\epsilon \log Z^\epsilon(X^\epsilon, Y^\epsilon)$$
$$= F^\epsilon(X^\epsilon, Y^\epsilon) - \int_0^T \sqrt{\epsilon} Y_t^\epsilon\left(\frac{\delta h}{\delta x}\right)(X_t^\epsilon)\, dW_t - \frac{1}{2}\int_0^T \left(Y_t^\epsilon\left(\frac{\delta h}{\delta x}\right)(X_t^\epsilon)\right)^2 dt$$

with

$$V^\epsilon(x,y) = \frac{1}{2}h^2(x) + y\left(f(x)\frac{\delta h}{\delta x}(x) + \frac{1}{2}\epsilon\frac{\delta^2 h}{\delta x^2}(x)\right) - \frac{1}{2}\left(y\frac{\delta h}{\delta x}(x)\right)^2$$
$$F^\epsilon(\theta, \eta) = h(\theta_T)\eta_T - \int_0^T V^\epsilon(\theta_t, \eta_t)\, dt.$$

Using Girsanov's theorem we obtain

$$E\left(\varphi\left(X^{\epsilon}\right)/\mathcal{F}_{T}^{Y^{\epsilon}}\right) = \frac{\int_{\Omega_{T}} \varphi\left(\theta\right) \; exp\left(\frac{1}{\epsilon}F^{\epsilon}\left(\theta,Y^{\epsilon}\right)\right) \; \check{P}_{[Y^{\epsilon}]}^{\epsilon}(d\theta)}{\int_{\Omega_{T}} exp\left(\frac{1}{\epsilon}F^{\epsilon}\left(\theta,Y^{\epsilon}\right)\right) \; \check{P}_{[Y^{\epsilon}]}^{\epsilon}(d\theta)}$$

where $\check{P}_{[Y^{\epsilon}]}^{\epsilon}(d\theta)$ is the distribution on Ω_{T} of the diffusion:

$$d\check{X}_{t}^{\epsilon} = \left(f\left(\check{X}_{t}^{\epsilon}\right) - Y_{t}^{\epsilon}\frac{\delta h}{\delta x}\left(\check{X}_{t}^{\epsilon}\right)\right) dt + \sqrt{\epsilon}dW_{t} \qquad \text{with} \quad \check{X}_{0}^{\epsilon} = X_{0}^{\epsilon}.$$

The term F^{ϵ} does not involve stochastic integration and thus is well defined for all $\eta \in \Omega_{0}$ and not only on a set of ϵ-Wiener measure equal to 1, so that the final estimates can be made uniform in the parameter η over a compact subset in Ω_{0}. Futher it depends continuously on $\eta \in \Omega_{0}$. These properties are inherited by the measures defined as follows ($A \in \mathbf{B}\left(\Omega_{T}\right)$, $B \in \mathbf{B}\left(\mathbb{R}\right)$, $\eta \in \Omega_{0}$, $t \in [0,T]$):

$$\sum{}^{\epsilon}\left(A\right)\left(\eta\right) = \int_{A} e^{\frac{1}{\epsilon}F^{\epsilon}\left(\theta,\eta\right)} \check{P}_{[\eta]}^{\epsilon}(d\theta) \qquad \sigma_{t}^{\epsilon}(B)(\eta) = \sum{}^{\epsilon}\left(\{\theta \in \Omega_{T} : \theta_{t} \in B\}\right)(\eta)$$

$$\prod{}^{\epsilon}\left(A\right)\left(\eta\right) = \frac{\sum{}^{\epsilon}\left(A\right)\left(\eta\right)}{\sum{}^{\epsilon}\left(\Omega_{T}\right)\left(\eta\right)} \qquad \pi_{t}^{\epsilon}(B)(\eta) = \frac{\sigma_{t}^{\epsilon}(B)(\eta)}{\sigma_{t}^{\epsilon}\left(\mathbb{R}\right)\left(\eta\right)}$$

Let $(\Omega,\mathcal{F}_{T},\mathbb{P})$ be a performance space with a filtration $(\mathcal{F}_{t})_{t\in[0,T]}$ on which are defined two real-valued and $\mathbb{P}$-independent optimization processes U^{X} and U^{Y}. We also define a real-valued optimization variable X_{0} with performance distribution $\mathbb{p}_{0}^{X} = S_{0}$. Assume, for every $(x,u,v) \in \mathbb{R} \times \Omega_{T} \times \Omega_{T}$

$$\mathbb{p}^{X_{0},U^{X},U^{Y}}\left(x,u,v\right) = \begin{cases} S_{0}(x) - \frac{1}{2}\int_{0}^{T}\|u_{t}\|^{2}\,dt - \frac{1}{2}\int_{0}^{T}\|v_{t}\|^{2}\,dt \\ \quad \text{(whenever the integrals exist)} \\ \mathbb{0} \qquad \text{(otherwise)}. \end{cases}$$

We consider also the pair of real-valued optimization processes defined by:

$$\dot{X}_{t} = f\left(X_{t}\right) + U_{t}^{X}, \quad X(0) = X_{0} \qquad \text{and} \qquad \dot{Y}_{t} = h\left(X_{t}\right) + U_{t}^{Y}, \quad Y(0) = 0.$$

Let $\mathbb{p}^{X,Y}$ be the performance density on $(\Omega_{T} \times \Omega_{0},\mathbf{B}\left(\Omega_{T}\right) \otimes \mathbf{B}\left(\Omega_{0}\right))$ of the pair of optimization variables (X,Y). For every $(x,y) \in \Omega_{T} \times \Omega_{0}$

$$\mathbb{p}^{X,Y}\left(x,y\right) = S_{0}(x) - \frac{1}{2}\int_{0}^{T}\|\dot{x}_{t} - f\left(x_{t}\right)\|^{2}\,dt - \frac{1}{2}\int_{0}^{T}\|\dot{y}_{t} - h\left(x_{t}\right)\|^{2}\,dt$$

if x and y are absolutely continuous, $\mathbb{0}$ otherwise. One can check that $\mathbb{p}^{X,Y}$ is upper-semicontinuous over $(\Omega_{T} \times \Omega_{0})$ and

$$\mathbb{p}^{X}\left(x\right) = S_{0}\left(x_{0}\right) - \frac{1}{2}\int_{0}^{T}\|\dot{x}_{t} - f\left(x_{t}\right)\|^{2}\,dt, \qquad \mathbb{p}^{Y}\left(y\right) = \sup_{x\in\Omega^{n}X} \mathbb{p}^{X,Y}\left(x,y\right)$$

$$\mathbb{p}^{Y/X}\left(y/x\right) = \begin{cases} \frac{\mathbb{p}^{X,Y}\left(x,y\right)}{\mathbb{p}^{X}\left(x\right)}{}_{\mathbb{0}} = -\frac{1}{2}\int_{0}^{T}\|\dot{y}_{t} - h\left(x_{t}\right)\|^{2}\,dt & \text{if } \mathbb{p}^{X}\left(x\right) > \mathbb{0} \\ \mathbb{0} & \text{otherwise}. \end{cases}$$

Finally, for every bounded and measurable function φ from Ω_T into $\mathcal{A}$, one has $\mathbb{P}$-a.e.

$$\mathbb{E}\left(\varphi\left(X\right)\big/\mathcal{F}_T^Y\right) \;=\; \frac{\int^{\oplus} \varphi(\theta)\odot \mathbb{p}^{Y/X}\left(Y/\theta\right)\odot \mathbb{p}^X(\theta)\odot d\theta}{\int^{\oplus}\odot\, \mathbb{p}^{Y/X}\left(Y/\theta\right)\odot \mathbb{p}^X(\theta)\odot d\theta}$$

$$=\; \frac{\mathbb{E}_0\left(\varphi\left(X\right)\odot Z\left(X,Y\right)\big/\mathcal{F}_T^Y\right)}{\mathbb{E}_0\left(Z\left(X,Y\right)\big/\mathcal{F}_T^Y\right)}$$

where $\mathbb{E}_0(.)$ denotes the Maslov expectation associated with the performance measure on $(\Omega, \mathcal{F}_T)$ defined by

$$\frac{d\mathbb{P}}{d\mathbb{P}_0}\big/_{\mathcal{F}_T} = Z\left(X,Y\right) = \int_0^T h\left(X_t\right)\dot{Y}_t dt - \frac{1}{2}\int_0^T h^2\left(X_t\right)dt.$$

By the same line of argument as before, we use these formula to define several measures as follows ($A\in \mathbf{B}\left(\Omega_T\right)$, $B\in \mathbf{B}\left(\mathbb{R}\right)$, $\eta\in \Omega_0$, $t\in [0,T]$):

$$\sum(A)(\eta) \;=\; \int_A^{\oplus} Z(\theta,\eta)\odot \mathbb{p}^X(\theta)\odot(d\theta)$$

$$\sigma_t(B)(\eta) \;=\; \sum(\{\theta\in \Omega_T\,:\,\theta_t\in B\})(\eta)$$

$$\prod(A)(\eta) \;=\; \frac{\sum(A)(\eta)}{\sum(\Omega_T)(\eta)}$$

$$\pi_t(B)(\eta) \;=\; \frac{\sigma_t(B)(\eta)}{\sigma_t(\mathbb{R})(\eta)}$$

The classical theorems on large deviations have the following formulation.

Theorem 6 $\forall A\subset \Omega_T$, A closed, $\forall \mathcal{O}\subset \Omega_T$, $\mathcal{O}$ open, $\forall x\in \mathbb{R}$

$$\limsup_{\epsilon\to 0} \epsilon\log P^{X^\epsilon/X_0^\epsilon}\left(A/x\right) \;\leq\; \mathbb{P}^{X/X_0}\left(A/x\right)$$

$$\liminf_{\epsilon\to 0}\epsilon\log P^{X^\epsilon/X_0^\epsilon}\left(\mathcal{O}/x\right) \;\geq\; \mathbb{P}^{X/X_0}\left(\mathcal{O}/x\right).$$

Theorem 7 *Let* $(P^\epsilon)_{\epsilon>0}$ *be a sequence of probability measures over* $(\Omega_T, \mathbf{B}\left(\Omega_T\right))$ *and* $\mathbb{P}$ *a Maslov performance measure over* $(\Omega_T, \mathbf{B}\left(\Omega_T\right))$ *such that for every open subset* $\mathcal{O}$ *and for every closed subset* A *in* Ω_T

$$\limsup_{\epsilon\to 0}\epsilon\log P^\epsilon(A)\leq \mathbb{P}(A)\qquad \liminf_{\epsilon\to 0}\epsilon\log P^\epsilon(\mathcal{O})\geq \mathbb{P}(\mathcal{O}).$$

If $(F^\epsilon)_{\epsilon>0}$ *is a sequence of functions from* Ω_T *into* $\mathbb{R}$ *which uniformly converges to a function* F *when* ϵ *goes to 0, then*

$$\limsup_{\epsilon\to 0}\epsilon\log\left(\int_A e^{\frac{1}{\epsilon}F^\epsilon(x)}P^\epsilon(dx)\right) \;\leq\; \int_A^{\oplus}F(x)\odot \mathbb{P}(dx) \qquad (6.2)$$

$$\liminf_{\epsilon\to 0}\epsilon\log\left(\int_{\mathcal{O}} e^{\frac{1}{\epsilon}F^\epsilon(x)}P^\epsilon(dx)\right) \;\geq\; \int_{\mathcal{O}}^{\oplus}F(x)\odot \mathbb{P}(dx). \qquad (6.3)$$

If we combine these theorems with the previous study we obtain:

1. For every closed subset A and open subset $\mathcal{O}$ in Ω_T

$$\limsup_{\epsilon \to 0} \epsilon \log P^{X^\epsilon}(A) \leq \mathbb{P}^X(A) \qquad \liminf_{\epsilon \to 0} \epsilon \log P^{X^\epsilon}(\mathcal{O}) \geq \mathbb{P}^X(\mathcal{O}).$$

2. For every closed subset A and open subset $\mathcal{O}$ in $\Omega_T \times \Omega_0$

$$\limsup_{\epsilon \to 0} \epsilon \log P^{X^\epsilon, Y^\epsilon}(A) \leq \mathbb{P}^{X,Y}(A) \qquad \liminf_{\epsilon \to 0} \epsilon \log P^{X^\epsilon, Y^\epsilon}(\mathcal{O}) \geq \mathbb{P}^{X,Y}(\mathcal{O}).$$

According to the definition of the probability measure $\left(\check{P}^\epsilon_{[\eta]} \right)_{\epsilon > 0, \eta \in \Omega_0}$, we have

1. For every closed subset A and open subset $\mathcal{O}$ in Ω_T, $\eta \in \Omega_0$,

$$\limsup_{\epsilon \to 0} \epsilon \log \check{P}^\epsilon_{[\eta]}(A) \leq \mathbb{P}^{\check{X}}_{[\eta]}(A) \qquad \liminf_{\epsilon \to 0} \epsilon \log \check{P}^\epsilon_{[\eta]}(\mathcal{O}) \geq \mathbb{P}^{\check{X}}_{[\eta]}(\mathcal{O})$$

where $\mathbb{P}^{\check{X}}_{[\eta]}$ is the performance measure on Ω_T of the optimization process

$$\dot{\check{X}}_t = f\left(\check{X}_t \right) - \eta_t \frac{\delta h}{\delta x}\left(\check{X}_t \right) + U^X_t \qquad \text{with} \quad \check{X}_0 = X_0.$$

For every $\eta \in \Omega_0$, one can check that the sequence $(F^\epsilon(.,\eta))_{\epsilon > 0}$ satisfies the condition of the theorem with

$$F(\theta, \eta) \overset{def}{=} h(\theta_T)\eta_T$$
$$- \int_0^T \left(\frac{1}{2}h^2(\theta_t) + \eta_t f(\theta_t)\frac{\delta h}{\delta x}(\theta_t) - \frac{1}{2}\left(\eta_t \frac{\partial h}{\partial x}(\theta_t) \right)^2 \right) dt.$$

2. For every closed subset A and open subset $\mathcal{O}$ in Ω_T, and $\eta \in \Omega_0$,

$$\limsup_{\epsilon \to 0} \epsilon \log \sum^\epsilon(A) \leq \sum(A)(\eta) \qquad \liminf_{\epsilon \to 0} \epsilon \log \sum^\epsilon(\mathcal{O}) \geq \sum(\mathcal{O})(\eta).$$

Consequently, for every closed subset A and every open subset $\mathcal{O}$ in Ω_T, $\eta \in \Omega_0$,

$$\limsup_{\epsilon \to 0} \epsilon \log \prod^\epsilon(A) \leq \prod(A)(\eta) \qquad \liminf_{\epsilon \to 0} \epsilon \log \prod^\epsilon(\mathcal{O}) \geq \prod(\mathcal{O})(\eta).$$

In particular, for every closed subset A and every open subset $\mathcal{O}$ in $\mathbb{R}$, and $\eta \in \Omega_0$, $t \in [0, T]$

$$\limsup_{\epsilon \to 0} \epsilon \log \pi^\epsilon_t(A) \leq \pi(A)_t(\eta) \qquad \liminf_{\epsilon \to 0} \epsilon \log \pi^\epsilon_t(\mathcal{O}) \geq \pi_t(\mathcal{O})(\eta).$$

Assume that for every $\eta \in \Omega_0$ and $t \in [0, T]$ there exists a unique conditional optimal state $op(X_t/\eta)$. In other words, for every $\gamma > 0$

$$\pi_t(\{x \in \mathbb{R} : \|x - op(X_t/\eta)\| \leq \gamma\})(\eta) = 1$$

$$\pi_t(\{x \in \mathbb{R} : \|x - op(X_t/\eta)\| > \gamma\})(\eta) < 1.$$

Then using the Prohorov topology, we obtain that $\pi^\epsilon_t(\eta)$ weakly converges to $\delta_{op(X_t/\eta)}$ when ϵ goes to 0.

6.2   The Log-Exp transform

We briefly recall the Log-Exp transform (for details see [7]). This mapping leads to useful conclusions because it makes explicit the relationship between the performance and the probability measure of an event. For $\nu > 0$, $d \geq 1$, we denote by D_d^ν the class of probability densities p on $\mathbb{R}^d$ such that

$$\int^{\oplus} \log\left(p(x)^\nu\right) \odot dx \stackrel{def}{=} \mathbb{N}_\nu(p) > \mathbb{0}$$

and by $\mathbb{D}_d^\nu$ the class of performance measures $\mathbb{p}$ on $\mathbb{R}^d$ such that

$$\int \exp\left(\frac{\mathbb{p}(x)}{\nu}\right) dx \stackrel{def}{=} N_\nu(\mathbb{p}) > 0.$$

We use the conventions, when discrete events are embedded in a continuous fashion:

$$\log\left(\sum_{n \geq 0} p_n \, \delta_{z_n}\right) = \bigoplus_{n \geq 0} \log\left(p_n\right) \odot \mathbb{1}_{z_n} \quad \exp\left(\bigoplus_{n \geq 0} \mathbb{p}_n \odot \mathbb{1}_{z_n}\right) = \sum_{n \geq 0} \exp\left(\mathbb{p}_n\right) \delta_{z_n}.$$

These spaces are in a one to one correspondence by the following transformations:

$$\mathbb{Exp}_\nu(\mathbb{p}) \stackrel{def}{=} \frac{e^{\frac{1}{\nu}\mathbb{p}}}{N_\nu(\mathbb{p})} \qquad \mathbb{Log}_\nu(p) \stackrel{def}{=} \frac{\log p^\nu}{\mathbb{N}_\nu(p)} \qquad \mathbb{Exp}_\nu = \mathbb{Log}_\nu^{-1}. \qquad (6.4)$$

Let $(\Omega, \mathcal{F}, \mathbb{P})$ be an optimization basis, and $\mathbf{F} \stackrel{def}{=} (\mathcal{F}_k)_{k \geq 0}$ an increasing filtration of $\mathcal{F}$ on which are defined two real-valued and independent optimisation processes U and V with values in $\mathbb{R}^{n_X}$ and $\mathbb{R}^{n_Y}$, $n_X, n_Y \geq 1$. Let us now define on $(\Omega, \mathcal{F}, \mathbb{P})$ the following optimization processes, for every $k \geq 0$:

$$\mathcal{O}(X/Y): \quad X_k = F(X_{k-1}, U_k) \quad X_0 = U_0, \text{ and } Y_k = H(X_k) + V_k$$

with F a measurable function from $\mathbb{R} \times \mathbb{R}$ into $\mathbb{R}$ and H a measurable function from $\mathbb{R}$ into $\mathbb{R}$. From the above,

$$\mathbb{p}^{U_k, Y_k}(u, y) = \mathbb{p}^{U_k}(u) \odot \mathbb{p}^{V_k}(y - H(\phi(u)))$$

where $\left(U_{\underline{k}}, V_{\underline{k}}\right) \stackrel{def}{=} (U_0, U_1, \ldots, U_k, V_0, V_1, \ldots, V_k)$ and $\phi(u) = X_{\underline{k}}$, the state path associated with the value $U = u_{\underline{k}}$. It is now straightforward to apply the Log-Exp transform. For every $\nu > 0$ such that

$$\forall k \geq 0 \quad N_\nu(k) \stackrel{def}{=} \int_{(\mathbb{R} \times \mathbb{R})^{[0,k]}} \exp\left(\frac{1}{\nu}\mathbb{p}^{U_{\underline{k}}, Y_{\underline{k}}}(u, y)\right) du \, dy > 0$$

the measure $p^{W_{\underline{k}}^\nu, Y_{\underline{k}}^\nu} \stackrel{def}{=} \mathbb{Exp}_\nu\left(\mathbb{p}^{U_{\underline{k}}, Y_{\underline{k}}}\right) = \frac{1}{N_\nu(k)} \exp\left(\mathbb{p}^{U_{\underline{k}}, Y_{\underline{k}}}\right)$ is the probability measure associated with the filtering problem $\mathcal{F}^\nu$ defined for every $k \geq 0$ by

$$\mathcal{F}^\nu(X/Y): \quad X_k^\nu = F\left(X_{k-1}^\nu, W_k^\nu\right) \quad X_0^\nu =, W_0^\nu, \text{ and } Y_k^\nu = H(X_k^\nu) + V_k^\nu$$

where W^ν, V^ν are two P-independent stochastic processes with probability measures $\mathrm{Exp}_\nu\left(\mathbb{p}^U\right)$ and $\mathrm{Exp}_\nu\left(\mathbb{p}^V\right)$.

Example 1 *Let $H(x) = x$, $\tau \in [0, k]$, $0 < \lambda < 1$, c a fixed real value.*

1. $\mathbb{p}^{V_\tau}(u) = -\frac{1}{2}v^2 \implies \mathbb{p}^{V_k}\left(c - H\left(\phi(u)\right)\right) = -\frac{1}{2}\sum_{\tau=0}^{k}\left(c - \phi_\tau(u)\right)^2.$

2. $\mathbb{p}^{U_\tau}(u) = -\frac{1}{2}u^2 \implies p^{W_\tau^\nu}(u) \overset{def}{=} \mathrm{Exp}_\nu\left(\mathbb{p}^{U_\tau}\right)(u) = \frac{1}{\sqrt{2\nu\pi}}e^{-\frac{1}{2\nu}u^2}.$

3. $\mathbb{p}^{U_\tau}(u) = \log\left(\frac{\lambda}{\lambda \oplus (1-\lambda)}\right) \odot \mathbb{1}_1 \oplus \log\left(\frac{1-\lambda}{\lambda \oplus (1-\lambda)}\right) \odot \mathbb{1}_0$

   $\implies p^{W_\tau^\nu}(u) \overset{def}{=} \mathrm{Exp}_\nu\left(\mathbb{p}^{U_\tau}\right)(u) = \frac{\lambda^{\frac{1}{\nu}}}{\lambda^{\frac{1}{\nu}}+(1-\lambda)^{\frac{1}{\nu}}}\delta_1 + \left(1 - \frac{\lambda^{\frac{1}{\nu}}}{\lambda^{\frac{1}{\nu}}+(1-\lambda)^{\frac{1}{\nu}}}\right)\delta_0.$

4. *Initial constraint:* $\mathbb{p}^{U_0} = \mathbb{1}_{x_0} \implies p^{W_0^\nu} \overset{def}{=} \mathrm{Exp}_\nu\left(\mathbb{p}^{U_0}\right) = \delta_{x_0}.$

5. *Final constraint:* $\mathbb{p}^{V_k} = \mathbb{1}_0 \implies \mathbb{p}^{V_k}\left(c - H\left(\phi_k(u)\right)\right) = \mathbb{1}_c\left(\phi_k(u)\right) \implies p^{V_k^\nu} \overset{def}{=} \mathrm{Exp}_\nu\left(\mathbb{p}^{U_k}\right) = \delta_0.$

In other words one may regard the regulation problem $\mathcal{O}(X/Y)$ as the maximum likelihood estimation problem associated with a filtering problem $\mathcal{F}^\nu, \nu > 0$.

6.3 The Cramér transform

As is well known, the Cramér transform is defined by $\mathcal{C} \overset{def}{=} \mathcal{F} \circ \log \circ \mathcal{L}$, where $\mathcal{L}$ is the Laplace transform and $\mathcal{F}$ the Fenchel transform. An excellent account of this transform can be found in Stroock [35]. This transform maps the set of probability measures into the set of upper-semicontinuous performance densities. It also converts probability convolutions into Maslov convolutions and the classical expectation of a random variable into the optimal state of the induced optimization variable. More details were developed in Huillet–Salut [20], [21] and Huillet–Rigal–Salut [22]. Later results have been obtained by Quadrat in [2] and Akian–Quadrat–Viot [1].

References

[1] M. AKIAN, J.P. QUADRAT, M. VIOT, *Bellman processes*, 11eme Conférence Internationale sur l'Analyse et l'Optimisation des Systèmes. Ecole des Mines Sophia-Antipolis, France, 15–17 June 1994. Lecture Notes in Control and Information Sciences 199. Springer-Verlag.

[2] F. BACCELLI, G. COHEN, G.J. OLSDER, J.P. QUADRAT, *Synchronization and Linearity*, Wiley, 1992.

[3] F. BELLALOUNA, *Un point de vue linéaire sur la programmation dynamique. Détecteur de ruptures dans le cadre des problèmes de fiabilité.* Thèse Université Paris Dauphine, Septembre 1992.

[4] R.E. BELLMAN, S.E. DREYFUS, *Dynamic Programing and Applications*, Dunod, Paris, 1965.

[5] H. COX, *Estimation of state variables via dynamic programming*, Proc. 1964 Joint Automatic Control Conf., pp 376–381, Stanford, California.

[6] P. DEL MORAL, T. HUILLET, G. RIGAL, G. SALUT, *Optimal versus random processes: the non linear case*, LAAS report no. 91131, April 1991.

[7] P. DEL MORAL, *Résolution particulaire des problèmes d'estimation et d'optimisation non linéaires*, Thèse Université Paul Sabatier, Toulouse, 1994.

[8] P. DEL MORAL, *Maslov Optimization theory: Topological Aspects*, April 1994, LAAS-CNRS report no. 94 110.

[9] P. DEL MORAL, G. RIGAL, G. SALUT, *Estimation et commande optimale non linéaire: Méthodes de résolutions particulaires* Rapports de contrat DIGILOG-DRET no. 89.34.553.00.470.75.01, October 1992, January 1993.

[10] P. DEL MORAL, G. SALUT, *Maslov Optimization Theory: Optimality Versus Randomness* LAAS report no. 94211, May 1994.

[11] P. DEL MORAL, *An introduction to the theory of optimization processes*, LAAS report no. 94127, April 1994.

[12] P. DEL MORAL, *Optimization processes: Hamilton-Liapunov stability,* LAAS report no. 94207, May 1994.

[13] P. DEL MORAL, J.C. NOYER, G. SALUT, *Maslov Optimization Theory: Stochastic Interpretation, Particle Resolution,* 11eme Conférence Internationale sur l'Analyse et l'Optimisation des Systèmes. Ecole des Mines Sophia-Antipolis, France, 15-16-17 Juin 1994. Lecture Notes in Control and Information Sciences 199. Springer-Verlag.

[14] P. DEL MORAL, G. SALUT, *Maslov Optimization Theory*, Russian Journal of Mathematical Physics, John Wiley and Sons (1995).

[15] P. DEL MORAL, G. SALUT, *Particle Interpretation of Non-Linear Filtering and Optimization*, Russian Journal of Mathematical Physics, John Wiley and Sons (1995).

[16] P.I. DUDNIKOV, S.N. SAMBORSKII, *Endomorphisms of semimodules over semirings with an idempotent operation*, Preprint $N°$ 87-48, Inst. Mat. Akad. Nauk Ukrain., Kiev, 1987 (Russian).

[17] O. HIJAB, *Minimum energy estimation*, Doctoral Disseration, University of California, Berkeley, 1980.

[18] O. HIJAB, *Asymptotic Bayesian Estimation of a first order Equation with small diffusion*, Annals of Probability, 12 (1984), 890–902.

[19] A. H. JAZWINSKI, *Stochastic Processes and Filtering Theory*, Academic Press, New York, 1970.

[20] T. HUILLET, G. SALUT, *Stochastic processes and Optimal processes*, LAAS-CNRS report $N°$ 89025, January 1989.

[21] T. HUILLET, G. SALUT, *Optimal versus Random processes: the regular case*. Rapport LAAS $N°$ 89189, Juin 1989.

[22] T. HUILLET, G. RIGAL, G. SALUT, *Optimal versus Random processes: a general framework*, LAAS report no 89251, July 1989.

[23] V. MASLOV, *Méthodes opératorielles*, Editions Mir, 1987.

[24] V.N. KOLOKOLTSOV, *Semiring analogs of linear equivalent spaces*, Abstr. Conf. on Optimal Control, Geometry, Kemerovo, 1988, p. 30 (Russian).

[25] V.N. KOLOKOLTSOV, V. MASLOV, *Idempotent calculus as the apparatus of optimization theory*. I, Funktsional. Anal. i Prilozhen. 23 (1989), no. 1, 1-14; II, 23, no. 4, 53-62; English transl. in Functional Anal. Appl. 23 (1989).

[26] V.N. KOLOKOLTSOV, V.P. MASLOV, *The general form of endomorphisms in the space of continuous functions with values in a numerical semiring with idempotent addition.*, Dokl. Akad. Nauk SSSR 295 (1987), no. 2, 283–287; English transl. Soviet Math. Dokl. 36 (1988), no. 1, 55–59.

[27] G. LITVINOV, V.P. MASLOV, Correspondence principle for idempotent calculus and some computer applications. Institut des Hautes Etudes Scientifiques, IHES-M-95-33.

[28] V. MASLOV, *Quasilinear systems which are linear in some semimoduli* Congrès international sur les problèmes hyperboliques, 13–17 January 1986, Saint Etienne, France.

[29] V. MASLOV, S.N. SAMBORSKII, *Idempotent Analysis*, Advances in Soviet Mathematics, 13, Amer. Math Soc., Providence (1992).

[30] V. MASLOV, S.N. SAMBORSKII, *Stationary Hamilton–Jacobi and Bellman equations*, Advances in Soviet Mathematics 13 AMS (1992) 119-133.

[31] L.S. PONTRYAGIN, A.A. ANDRONOV, A.A. VITT, *O statisticheskom rassmotrenii dinamicheskikh sistem*, Zh. Eksper. Teoret. Fiz., 3(1933)

[32] J.P. QUADRAT, *Théorèmes asymptotiques en programmation dynamique*, Comptes Rendus à l'Académie des Sciences, 311:745-748,1990.

[33] R.T. ROCKAFELLAR, Convex Analysis, Princeton University, 1970.

[34] R.T. ROCKAFELLAR, Convex functions and duality in optimization problems and dynamics, Mathematical Systems Theory and Economics, I. H.W. Kuhn, G.P. Szergo editors, Lecture Notes in ORME, Springer-Verlag, 1969.

[35] D.W. STROOCK, *An Introduction to the Theory of Large Deviations*, Springer-Verlag, AMS 60F10, 1984.

[36] M.I. FREIDLIN, A.D. WENTZELL, *Random Perturbations of Dynamical Systems*, Springer Verlag 1979, AMS Classifications: 60HXX, 58G32.

Random Particle Methods in (max,+) Optimization Problems

Pierre Del Moral and Gérard Salut

Introduction

We assume the reader is familiar with basic facts about Maslov optimization theory and non-linear filtering. We first recall some standard notations. Let $\mathcal{A} \overset{def}{=} [-\infty, +\infty[$ be the semi-ring of real numbers endowed with the commutative semigroup laws $\oplus$, $\odot$, the neutral elements $\mathbb{0}$, $\mathbb{1}$ and the exponential metric ρ such that

$$
\begin{aligned}
\mathbb{0} &= -\infty & a \oplus b &= \sup(a,b) & \rho(a,b) &= \left| e^a - e^b \right| \\
\mathbb{1} &= 0 & a \odot b &= a + b & \tfrac{a}{b} &= a - b & (b \neq \mathbb{0}).
\end{aligned}
$$

In this paper we consider discrete time non-linear filtering and deterministic optimization problems. In [5] we show that the Log-Exp transform is a bijective correspondence between Maslov performances and deterministic optimization problems on one hand, and Markov probabilities and filtering problems on the other hand. This stochastic interpretation of deterministic optimization problems provides a firm basis for transposing the recently developed particle procedures for non-linear filtering to optimization problems, because they make explicit the links between the performance and probability measures of an event. This is a most powerful approach to the numerical solution of infinite-dimensional problems arising where non-linearities do not allow the use of analysis and non-stationarities do not allow the application of fixed discretization schemes. Like all particle methods, they are based on a Dirac comb representation of the performance/probability measure in question, but the "teeth" of this comb depend, in both mass and position, on the flow of the system and partial observations or its desired reference path. This method, which has revealed its efficiency in radar and G.P.S. signal processing ([3], [6], [8]), may also be used in solving non-linear filtering problems and in deterministic optimization of discrete event dynamical systems, such as the determination of communication networks or manufacturing systems from partial observations or from a reference trajectory. We briefly review some basic facts about Monte-Carlo principles in this paper. We show in [9], [4] and [7] that these principles are a powerful tool to study the conditional mean of a random variable as well as the conditional optimal state of an optimization variable. We introduce some recursive distributions which can be used to explore the performance/probability space in a way closely associated with the performance criterion or, respectively, likelihood function. These distributed weights are related to the performance likelihood of each exploration path of the finite N particles. It is important to notice that they are *a priori* time degenerative, for individual paths have divergent likelihood/performance. This degeneracy of weights is eliminated by a regularization of the problem ([4], [7], [8]).

1   Non-linear Filtering and Deterministic Optimization

In [6] we state that the Bellman optimality principle may be viewed as a basic definition of optimization processes like the Markov property rather than a deductive conclusion. Indeed, Maslov integration theory allows an optimization theory to be derived at the same level of generality as stochastic process theory. Maslov optimization processes, such as encountered in maximum likelihood problems, are introduced in forward time; reversal of time yields optimal control problems of regulation type. Maslov optimization processes and Markov stochastic processes can be mapped into each other via various transformations. The Log-Exp transformation is a powerful tool to study the stochastic interpretation of a Maslov performance. To simplify the notations, to any real valued deterministic optimization problem, on $[0,T]$, denoted by $\mathcal{O}(\mathcal{X}/\mathcal{Y})$ and defined on a performance space $(\Omega, \mathcal{F}, \mathbb{P})$, we associate the real-valued nonlinear filtering problem, on $[0,T]$, denoted by $\mathcal{F}(\mathcal{X}/\mathcal{Y})$ and defined on some convenient probability space $(\Omega, \mathcal{F}, P)$ as follows:

$\mathcal{O}(\mathcal{X}/\mathcal{Y})$	$\mathcal{F}(\mathcal{X}/\mathcal{Y})$
$\begin{cases} X_n = \phi(X_{n-1}, U_n), & X_0 = U_0 \\ Y_n = H(X_n) + V_n \end{cases}$	$\begin{cases} X_n^e = \phi(X_{n-1}^e, U_n^e), & X_0^e = U_0^e \\ Y_n^e = H(X_n^e) + V_n^e \end{cases}$
U, V two $\mathbb{P}$-independent optimization processes	U^e, V^e two P-independent stochastic processes, with:
$\left(\text{Log } p^{U^e, V^e} = \mathbb{p}^{U,V}\right)$	$\text{Exp } \mathbb{p}^{U,V} = p^{U^e, V^e}$

where ϕ and H denote two real-valued measurable functions. After a careful examination of these problems, several comments are in order. Note that the random variable V^e and the optimization variable V completely describe the Bayesian factor which produces the conditional probability or performance measure. The following examples suggest how these results may be useful in analysing some deterministic optimization problems. Let $H(x) = x$, $n \in [0, T]$, $0 < \lambda < 1$, c a fixed real value.

1. $\mathbb{p}(v_n) = -\frac{1}{2}v_n^2 \Longrightarrow \mathbb{p}^V (c - H(\phi(u))) = -\frac{1}{2}\sum_{n=0}^{T} (c - \phi_\tau(u))^2$.

2. $\mathbb{p}(u_n) = -\frac{1}{2}u_n^2 \Longrightarrow p(u_n^e) = \frac{1}{\sqrt{2\pi}}e^{-\frac{1}{2}(u_n^e)^2}$.

3. Poisson processes are also realistic models for a large class of point processes: photon counting, electron emission, telephone calls, data communications, servicing, etc. Assume U^e is a Poisson counting process with a non-homogeneous intensity function λ. Its sample function is given for every piecewise constant path u^e, such that $\Delta u_n^e \in \{0, 1\}$, by

$$p(u^e) = \exp\left(-\int_0^T \lambda_\tau \, d\tau + \int_0^T \log(\lambda_\tau) \, du_\tau^e\right)$$

and $\mathbb{p}(u) = \int_0^T \log(\lambda_\tau) \, du_\tau$.

4. Initial constraint:

$$\mathbb{p}(u_0) = \mathbb{1}_{x_0} \Longrightarrow p(u_0^e) = \delta_{x_0}(u_0^e).$$

5. Final constraint:

$$\mathbb{p}(v_T) = \mathbb{1}_0(v_T) \quad \Longrightarrow \quad \mathbb{p}^{V_T^e}(c - H(\phi_T(u))) = \mathbb{1}_c(\phi_T(u))$$
$$\Longrightarrow \quad p(v_T^e) = \delta_0(v_T^e).$$

Clearly these results can be generalized to the vector case. As a matter of fact, the differential equations introduced in $\mathcal{F}(\mathcal{X}/\mathcal{Y})$ or $\mathcal{O}(\mathcal{X}/\mathcal{Y})$ are usually constructed using the conventional addition and multiplication. We end this section by recalling that **non-linear filtering and deterministic optimization may be useful in optimization and control of communication networks and manufacturing systems**. The time behaviour of such systems might be described in terms of $(\oplus, \odot)$-differential equations as a physical phenomenon (the reader is referred to [1] and [2] for $(\oplus, \odot)$-linear systems). In such areas, **the functions ϕ and H are constructed using the operations min, max, conventional addition and multiplication.**

2 Monte-Carlo Principles

In this short section we briefly review some basic facts about Monte-Carlo principles and we show that these principles might be a powerful tool to study the mean of a random variable as well as the optimal state of an optimization variable. Probability results from a deductive argument in which we estimate the chance of some event realization. When the event is associated with a random error of some approximation, this measure evaluates the chance of achieving a given precision. In what follows particle algorithms will be studied in this way. Independence between random variables means that the realization of any variable is not altered by the realizations of the others. This concept is fundamental, in fact it justifies the mathematical development of probability not merely as a topic in measure theory, but as a separate discipline. The significance of independence arises in the context of repeated trials. We will denote in the sequel, for every sequence of real numbers u and $n \in [0, T]$:

$$u_{\underline{n}} = (u_0, \dots, u_n) \qquad \|u_{\underline{n}}\|_2^2 = \sum_{m=0}^{n} u_m^2.$$

As is well known, the sensor H may always be chosen as a linear function, through a suitable state-space basis, so that the conditions for L^0-convergence have a simplest form. By the same line of argument as before, with some obvious extension of the notation, if $(U^i)_{i \geq 1}$ is a sequence of independent random variables with the same probability law as U^e and defined on the same probability space (Ω, σ, P), then for every $N \geq 1$, $1 \leq n \leq T$ and $\epsilon > 0$, one has ([4]):

Conditional expectation estimate:

$$P\left(\left| E_N(\phi_n(U^e)/Y_{\underline{n}}^e) - E(\phi_n(U^e)/Y_{\underline{n}}^e) \right| > \epsilon \right)$$
$$\leq \frac{C_T}{N\epsilon^2} \, E\left(\left(\phi_n(U^1) - E(\phi_n(U^e)/Y_{\underline{n}}^e) \right)^2 \right)$$

where $E_N(\phi_n(U^e)/Y_{\underline{n}}^e) = \sum_{i=1}^{N} \dfrac{p(Y_{\underline{n}}^e/U_{\underline{n}}^i)}{\sum_{j=1}^{N} p(Y_{\underline{n}}^e/U_{\underline{n}}^j)} \, \phi_n(U^i)$ and $C_T > 0$.

Conditional optimization estimate:
Let Y be a reference value for which the conditional performance $\mathbb{p}(u_{\underline{n}}/Y_{\underline{n}})$ satisfies some regularity conditions ([4]). After some algebraic manipulations one finds that, for every $\epsilon > 0$, there exists some $\eta > 0$ such that

$$P\left(\left\| op_N(U_{\underline{n}}/Y_{\underline{n}}) - op(U_{\underline{n}}/Y_{\underline{n}}) \right\|_2 > \epsilon \right) \leq \left(1 - P\left(\left\| U_{\underline{n}}^1 - op(U_{\underline{n}}/Y_{\underline{n}}) \right\|_2 \leq \eta \right) \right)^N$$

where $\Omega_N = \{ U_{\underline{n}}^1, \dots, U_{\underline{n}}^N \}$ and $op_N(U_{\underline{n}}/Y_{\underline{n}}) = Arg \sup_{u \in \Omega_N} \mathbb{p}\left(u, Y_{\underline{n}} \right) \stackrel{def}{=} A \bigoplus_{i=1}^{N} \mathbb{p}\left(U_{\underline{n}}^i, Y_{\underline{n}} \right)$.

3   Particle Interpretations

Particle algorithms are based on a Dirac comb which depends on the flow of the system and partial observations or reference values in both mass and position. In order to clarify the notation, the symbol $(.)^e$ will be omitted and the random variables V will be taken as centred Gaussian variables with zero mean and variance function R. All the stochastic processes defined in what follows are assumed to be carried by some probability space. We now describe some time-recursive explorative distributions which will be used in the sequel. Using the above, we emphasize that these distributions are exhibited by some natural change of probability distributions. The detailed assumptions under which these convergences are uniform in time are studied in [4]. The following change of probability is a simple consequence of Bayes' formula.

A priori exploration

$$p\left(u_{\underline{n}}, y_{\underline{n}}\right) = Z_n^0(u, y) \; p_0\left(u_{\underline{n}}, y_{\underline{n}}\right) = Z_n^0(u, y) \; p_0\left(u_{\underline{n}}/y_{\underline{n}}\right) \; p_0\left(y_{\underline{n}}\right)$$

with

$$\left\{ \begin{array}{rclcrcl} p_0\left(y_{\underline{n}}\right) & = & G_n(y), & & p_0\left(u_{\underline{n}}, y_{\underline{n}}\right) & = & p_0\left(u_{\underline{n}}\right) = p\left(u_{\underline{n}}\right) \\ G_n(y) & = & g_n(y) \, G_{n-1}(y), & & g_n(y) & = & p^{V_n}(y_n) \\ Z_n^0(u,y) & = & z_n^0(u,y) \, Z_{n-1}^0(u,y), & & z_n^0(u,y) \, g_n(y) & = & p\left(y_n/u_{\underline{n}}\right) \end{array} \right.$$

Let $\breve{U}_{\underline{n}}$ be a *generic* explorative stochastic process P-independent of U and V, with distribution $p(u_{\underline{n}})$, and let $(U^i)_{i \geq 1}$ be a sequence on independent copies of $\breve{U}$. Then ([4])

$$E_N(\phi_n(U)/Y_{\underline{n}}) \xrightarrow[N \to +\infty]{\mathbf{L}^0} E(\phi_n(U)/Y_{\underline{n}})$$

and

$$op_N(U_{\underline{n}}/Y_{\underline{n}}) \xrightarrow[N \to +\infty]{\mathbf{L}^0} op(U_{\underline{n}}/Y_{\underline{n}})$$

where

$$E_N(\phi_n(U)/Y_{\underline{n}}) = \sum_{i=1}^{N} \frac{Z_n^0(U^i, Y)}{\sum\limits_{j=1}^{N} Z_n^0(U^j, Y)} \; \phi_n(U^i)$$

and

$$op_N(U_{\underline{n}}/Y_{\underline{n}}) = A \bigoplus_{i=1}^{N} \mathbb{p}\left(U_{\underline{n}}^i, Y_{\underline{n}}\right) \tag{3.1}$$

Conditional exploration

$$p\left(u_{\underline{n}}, y_{\underline{n}}\right) = Z_n^1(u, y) \; p_1\left(u_{\underline{n}}, y_{\underline{n}}\right) = Z_n^1(u, y) \; p_1\left(u_{\underline{n}}/y_{\underline{n}}\right) \; p_1\left(y_{\underline{n}}\right)$$

with $p_1\left(u_{\underline{n}}/y_{\underline{n}}\right) = p\left(u_n/u_{\underline{n-1}}, y_{\underline{n}}\right) \; p_1\left(u_{\underline{n-1}}/y_{\underline{n-1}}\right)$ and

$$\left\{ \begin{array}{rclcrcl} p_1\left(y_{\underline{n}}\right) & = & G_n(y), & & & & \\ G_n(y) & = & g_n(y) \, G_{n-1}(y), & & g_n(y) & = & p^{V_n}(y_n) \\ Z_n^1(u,y) & = & z_n^1(u,y) \, Z_{n-1}^1(u,y), & & z_n^1(u,y) \, g_n(y) & = & p\left(y_n/u_{\underline{n-1}}\right). \end{array} \right.$$

Let $\check{U}_{\underline{n}}$ be a *generic* explorative stochastic process P-independent of U and V, with Y-conditional distribution $p_1\left(u_{\underline{n}}/y_{\underline{n}}\right)$, and let $(U^i)_{i\geq 1}$ be a sequence on Y-conditionally independent copies of $\check{U}$. We obtain ([4])

$$E_N(\phi_n(U)/Y_{\underline{n}})\xrightarrow[N\to+\infty]{\mathbf{L}^0} E(\phi_n(U)/Y_{\underline{n}})$$

and

$$op_N(U_{\underline{n}}/Y_{\underline{n}})\xrightarrow[N\to+\infty]{\mathbf{L}^0} op(U_{\underline{n}}/Y_{\underline{n}})$$

where

$$E_N(\phi_n(U)/Y_{\underline{n}}) = \sum_{i=1}^{N} \frac{Z_n^1(U^i,Y)}{\sum\limits_{j=1}^{N} Z_n^1(U^j,Y)}\ \phi_n(U^i)$$

and

$$op_N(U_{\underline{n}}/Y_{\underline{n}}) = A\bigoplus_{i=1}^{N}\mathbb{p}\left(U_{\underline{n}}^i,Y_{\underline{n}}\right).$$

In the following example the non-linear structure of the problem in hand can be directly exploited. If $\phi(x,u) = F(x)+u$, $H(x) = C.x$ and U is a discrete time Gaussian process with zero mean and variance function Q, then, with $S_n^{-1} = Q_n^{-1} + C\,R_n^{-1}\,C$

$$p\left(u_n/u_{n-1},y_{\underline{n}}\right) =$$

$$\frac{1}{\sqrt{2\,\pi\,|S_n|}}\exp\left(-\frac{1}{2\,|S_n|}\left(u_n - S_n\,C\,R_n^{-1}\left(y_n - C\,F(\phi_{n-1}(u))\right)\right)^2\right).$$

The conditional exploration transitions may depend on several observation values. We show in [4] that through a suitable state space basis this case may be reduced to the latter. Moreover, for the explicit determination of the conditional exploration transitions we need, in general, another particle approximation scheme (see [4]).

Sampling/Resampling To conclude this paper we now propose a strategy to accelerate the exploration of the performance/probability space. The following algorithms are an extension of the well known sampling/resampling (S/R) principles introduced in [13], [10], [3] and more recently [14], [11], [4]. Other interesting particle schemes based on birth and death principles have been introduced in [12]. The sampling/resampling approach differs from the others in the way it stores and updates the information that is accumulated through the resampling of the positions. The basic idea is to build up, iteratively, a pure Dirac comb approximation (without weights) of the conditional sample functions $p(u_{\underline{n}}/y_{\underline{n}})$, that is to construct a discrete-time stochastic processes $\widehat{U}_{\underline{n}}^n = \left(\widehat{U}_0^n,\ldots,\widehat{U}_n^n\right)$ such that, in a sense to be defined,

$$\frac{1}{N}\sum_{i=1}^{N}\delta_{\widehat{U}_{\underline{n}}^{n,i}}\xrightarrow[N\to+\infty]{} p(u_{\underline{n}}/y_{\underline{n}})$$

where $\widehat{U}_{\underline{n}}^{n,i}$ is a sequence of independent processes having the same law as $\widehat{U}_{\underline{n}}^{n}$. The symbol $\widehat{(.)}^{n}$ means that the conditional sample function of the process depends on the observation path $y_{\underline{n}}$. We initialize the sampling/resampling algorithm by first introducing a sequence of independent stochastic processes $U_{\underline{n}}^{i} = (U_{0}^{i}, \ldots, U_{n}^{i})$ having the same law as $U_{\underline{n}}$. By the same line of argument as before

$$(1) \qquad p_0(u_0) \overset{def}{=} \frac{1}{N}\sum_{i=1}^{N} \delta_{U_0^i} \xrightarrow[N \to +\infty]{} p(u_0)$$

$$(2) \qquad p_0(u_0/y_0) \overset{def}{=} \frac{p(y_0/u_0)}{\int p(y_0/u_0)\, dp_0(u_0)} p_0(u_0) \xrightarrow[N \to +\infty]{} p(u_0/y_0)$$

We denote by $\widehat{U}_0^0$ the random variable whose law is given by $p_0(u_0/y_0)$. and by $\widehat{U}_0^{0,i}$ a sequence of independent variables with the same law as $\widehat{U}_0^0$. For every $n \geq 1$, the S/R processes $\widehat{U}_{\underline{n}}^n$ is defined recursively as follows

Let $p_{n-1}(u_{\underline{n-1}}/y_{\underline{n-1}}) \xrightarrow[N \to +\infty]{} p(u_{\underline{n-1}}/y_{\underline{n-1}})$ be the sampling function of the process $\widehat{U}_{\underline{n-1}}^{n-1}$. Then

$$(1)\ \ p_n(u_{\underline{n}}/y_{\underline{n-1}}) \overset{def}{=} \frac{1}{N}\sum_{i=1}^{N} \delta_{\widehat{U}_{\underline{n}}^{n-1,i}} \xrightarrow[N \to +\infty]{} p(u_n)\, p(u_{\underline{n-1}}/y_{\underline{n-1}}) =$$
$$p(u_{\underline{n}}/y_{\underline{n-1}}) \ \ (n^{th}\ S/R\ update)$$

with $\ \ \widehat{U}_{\underline{n}}^{n-1,i} \overset{def}{=} (\widehat{U}_{\underline{n-1}}^{n-1,i}, U_n^i)$

$$(2)\ \ p_n(u_{\underline{n}}/y_{\underline{n}}) \overset{def}{=} \frac{p(y_n/u_{\underline{n}})}{\int p(y_n/u_{\underline{n}})\, dp_n(u_{\underline{n}}/y_{\underline{n-1}})} p_n(u_{\underline{n}}/y_{\underline{n-1}}) \xrightarrow[N \to +\infty]{} p(u_{\underline{n}}/y_{\underline{n}})$$
$$(n^{th}\ Bayesian\ correction)$$

Then we denote by $\widehat{U}_{\underline{n}}^n$ the stochastic process whose sampling function is $p_n(u_{\underline{n}}/y_{\underline{n}})$. This S/R principles can be summarized as follows:

$$\widehat{U}_{\underline{n}}^{n-1} \xrightarrow{\hspace{3cm}} \widehat{U}_{\underline{n}}^{n-1} \xrightarrow{\hspace{3cm}} \widehat{U}_{\underline{n}}^{n}$$
$$N\ \textbf{a priori samplings} \qquad N\ \textbf{conditional resamplings}$$

When we use the above approximations we have, in a sense to be defined,

$$\frac{1}{N}\sum_{i=1}^{N} \delta_{\widehat{U}_{\underline{n}}^{n,i}} \xrightarrow[N \to +\infty]{} p(u_{\underline{n}}/y_{\underline{n}})$$

and

$$E_N\left(\phi_n(U)/Y_{\underline{n}}\right) \overset{def}{=} \frac{1}{N}\sum_{i=1}^{N} \phi_n(\widehat{U}_{\underline{n}}^{n,i}) \xrightarrow[N \to +\infty]{} E\left(\phi_n(U)/Y_{\underline{n}}\right).$$

The analysis of this convergence necessarily involves the study of all the approximations which lead to $p_n(u_{\underline{n}}/y_{\underline{n}})$. One open problem is to find sufficient conditions for these S/R particle schemes to $\mathbf{L}^0$-converge uniformly in time to the conditional expectation. Some local proofs can be found in [4]. As usual these S/R particle schemes are also applicable to **deterministic optimization problems**. Indeed, if Y is a reference path which satisfies certain regularity conditions ([4]), then, for every $\epsilon > 0$, there exists some $\eta > 0$ such that

$$P\left(\left\|op_N(U_{\underline{n}}/Y_{\underline{n}}) - op(U_{\underline{n}}/Y_{\underline{n}})\right\|_2 > \epsilon\right) \leq \left(1 - P\left(\left\|\widehat{U}_{\underline{n}}^n - op(U_{\underline{n}}/Y_{\underline{n}})\right\|_2 \leq \eta\right)\right)^N$$

where $op_N(U_{\underline{n}}/Y_{\underline{n}}) = A \bigoplus_{i=1}^{N} \mathbb{p}\left(\widehat{U}_{\underline{n}}^{n,i}, Y_{\underline{n}}\right)$. More generally, the resampling updates may be given by some timing sequence schedule t_n. The recent literature ([14], [11], [3] and [4]) describes several different schemes for choosing $\Delta t_n = t_n - t_{n-1}$, none of which, in our view, is completely reliable. It is our opinion that the choice of the control parameters Δt_n and the assignment of the schedule must require physical insight and/or trial and error. To clarify the presentation we have restricted the study to the case $\Delta t_n = 1$. In fact, a suitable state-space augmentation, allows the more general case to be reduced to the above.

References

[1] F. BACELLI, G. COHEN, G.J. OLSDER, J.P. QUADRAT, *Synchronization and Linearity, An Algebra for Discrete Event Systems*, John Wiley and Sons, 1992

[2] R. CUNINGHAME-GREEN, *Minimax Algebra*, Lectures Notes in Economics and Mathematical Systems, Springer-Verlag, 166, 1979.

[3] P. DEL MORAL, G. RIGAL, G. SALUT, *Filtrage non linéaire non-gaussien appliqué au recalage de plates-formes inertielles-Mise en équations spécifiques*, Contract DIGILOG-S.T.C.A.N., no. A.91.77.013, Report no. 1, September 1991.

[4] P. DEL MORAL *Résolution particulaire des problèmes d'estimation et d'optimisation non linéaires*, Thesis dissertation, University Paul Sabatier, Toulouse, June 1994.

[5] P. DEL MORAL *Maslov Optimization theory: Topological Aspects*, April 1994, LAAS-CNRS report no. 94 110 and this volume.

[6] P. DEL MORAL, G. RIGAL, G. SALUT, *Estimation et commande optimale non linéaire: Méthodes de résolutions particulaires* Reports no. 3, 4, Contract DIGILOG-DRET no. 89.34.553.00.470.75.01, October 1992, January 1993.

[7] P. DEL MORAL, G. SALUT, *Particle Interpretation of Non-Linear Filtering and Optimization*, Russian Journal of Mathematical Physics, John Wiley and Sons 1995.

[8] P. DEL MORAL, G. SALUT,*Non-linear filtering: Monte Carlo particle methods*, Comptes Rendus de l'Académie des Sciences. Note communiquée par A. Bensoussan, 1995.

[9] P. DEL MORAL, J.C. NOYER, G. SALUT, *Maslov Optimization Theory: Stochastic Interpretation, Particle Resolution*, 11eme Conférence Internationale sur l'Analyse et l'Optimisation des Systèmes. Ecole des Mines Sophia-Antipolis, France, 15-16-17 June 1994. Lecture Notes in Control and Information Sciences 199. Springer-Verlag.

[10] B. EFRON, *The Bootstrap, Jackknife and Other Resampling Plans*, Philadelphia: Society of Industrial and Applied Mathematics, 1982.

[11] N.J. GORDON, D.J. SALMOND, A.F.M. SMITH, *Novel approach to nonlinear/non-Gaussian Bayesian State estimation*, IEEE, 1993.

[12] P. MULLER, *Monte Carlo integration in general dynamic models*, Contemp. Math., 1991, 115, pp. 145–163.

[13] D.B. RUBIN, *Using the SIR Algorithm to simulate Posterior Distributions in Bayesian Statistics*, eds. J.M. Bernardo, M.H. DeGroot, D.V. Lindley, and A.F.M. Smith, Cambridge, MA: Oxford University Press, pp. 395–402, 1988.

[14] A.F.M. SMITH, A.E. GELFAND, *Bayesian Statistics Without Tears: A sampling-Resampling Perspective*, The American Statistician, May 1992, vol. 46, Number 2.

The Geometry of Finite Dimensional Pseudomodules

Edouard Wagneur

Abstract *A semimodule M over an idempotent semiring P is also idempotent. When P is linearly ordered and conditionally complete, we call it a pseudoring, and we say that M is a pseudomodule over P. The classification problem of the isomorphism classes of pseudomodules is a combinatorial problem which, in part, is related to the classification of isomorphism classes of semilattices. We define the structural semilattice of a pseudomodule, which is then used to introduce the concept of torsion. Then we show that every finitely generated pseudomodule may be canonically decomposed into the "sum" of a torsion free sub-pseudomodule, and another one which contains all the elements responsible for the torsion of M. This decomposition is similar to the classical decomposition of a module over an integral domain into a free part and a torsion part. It allows a great simplification of the classification problem, since each part can be studied separately. For sub-pseudomodules of the free pseudomodule over m generators, we conjecture that the torsion free part, also called semiboolean, is completely characterized by a weighted oriented graph whose set of vertices is the structural semilattice of M. Partial results on the classification of the isomorphism class of a torsion sub-pseudomodule of P^m with m generators will also be presented.*

1 Pseudorings and Pseudomodules

A pseudoring $(P, +, \cdot)$ is an idempotent, commutative, completely ordered, and conditionally complete semiring with minimal element 0, such that $(P \setminus \{0\}, \cdot)$ is an integrally closed commutative group. For simplicity, the reader may think of P as $R \cup \{-\infty\}$ (or $R \cup \{\infty\}$) with $+$ the max (resp. the min) operator, and $\cdot$ the usual addition. The neutral element of $\cdot$ will be written 1 (this corresponds to the real number zero), and the 'multiplication' symbol $\cdot$ will usually be omitted. The structure of a pseudomodule (p.m. for short) over a pseudoring is defined in a similar way to a module over a ring ([1], [5], see also [3] where slightly more general structures are studied). Addition in a pseudomodule will be written additively, and **0** will also stand for its neutral element. The context will usually eliminate the risk of ambiguity.

We say that $X = \{x_1, \ldots, x_n\}$ **generates** M, or that M is the **span** of X, if any $x \neq \mathbf{0}$ in M may be written as a finite linear combination $x = \sum_{i=1}^{k} \lambda_i x_i$.

Also X is **independent** if

$$\forall Y \subset X, \quad x_i \notin Y \Rightarrow M_Y \bigcap M_{x_i} = \{\mathbf{0}\}, \tag{1.1}$$

where M_Y (M_{x_i}) is the span of Y ($\{x_i\}$).

Remark 1.1 Independence in a vector space may be defined as usual, but also by

$$\forall Y \subset X, \forall Z \subset X, \ M_Y \bigcap M_Z = M_{Y \cap Z}. \tag{1.2}$$

The free pseudomodule is defined similarly to free modules. It has been shown by Marczewski [4], that condition (1.2) characterizes the free pseudomodule. Also, a slightly weaker condition is given by

$$\forall Y \subset X, \forall Z \subset X \ Y \bigcap Z = \vee \Rightarrow M_Y \bigcap M_Z = \{\mathbf{0}\}. \tag{1.3}$$

In [1], the authors define independence by (1.3). It is easy to see that (1.1) does not imply (1.2). On the other hand, if $X = \{x_1, \ldots, x_n\}$ is a subset of the free pseudomodule P^n satisfying (1.3), then the matrix (x_{ij}) with jth column x_j, $j = 1, \ldots, n$ is nonsingular, and conversely, i.e.

$$\sum_{\sigma_e} x_{i_1 \sigma(i_1)} x_{i_2 \sigma(i_2)} \cdots x_{i_n \sigma(i_n)} \neq \sum_{\sigma_o} x_{i_1 \sigma(i_1)} x_{i_2 \sigma(i_2)} \cdots x_{i_n \sigma(i_n)},$$

where the sum runs over all even permutations σ_e on the l.h.s., and over all odd permutations σ_o on the r.h.s. Thus (1.3) relates to the regularity of the matrix whose columns are the elements of X, while the weaker concept (1.2) relates to the existence of a minimal set of generators. For example, it is easy to see that

$$X = \{(1, 1, 0, 0), \ (1, 0, 1, 0), \ (0, 1, 0, 1), \ (0, 0, 1, 1)\},$$

then (1.1) holds, while (1.3) does not.

M is partially ordered by $\leq$, where $x \leq y$ iff $x + y = y$. It follows that $(M, \leq)$ is a (sup) semilattice with least element $\mathbf{0}$, the neutral element of $+$. For $X \subset M$, the set X^+ of finite sums $\{\Sigma_i x_i \,|\, x_i \in X\}$ is a (sup) sub-semilattice of M_X. Moreover, the set of irreducible elements of X generates M_X. A *basis* of M_X (i.e. a minimal set of independent generators) is then extracted from the set of irreducible elements of X by deleting all but one of the elements of the form λx, $\lambda \in P$ ($\lambda \neq 0$). We have a first classification result.

Theorem 1.1 The free pseudomodule with n generators is isomorphic to P^n.   □

The free pseudomodule over n generators is generated by $E_n = (e_i)_{i=1}^n$, with $e_i = (\delta_{i1}, \ldots, \delta_{in})$, where δ_{ij} is the Kronecker symbol. The free semilattice E_n^+ is isomorphic to $2^n \setminus \emptyset$.

Let X, Y be two bases of M and $\Phi\colon M \to M$ an automorphism. It is well-known that Φ is isotone, i.e. $\forall x_i, x_j \in X$, $x_i \leq x_j \Rightarrow \Phi(x_i) \leq \Phi(x_l)$. Since a change of basis has the form $x_i \mapsto \lambda_i x_i$, this means that, except in the case of the free pseudomodule, the "extension by linearity" of a change of basis **is not an automorphism in general**. The following example throws some light on this unexpected situation.

Example 1.3 If $X = \{(0,1), (1,1)\}$, and $\lambda > 1$, the change of basis of M_X defined by $\varphi(x_1) = \lambda x_1$, $\varphi(x_2) = x_2$ is not a morphism. Indeed, we have $x_1 \leq x_2$, while
$\varphi(x_1) \not\leq \varphi(x_2)$. In particular, since M_X may be represented by $\mathrm{Im}A$, where $A = \begin{pmatrix} 0 & 1 \\ 1 & 1 \end{pmatrix}$, and $M_{\varphi X}$ by $\mathrm{Im}B$, with $B = \begin{pmatrix} 0 & 1 \\ \lambda & 1 \end{pmatrix}$, then this means that $B = \Phi A$ holds for no matrix Φ.

Remark 1.4 In the previous example, we argued that the matrix equation

$$\Phi \begin{pmatrix} 0 & 1 \\ 1 & 1 \end{pmatrix} = \begin{pmatrix} 0 & 1 \\ \lambda & 1 \end{pmatrix}$$

has no solution. This is equivalent to saying that the 2×2 matrix A (which has rank two) is not invertible. Again, this is a consequence of the order properties of morphisms. For A maps the antichain $\{(0,1), (1,0)\}$ onto x_1, x_2, which is a chain. Thus the inverse of A would have to map a chain onto an antichain, which is impossible.

An n-dimensional sub-pseudomodule of P^m can always be represented by a morphism $\varphi \in \mathrm{Hom}(P^n, P^m)$ with "source rank" n, i.e. by an $m \times n$ matrix of column rank n. An equivalence class of such morphisms will then correspond to a factorization of the form

$$\begin{array}{ccc} P^n & \xrightarrow{\Phi_1} & P^n \\ \varphi \downarrow & & \check{\varphi} \downarrow \\ P^m & \xleftarrow{\Phi_2} & P^m \end{array}$$

where Φ_i, $i = 1, 2$ is the product of a permutation matrix and a diagonal matrix, and $\check{\varphi}$ some "canonical" matrix whose structure has to be specified.

This type of equivalence between matrices can be decomposed into elementary steps as follows:

- multiplication of a row (column) by a scalar
- permutation of two rows (columns).

This reduction of matrices may be used to define the "standard" form of a matrix, hence of a basis, which yields an easy way to recognize the type of p.m. defined by a given map. Another problem is to be able to determine when two pseudomodules M and N are isomorphic. In Section 2 below, we introduce torsion and bending in a p.m., and recall a decomposition theorem, which may be seen as the counterpart of the classical decomposition of a module over a principal ideal domain into a biproduct of a free module and a torsion module. Some examples are provided in Section 3. Then, in Section 4, we study the classification problem for finite dimensional sub-pseudomodules of P^n.

2 Torsion and Bending in a Pseudomodule

We briefly recall the main results of [5]. As soon as $n > 1$, there are infinitely many nonisomorphic pseudomodules of dimension n. Let X be a finite basis of M. The relation $x \prec y \iff \exists \lambda \in P$ such that $x \leq \lambda y$ defines a quasi-order on M, hence also on X and X^+. It is easy to see that $x \simeq y \iff x \prec y$ and $y \prec x$ is a congruence relation with respect to the semilattice structure of M. For every $x \in M$, let c_x stand for the equivalence class of x. Clearly $\prec$ induces an order relation on the set of equivalence classes (since the context will usually limit the risk of ambiguity, we will also write $\leq$ for this order relation). Moreover, since X is unique up to a rescaling map, the graph of $(X, \prec)$ is independent of the choice of basis. Also $\{c_x \mid x \in X\}$ is the set of irreducible elements of the semilattice it generates, and this semilattice coincides with the set $S(X^+)$ of *mod* $\sim$ equivalence classes of X^+. This is summarized in the following statement ([6]).

Theorem 2.1 For any basis X, the quotient map $\pi_X : X^+ \to S(X^+) = X^+|_{\simeq}$, $x \mapsto c_x$ is an epimorphism of (join) semilattices, and $S(X^+)$ is independent of the particular choice of basis X of M. $\square$

Theorem 2.1 states in particular that the semilattice $\pi(X^+)$ is an intrinsic invariant of M_X. Thus, we will write $S(M)$ for this semilattice. Many examples show that, although necessary, it is not sufficient for the characterization of the isomorphism class of M (e.g. Examples 3.3 and 3.7 below).

Definition 2.2 The semilattice $S(M)$ is called the *structural* semilattice of M.

It is easy to see that for every basis X the map $\pi_X : X^+ \to S(M)$ satisfies: $\forall c \in S(M)$, $\pi_X^{-1}(c)$ is a semilattice. We call it the *fiber over* c.

Definition 2.3 We say that M is *semiboolean* if there is a basis X such that π is an isomorphism of semilattices.

In a general p.m., we say that $x \in S(M)$ is a *torsion element* if, for every basis X, $\pi_X^{-1}(x)$ contains at least two elements, and the fiber $\pi_X^{-1}(x)$ over x is called a **torsion cycle**.

For a given pseudomodule M, with basis X, we would like to be able to write $X = X_1 \cup X_2$ in such a way that if $M_i = M_{X_i}$, $i = 1, 2$, then M may be written as some sort of composition of M_1 and M_2. Note that if X is a chain, then its decomposition into subchains yields no trivial (i.e. direct sum) decomposition of M_X. In [6], we introduce the *sub-direct sum* of semilattices and of pseudomodules, written $M_1 \dot\oplus M_2$. Then we prove the following statement.

Theorem 2.4 Every finite dimensional pseudomodule over a pseudoring P is isomorphic to the subdirect sum $M_1 \dot\oplus M_2$ of a semiboolean pseudomodule M_1, and a torsion pseudomodule M_2. ☐

Definition 2.5 We say that M_1 (resp. M_2) is the semiboolean (resp. torsion) part of M.

Definition 2.6 A basis X is said to be **regular** if $\forall \mathbf{c} \in S(M)$, $\pi_X^{-1}(\mathbf{c})$ is minimum, i.e. for every basis Y, we have $\pi_X^{-1}(\mathbf{c}) \leq \pi_Y^{-1}(\mathbf{c})$.

The proof of the existence of regular bases is straightforward.

Let X be a regular basis of M, and let $(X^+, \prec)$ be the graph of $\prec$ on X^+. This is an oriented graph, and there is an arc (i, j) iff

i) $x_i \prec x_j$,   and

ii) $x_i \prec x_{i_1} \prec \cdots \prec x_{i_k} \prec x_j \Rightarrow \lambda_{ij} < \lambda_{i_1 i_2} < \cdots < \lambda_{i_k j}$

where $\lambda_{ij} = inf\{\lambda \,|\, x_i \leq \lambda x_j\}$. Without loss of generality, we may always assume that X is such that the number of arcs in $(X^+, \prec)$ is minimum.

Definition 2.7 We say that M is *bent* if for every basis there is an arc (i, j) in $(X^+, \prec)$ together with, for some $k \geq 1$, a $\prec$-chain $x_i \prec x_{i_1} \prec \cdots \prec x_{i_k} \prec x_j$, with $x_i \neq x_{i_1} \neq x_j$.

Example 3.6 below shows that a semiboolean p.m. may be bent, while Examples 3.9–3.10 exhibit bending in a torsion p.m. Hence both the semiboolean and the torsion pseudomodules can be decomposed into two subclasses. We will say that a semiboolean p.m. with no bending is Boolean. The definition can be made more precise as follows. Let X be a regular basis as above. The semilattice morphism $\pi_X \colon X^+ \to S(M)$ extends to a morphism of oriented graphs $\pi_{G(X)} \colon (X^+, \prec) \to (S(M), \leq)$.

Definition 2.8

We say that M is *Boolean* or *flat* if, for some regular basis X $\pi_{G(X)} \colon (X^+, \prec) \to (S(M), \leq)$ is an isomorphism of oriented graphs.

In the next section, we provide some examples, both for illustration purposes, and also to help the reader get more insight into the combinatorially complex structure of pseudomodules.

3 Examples

For a pseudomodule $M \subset P^m$ with a (finite) basis $X = \{x_1, \ldots, x_n\}$ (alternatively, for a matrix A with independent columns $x_1, x_2, \ldots, x_n$), we construct the weighted oriented graph $\Gamma_X(M) = (X^+, \lambda_{ij})$ with vertices x_i and arcs (i,j) in $(X^+, \prec)$. The weight λ_{ij} of (i,j) is defined by $\lambda_{ij} = \inf\{\lambda \in P \,|\, x_i \leq \lambda x_j\}$. Since P is conditionally complete with least element 0, these weights are well-defined.

For $\dim M = 2$, let $A = \begin{pmatrix} a & c \\ b & d \end{pmatrix}$. W.l.o.g., we may assume that $a = 1$.

Assume first that $b = 0$. Since $x_1 = \begin{pmatrix} 1 \\ 0 \end{pmatrix}$ and $x_2 = \begin{pmatrix} c \\ d \end{pmatrix}$ are independent, we necessarily have $d \neq 0$. Hence there are only two cases, $c = 0$ and $c \neq 0$.

Example 3.1 If $c = 0$, then, by an elementary column transformation (multiply x_2 by d^{-1}), we get a matrix $B = \begin{pmatrix} 1 & 0 \\ 0 & 1 \end{pmatrix} \simeq A$. Thus M is (isomorphic to) the free pseudomodule P^2. Also, $\Gamma_X(M) = X^+$ is the free semilattice $\mathbf{2}^2$.

Example 3.2 If $c \neq 0$, then, by elementary transformations (multiply x_2 by c^{-1}, and row 2 by cd^{-1}), we get a matrix $B = \begin{pmatrix} 1 & 1 \\ 0 & 1 \end{pmatrix} \simeq A$. Therefore, $\Gamma_X(M) = X^+$, with $x_1 + x_2 = x_2$, and $\lambda_{12} = 1$.

In the previous Examples, M is Boolean. In Example 3 below, M is a torsion p.m.

Example 3.3 If $b \neq 0$, then ($c \neq 0$, $d \neq 0$, and) by elementary transformations, we get a matrix $B = \begin{pmatrix} 1 & 1 \\ 1 & \lambda \end{pmatrix} \simeq A$, where we may assume w.l.o.g. that $\lambda > 1$ (for if $\lambda < 1$, just multiply the second row by λ^{-1}, and interchange the two columns of the matrix). Clearly, if $C = \begin{pmatrix} 1 & 1 \\ 1 & \mu \end{pmatrix} \simeq B$, then we must have $\mu = \lambda$. We associate with B a graph $\Gamma_X(M)$ with vertices x_1, x_2, and weights $\lambda_{12} = 1$, $\lambda_{21} = \lambda > 1$. Clearly, if $N \simeq M$, then $\Gamma_X(M) \simeq \Gamma_Y(N)$ (in the graph theoretic sense). Conversely, given $\Gamma = (x_1, x_2; \lambda_{12} = 1, \lambda_{21} = \lambda)$, we readily construct a matrix B as above, hence (a basis X and) a torsion sub-pseudomodule $M_X \subset P^2$ with $\Gamma_X(M) = \Gamma$.

In a geometrical representation, where P^2 is identified with the positive cone of the Euclidean plane, we have $\lambda = (1 + tg\alpha)/(1 - tg\alpha)$, with α the angle defined by the generators $(1, 1)$ and $(\lambda, 1)$. We may say that λ is a measure for this angle, and conclude that, geometrically, sub-pseudomodules of P^2 with this type of graph are classified by the angle of the cone defined by their generators.

The case $n = 3$ already becomes combinatorially complicated. Recall the 7 nonisomorphic structures whose graphs $\Gamma(M)$ are subsemilattices of $\mathbf{2}^3$ [5]. In addition we have the following.

Example 3.4 Let $A = \begin{pmatrix} 1 & 1 & 0 \\ 1 & 0 & 1 \\ 0 & 1 & 1 \\ 0 & 0 & 1 \end{pmatrix}$, and let $x_i \in P^4$ be given by the

ith column of A ($i = 1, 2, 3$). We have $x_1 + x_2 < x_1 + x_3 = x_2 + x_3$, which is impossible in E_3^+ when $\{x_1, x_2, x_3\}$ is an antichain. Note that $\Gamma_X(M) = X^+$.

Example 3.5 Let $A = \begin{pmatrix} 1 & 0 & 0 \\ 1 & 1 & 0 \\ 0 & 1 & 1 \\ 0 & 0 & 1 \end{pmatrix}$, with x_i, $i = 1, 2, 3$ as above. We

have $x_1 + x_2 < x_1 + x_3$, and $x_2 + x_3 < x_1 + x_3$, which is also impossible for an antichain $\{x_1, x_2, x_3\}$ in E_3^+. Also $\Gamma_X(M) = X^+$, as above.

Examples 3.4 and 3.5 show that (at least) two 3-dimensional Boolean pseudomodules cannot be embedded in P^3.

Example 3.6 Let $A = \begin{pmatrix} 1 & 1 & \xi^{-1} \\ 0 & 1 & 1 \\ 0 & 0 & 1 \end{pmatrix}$, where $\xi < 1$. With the same

notation as above, we have $x_1 \leq x_2 \leq x_3$, and $x_1 \leq \xi x_3$, i.e. the triangle inequality holds strictly, and M is a bent semi-boolean p.m. It is easy to see that, as long as we constrain $\Gamma_X(M)$ to have $\lambda_{12} = \lambda_{23} = 1$, we cannot eliminate the bending coefficient ξ from $\Gamma_X(M)$ (alternatively, there is no sequence of elementary column and row transformations leading to a matrix $B \simeq A$ with the property that $inf\{\lambda \in P \,|\, y_1 \leq \lambda y_2\} = inf\{\lambda \in P \,|\, y_2 \leq \lambda y_3\} = 1$, and $inf\{\lambda \in P \,|\, y_1 \leq \lambda y_3\} \neq \xi$, where the y_i correspond to the columns of B). Hence every pseudomodule defined by a matrix such as

$$C = \begin{pmatrix} 1 & 1 & \nu \\ 0 & \lambda & \lambda \\ 0 & 0 & \mu \end{pmatrix} \simeq \begin{pmatrix} 1 & 1 & \nu \\ 0 & 1 & 1 \\ 0 & 0 & 1 \end{pmatrix}$$

is isomorphic to M iff $\nu = \xi^{-1}$. Note that for $\dim M = 3$, and M a subpseudomodule of P^3, Example 3.6 exhibits the sole family of semiboolean bent p.m.'s. This family is characterized by the graph structure $(X^+, \prec)$, and contains infinitely many isomorphism classes, each of which is determined by the bending coefficient $\xi < 1$.

We now describe the general case of a torsion cycle in P^3.

Example 3.7 Let $A = \begin{pmatrix} a_{11} & a_{12} & a_{13} \\ a_{21} & a_{22} & a_{23} \\ a_{31} & a_{32} & a_{33} \end{pmatrix}$. By elementary column and row

transformations, we get $B = \begin{pmatrix} 1 & 1 & c \\ 1 & a & d \\ 1 & b & e \end{pmatrix} \simeq A$, with $a = a_{11}a_{22}(a_{12}a_{21})^{-1}$,

$b = a_{11}a_{32}(a_{12}a_{31})^{-1}$, $c = a_{11}^{-1}a_{13}$, $d = a_{21}^{-1}a_{23}$, and $e = a_{31}^{-1}a_{33}$. W.l.o.g., we may assume that $1 \leq a \leq b$, $1 \leq c$, $a \leq d$, and $b \leq e$.

The following example shows that there exists a torsion pseudomodule $M \subset P^3$ of arbitrary dimension $n > 1$ ($n = \infty$ included).

Example 3.8 Let $x_i = (1, i, i^2)'$, $i = 1, 2, \ldots$. The x_i are independent. Indeed, if x_j were dependent, then it could be expressed as a nonredundant linear combination of three of the other x_i's only (since there are only 3 components): $x_j = \lambda_i x_i + \lambda_k x_k + \lambda_\ell x_\ell$, say. Then we would have

$$\lambda_i + \lambda_k + \lambda_\ell = 1, \tag{2.1}$$

$$i\lambda_i + k\lambda_k + \ell\lambda_\ell = j, \tag{2.2}$$

$$i^2\lambda_i + k^2\lambda_k + \ell^2\lambda_\ell = j^2. \tag{2.3}$$

From (2.1), we must have $\lambda_p \leq 1$, $p = i, k, \ell$, with equality in at least one case. W.l.o.g. we may assume $\lambda_i = 1$, therefore $i < j$, and the system becomes:

$$k\lambda_k + \ell\lambda_\ell = j, \tag{2.2'}$$

$$k^2\lambda_k + \ell^2\lambda_\ell = j^2. \tag{2.3'}$$

W.l.o.g., we may assume that $k\lambda_k = j$. But then $k^2\lambda_k = kj \leq j^2$, by (2.3). Hence $k \leq j$. Since $k \neq j$, we necessarily have $k < j$, and (since $\lambda_k \leq 1$ by (2.1)) we get the contradiction $k\lambda_k < j$.

To get a geometric insight into such pseudomodules, just take a polygon Q with vertices $x_1, \ldots, x_n$ in general position in P^3. Then M is generated by the vertices of Q.

Our last two examples show that the generators of a torsion p.m. need not belong to a torsion cycle.

Example 3.9 Let M be given by $A = \begin{pmatrix} 1 & 1 & 0 \\ 1 & 0 & 1 \\ 0 & 1 & \lambda \end{pmatrix}$, with $\lambda > 1$. The graph $(X^+, \leq)$ has 5 vertices:

$$x_1 = (1, 1, 0),$$

$$x_2 = (1, 0, 1),$$

$$x_3 = (0, 1, \lambda),$$

$$x_4 = x_1 + x_2,$$

$$x_5 = x_1 + x_3 = x_2 + x_3,$$

with arc weights

$$\lambda_{14} = \lambda_{15} = \lambda_{24} = \lambda_{25} = \lambda_{35} = \lambda_{45} = 1, \text{ and } \lambda_{54} = \lambda > 1.$$

A torsion cycle binds the vertices x_4 and x_5.

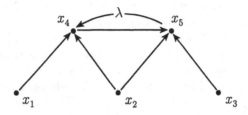

Example 3.10 Let $A = \begin{pmatrix} 1 & 0 & 0 & 1 \\ 0 & 1 & 1 & 0 \\ 1 & 0 & 1 & 0 \\ 0 & 1 & 0 & \lambda \end{pmatrix}$, with $\lambda > 1$. The generators are

an antichain, as well as $x_1 + x_3 = (1, 1, 1, 0)$, $x_2 + x_3 = (0, 1, 1, 1)$, $x_1 + x_4 = (1, 0, 1, \lambda)$ and $x_2 + x_4 = (1, 1, 0, \lambda)$, while $x_1 + x_2 = (1, 1, 1, 1)$ and $x_3 + x_4 = (1, 1, 1, \lambda)$ belong to a torsion cycle, i.e. $x_1 + x_2 < x_3 + x_4 < \lambda x_1 + x_2$.

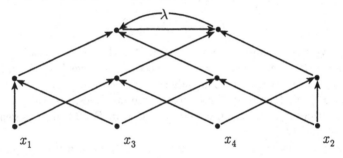

4 Classification

4.1 Finite dimensional semi-boolean pseudomodules

In [6], we showed how an arbitrary p.m. splits into a sub-direct sum of a semiboolean p.m. and a torsion p.m. by first eliminating irreducible torsion elements, then pairs of irreducible elements which generate torsion elements, and so on.

Similarly, in a semiboolean p.m., for $k = 1, \ldots$, we eliminate successively the irreducible elements $c_{i_j} \in S(M)$ such that $\sum_{j=1}^{k} c_{i_j} \in S(M)$ with $\pi_X^{-1}(\sum_{j=1}^{k} c_{i_j})$ the first or last element of a bent chain $x_1 \leq x_2 \leq \ldots \leq x_\ell$, with $x_1 \leq \lambda_{1\ell} x_\ell$, $\lambda_{1\ell} < 1$, and $\lambda_{ij} = 1$ whenever $(ij) \neq (1\ell)$. After a finite number of steps the remaining set $\{c_1, c_2, \ldots, c_m\}$ of irreducible elements of $S(M)$, if not empty, generates a semilattice $S_1(M)$. Then, if

$X_1 = \{\pi_X^{-1}(c_1), \pi_X^{-1}(c_2), \ldots, \pi_X^{-1}(c_m)\}$, the graph $(X_1^+, \prec)$ is isomorphic to $(S_1(M), \leq)$. In other words, the p.m. generated by X_1 is Boolean. It follows that $M = M_1 \dot{\oplus} M_2$, with M_1 a Boolean p.m., called the flat part of M. Accordingly, M_2 is called the bent part of M.

We state the following conjecture.

Conjecture 4.1

A. Flat sub-pseudomodules of P^n are classified by semilattices.

B. The classification of bendt p.m.'s is given by weighted oriented graphs $(X^+, \prec, \lambda_{ij})$, where X is a regular basis for which the number of arcs in $(X^+, \prec)$ is minimum, and where the λ_{ij} are the least weights satisfying $x_i \leq \lambda_{ij} x_j$.

Just as in the case of semilattices, the classification of weighted oriented graphs such as $(X^+, \prec, \lambda_{ij})$ is a combinatorial problem which requires further investigation. We have the following lemma.

Lemma 4.2 In a bent chain $x_1 \leq x_2 \leq \ldots \leq x_k$, the sequence λ_{ij} of bending coefficients is nondecreasing in the first argument, and nonincreasing in the second.

Proof It is enough to show that the statement holds for an arbitrary bended sub-chain $i_1 < i_2 < i_3 < i_4$, with $x_{i_1} \leq x_{i_2} \leq x_{i_3} \leq x_{i_4}$. We have $x_{i_1} \leq x_{i_2} \leq \lambda_{i_2 i_4} x_{i_4} \Rightarrow \lambda_{i_1 i_4} \leq \lambda_{i_2 i_4}$. Similarly, $x_{i_1} \leq \lambda_{i_1 i_3} x_{i_3} \leq \lambda_{i_1 i_3} x_{i_4}$, hence, we necessarily have $\lambda_{i_1 i_4} \leq \lambda_{i_1 i_3}$. $\square$

Example 4.3 Let M be generated by the columns of $A = \begin{pmatrix} 1 & 1 & \lambda & \xi \\ 0 & 1 & 1 & \mu \\ 0 & 0 & 1 & 1 \\ 0 & 0 & 0 & 1 \end{pmatrix}$,

$\lambda < \xi$. Writing x_i for the ith column of A, we have

$$x_1 \leq \lambda^{-1} x_3, \; x_1 \leq \xi^{-1} x_4 \text{ with } \xi^{-1} = \lambda_{14} < \lambda_{13} = \lambda^{-1}, \text{ and}$$

$$x_2 \leq (\xi^{-1} + \mu^{-1}) x_4 \text{ with } \lambda_{14} = \xi^{-1} \leq (\xi^{-1} + \mu^{-1}) = \lambda_{24}.$$

4.2 The case of torsion pseudomodules

We restrict to the case of a torsion cycle consisting of n irreducible elements in P^n. We show that such a torsion cycle is defined by a set of $(n-1)^2$ sufficient parameters. These, in turn, determine $(n-1)^2$ necessary invariants. Then we are led to the question: When are the necessary conditions also sufficient? This problem will be given a purely algebraic content below. More precisely, we show that it is equivalent to the uniqueness of the solution to a set of quadratic equations over P. Finally, we give an extensive development of the case $n = 3$, and provide an example showing that the necessary conditions are not sufficient in general.

Let A be a matrix with entries $a_{ij} \neq 0$, $1 \leq i, j \leq n$. Write $x_{\sigma(i)}$ for the columns of A, where $\sigma \in S_n$ has to be determined. We normalize A in such a way that, for the new columns $\xi_{\sigma(i)}x_{\sigma(i)}$, we have:

$$\xi_{\sigma(i)}x_{\sigma(i)} \leq \xi_{\sigma(i+1)}x_{\sigma(i+1)}, \tag{4.1}$$

and

$$\xi_{\sigma(i+1)}x_{\sigma(i+1)} \leq \lambda_{\sigma(i+1),\sigma(i)}\xi_{\sigma(i)}x_{\sigma(i)}, \tag{4.2}$$

with $1 < \lambda_{21} \leq \cdots \leq \lambda_{i+1,i} \leq \cdots \leq \lambda_{n,n-1}$. By (4.1), we have $a_{j\sigma(i)} \leq \xi_{\sigma(i)}^{-1}\xi_{\sigma(i+1)}a_{j\sigma(i+1)}$, $j = 1, \ldots, n$, hence

$$\xi_{\sigma(i+1)} = \xi_{\sigma(i)}\sum_{j=1}^{n} a_{j\sigma(i)}a_{j\sigma(i+1)}^{-1}. \tag{4.3}$$

Similarly, (4.2) yields

$$\lambda_{\sigma(i+1),\sigma(i)} = \xi_{\sigma(i)}^{-1}\xi_{\sigma(i+1)}\sum_{j=1}^{n} a_{j\sigma(i+1)}a_{j\sigma(i)}^{-1}. \tag{4.4}$$

It follows that

$$\lambda_{\sigma(i+1),\sigma(i)} = \sum_{j=1}^{n} a_{j\sigma(i+1)}a_{j\sigma(i)}^{-1}\sum_{k=1}^{n} a_{k\sigma(i)}a_{k\sigma(i+1)}^{-1}$$

$$= \sum_{j,k=1}^{n} a_{j\sigma(i)}a_{k\sigma(i+1)}a_{j\sigma(i+1)}^{-1}a_{k\sigma(i)}^{-1}$$

$$= \sum_{j,k=1}^{n} \begin{vmatrix} a_{j\sigma(i)} & a_{j\sigma(i+1)} \\ a_{k\sigma(i)} & a_{k\sigma(i+1)} \end{vmatrix}, \tag{4.5}$$

where $\begin{vmatrix} a & c \\ b & d \end{vmatrix} = ad(bc)^{-1}$ is called the *pseudodeterminant* of $\begin{pmatrix} a & c \\ b & d \end{pmatrix}$. Then σ will be the permutation which yields a minimal element w.r.t. the lexicographic order, in the set of vectors of torsion and bending coefficients:
$$(\lambda_{21}, \ldots, \lambda_{n,n-1}, \lambda_{31}, \ldots, \lambda_{n-1,n-3}, \ldots\ldots, \lambda_{i+k,i}, \ldots, \lambda_{n1};$$
$$\lambda_{13}, \ldots, \lambda_{1n} \ldots\ldots \lambda_{i,i+k}, \ldots, \lambda_{n-2,n}).$$

Assume that the columns of A (i.e. the generators of M) are ordered as specified above. Then it is easy to see that A may be assumed to have the form

$$A = \begin{pmatrix} 1 & 1 & a_{13} & \cdots & a_{1n} \\ 1 & a_{22} & a_{23} & \cdots & a_{2n} \\ \cdot & \cdot & \cdot & \cdots & \cdot \\ \cdot & \cdot & \cdot & \cdots & \cdot \\ 1 & a_{n2} & a_{n3} & \cdots & a_{nn} \end{pmatrix},$$

with $1 \leq a_{22} \leq a_{32} \leq \ldots \leq a_{n2}$. Also, since for $j = 1, \ldots, n-1$, the columns of A satisfy $x_j \leq x_{j+1}$, with $1 = inf\{\lambda \mid x_j \leq \lambda x_{j+1}\}$, we necessarily have $a_{ij+1} = a_{ij}$ for at least one i ($1 \leq i \leq n$). Note that we can always achieve this in an arbitrary torsion matrix by multiplying column $j+1$ by $\sum_{i=1}^{n} a_{ij} a_{ij+1}^{-1}$.

Let X be the set of columns of A, and $M = M_X$.

Definition 4.4 We say that X is a *canonical* basis of M, and A is written in *canonical* form.

Let A be given in canonical form: then $\lambda_{jk} = \sum_{i=1}^{n} a_{ik}^{-1} a_{ij}$, $1 \leq j, k \leq n$, and we have

$$A^t A^- = \Lambda_A, \tag{4.6}$$

where $A^- = (a_{ij}^{-1})$. Then

$$\Lambda_A = \begin{pmatrix} 1 & 1 & \lambda_{13} & \ldots & \lambda_{1n} \\ \lambda_{21} & 1 & 1 & \ldots & \lambda_{2n} \\ \ldots & \ldots & \ldots & \ldots & \ldots \\ \lambda_{n-11} & \lambda_{n-12} & \ldots & 1 & 1 \\ \lambda_{n1} & \lambda_{n2} & \ldots & \ldots & 1 \end{pmatrix}. \tag{4.7}$$

Problem 4.5 Assume Λ is given by (4.7). Then when does (4.6) have a unique solution A?

Assume that the solution to (4.6) is unique. This means that the graph $\Gamma_X(M)$ characterizes the isomorphism class of M. Hence the statements of Section 4.1 above can be extended to torsion p.m.'s. For $n = 3$, we have $a_{13} = 1$, or $a_{23} = a_{22}$, or $a_{33} = a_{32}$, and the three matrices

$$\begin{pmatrix} 1 & 1 & 1 \\ 1 & a & \alpha \\ 1 & b & \beta \end{pmatrix}, \begin{pmatrix} 1 & 1 & \alpha \\ 1 & a & a \\ 1 & b & \beta \end{pmatrix}, \begin{pmatrix} 1 & 1 & \alpha \\ 1 & a & \beta \\ 1 & b & b \end{pmatrix}$$

yield 11 torsion matrices, and associated torsion graphs as follows:

$$\begin{pmatrix} 1 & 1 & \lambda_{13} \\ b & 1 & 1 \\ \lambda_{31} & \lambda_{32} & 1 \end{pmatrix}, \quad \text{and} \quad$$

Depending on the relative values of the entries, there are three cases: $a\beta(ba)^{-1} \geq 1$, and $a\beta(ba)^{-1} < 1$ with $\beta < \alpha$ or $\alpha < \beta$.

For $a_{11} = 1$

$$\begin{pmatrix} 1 & 1 & 1 \\ b & 1 & 1 \\ \beta & b^{-1}\beta & 1 \end{pmatrix} \qquad \begin{pmatrix} 1 & 1 & 1 \\ b & 1 & 1 \\ \alpha & a^{-1}\alpha & 1 \end{pmatrix} \qquad \begin{pmatrix} 1 & 1 & 1 \\ b & 1 & 1 \\ \beta & a^{-1}\alpha & 1 \end{pmatrix}$$

case1 : $a\beta \geq b\alpha$ case2 : $a\beta < b\alpha$ case3 : $a\beta < b\alpha$
with $\beta < \alpha$ with $\alpha < \beta$.

For $a_{23} = a_{22}$

$$\begin{pmatrix} 1 & 1 & \alpha^{-1} \\ b & 1 & 1 \\ \beta & b^{-1}\beta & 1 \end{pmatrix} \qquad \begin{pmatrix} 1 & 1 & a^{-1} \\ b & 1 & 1 \\ \beta & b^{-1}\beta & 1 \end{pmatrix} \qquad \begin{pmatrix} 1 & 1 & a^{-1} \\ b & 1 & 1 \\ \alpha & \alpha & 1 \end{pmatrix}$$

case4 : $\alpha < a$ case5 : $a < \alpha, b\alpha < \beta$ case6 : $a < \alpha, \beta < b\alpha$

For $a_{33} = a_{32}$

$$\begin{pmatrix} 1 & 1 & \alpha^{-1} \\ b & 1 & 1 \\ \beta & a^{-1}\beta & 1 \end{pmatrix} \qquad \begin{pmatrix} 1 & 1 & \beta^{-1} \\ b & 1 & 1 \\ \alpha & \alpha & 1 \end{pmatrix} \qquad \begin{pmatrix} 1 & 1 & b^{-1} \\ b & 1 & 1 \\ \alpha & \alpha & 1 \end{pmatrix}$$

case7 : $\alpha < b < a^{-1}\beta$ case8 : $\beta < b < \alpha$ case9 : $b < \beta < \alpha$

together with, when $b < \alpha < \beta$:

$$\begin{pmatrix} 1 & 1 & b^{-1} \\ b & 1 & 1 \\ \beta & \alpha & 1 \end{pmatrix} \qquad \begin{pmatrix} 1 & 1 & b^{-1} \\ b & 1 & 1 \\ \beta & a^{-1}\beta & 1 \end{pmatrix}$$

case10 : $\beta < a\alpha$ case11 : $a\alpha < \beta$

The following statement shows that Problem 4.5 has no elementary solution.

Theorem 4.6 If M is a nonbent three-dimensional torsion pseudomodule, then the solution to (4.6) is not unique in general.

Proof Let

$$A = \begin{pmatrix} 1 & 1 & 1 \\ 1 & a & c \\ 1 & b & d \end{pmatrix}, \quad B = \begin{pmatrix} 1 & 1 & 1 \\ 1 & a & c' \\ 1 & b & d \end{pmatrix},$$

where $a < b < d$, $a < c < c' < d$, and $bc < ad$, the M_A and M_B are not isomorphic, while $\Gamma_A = \Gamma_B$. This last equality is easy. Now, take $\xi_2, \xi_3 \leq \xi_1$, with $a^{-1}c\xi_3 \leq \xi_2 \, le \, b^{-1}d\xi_3$, and $a\xi_2\xi_3^{-1} < c'$. Then if $\varphi: M_A \to M_B$ were an isomorphism, we would necessarily have $\varphi(a_i) = b_i$, where a_i (b_i) stands for the ith column of A (B). But it is easy to see that $x = \sum_{i=1}^{3} \xi_i a_i$

is a nonredundant combination of the a_i, while $\varphi(x) = \xi_1 b_1 + \xi_3 b_3 = \varphi(\xi_1 a_1 + \xi_3 a_3)$. Hence φ is not injective. □

Although Theorem 4.6 shows that there is no elementary solution to our Problem 4.5, there are cases where the problem can be solved (in particular for $n = 2$). For a complete solution to the problem, i.e. the conditions required for the uniqueness of the solution to (4.6), more analysis has to be done, in particular the cases listed above for $n = 3$ have to be further investigated.

Acknowledgements This research has been supported by the ESPRIT-III basic research project SESDIP.

BIBLIOGRAPHY

[1] COHEN, G., MOLLER, P., QUADRAT, J.P., VIOT, M., Une théorie linéaire des systèmes à événements discrets. Rapport de recherche IN-RIA # 362. Le Chesnay, 1985.

[2] CUNINGHAME-GREEN, R. A., Minimax Algebra, Lecture Notes in Economics and Mathematical Systems, 83, SpringerVerlag, 1979.

[3] GAUBERT, S., Théorie des Systèmes linéaires dans les Dioïdes. Thèse. Ecole nationale Supérieure des Mines de Paris, juillet 1992.

[4] MARCZEWSKI, E., A General Scheme of the Notion of Independence, Bull. Acad. pol. Sc., Math., Astr. et Phys. Vol. VI, No. 12, 1958.

[5] WAGNEUR, E., Moduloids and Pseudomodules. 1. Dimension Theory. Discr. Math., 98, 57–73, 1991.

[6] WAGNEUR, E., Subdirect sum decomposition of finite dimensional pseudomodules. 11th Int. Conf. on Anal. and Sys. Opt. , Discr. Event. Sys. Ecole des Mines, Sophia-Antipolis, 15–17 June 1994, pp 322–328, Springer-Verlag.

A General Linear Max-Plus Solution Technique

Elizabeth A. Walkup and Gaetano Borriello

1   Introduction and Background

This paper gives a general method for solving linear max-plus systems and optimizing linear max-plus expressions subject to a linear max-plus system. Such systems are of interest both as mathematical objects and for their ability to represent timing relationships between systems of synchronizing events. Previous solution techniques for linear max-plus systems center upon the use of **max-plus balances** [1], a symmetrization of the max-plus algebra that allows square systems to be solved with the max-plus analog of Cramer's rule. Gaubert [2] extends the balance technique for use with non-square systems, but the technique is non-intuitive, and also computationally expensive since for a system over n variables it requires solving all square subsystems of size $n - 1$.

As with Gaubert's method, the method presented in this paper solves sub-systems of the original linear max-plus system, but it is much simpler in that it relies upon the more intuitive max-plus closure rather than balances to solve the problem, and solves a series of subsystems which result in decreasing bounds on the maximum solution to the original system. Space considerations prohibit us from providing the full background required for this paper, so the reader is referred to Baccelli *et al. 's* text [1], as well as other sources referenced here.

The max-plus algebra is the dioid $(\mathcal{R} \cup \epsilon, \oplus, \otimes)$ where $\oplus$ is the binary maximum operator, $\otimes$ is the familiar addition, and $\epsilon = -\infty$. This work is concerned with **linear max-plus systems**, namely those which can be expressed as a matrix equation

$$[A \otimes x] \oplus b = [C \otimes x] \oplus d, \qquad (1.1)$$

representing m equations over the n variables, $\{x_0, \ldots, x_{n-1}\}$. The ith such equation is

$$\left[\bigoplus_{j=0}^{n-1} A_{i,j} \otimes x_j \right] \oplus b_i = \left[\bigoplus_{j=0}^{n-1} C_{i,j} \otimes x_j \right] \oplus d_i. \qquad (1.2)$$

A and C are $m \times n$ matrices, and b and d are m element column vectors, all with entries in $\mathcal{R} \cup \epsilon$, and x is a column vector of n variables. Matrix operations are built from the primitive operations as expected. **Linear max-plus expressions** have the form

$$\left[\bigoplus_{j=0}^{n-1} a_j \otimes x_j\right] \oplus b, \tag{1.3}$$

where b and the a_j's are elements of $\mathcal{R} \cup \epsilon$.

We assume the reader is familiar with max-plus **matrix closure**, denoted by A^* for the matrix A, and defined as the minimum solution for the square matrix X in the equation

$$X = [A \otimes X] \oplus E_n, \tag{1.4}$$

where E_n is the $n \times n$ identity matrix for the $\otimes$ operation, and consists of entries 0 along its diagonal, and ϵ everywhere else. A recursive formulation for closure is derived by Baccelli *et al.* [1] as

$$\begin{bmatrix} W & X \\ Y & Z \end{bmatrix}^* = \begin{bmatrix} W^* \oplus W^*X(YW^*X \oplus Z)^*YW^* & W^*X(YW^*X \oplus Z)^* \\ (YW^*X \oplus Z)^*YW^* & (YW^*X \oplus Z)^* \end{bmatrix}. \tag{1.5}$$

Intuitively, if each matrix entry $A_{i,j}$ is the length of a directed edge to node i from node j in a graph, then the length of the longest path to i from j is $A_{i,j}^*$ [3, 1].

2   The Upper Bound Technique

In this section, we show how to find the maximum solution to an arbitrary system of linear max-plus equations using the following three steps:

- translate each linear max-plus equation into a small set of **upper bound constraints**, each of which bounds the value of a single variable from above (Section 2.1),

- employ the max-plus closure operation to find the maximum solution to a special subset of the upper bound constraints (Section 2.2), and

- use that subset's maximum solution to guide the choice of a new constraint subset which will have a smaller maximum solution (Section 2.3).

The last two steps above are repeated until either the process converges upon a solution which meets all the upper bound constraints, or it is found that the system is **infeasible** because some variable has a maximum solution of ϵ. Section 2.4 addresses the problem of finding an appropriate initial UBC subset.

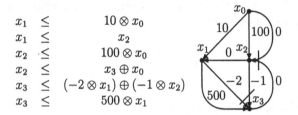

$$
\begin{aligned}
x_1 &\le 10 \otimes x_0 \\
x_1 &\le x_2 \\
x_2 &\le 100 \otimes x_0 \\
x_2 &\le x_3 \oplus x_0 \\
x_3 &\le (-2 \otimes x_1) \oplus (-1 \otimes x_2) \\
x_3 &\le 500 \otimes x_1
\end{aligned}
$$

Figure 1: A set of UBCs and the graph they induce.

2.1   Upper bound constraints

Definition 2.1 *An* **upper bound constraint** *or* **UBC** *has the form*

$$
x_{\tau(i)} \le \bigoplus_{j=0}^{n-1} A_{i,j} \otimes x_j, \tag{2.1}
$$

where $\tau(i)$ is used to index the variable on the left hand side of the ith such UBC. A single x_i may be the **target**, *or left hand side, of several UBCs.*

Lemma 2.1 *In any system of UBCs as given in Equation 2.1 either the system is infeasible, or each such equation may be written so that $A_{i,\tau(i)} = \epsilon$.*

Proof Suppose that the system is feasible but there is at least one i for which $A_{i,\tau(i)} \ne \epsilon$. We may then write Equation 2.1 as

$$
x_{\tau(i)} \le a \oplus \left(A_{i,\tau(i)} \otimes x_{\tau(i)} \right) \tag{2.2}
$$

where the expression a contains no $x_{\tau(i)}$ terms. Then if $A_{i,\tau(i)} \ge 0$, the ith UBC may be removed from the system since it is always satisfied. Alternatively, if $A_{i,\tau(i)} < 0$ then $A_{i,\tau(i)} \otimes x_{\tau(i)}$ will never be the maximum term in $a \oplus (A_{i,\tau(i)} \otimes x_{\tau(i)})$, since that maximum must be at least as large as $x_{\tau(i)}$. We may then write the UBC as $x_{\tau(i)} \le a$. Should any UBC be reduced to $x_{\tau(i)} \le \epsilon$, the system must be infeasible.   □

Figure 1 gives an example of a convenient graphical representation of a system of UBCs. For every system variable, there is a node labeled with the variable's name. For every UBC there is a set of arcs, one to represent each non-ϵ term $A_{i,j}$ of the UBC's right hand side. Such an arc points from x_j to $x_{\tau(i)}$, and all arcs derived from the same UBC are bundled together with a line drawn through them near their arrow end. This indicates which arcs represent terms in the same constraint. A set of UBCs **induces** such a graph.

A system of UBCs may also be expressed in matrix form as follows. We first take note of the fact that for the max-plus algebra the equations $a \le b$

and $a \oplus b = b$ are equivalent, and thus each equation of the form given in Equation 2.1 may also be written as

$$x_{\tau(i)} \oplus \left[\bigoplus_{j=0}^{n-1} A_{i,j} \otimes x_j\right] = \bigoplus_{j=0}^{n-1} A_{i,j} \otimes x_j. \qquad (2.3)$$

We can combine all such UBC representations for a given system into the single matrix equation

$$(J \oplus A) \otimes x = A \otimes x \qquad (2.4)$$

where A is the $m \times n$ matrix of entries $A_{i,j}$ and J is the $m \times n$ matrix with entries $J_{i,\tau(i)} = 0$ and all other entries equal to ϵ. For the system given in Figure 1 the J and A matrices are

$$J = \begin{bmatrix} \epsilon & 0 & \epsilon & \epsilon \\ \epsilon & 0 & \epsilon & \epsilon \\ \epsilon & \epsilon & 0 & \epsilon \\ \epsilon & \epsilon & 0 & \epsilon \\ \epsilon & \epsilon & \epsilon & 0 \\ \epsilon & \epsilon & \epsilon & 0 \end{bmatrix} \text{ and } A = \begin{bmatrix} 10 & \epsilon & \epsilon & \epsilon \\ \epsilon & \epsilon & 0 & \epsilon \\ 100 & \epsilon & \epsilon & \epsilon \\ 0 & \epsilon & \epsilon & 0 \\ \epsilon & 500 & \epsilon & \epsilon \\ \epsilon & -2 & -1 & \epsilon \end{bmatrix}. \qquad (2.5)$$

2.1.1   Expressing arbitrary linear max-plus equations with UBCs

We can express any linear max-plus system with upper bound constraints in the following manner:

- Create a new set of variables $\mathcal{X} = \{x_0, \ldots, x_{n-1}\} \cup \{\hat{x}_0, \ldots, \hat{x}_{m-1}\} \cup \{x_\emptyset\}$, where the x_i terms are as in the original system, and there is a $\hat{x}_i$ term for each equation of the system. The variable $x_\emptyset$ represents 0.

- For each row i of a system as given above in Equation 1.2 we create upper bound constraints which will force each $\hat{x}_i$ to equal both sides of the equation it represents. To bound each $\hat{x}_i$ from above we add:

$$\hat{x}_i \leq \left[\bigoplus_{j=0}^{n-1} A_{i,j} \otimes x_j\right] \oplus [b_i \otimes x_\emptyset], \text{ and } \hat{x}_i \leq \left[\bigoplus_{j=0}^{n-1} C_{i,j} \otimes x_j\right] \oplus [d_i \otimes x_\emptyset].$$

To bound each $\hat{x}_i$ from below we note that if $\hat{x}_i$ is greater than the maximum of several terms, it must be greater than each of them, and so we add:

$$\begin{aligned} x_j &\leq -A_{i,j} \otimes \hat{x}_i, \text{ for all } j \text{ such that } A_{i,j} > \epsilon \\ x_j &\leq -C_{i,j} \otimes \hat{x}_i, \text{ for all } j \text{ such that } C_{i,j} > \epsilon \\ x_\emptyset &\leq -b_i \otimes \hat{x}_i \text{ for all } i \text{ such that } b_i > \epsilon \\ x_\emptyset &\leq -d_i \otimes \hat{x}_i \text{ for all } i \text{ such that } d_i > \epsilon. \end{aligned}$$

Here a unary minus sign indicates the standard additive inverse, which is the $\otimes$ inverse in this scheme. The variables are then reordered so that x_0 takes on the role of $x_\emptyset$. The maximum solution to the original system can then be found by finding the maximum solution to the new system when $x_0 = 0$.

2.2   Bounding targeting subsets

Definition 2.2 *A **targeting subset** of a system of UBCs over n variables is a set of $n - 1$ UBCs such that each variable $x_i \neq x_0$ is the target of exactly one UBC. For a targeting subset, the corresponding **target matrix** is the $n \times n$ matrix $\mathcal{P}A$ in which the entry $(\mathcal{P}A)_{i,j}$ is the length of the arc to x_i from x_j in the targeting subset's induced graph. When no such arc exists, $(\mathcal{P}A)_{i,j} = \epsilon$.*

Definition 2.3 *A targeting subset is called a **safe targeting subset** if all cycles in the graph induced by the targeting subset have negative weight.*

Theorem 2.1 *(Baccelli et al. [1], Theorem 3.17) For a matrix A with induced graph $\mathcal{G}(A)$, if the cycle weights in $\mathcal{G}(A)$ are all negative then there is a unique solution to the equation $x = Ax \oplus b$, which is given by $x = A^*b$.*

Lemma 2.2 *Given a safe targeting subset of upper bound constraints,*

$$x_i \leq \bigoplus_{j=0}^{n-1} (\mathcal{P}A)_{i,j} \otimes x_j \qquad (2.6)$$

the vector $l = [l_0, l_1, \ldots, l_{n-1}]^T$ where $l_i = (\mathcal{P}A)^*_{i,0}$ is the maximum solution to the constraint subset when $x_0 = 0$

Proof By Theorem 2.1 above, the values $l_i = (\mathcal{P}A)^*_{i,0}$ are the unique solution to the equations $x_0 = 0$ and, for $i \neq 0$,

$$x_i = \bigoplus_{j=0}^{n-1} (\mathcal{P}A)_{i,j} \otimes x_j. \qquad (2.7)$$

Since this set of equations is stronger than the constraints given in Equation 2.6 we know that the vector l gives a solution to our constraint subset. Furthermore, any solution to the system of Equation 2.6 in which any l_i is strictly less than its upper bound of

$$\bigoplus_{j=0}^{n-1} (\mathcal{P}A)_{i,j} \otimes l_j \qquad (2.8)$$

is not a maximum solution since increasing the value of l_i to equal its upper bound as given in Equation 2.8 cannot cause any other UBC in the safely targeting subset to be violated. $\square$

2.3   Choosing converging subsets

As we showed in Section 2.2, it is easy to calculate the maximum solution to a safe targeting subset of UBCs. Since such a solution is derived from a subset of the total system constraints, it is not guaranteed to be a solution to the entire UBC system. However, since a solution to the whole system must satisfy any subset of the constraints, the maximum solution to a safe targeting subset must bound from above any solution to the system proper. In this section we show how given one safe targeting subset and its associated maximum solution, we may find another safe targeting subset whose maximum solution is smaller.

If l, the maximum solution to the current safe targeting subset, is not a solution to the entire UBC system, then there must be at least one UBC, u_i, which is not currently satisfied and is not in the current targeting subset. If constraint u_i is as given in Equation 2.1 then we may determine that u_i is currently unsatisfied if

$$l_{\tau(i)} > \bigoplus_{j=0}^{n-1} A_{i,j} \otimes l_j. \tag{2.9}$$

The next two subsections show that if we replace the UBC currently targeting $x_{\tau(i)}$ with u_i, the resulting targeting subset is safe (Section 2.3.1), and will yield a new maximum solution, l', for which $l_j \geq l'_j$ for all j and $l_{\tau(i)} > l'_{\tau(i)}$ (Section 2.3.2). Section 2.4 gives the method for finding an initial safe targeting subset.

2.3.1   The new target subset is safe

Definition 2.4 *If $\mathcal{U}$ is the set of all UBCs in a system, then given a UBC, $u_i \in \mathcal{U}$,*

$$u_i : x_{\tau(i)} \leq \bigoplus_{j=0}^{n-1} A_{i,j} \otimes x_j, \tag{2.10}$$

*and a vector $l = [l_0, \ldots, l_{n-1}]^T$ of values bounding the variables $\mathcal{X}$ of a system, $u_i(l)$, the **value of u_i subject to l**, is the value*

$$u_i(l) = \bigoplus_{j=0}^{n-1} A_{i,j} \otimes l_j. \tag{2.11}$$

Lemma 2.3 *Given a safe targeting subset $\mathcal{S}$ of $\mathcal{U}$, with maximum solution l, a UBC $v \in \mathcal{S}$ targeting $x_{\tau(i)}$ and an additional UBC $u \in \mathcal{U}$ also targeting $x_{\tau(i)}$, if $u(l) < l_{\tau(i)}$ then $\mathcal{S}' = \mathcal{S} \cup \{u\} - \{v\}$ is also a safe targeting subset.*

Proof Clearly $\mathcal{S}'$ is a targeting subset since one UBC targeting $x_{\tau(i)}$ has been exchanged for another. $\mathcal{S}'$ is safe so long as the total path weights of all cycles in its induced subgraph are less than 0. Thus, we assume that a "bad" cycle

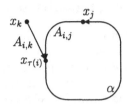

Figure 2: Diagram supporting the argument for Lemma 2.3.

of weight 0 or greater does exist and then show this contradicts the hypothesis that $u(l) < l_{\tau(i)}$. The diagram in Figure 2 accompanies the argument below.

If the exchange of u for v introduces a cycle of weight 0 or greater into the induced graph, it must be the case that any such cycle includes one of the arcs induced by constraint u. Let x_j be at the tail end of such an arc, and let $A_{i,j}$ be the weight of that arc. Furthermore, let α be the $\otimes$ of all edge weights of the path from $x_{\tau(i)}$ to x_j along the corresponding "bad" cycle. All the arcs which contribute to the value of α are present in the graphs induced by both S and S', since only the edges directed to $x_{\tau(i)}$ have changed. Since the cycle in S' has weight 0 or greater, it must be the case that

$$\alpha \otimes A_{i,j} \geq 0. \tag{2.12}$$

Consider now what this means for the values in l. The longest path from $x_{\tau(i)}$ to x_j in the graph induced by S must have weight $\geq \alpha$ since all directed edges along that α-weight path are in S. Therefore

$$l_j \geq \alpha \otimes l_{\tau(i)}. \tag{2.13}$$

We also know that

$$u(l) \geq A_{i,j} \otimes l_j \tag{2.14}$$

since $u(l)$ was the $\oplus$ of several such terms. Together, Equations 2.13 and 2.14 give us

$$u(l) \geq A_{i,j} \otimes \alpha \otimes l_{\tau(i)}. \tag{2.15}$$

Now since $\alpha \otimes A_{i,j} \geq 0$ this means that $u(l) \geq l_{\tau(i)}$, which directly contradicts the assumption that $u(l) < l_{\tau(i)}$, and so S' must be a safe targeting subset of $\mathcal{U}$. $\qquad\square$

2.3.2 Proof of converging subsets

Lemma 2.4 *Given a safe targeting subset S of $\mathcal{U}$, with maximum solution l, a UBC $v \in S$ targeting $x_{\tau(i)}$ and an additional UBC $u \in \mathcal{U}$ also targeting $x_{\tau(i)}$, such that $u(l) < l_{\tau(i)}$, and l' the maximum solution to $S' = S \cup \{u\} - \{v\}$ it must be the case that $l_{\tau(i)} > l'_{\tau(i)}$ and that $l_j \geq l'_j$ for all j.*

Proof Let C and C' be the safe target matrices corresponding to S and S'. Assuming that C and C' are $n \times n$ matrices and that $\tau(i) = n - 1$ we note that matrices C and C' can be divided up so that

$$C = \begin{bmatrix} W & X \\ Y & Z = \epsilon \end{bmatrix} \text{ and } C' = \begin{bmatrix} W & X \\ Y' & Z' = \epsilon \end{bmatrix}, \quad (2.16)$$

where W is an $n - 1 \times n - 1$ matrix, and the ϵ terms follow from Lemma 2.1.

Now, since the graphs corresponding to both C and C' contain only negative weight cycles, the terms $(YW^*X \oplus Z)^*$ and $(Y'W^*X \oplus Z')^*$, which represent the longest paths from x_{n-1} to itself in C and C', must both have value 0, and the closure (Equation 1.5) for C and C' simplifies to

$$C^* = \begin{bmatrix} W^* \oplus W^*XYW^* & W^*X \\ YW^* & 0 \end{bmatrix} \text{ and } C'^* = \begin{bmatrix} W^* \oplus W^*XY'W^* & W^*X \\ Y'W^* & 0 \end{bmatrix}.$$
$$(2.17)$$

Since our two solutions, l and l', come from the first columns of C^* and C'^*, we define an $n - 1 \times 1$ matrix $\mathcal{F} = [0, \epsilon, \dots, \epsilon]^T$ to select the first column from C^* and C'^*. Showing that l' provides a tighter bound than l can be done by showing that $l_{\tau(i)} > l'_{\tau(i)}$, which may be expressed as

$$YW^*\mathcal{F} > Y'W^*\mathcal{F}, \quad (2.18)$$

and that for $j \neq \tau(i)$, $l_j \geq l'_j$, which is equivalent to showing that entry-for-entry

$$(W^* \oplus W^*XYW^*)\mathcal{F} \geq (W^* \oplus W^*XY'W^*)\mathcal{F}. \quad (2.19)$$

We begin by proving Equation 2.18 holds. We know that the constraint u targeting x_{n-1} in C' was chosen because the value of u subject to l is smaller than l_{n-1}. This is equivalent to saying

$$YW^*\mathcal{F} > Y'(W^* \oplus W^*XYW^*)\mathcal{F}. \quad (2.20)$$

Multiplying out the right hand side of this inequality and taking advantage of the fact that performing a $\oplus$ operation on something cannot decrease its value, we get

$$YW^*\mathcal{F} > Y'W^*\mathcal{F} \oplus Y'W^*XYW^*\mathcal{F} \geq Y'W^*\mathcal{F}, \quad (2.21)$$

thus proving Equation 2.18 holds. Multiplying out Equation 2.19, we find that we must demonstrate that

$$W^*\mathcal{F} \oplus W^*XYW^*\mathcal{F} \geq W^*\mathcal{F} \oplus W^*XY'W^*\mathcal{F}, \quad (2.22)$$

which follows immediately from Equation 2.18 if we multiply both sides of the inequality by W^*X and $\oplus$ each side with $W^*\mathcal{F}$. $\quad\square$

2.4   Generating an initial safe targeting subset

It now remains to show that an initial safe targeting subset can be found. We begin by assuming that the values of all of the x_i are finite in the maximum solution to the UBC system. Let $\mathcal{V}$ be larger than the $\oplus$ of the values of all n of the x_i's maximum solution. Clearly, if we augment the initial UBC system with $n - 1$ constraints of the form $x_i \leq x_0 \otimes \mathcal{V}$, then the addition of this set of constraints does not change the maximum solution of the system, and is a safe targeting subset.

Unfortunately, we do not know a priori a suitable value for $\mathcal{V}$. Instead we treat $\mathcal{V}$ as a finite but very large element of $\mathcal{R}$, for which $\mathcal{V} \otimes a > b$ for all $a, b \in \mathcal{R}$ with $a \neq \epsilon$. Unlike the other elements of $\mathcal{R}$, $\mathcal{V}$ has no $\otimes$-inverse, and $\mathcal{V} \otimes \mathcal{V}$ is undefined.

Fortunately, the max-plus closure operation does not require $\otimes$ inverses, and we never need to calculate $\mathcal{V} \otimes \mathcal{V}$. If we begin with the initial target matrix V, we get the corresponding closure matrix V^* as shown below:

$$V = \begin{bmatrix} \epsilon & \epsilon & \cdots & \epsilon \\ \mathcal{V} & \epsilon & \cdots & \epsilon \\ \vdots & \vdots & \ddots & \vdots \\ \mathcal{V} & \epsilon & \cdots & \epsilon \end{bmatrix} \text{ and } V^* = \begin{bmatrix} 0 & \epsilon & \cdots & \epsilon \\ \mathcal{V} & 0 & \cdots & \epsilon \\ \vdots & \vdots & \ddots & \vdots \\ \mathcal{V} & \epsilon & \cdots & 0 \end{bmatrix}. \tag{2.23}$$

As we replace equations $x_i \leq \mathcal{V} \otimes x_0$ with equations not containing a $\mathcal{V}$ term, we need to calculate the closure of matrices which can be written

$$\begin{bmatrix} V & \bar{\epsilon} \\ Y & Z \end{bmatrix} \tag{2.24}$$

where, if V is a $k \times k$ sub-matrix, V has entries $V_{i,0} = \mathcal{V}$ for $1 \leq i < k$, and ϵ everywhere else, while the sub-matrix denoted $\bar{\epsilon}$ above has all entries equal to ϵ. The closure of this matrix can then be seen to be

$$\begin{bmatrix} V^* & \bar{\epsilon} \\ Z^*YV^* & Z^* \end{bmatrix}, \tag{2.25}$$

where V and V^* are related as in Equation 2.23.

2.5   Convergence

Since there are only a finite number of safe targeting subsets, each with a single maximum solution, the process must converge in a finite number of steps. Whenever an upper bound of ϵ is found for a system variable, we may stop and declare the system inconsistent. In addition, should we ever find a UBC u targeting x_0 such that $u(l) < 0$, we must declare the system inconsistent. Should the process converge while some x_i are still bound with an equation $x_i \leq \mathcal{V} \otimes x_0$ then there must be no maximum possible value for any x_i whose current bound, l_i, contains a $\mathcal{V}$ term.

3 Optimization

While maximizing the value of any linear max-plus expression subject to a set of linear max-plus equations is quite easy – the maximum solution to the expression must occur at the maximum solution to the system itself since decreasing a system variable's value cannot increase the expression's value – minimizing a linear max-plus expression is also quite easy. We simply augment the system with a new variable, x_{min}, which is set equal to the expression to be minimized, and find the maximum value of x_0 when $x_{min} = 0$. The resulting value is the $\otimes$-inverse of the minimum value of the expression.

4 Conclusion

We have presented a general technique for solving linear max-plus systems and optimizing linear max-plus expressions subject to the constraints of a linear max-plus system. The new technique is distinct from the solution of Gaubert [2] in that the new technique uses for its basic solution component the max-plus closure operation rather than the more complicated max-plus balances [1] and that it solves a series of subsystems with decreasing maximum solutions instead of solving all square subsystems with one fewer variable than appears in the original system. The technique is also the basis for a practical algorithm for verifying timing relationships of interface glue logic [4]. A more thorough description of the technique may be found in [5].

References

[1] F. Baccelli, G. Cohen, G.J. Olsder, and J.-P. Quadrat (1992), *Synchronization and Linearity*, John Wiley and Sons, Chichester, UK.

[2] S. Gaubert (July 1992), *Théorie des systèmes linéaires dans les dioïdes*. Thèse, École des Mines de Paris.

[3] M. Gondran and M. Minoux (1977, English edition 1984), *Graphs and Algorithms*, John Wiley and Sons.

[4] E. A. Walkup and G. Borriello (June 1994), *Interface Timing Verification with Application to Synthesis*. Proceedings of the 31st Design Automation Conference.

[5] E. A. Walkup (September 1995), *Optimization of Linear Max-Plus Systems with Application to Timing Analysis*. PhD Thesis, University of Washington.

Axiomatics of Thermodynamics and Idempotent Analysis

Victor P. Maslov

First let us consider an important simple example of "basic" thermodynamics, namely, the thermodynamics containing four main thermodynamic variables: pressure P, specific volume v, temperature T, specific entropy s. Thus we consider the 4-dimensional space $\mathbf{R}^4$ of these variables. For these variables there are two relations, namely, two state equations. This means that we consider a 2-dimensional surface $\Lambda^2 \subset \mathbf{R}^4$.

From the physical viewpoint, the pairs of variables: pressure – specific volume and temperature – specific entropy are dual in a certain sense. Therefore we consider the space $\mathbf{R}^4$ as a phase space (q_1, q_2, p_1, p_2), where the pressure and temperature play the part of coordinates, and the specific variables play the part of corresponding momenta:

$$q_1 = P, \qquad q_2 = T,$$

$$p_1 = V, \qquad p_2 = -s.$$

(We take the entropy with the sign "minus" in order to introduce a (symplectic) structure corresponding to both Gibbs' thermodynamics and the usual phase space).

We shall assume that

1) Λ^2 is a smooth surface,

2) $\oint (p_1 \, dq_1 + p_2 \, dq_2) = 0$ along any closed path[1].

The latter relation implies that the function

$$\Phi(\alpha) = -\int_{\alpha^0}^{\alpha} q \, dp \quad (\alpha^0 \text{ is a fixed point on } \Lambda^2), \quad q = (q_1, q_2), \quad p = (p_1, p_2)$$

is defined on the surface Λ^2.

Actually, in his famous work [1] Gibbs assumed that the surface Λ^2 is diffeomorphically projected on the p-plane. This is equivalent to the assumption that $\Phi(\alpha)$ is a smooth function of momenta p_1, p_2.

Precisely this latter assumption was made in Gibbs' work. In thermodynamics the function $\Phi(p_1, p_2) = u(v, s)$ is the so-called *internal energy*.

---

[1] This is the First Law of Thermodynamics.

Gibbs made the following geometric construction that differs from the above. In the 3-dimensional space of variables u, v, s he considered the surface

$$u = u(v, s). \tag{1}$$

According to Gibbs, the convex hull of this surface corresponds to the stable thermodynamics. The points of the surface (1) that do not belong to the convex hull are unstable or metastable.

We shall present another geometric interpretation of unstable and metastable points [3], [4]. This interpretation coincides with our previous concept of the surface Λ^2 in the phase space $\mathbf{R}^4$.

According to Gibbs, the surface Λ^2 is determined by the equations

$$q_1 = -\frac{\partial \Phi}{\partial p_1}, \qquad q_2 = -\frac{\partial \Phi}{\partial p_2}.$$

We introduce the function $\Phi_1(q_1, q_2)$ (Gibbs' potential $G(P, T)$) as the Legendre transformation of $\Phi(p_1, p_2)$:

$$\Phi_1(q_1, q_2) = \min_{p_1, p_2}\{(p, q) + \Phi(p_1, p_2)\}, \qquad (p, q) = p_1 q_1 + p_2 q_2. \tag{2}$$

By our "axiom", the points $\alpha \in \Lambda^2$ (or $p_1, p_2 \in \Lambda^2$) that do not minimize the right hand side of (2) are unstable or metastable. In other words, among all points $\alpha \in \Lambda^2$ corresponding to the coordinates q_1, q_2 (i.e., among the points projected from Λ^2 onto the point (q_1, q_2)), only the points that minimize the action

$$\Phi_1(\alpha) = \int_{\alpha^0}^{\alpha} p\, dq = \int_{\alpha^0}^{\alpha}(p_1\, dq_1 + p_2\, dq_2)$$

are stable. These conditions are close to the concept of idempotent analysis, i.e., to the concept of spaces with values in a semiring, where addition is defined as minimization, and multiplication is defined as addition. This concept yields the following situation: if we have two different surfaces Λ_1^q and Λ_2^q, then, as the action corresponding to the pair $\{\Lambda_1^q, \Lambda_2^q\}$, we must take a linear combination in the sense of the above idempotent space. For example, if ahead of a shock wave the relations between pressure, temperature, entropy and volume give the surface Λ_1^q and behind the shock wave they give the surface Λ_2^q, then on the shock wave itself, as Gibbs' thermodynamic potential, we must take a linear combination in the space (min, +).

This concept corresponds to the well-known Mott–Smith model [2].

Actually, the above considerations give a certain new interpretation of Gibbs' and Mott–Smith's theories. However, the above concept can be generalized in a nontrivial way.

First of all, it is not necessary to demand that Λ^2 be uniquely projected on the p-plane. Then, and this is the principal point, we can generalize our construction to the n-dimensional case.

Let $\mathbf{R}^{18}$ be an 18-dimensional phase space. There are intensive thermodynamic variables (9 coordinates $(q_1, q_2, \ldots, q_9)$): temperature T, pressure P, electric intensity $\bar{E}$, magnetic intensity $\bar{H}$, electrostatic potential φ; and extensive thermodynamic specific variables (9 momenta $(p_1, p_2, \ldots, p_9)$): minus entropy s, volume v, dielectric polarization $\bar{\pi}$, magnetization $\bar{M}$, and charge e. There are 9 state equations.

The form

$$dU = T\, ds - P\, dv + (\bar{E}, d\bar{\pi}) + (\bar{H}, d\bar{M}) + \varphi\, de$$

defines the thermodynamic lagrangian manifold Λ (dim $= 9$) in $\mathbf{R}^{18}$. We have

Axiom I *The lagrangian manifold Λ is twice differentiable.*

We shall define the action

$$S(q) \overset{\text{def}}{=} \int p\, dq.$$

A point $r_1 \in \Lambda$ is called *unessential*, if there exists another point $r_2 \in \Lambda$ such that $q(r_1) = q(r_2)$ is the projection on $\mathbf{R}_q^n$, and

$$S|_{r=r_1} > S|_{r=r_2}.$$

Then we have

Axiom II *Unessential points correspond to metastable or unstable states of the system.*

Remark For given "coordinate" q, the points r_1 and r_2 correspond to two phases. As follows from physics there always exists a domain of coordinates that corresponds to only one phase. If we accept this assumption, then from Axiom I we find that the number of all phases – stable, metastable, and unstable – is always odd. One may be convinced that this "law" works in all known thermodynamic systems.

The axiomatics can trivially be transferred to the n-dimensional case.

The second kind of phase transition occurs (by definition) at the point where the volume and the entropy are continuous, but the derivatives $\partial V/\partial p$ and $\partial S/\partial t$ are discontinuous. By Axiom I, this is impossible for the "basic" 4-dimensional thermodynamics. It is necessary that (1) the phase space be of dimension ≥ 6 and (2) a first-kind discontinuity of one of the "non-basic" thermodynamic extensive variables occurs at this point. It is not always possible to give physical meaning to new thermodynamic variables. However, once introduced, they dramatically simplify and clarify the situation. It is another thermodynamic "law" that follows from the proposed axioms.

Our generalization is not trivial since it describes second-kind phase transitions. Moreover, Axiom I is not undeniable. However, all effects of second kind phase transitions known to the author can be interpreted by using our axioms.

References

[1] J.W. Gibbs (1964), *Papers on Thermodynamics*, GITTL, Moscow–Leningrad. (Russian.)

[2] J.O. Hirschfolder, S.F. Curtiss, and R.B. Bird (1954), *Molecular Theory of Gases and Liquids.*

[3] V.P. Maslov (1994), 'Analytic continuation of asymptotic formulas and axiomatics of thermodynamics and semithermodynamics', *Funktsional. Anal. i Prilozhen.*, **28** 28–42. (Russian.)

[4] V.P. Maslov (1994), 'Geometric quantization of thermodynamics. Phase transitions and asymptotics at critical points', *Matem. Zametki*, **56** 155–156. (Russian.)

The Correspondence Principle for Idempotent Calculus and some Computer Applications

Grigori L. Litvinov and Victor P. Maslov

1 Introduction

This paper is devoted to heuristic aspects of the so-called idempotent calculus. There is a correspondence between important, useful and interesting constructions and results over the field of real (or complex) numbers and similar constructions and results over idempotent semirings, in the spirit of N. Bohr's correspondence principle in Quantum Mechanics. Idempotent analogs for some basic ideas, constructions and results in Functional Analysis and Mathematical Physics are discussed from this point of view. Thus the correspondence principle is a powerful heuristic tool to apply unexpected analogies and ideas borrowed from different areas of Mathematics and Theoretical Physics.

It is very important that some problems nonlinear in the traditional sense (for example, the Bellman equation and its generalizations and the Hamilton–Jacobi equation) turn out to be linear over a suitable semiring; this linearity considerably simplifies the explicit construction of solutions. In this case we have a natural analog of the so-called superposition principle in Quantum Mechanics (see [1]–[3]).

The theory is well advanced and includes, in particular, new integration theory, new linear algebra, spectral theory and functional analysis. Applications include various optimization problems such as multicriteria decision making, optimization on graphs, discrete optimization with a large parameter (asymptotic problems), optimal design of computer systems and computer media, optimal organization of parallel data processing, dynamic programming, discrete event systems, computer science, discrete mathematics, mathematical logic and so on. See, for example, [4]–[64]. Let us indicate some applications of these ideas in mathematical physics and biophysics [65]–[70].

In this paper the correspondence principle is used to develop an approach to object-oriented software and hardware design for algorithms of idempotent calculus and scientific calculations. In particular, there is a regular method for constructing back-end processors and technical devices intended for an implementation of basic algorithms of idempotent calculus and mathematics of semirings. These hardware facilities increase the speed of data processing. Moreover this approach is useful for software and hardware design in the general case of algorithms which are not "idempotent" [72].

The paper contains a brief survey of the subject but our list of references is not complete. Additional references could be found in [4]–[9], [11], [14], [15], [17], [19]–[24], [27]–[29], [47], [53], [63]; the corresponding lists of references are also incomplete but very useful.

The authors are grateful to I. Andreeva, B. Doubrov, M. Gromov, J. Gunawardena, G. Henkin, V. Kolokoltsov, G. Mascari, P. Del Moral, A. Rodionov, S. Samborski, G. Shpiz, and A. Tarashchan for discussions and support.

The work was supported by the Russian Fund for Fundamental Investigations (RFFI), Project 96–01–01544.

2 Idempotent Quantization and Dequantization

Let $\mathbb{R}$ be the field of real numbers, $\mathbb{R}_+$ the subset of all nonnegative numbers. Consider the change of variables

$$u \mapsto w = h \ln u, \tag{2.1}$$

where $u \in \mathbb{R}_+$, $h > 0$; thus $u = e^{w/h}$, $w \in \mathbb{R}$. We have a natural map

$$D_h : \mathbb{R}_+ \to A = \mathbb{R} \cup \{-\infty\} \tag{2.2}$$

defined by the formula (2.1). Denote by $\mathbb{0}$ the "additional" element $-\infty$ and by $\mathbb{1}$ the zero element of A (that is $\mathbb{1} = 0$); clearly $\mathbb{0} = D_h(0)$ and $\mathbb{1} = D_h(1)$. Denote by A_h the set A equipped with the two operations $\oplus$ (generalized addition) and $\odot$ (generalized multiplication) borrowed from the usual addition and multiplication in $\mathbb{R}_+$ by the map D_h; thus $w_1 \odot w_2 = w_1 + w_2$ and $w_1 \oplus w_2 = h \ln(e^{w_1/h} + e^{w_2/h})$. Clearly, $D_h(u_1 + u_2) = D_h(u_1) \oplus D_h(u_2)$ and $D_h(u_1 u_2) = D_h(u_1) \odot D_h(u_2)$. It is easy to prove that $w_1 \oplus w_2 = h \ln(e^{w_1/h} + e^{w_2/h}) \to \max\{w_1, w_2\}$ as $h \to 0$.

Let us denote by $\mathbb{R}_{\max}$ the set $A = \mathbb{R} \cup \{-\infty\}$ equipped with operations $\oplus = \max$ and $\odot = +$; put $\mathbb{0} = -\infty$, $\mathbb{1} = 0$. Algebraic structures in $\mathbb{R}_+$ and A_h are isomorphic, so $\mathbb{R}_{\max}$ is a result of a deformation of the structure in $\mathbb{R}_+$. There is an analogy to the quantization procedure, and h is an analog for the Planck constant. Thus $\mathbb{R}_+$ (or $\mathbb{R}$) can be treated as a "quantum object" with respect to $\mathbb{R}_{\max}$, and $\mathbb{R}_{\max}$ can be treated as a "classical" or "semiclassical" object and as a result of a "dequantization" of this quantum object.

Similarly denote by $\mathbb{R}_{\min}$ the set $\mathbb{R} \cup \{+\infty\}$ equipped with operations $\oplus = \min$ and $\odot = +$; in this case $\mathbb{0} = +\infty$ and $\mathbb{1} = 0$. Of course, the change of variables $u \mapsto w = -h \ln u$ generates the corresponding dequantization procedure for this case.

The set $\mathbb{R} \cup \{+\infty\} \cup \{-\infty\}$ equipped with the operations $\oplus = \min$ and $\odot = \max$ can be obtained as a result of a "second dequantization" with respect to $\mathbb{R}$ (or $\mathbb{R}_+$). In this case $\mathbb{0} = \infty$, $\mathbb{1} = -\infty$ and the dequantization procedure can be applied to the subset of negative elements of $\mathbb{R}_{max}$ and the corresponding change of variables is $w \mapsto v = h \ln(-w)$.

3 Semirings

It is easy to check that for these constructed operations $\oplus$ and $\odot$ the following basic properties are valid for all elements a, b, c:

$$(a \oplus b) \oplus c = a \oplus (b \oplus c); \qquad (a \odot b) \odot c = a \odot (b \odot c); \qquad (3.1)$$

$$\mathbb{0} \oplus a = a \oplus \mathbb{0} = a; \qquad \mathbb{1} \odot a = a \odot \mathbb{1} = a; \qquad (3.2)$$

$$\mathbb{0} \odot a = a \odot \mathbb{0} = \mathbb{0}; \qquad (3.3)$$

$$a \odot (b \oplus c) = (a \odot b) \oplus (a \odot c); \qquad (b \oplus c) \odot a = (b \odot a) \oplus (c \odot a); \qquad (3.4)$$

$$a \oplus b = b \oplus a; \qquad (3.5)$$

$$a \oplus a = a; \qquad (3.6)$$

$$a \odot b = b \odot a. \qquad (3.7)$$

A set A equipped with binary operations $\oplus$ and $\odot$ and having distinguished elements $\mathbb{0}$ and $\mathbb{1}$ is called a *semiring*, if the properties (axioms) (3.1)–(3.5) are fulfilled. We shall suppose that $\mathbb{0} \neq \mathbb{1}$.

This semiring is *idempotent* if (3.6) is valid. Idempotent semirings are often called dioids. A semiring (not necessarily idempotent) is called *commutative* if (3.7) is valid. Note that different versions of this axiomatics are used see, for example, [4]–[9], [14], [20]–[24], [27], [28] and literature indicated in [53].

Example 3.1 The set $\mathbb{R}_+$ of all nonnegative real numbers endowed with the usual addition and multiplication is a commutative (but not idempotent) semiring. Of course, the field $\mathbb{R}$ of all real numbers is also a commutative semiring.

Example 3.2 $\mathbb{R}_{max}$ and $\mathbb{R}_{min}$ are isomorphic commutative idempotent semirings.

Example 3.3 $A = \mathbb{R}_+$ with the operations $\oplus = \max$ and $\odot = \cdot$ (the usual multiplication); $\mathbb{0} = 0$, $\mathbb{1} = 1$. This idempotent semiring is isomorphic to $\mathbb{R}_{max}$ by the mapping $x \mapsto \ln(x)$.

Example 3.4 $A = [a, b] = \{x \in \mathbb{R} \mid a \leq x \leq b\}$ with the operations $\oplus = \max$, $\odot = \min$ and the neutral elements $\mathbb{0} = a$ and $\mathbb{1} = b$ (the cases $a = -\infty$, $b = +\infty$ are possible).

Semirings similar to these examples are the closest to the initial "quantum" object $\mathbb{R}_+$ and can be obtained by dequantization procedures. However, there are many important idempotent semirings which are unobtainable by means of these procedures. Note that there exist important quantum mechanical systems which cannot be obtained from classical systems by quantization (for example, particles with spin and systems consisting of identical particles). Thus our analogy fits the natural situation well enough.

Example 3.5 Let $\mathrm{Mat}_n(A)$ be the set of $n \times n$ matrices with entries belonging to an idempotent semiring A. This set forms a noncommutative idempotent semiring with respect to matrix addition $\oplus$ and matrix multiplication $\odot$, that is

$$(X \oplus Y)_{ij} = X_{ij} \oplus Y_{ij} \quad \text{and} \quad (X \odot Y)_{ij} = \bigoplus_{k=1}^{n} X_{ik} \odot Y_{kj}.$$

Clearly, $(\mathbb{0})_{ij} = \mathbb{0} \in A$, and $(\mathbb{1})_{ij} = \mathbb{0} \in A$ if $i \neq j$, and $(\mathbb{1})_{ii} = \mathbb{1} \in A$.

Example 3.6 $A = \{0, 1\}$ with the operations $\oplus = \max$, $\odot = \min$, $\mathbb{0} = 0$, $\mathbb{1} = 1$. This is the well-known *Boolean* semiring (or Boolean algebra).

Note that every bounded distributive lattice is an idempotent semiring.

Example 3.7 $A = \{\mathbb{0}, \mathbb{1}, a\}$, where $\{\mathbb{0}, \mathbb{1}\}$ is a Boolean semiring, $\mathbb{0} \oplus a = a$, $\mathbb{0} \odot a = \mathbb{0}$, $\mathbb{1} \odot a = a$, $\mathbb{1} \oplus a = \mathbb{1}$, $a \oplus a = a$, $a \odot a = a$. This example can be treated as a three-valued logic.

There are many finite idempotent semirings; a classification of commutative idempotent semirings consisting of two, or three, or four elements is presented in [52].

Example 3.8 Let A be the set of all compact convex subsets of $\mathbb{R}^n$ (or of any closed convex cone in $\mathbb{R}^n$); this set is an idempotent semiring with respect to the following operations:

$$\alpha \oplus \beta = \text{convex hull of } \alpha \text{ and } \beta;$$
$$\alpha \odot \beta = \{a + b \mid a \in \alpha, b \in \beta\}$$

for all $\alpha, \beta \in A$; $\mathbb{0} = \emptyset$, $\mathbb{1} = \{0\}$. This idempotent semiring is used in mathematical economics and in the multicriterial optimization problem (evolution of the so-called Pareto sets; see, for example [35], [74]).

Example 3.9 If A_1 and A_2 are idempotent semirings, then $A = A_1 \times A_2$ is also an idempotent semiring with respect to the natural component-wise operations of the direct product; in this case $(\mathbb{0}, \mathbb{0})$ and $(\mathbb{1}, \mathbb{1})$ are the corresponding neutral elements. A similar (and natural, see [52]) construction turns $(A_1 \backslash \{\mathbb{0}\}) \times (A_2 \backslash \{\mathbb{0}\}) \cup \mathbb{0}$ into an idempotent semiring.

Probably the first interesting and nontrivial idempotent semiring of all languages over a finite alphabet was examined by S. Kleene [73] in 1956. This noncommutative semiring was used for applications to compiling and syntax analysis; see also [6], [7]. There are many other interesting examples of idempotent semirings (including the so-called "tropical" semirings; see, for example, [47], [48], [60], [63], [64]) with applications to theoretical computer science (linguistic problems, finite automata, discrete event systems and Petri nets, stochastic systems, computational problems etc.), algebra (semigroups of matrices over semirings), logic, optimization etc.; in particular, see also [5]–[7], [9], [11], [12], [15]–[17], [19]–[24], [26]–[29], [32], [33], [35], [53], [63]–[66].

There is a naturally defined *partial order* (i.e. partial ordering relation) on any idempotent semiring (and on any idempotent semigroup); by definition, $a \preceq b$ if and only if $a \oplus b = b$. For this relation the reflexivity is equivalent to the idempotency of the (generalized) addition, and the transitivity and the antisymmetricity follow, respectively, from the associativity and commutativity of this operation. This ordering relation on $\mathbb{R}_{\max}$ (and on the semirings described in examples 3.3 and 3.4) coincides with the natural one but for $\mathbb{R}_{\min}$ it is opposite to the natural ordering relation on the real axis.

Every element a of an idempotent semiring A is "nonnegative": $\mathbb{0} \preceq a$; indeed, $\mathbb{0} \oplus a = a$ because of (3.2). Similarly, for all $a, b, c \in A$ such that $a \preceq b$ we have $a \oplus c \preceq b \oplus c$ and $a \odot c \preceq b \odot c$.

Using this standard partial order it is possible to define in the usual way the notions of upper and lower bounds, bounded sets, $\sup M$ and $\inf N$ for upper/lower bounded sets M and N etc. On the basis of these concepts an algebraic approach to the subject is developed; see, for example, [4]–[9], [17], [19]–[24], [27], [32], [33], [52], [53].

An idempotent semiring can be a metric or topological space with natural correlations between topological and algebraic properties. For example, for $\mathbb{R}_{\min}$ there is a natural metric $\rho(x, y) = |e^{-x} - e^{-y}|$, and for the semiring from Example 3.4 it is convenient to use the metric $\rho(x, y) = |\arctan x - \arctan y|$ if $a = -\infty$, $b = +\infty$. The corresponding "topological" approach was developed, in e.g. [14], [15], [19]–[24], [39], [42], [61]–[63], [66]–[70].

4 Semirings with Special Properties

It is convenient to treat some special classes of semirings for which some additional conditions are fulfilled. Let us discuss some conditions of this type.

Suppose A is an arbitrary semiring. The so-called *cancellation condition* is fulfilled for A if $b = c$ whenever $a \odot b = a \odot c$ and $a \neq \mathbb{0}$. If the multiplication in A is invertible on $A \backslash \{\mathbb{0}\}$, then A is called *a semifield*. Clearly, the cancellation condition is fulfilled for all semifields. For example, $\mathbb{R}_{\max}$ is a semifield. Idempotent semirings with the cancellation condition or with an idempotent multiplication are especially interesting.

For arbitrary commutative idempotent semirings with the cancellation condition the following version of Newton's binomial formula is valid:

$$(a \oplus b)^n = a^n \oplus b^n, \tag{4.1}$$

see [32], [33]. However, this formula is valid also for the semirings from Example 3.4, which lack the cancellation condition. It is easily proved (by induction) that for arbitrary commutative idempotent semirings the binomial formula has the form

$$(a \oplus b)^n = \bigoplus_{i=0}^{n} a^{n-i} \odot b^i. \tag{4.2}$$

Suppose A is an arbitrary idempotent semiring. Applying (4.2) to the semiring generated by the elements $\mathbb{1}, a \in A$, we deduce the following formula:

$$(\mathbb{1} \oplus a)^n = \mathbb{1} \oplus a \oplus a^2 \oplus \cdots \oplus a^n. \tag{4.3}$$

Now let A be an arbitrary semiring (not necessarily idempotent) and suppose that the infinite sum

$$a^* = \bigoplus_{i=0}^{\infty} a^i = \mathbb{1} \oplus a \oplus a^2 \oplus \cdots \oplus a^n \cdots \tag{4.4}$$

is well-defined for an element $a \in A$. For particular semirings a^* may be defined as e.g. $\sup_n \{(\mathbb{1} + a)^n\}$ or $\lim_{n \to \infty} (\mathbb{1} \oplus a)^n$. This important star operation $a \mapsto a^*$ was introduced by S. Kleene [73]; the element a^* is called the *closure* of a.

It is natural to set $a^* = (\mathbb{1} - a)^{-1}$ if A is a field and $a \neq \mathbb{1}$. It is easy to prove that $a^* = \mathbb{1}$, if A is an idempotent semiring and $a \preceq \mathbb{1}$. For $\mathbb{R}_{\max}$ the closure a^* is not defined if $\mathbb{1} \prec a$. The situation can be corrected if we add an element ∞ such that $a \oplus \infty = \infty$ for all $\in \mathbb{R}_{\max}$, $\mathbb{0} \odot \infty = \mathbb{0}$, $a \odot \infty = \infty$ for all $a \neq \mathbb{0}$. For this new semiring $\overline{\mathbb{R}}_{\max} = \mathbb{R}_{\max} \cup \{\infty\}$ we have $a^* = \infty$ if $\mathbb{1} \prec a$, see e.g. [18], [30]. For all semirings described in Examples 3.4, 3.6 and 3.7 we have $a^* = \mathbb{1}$ for any element a.

An idempotent semiring A is *algebraically closed* (with respect to the operation $\odot$) if the equation $x^n = a$ (where $x^n = x \odot \cdots \odot x$) has a solution $x \in A$ for any $a \in A$ and any positive integer n, see [32], [33]. It is remarkable and important that the semiring $\mathbb{R}_{\max}$ is algebraically closed in this sense. However, the equation $x^2 \oplus \mathbb{1} = \mathbb{0}$ has no solutions.

5   The Correspondence Principle

The analogy with Quantum Mechanics discussed in Section 2 leads to the following *correspondence principle* in idempotent calculus:

There is a (heuristic) correspondence, in the spirit of the correspondence principle in Quantum Mechanics, between important, useful and interesting constructions and results over the field of real (or complex) numbers (or the semiring of all nonnegative numbers) and similar constructions and results over idempotent semirings.

Example 5.1. Semimodules (see e.g. [4]–[12], [17], [22]–[24], [27], [32], [33], [50]–[53]). A set V is called a *semimodule over a semiring A* (or an A-semimodule), if there is a commutative associative addition operation $\oplus$ in V with neutral element $\mathbb{0}$, and a multiplication $\odot$ of elements from V by elements of A is defined, and the following properties are fulfilled:

$$(\lambda \odot \mu) \odot v = \lambda \odot (\mu \odot v) \qquad \text{for all } \lambda, \mu \in A, \ v \in V;$$
$$\lambda \odot (v_1 \oplus v_2) = \lambda \odot v_1 \oplus \lambda \odot v_2 \qquad \text{for all } \lambda \in A, \ v_1, v_2 \in V;$$
$$\mathbb{0} \odot v = \lambda \odot \mathbb{0} = \mathbb{0} \qquad \text{for all } \lambda \in A, \ v \in V.$$

The addition $\oplus$ in V is assumed to be idempotent if A is an idempotent semiring (i.e. $v \oplus v = v$ for all $v \in V$). Then we assume that

$$\sup_{\alpha}\{\lambda_\alpha\} \odot v = \sup_{\alpha}\{\lambda_\alpha \odot v\}, \qquad \text{if } v \in V \text{ and } \sup_{\alpha}\{\lambda_\alpha\} \in A.$$

Roughly speaking, semimodules are "linear spaces" over semirings. The simplest A-semimodule is the direct sum (product) $A^n = \{(a_1, a_2, \cdots, a_n) : a_j \in A\}$. The set of all endomorphisms $A^n \to A^n$ coincides with the semiring $\mathrm{Mat}_n(A)$ of all A-valued matrices (see Example 3.5).

The theory of A-valued matrices is an analog of the well-known O. Perron–G. Frobenius theory of nonnegative matrices, see e.g. [75]. For example, suppose A is an algebraically closed commutative idempotent semiring with the cancellation condition and the sequence $a^n \oplus b$ stabilizes for any $a \preceq \mathbb{1}$ and $b \neq \mathbb{0}$, $a, b \in A$. Then for every endomorphism K of A^n ($n \geq 1$) there exists a nontrivial subsemimodule $S \subset A^n$ (an "eigenspace") and $\lambda \in A$ (an "eigenvalue") such that $Kv = \lambda \odot v$ for all $v \in S$; this element λ is unique if K is irreducible, see [32], [33]. In particular, this result is valid if $A = \mathbb{R}_{\max}$ (or $\mathbb{R}_{\min}$). Similar results can be proved for semimodules of bounded functions and continuous functions, see [32], [33], [22]–[24].

Idempotent analysis deals with functions taking values in idempotent semirings and with the corresponding function spaces (semimodules). Let X be a set and A an idempotent semiring. Let us denote by $B(X, A)$ the set of all bounded mappings (functions) $X \to A$ (i.e. mappings with order-bounded images) equipped with the natural structure of an A-semimodule. If X is finite, $X = \{x_1, \ldots, x_n\}$, then $B(X, A)$ can be identified with the semimodule A^n (see Example 5.1 above). Actually $B(X, A)$ is an idempotent semiring with respect to the corresponding pointwise operations.

Let A be a metric semiring; then there is the corresponding uniform metric on $B(X, A)$. Suppose that X is a topological space and then denote by $C(X, A)$ the subsemimodule of continuous functions in $B(X, A)$.

Suppose now that the space X is locally compact and denote by $C_0(X, A)$ the A-semimodule of continuous A-valued functions with compact supports endowed with a natural topology (see [19]–[24] for details).

These spaces (and some other spaces of this type) are examples of "idempotent" function spaces. Many basic ideas, constructions and results can be extended to idempotent analysis from the usual analysis and functional analysis.

Example 5.2 Idempotent integration and measures. For the sake of simplicity set $A = \mathbb{R}_{\max}$ and let X be a locally compact space. An idempotent analog of the usual integration can be defined by the formula

$$\int_X^{\oplus} \varphi(x)\, dx = \sup_{x \in X} \varphi(x), \tag{5.1}$$

if φ is a continuous or upper semicontinuous function on X. The set function

$$m_\varphi(B) = \sup_{x \in B} \varphi(x), \tag{5.2}$$

where $B \subset X$, is called an A-measure on X and $m_\varphi(\bigcup B_\alpha) = \bigoplus_\alpha m_\varphi(B_\alpha) = \sup_\alpha m_\varphi(B_\alpha)$, so the function (5.2) is completely additive. An idempotent integral with respect to this A-measure is defined by the formula

$$\int_X^{\oplus} \psi(x)\, dm_\varphi = \int_X^{\oplus} \psi(x) \odot \varphi(x)\, dx = \sup_{x \in X} \psi(x) \odot \varphi(x). \tag{5.3}$$

It is obvious that this integration is "linear" over A and it is easy to see that (5.1) and (5.3) can be treated as limits of Riemann's and Lebesgue's sums. Clearly, if $\oplus = \min$ for the corresponding semiring A, then (5.3) turns into the formula

$$\int_X^{\oplus} \psi(x)\, dm_\varphi = \int_X^{\oplus} \psi(x) \odot \varphi(x)\, dx = \inf_{x \in X} \psi(x) \odot \varphi(x). \tag{5.4}$$

In this case, $\odot$ may coincide e.g. with max, or the usual addition or multiplication. See [14], [15], [19]–[24] for details.

Note that in (5.4) we mean inf (i.e. the greatest lower bound) with respect to the usual ordering of numbers. But if $\oplus = \min$, then this order is opposite to the standard partial order defined for any idempotent semiring (see Section 3 above). It is clear that (5.3) and (5.4) coincide from this point of view. In the general case A-measure and idempotent integral can be defined by (5.2) and (5.3), e.g. if the corresponding functions are bounded and A is *boundedly complete*, i.e. every bounded subset $B \subset A$ has a least upper bound $\sup B$.

There is a natural analogy between idempotent and probability measures. This analogy leads to a parallelism between probability theory and stochastic processes on the one hand, and optimization theory and decision processes on

Grigori L. Litvinov and Victor P. Maslov

the other hand. That is why it is possible to develop optimization theory at
the same level of generality as probability and stochastic process theory. In
particular, the Markov causality principle corresponds to the Bellman optimality principle; so the Bellman principle is an $\mathbb{R}_{\max}$-version of the Chapman–
Kolmogorov equation for Markov stochastic processes, see e.g. [43]–[46], [24],
[26], [29], [56], [63]. Applications to filtering theory can be found in [44], [46].

Example 5.3. Group idempotent (convolution) semirings. Let G be
a group, A an idempotent semiring; assume that A is boundedly complete.
Then the space $B(G, A)$ of all bounded functions $G \to A$ (see above) is an
idempotent semiring with respect to the following idempotent analog $\circledast$ of
convolution:

$$(\varphi \circledast \psi)(g) = \int_G^{\oplus} \varphi(x) \odot \psi(x^{-1} \cdot g) dx. \tag{5.5}$$

Of course, it is possible to consider other "function spaces" instead of
$B(G, A)$. In [23], [24] semirings of this type are referred to as *convolution
semirings*.

Example 5.4. Fourier–Legendre transform, see [14], [3], [19]–[24]. Let
$A = \mathbb{R}_{\max}$, $G = \mathbb{R}^n$ and let G be treated as a group. The usual Fourier–
Laplace transform is defined by the formula

$$\varphi(x) \mapsto \tilde{\varphi}(\xi) = \int_G e^{i\xi \cdot x} \varphi(x) dx, \tag{5.6}$$

where $e^{i\xi \cdot x}$ is a character of the group G, that is a solution of the functional
equation

$$f(x + y) = f(x)f(y).$$

The corresponding idempotent analog (for the case $A = \mathbb{R}_{\max}$) has the form

$$f(x + y) = f(x) \odot f(y) = f(x) + f(y),$$

so "idempotent characters" are linear functionals $x \mapsto \xi \cdot x = \xi_1 x_1 + \cdots + \xi_n x_n$.
Thus (5.6) turns into the following transform:

$$\varphi(x) \mapsto \tilde{\varphi}(\xi) = \int_G^{\oplus} \xi \cdot x \odot \varphi(x) dx = \sup_{x \in G}(\xi \cdot x + \varphi(x)). \tag{5.7}$$

This is the famous *Legendre transform*. Thus this transform is an $\mathbb{R}_{\max}$ version
of the Fourier–Laplace transform.

Of course, this construction can be generalized to different classes of groups
and semirings. Transformations of this type convert the generalized convolution to pointwise multiplication and possess analogs of some important properties of the usual Fourier transform. For the case of semirings of Pareto sets
the corresponding version of the Fourier transform reduces the multi-criterion
optimization problem to a family of single-criterion problems [35].

Examples 5.3 and 5.4 can be treated as fragments of an idempotent version of representation theory. In particular, idempotent representations of groups can be examined as representations of the corresponding convolution semirings (i.e. idempotent group semirings) in semimodules.

According to the correspondence principle, many important concepts, ideas and results can be converted from the usual functional analysis to idempotent analysis. For example, an idempotent scalar product can be defined by the formula

$$(\varphi, \psi) = \int_X^\oplus \varphi(x) \odot \psi(x) dx, \tag{5.8}$$

where φ, ψ are A-valued functions belonging to a certain idempotent function space. There are many interesting spaces of this type including $B(X, A)$, $C(X, A)$, $C_0(X, A)$, analogs of the Sobolev spaces and so on. There are analogs for the well-known theorems of Riesz, Hahn–Banach and Banach–Steinhaus; it is possible to treat dual spaces and operators, an idempotent version of the theory of distributions (generalized functions) etc.; see [19]–[24], [34], [36], [39], [40], [76] for details.

Example 5.5. Integral operators. It is natural to construct idempotent analogs *of integral operators* in the form

$$K : \varphi(y) \mapsto (K\varphi)(x) = \int_Y^\oplus K(x, y) \odot \varphi(y) dy, \tag{5.9}$$

where $\varphi(y)$ is an element of a space of functions defined on a set Y and taking their values in an idempotent semiring A, $(K\varphi)(x)$ is an A-valued function on a set X and $K(x, y)$ is an A-valued function on $X \times Y$. If $A = \mathbb{R}_{\max}$, then (5.9) turns into the formula

$$(K\varphi)(x) = \sup_{y \in Y} \{K(x, y) + \varphi(y)\}. \tag{5.10}$$

Formulas of this type are standard for optimization problems, see e.g. [77].

It is easy to see that the operator defined by (5.9) is linear over A, i.e. K is an A-endomorphism of the corresponding semimodule (function space). Actually every linear operator acting in an idempotent function space and satisfying some natural continuity-type conditions can be presented in the form (5.9). This is an analog of the well-known L. Schwartz kernel theorem. The topological version of this result in spaces of continuous functions was established in [78], [76]; see also [23], [24]. The algebraic version of the kernel theorem for the space of bounded functions can be found in [32], [33] and (in a final form) in [52].

6   The Superposition Principle

In Quantum Mechanics the correspondence principle means that the Schrö-dinger equation (which is basic for the theory) is linear. Similarly in idem-potent calculus the correspondence principle means that some important and basic problems and equations (e.g. optimization problems, the Bellman equa-tion and its generalizations, the Hamilton–Jacobi equation) that are nonlinear in the usual sense can be treated as linear over appropriate idempotent semi-rings, see [1]–[3], [19]–[24].

Example 6.1. Idempotent dequantization for the heat equation. Let us start with the heat equation

$$\frac{\partial u}{\partial t} = \frac{h}{2}\frac{\partial^2 u}{\partial x^2}, \tag{6.1}$$

where $x \in \mathbb{R}$, $t > 0$, and h is a positive parameter.

Consider the change of variables:

$$u \mapsto w = -h \ln u;$$

it converts (6.1) to the following (integrated) version of the Burgers equation:

$$\frac{\partial w}{\partial t} + \frac{1}{2}\left(\frac{\partial w}{\partial x}\right)^2 - \frac{h}{2}\frac{\partial^2 w}{\partial x^2} = 0. \tag{6.2}$$

This equation is nonlinear but it can be treated as linear over the following generalized addition $\oplus$ and multiplication $\odot$ (derived from the usual addition and multiplication by a change of variables):

$$w_1 \oplus w_2 = -h \ln(e^{-w_1/h} + e^{-w_2/h}), \tag{6.3}$$

$$w_1 \odot w_2 = w_1 + w_2. \tag{6.4}$$

So if w_1 and w_2 are solutions for (6.2), then their linear combination with respect to the operations (6.3) and (6.4) is also a solution for this equation. For $h \to 0$ (6.2) turns into a special case of the Hamilton–Jacobi equation:

$$\frac{\partial w}{\partial t} + \frac{1}{2}\left(\frac{\partial w}{\partial x}\right)^2 = 0. \tag{6.5}$$

This is the dequantization procedure described in Section 2 above. So it is clear that (6.3) and (6.4) turn into addition $\oplus = \min$ and multiplication $\odot = +$ in the idempotent semiring $\mathbb{R}_{\min}$, and equation (6.5) is linear over $\mathbb{R}_{\min}$; thus the set of solutions for (6.5) is an $\mathbb{R}_{\min}$-semimodule. This example was the starting point for the well-known Hopf method of vanishing viscosity.

In the general case the Hamilton–Jacobi equation has the form

$$\frac{\partial S(x,t)}{\partial t} + H\left(\frac{\partial S}{\partial x}, x, t\right) = 0, \tag{6.6}$$

where H is a smooth function on $\mathbb{R}^{2n} \times [0, T]$. Consider the Cauchy problem for (6.6): $S(x,0) = S_0(x)$, $0 \leq t \leq T$, $x \in \mathbb{R}^n$. Denote by U_t the resolving operator, i.e. the map that assigns to each given $S_0(x)$ the solution $S(x,t)$ of this problem at the moment of time t. Then the map U_t for each t is a linear (over $\mathbb{R}_{\min}$) integral operator in the corresponding $\mathbb{R}_{\min}$-semimodule.

The situation is similar for the Cauchy problem for the homogeneous Bellman equation

$$\frac{\partial S}{\partial t} + H\left(\frac{\partial S}{\partial x}\right) = 0, \quad S|_{t=0} = S_0(x),$$

where $H: \mathbb{R}^n \to \mathbb{R}$ is a (not strictly) convex first order homogeneous function

$$H(p) = \sup_{(f,g)\in V} (f \cdot p + g), \ f \in \mathbb{R}^n, \ g \in \mathbb{R},$$

and V is a compact set in $\mathbb{R}^{n+1}$. See [23], [24], [39], [76] for details.

It is well-known that the discrete version of the Bellman equation can be treated as linear over idempotent semirings. The so-called *generalized stationary* (finite dimensional) *Bellman equation* has the form

$$S = HS \oplus F, \tag{6.7}$$

where S, H, F are matrices with elements from an idempotent semiring A and the corresponding matrix operations are described in Example 3.5 above (for the sake of simplicity we write HS instead of $H \odot S$); the matrices H and F are given (specified) and it is necessary to determine S from the equation.

Equation (6.7) has the solution

$$S = H^*F, \tag{6.8}$$

where H^* is the closure of $H \in \mathrm{Mat}_n(A)$; see Section 4 and Example 3.5 above. Recall that

$$H^* = \mathbf{1} \oplus H \oplus H^2 \oplus \cdots \oplus H^k \oplus \cdots, \tag{6.9}$$

if the right-hand side of (6.9) is well-defined. In this case $H^* = \mathbf{1} \oplus HH^*$, so $H^*F = F \oplus HH^*F$; thus (6.8) is a solution of (6.7). For example, if the sequence $H^{(N)} = \sum_{k=0}^N H^k$ stabilizes (i.e. there exists N_0 such that $H^{(N)} = H^{(N_0)}$ for all $N \geq N_0$), then (6.9) is well-defined and can be calculated by means of a finite set of operations (steps).

This consideration and a version of the Gauss elimination method for solving (6.7) were presented by S. Kleene [73] in the case of the semiring of all languages over a finite alphabet. B.A. Carré [4] used semirings to show that many important problems for graphs can be formulated in a unified manner and are reduced to solving systems of algebraic equations. For example, Bellman's method of solving shortest path problems corresponds to a version of the Jacobi method for solving (6.7), and Ford's algorithm corresponds to a version of the Gauss–Seidel method. This subject is developed further in [4]–[18], [21]–[24], [27]–[31], [53], [65].

Let A be a semiring (not necessarily idempotent). For each square $n \times n$ matrix $H = (h_{ij}) \in \mathrm{Mat}_n(A)$ there is a standard way to construct a geometrical object called a *weighted directed graph*. This object consists of a set X of n elements $x_1, x_2, \ldots, x_n$ together with the subset Γ of all ordered pairs $(x_i, x_j) \in X \times X$ such that $h_{ij} \neq \mathbb{0}$ and the mapping $h \colon \Gamma \to A \backslash \{\mathbb{0}\}$ given by the correspondence $(x_i, x_j) \mapsto h_{ij}$. The elements of X are called *nodes*, and the members of Γ are called *arcs*; h_{ij} are arc *weights*.

Alternatively the quadruple $M(X, \Gamma, h, A)$ can be treated as a discrete medium with the points x_i, the set Γ of links and the so-called link characteristics h. This concept is convenient for analysis of parallel computations and for synthesis of computing media. Mathematical aspects of these problems are examined in [14]; the further development of the subject is presented e.g. in [15], [81]; see also [23], [24], [27], [29]–[31], [61]. For example, the operating period evaluation problem for parallel algorithms and digital circuits leads to shortest path problems for $M(X, \Gamma, h, A)$, where $A = \mathbb{R}_{\max}$.

Recall that a sequence of nodes and arcs of the form

$$p = (y_0, a_1, y_1, a_2, y_2, \ldots, a_k, y_k), \qquad (6.10)$$

where $k \geq 0$, y_i are nodes of the graph, and a_i are arcs satisfying $a_i = (y_{i-1}, y_i)$, is called a *path* (of order k) from the node y_0 to the node y_k in $M(X, \Gamma, h, A)$. The *weight* $h(p)$ *of the path* (6.10) is the product of the weights of its arcs:

$$h(p) = h(a_1) \odot h(a_2) \odot \ldots \odot h(a_k). \qquad (6.11)$$

The so-called *Algebraic Path Problem* is to find the matrix $D = (d_{ij})$:

$$d_{ij} \overset{\mathrm{def}}{=} \oplus_p h(p), \qquad (6.12)$$

where $i, j = 1, 2, \ldots, n$, and p runs through all paths from x_i to x_j. This problem does necessarily have a solution (the set of weights in (6.12) may be infinite). However, if there exists a closure H^* of the matrix $H = (h_{ij})$, then the matrix

$$D = (d_{ij}) = H^* = \mathbb{1} \oplus H \oplus H^2 \oplus \cdots \oplus H^k \oplus \cdots \qquad (6.13)$$

can be treated as a solution of this problem. Moreover, H^k corresponds to the value $\oplus_p h(p)$, where p contains exactly k arcs. For example, $h_{ij}^{(2)} =$

$\bigoplus_{k=1}^{n} h_{ik} \odot h_{kj}$ are elements (coefficients) of H^2, and each coefficient $h_{ij}^{(2)}$ corresponds to $\bigoplus_{p} h(p)$, where p runs through paths from x_i to x_j with exactly two arcs; similarly, $H^3 = H^2 \odot H$, etc.

Example 6.2. The shortest path problem. Let $A = \mathbb{R}_{\min}$, so h_{ij} are real numbers. In this case

$$d_{ij} = \bigoplus_{p} h(p) = \min_{p} h(p),$$

where (6.11) has the form

$$h(p) = h(a_1) + h(a_2) + \cdots + h(a_k).$$

Example 6.3. The relation closure problem. Let A be the Boolean semiring (see Example 3.6). In this case H corresponds to a relation $R \subset X \times X$, h_{ij} being 1 if and only if the relation holds between x_i and x_j. Then the transitive and reflexive closure R^* of the relation R corresponds to the matrix $D = H^*$.

Example 6.4. The maximal (minimal) width path problem. Let A be a semiring $\mathbb{R} \cup \{-\infty\} \cup \{\infty\}$ with the operations $\oplus = \max$ and $\odot = \min$ (see Example 3.4). Then

$$d_{ij} = \bigoplus_{p} h(p) = \max_{p} h(p),$$

where $h(p) = \min\{h(a_1), h(a_2), \ldots, h(a_k)\}$. If $h(a_i)$ is the width (or channel capacity) of a_i, then $h(p)$ is the possible width (or channel capacity) of p.

Example 6.5. The matrix inversion problem. Let A be the field $\mathbb{R}$ of real numbers (which is not an idempotent semiring). In this case

$$D = H^* = 1 + H + H^2 \cdots = (1 - H)^{-1},$$

if the series $\sum_{k=0}^{\infty} H^k$ converges; if the matrix $1 - H$ is invertible, then $(1 - H)^{-1}$ can be treated as a "regularized" sum of this series; here $H^0 = 1$ is the identity matrix.

Example 6.6. A simple dynamic programming problem. Let $A = \mathbb{R}_{\max}$, so h_{ij} are real numbers. Let us consider h_{ij} as the *profit* of moving from x_i to x_j, and suppose f_i is the *terminal prize* for the node x_i ($f_i \in \mathbb{R}$). Assume that p is a path of the form (6.10) and $y_0 = x_i$. Let M be the *total profit* for p, that is

$$M = h(a_1) + h(a_2) + \ldots + h(a_k) + f(y_k).$$

It is easy to see that $\max M = (H^k f)_i$, where f is a vector $\{f_i\}$, $H, H^k \in \mathrm{Mat}_n(A)$. So, the maximal value of the total profit for k steps is $(H^k f)_i$. It is clear that the maximal value of the total profit for paths of arbitrary order is $\max M = (H^* f)_i$.

See many other examples and details (including semiring versions of linear programming) in [4]–[17], [21]–[24], [27], [30], [73], [79], [80], [82]. The book [27] of F. L. Baccelli, G. Cohen, G. J. Olsder and J.-P. Quadrat is particularly useful.

7 The Correspondence Principle for Algorithms

Of course, the correspondence principle is valid for algorithms (and for their software and hardware implementations). Thus:

If we have an important and interesting numerical algorithm, then we have a good chance that its semiring analogs are important and interesting as well.

In particular, according to the superposition principle, analogs of linear algebra algorithms are especially important. Note that numerical algorithms for standard infinite-dimensional linear problems over semirings (i.e. for problems related to integration, integral operators and transformations, the Hamilton–Jacobi and generalized Bellman equations) deal with the corresponding finite-dimensional (or finite) "linear approximations". Nonlinear algorithms often can be approximated by linear ones. Recall that usually different natural algorithms for the same optimization problem correspond to different standard methods for solving systems of linear equations (like Gauss elimination method, iterative methods etc.).

It is well-known that algorithms of linear algebra are convenient for parallel computations (see e.g. [81]–[84]); similarly, their idempotent analogs can be parallelized. This is a regular way to use parallel computations for many problems including basic optimization problems.

Algorithms for the "scalar" (inner) product of two vectors, for matrix addition and multiplication do not depend on specific semirings. Algorithms to construct the closure H^* of an "idempotent" matrix H can be derived from standard methods for calculating $(1 - H)^{-1}$. For the Gauss–Jordan elimination method (via LU-decomposition) this trick was used in [30], and the corresponding algorithm is universal and can be applied both to the general algebraic path problem and to computing the inverse of a real (or complex) matrix $(1 - H)$. Computation of H^{-1} can be derived from this universal algorithm with some obvious transformations.

Note that numerical algorithms are combinations of basic operations. Usually these basic operations deal with "numbers". Abstractly these "numbers" are thought of as members of numerical *domains* (the real numbers, the integers, and so on). But every computer calculation deals with specific *models* (computer representations) of these numerical domains. For example, real numbers can be represented as ordinary floating point numbers, or as double precision floating point numbers, or as rational numbers etc. Differences between mathematical objects and their computer models lead to calculation errors. That is another reason to use universal algorithms which do not depend on a specific semiring and its computer model. Of course, one algorithm may be more universal than another algorithm of the same type. For example, numerical integration algorithms based on the Gauss–Jacobi quadrature formulas actually depend on computer models because they use finite precision constants. On the contrary, the rectangular formula and the trapezoid rule do not depend on models and in principle can be used even in the case of idempotent integration.

8   The Correspondence Principle for Hardware Design

A systematic application of the correspondence principle to computer calculations leads to a unifying approach to software and hardware design.

The most important and standard numerical algorithms have many hardware realizations in the form of technical devices or special processors. *These devices often can be used as prototypes for new hardware units generated by replacing the usual arithmetic operations by their semiring analogs and by adding tools for handling the neutral elements $\mathbb{0}$ and $\mathbb{1}$* (the latter usually is not difficult). Of course the case of numerical semirings whose elements (except perhaps for the neutral elements) are real numbers is the most simple and natural. Semirings of this type are presented in Examples 3.1–3.4. Semirings from Examples 3.6 and 3.7 can also be treated as numerical semirings. Note that for semifields (including $\mathbb{R}_{\max}$ and $\mathbb{R}_{\min}$) the operation of division is also defined.

Good and efficient technical ideas and decisions can be transposed from prototypes into new hardware units. Thus the correspondence principle generates a regular heuristic method for hardware design. Note that to get a patent it is necessary to present the so-called "invention formula", that is to indicate a prototype for the suggested device and the difference between the new and old devices. A survey of patents from the correspondence principle point of view is presented in [82].

Consider (as a typical example) the most popular and important algorithm for computing the scalar product of two vectors:

$$(x, y) = x_1 y_1 + x_2 y_2 + \ldots + x_n y_n. \tag{8.1}$$

The universal version of (8.1) for any semiring A is obvious:

$$(x, y) = (x_1 \odot y_1) \oplus (x_2 \odot y_2) \oplus \ldots \oplus (x_n \odot y_n). \tag{8.2}$$

In the case $A = \mathbb{R}_{\max}$ this formula turns into

$$(x, y) = \max\{x_1 + y_1, x_2 + y_2, \ldots, x_n + y_n\}. \qquad (8.3)$$

This calculation is standard for many optimization algorithms (see Section 6), so it is useful to construct a hardware unit for computing (8.3). There are many different devices (and patents) for computing (8.1) and every such device can be used as a prototype to construct a new device for computing (8.3) and even (8.2). Many processors for matrix multiplication and for other algorithms of linear algebra are based on computing scalar products and on the respective "elementary" devices.

Methods are available to make these new devices more universal than their prototypes. There is only a modest collection of possible operations for standard numerical semirings: max, min, and the usual arithmetic operations. So it is easy to construct programmable hardware processors with variable basic operations. Using modern technologies it is possible to construct cheap special-purpose multi-processor chips implementing examined algorithms. The so-called systolic processors are especially convenient for this purpose. A systolic array is a "homogeneous" computing medium consisting of elementary processors, where the general scheme and processor connections are simple and regular. Every elementary processor pumps data in and out performing elementary operations in a such way that the corresponding data flow is kept up in the computing medium; there is an analogy with the blood circulation and this is a reason for the term "systolic", see e.g. [83], [84].

Examples of systolic processors for the general algebraic path problem are presented in [30], [31]. In particular, there is a systolic array of $n(n + 1)$ elementary processors which performs computations of the Gauss–Jordan elimination algorithm and can solve the algebraic path problem within $5n - 2$ time steps. Of course, hardware implementations for important and popular basic algorithms increase the speed of data processing.

9   The Correspondence Principle for Software Design

Software implementations for universal semiring algorithms are not so efficient as hardware ones (with respect to the computation speed) but are much more flexible. Program modules can deal with abstract (and variable) operations and data types. Specific values for these operations and data types can be defined by input data types. In this case specific operations and data types are defined by means of additional program modules. For programs written in this manner it is convenient to use the special techniques of so-called object-oriented design, see e.g. [71]. Fortunately, powerful tools supporting the object-oriented software design have recently appeared including compilers for real and convenient programming languages (e.g. C^{++}).

A project has been set up to implement the correspondence principle approach to scientific calculations in the form of a powerful software system based on a unifying collection of universal algorithms. This approach ensures a reduction of working time for programmers and users because of software unification. Also, the accuracy and safety of numerical calculations can be ensured to whatever degree may be necessary [72].

The system contains several levels (including the programmer and user levels) and many modules. Roughly speaking it is divided into three parts. The first part contains modules that implement finite representations of basic mathematical objects (arbitrary-precision real and complex numbers, finite precision rational numbers, p-adic numbers, interval numbers, fuzzy numbers, basic semirings and rings etc.). The second part implements universal calculation algorithms (linear algebra, idempotent and standard analysis, optimization and optimal control, differential equations and so on). The third part contains modules implementing model-dependent algorithms (e.g. graphics, Gauss–Jacobi numerical integration, efficient approximation algorithms). The modules can be used in user programs written in C^{++}. See [72] for details.

References

[1] V. P. Maslov (1986), 'New superposition principle for optimization problems', *Seminaire sur les Equations aux Dérivées Partielles* 1985/86, Centre Math. École Polytech., Palaiseau, exposé **24**.

[2] V. P. Maslov (1987), 'A new approach to generalized solutions of nonlinear systems', *Soviet Math. Dokl.*, **42**, 1, 29–33.

[3] V. P. Maslov (1987), 'New superposition principle for optimization problems', *Russian Math. Surveys*, **42**.

[4] B. A. Carré (1971), 'An algebra for network routing problems', *J. Inst. Math. Appl.*, **7**, 273–294.

[5] B. A. Carré (1979), *Graphs and networks*, The Clarendon Press/Oxford Univ. Press, Oxford.

[6] A. V. Aho and J. D. Ullman (1973), *The theory of parsing, translation and compiling, Vol. 2: Compiling*, Prentice-Hall, Englewood Cliffs, N. J.

[7] A. V. Aho, J. E. Hopcroft and J. D. Ullman (1976), *The design and analysis of computer algorithms*, Addison-Wesley Publ. Co., Reading (Massachusetts) et al.

[8] M. Gondran (1975), 'Path algebra and algorithms', in *Combinatorial programming: methods and applications* (B. Roy, ed.), NATO Adv. Study Inst. Ser., Ser. C., **19**, 137–148.

[9] M. Gondran and M. Minoux (1979, 1988), *Graphes et algorithms*, Editions Eyrolles, Paris.

[10] D. J. Lehmann (1977), 'Algebraic structures for transitive closure', *Theor. Comp. Sci.*, **4**, 59–76.

[11] R. A. Cuningham-Green (1979), 'Minimax algebra', *Springer Lect. Notes in Economics and Mathematical Systems*, **166**, Berlin etc.

[12] R. A. Cuningham-Green (1991), 'Minimax algebra and its applications', *Fuzzy Sets and Systems*, **41**, 251–267.

[13] S. M. Avdoshin and V. V. Belov (1979), 'The generalized wave method for solving extremal problems on graphs' (Russian), *Zh. Vychisl. Mat. i Mat. Phyz.*, **19**, 739–755.

[14] S. M. Avdoshin, V. V. Belov and V. P. Maslov (1984), *Mathematical aspects of computing media synthesis* (Russian), MIEM Publ., Moscow.

[15] S. M. Avdoshin, V. V. Belov, V. P. Maslov and V. M. Piterkin (1987), *Optimization for flexible manufacturing systems* (Russian), MIEM Publ., Moscow.

[16] S. M. Avdoshin (1993), 'Linear programming in semirings with operators', *Russian J. Math. Phys.*, **1**, 127–130.

[17] U. Zimmermann (1981), 'Linear and combinatorial optimization in ordered algebraic structures', *Ann. Discrete Math.*, **10**, 1–380.

[18] B. Mahr (1984), 'Iteration and summability in semirings', *Ann. Discrete Math.*, **19**, 224–256.

[19] V. P. Maslov (1987), *Asymptotic methods for solving pseudodifferential equations* (Russian), Nauka Publ., Moscow.

[20] V. P. Maslov (1987), *Méthodes opératorielles*, Éditions MIR, Moscow.

[21] V. P. Maslov and K. A. Volosov, editors (1988), *Mathematical aspects of computer engineering*, MIR Publ., Moscow.

[22] V. P. Maslov and S. N. Samborskiĭ, editors (1992), *Idempotent analysis*, Adv. Sov. Math., **13**, Amer. Math. Soc., Providence, R. I.

[23] V. P. Maslov, V. N. Kolokoltsov (1994), *Idempotent analysis and its applications in optimal control* (Russian), Nauka Publ., Moscow.

[24] V. N. Kolokoltsov, V. P. Maslov (in press), *Idempotent analysis and applications*, Kluwer Acad. Publ., Dordrecht et al.

[25] G. Cohen, D. Dubois, J.-P. Quadrat and M. Voit (1983), 'Analyse du comportement périodique de systèmes de production par la théorie des dioïdes', *INRIA, Rapp. de Recherche*, **191**, Rocquencourt.

[26] J.-P. Quadrat (1990), 'Théorèmes asymptotiques en programmation dynamique', *Comptes Rendus Acad. Sci., Paris*, **311**, 745–748.

[27] F. L. Baccelli, G. Cohen, G. J. Olsder and J.-P. Quadrat (1992), *Synchronization and linearity: an algebra for discrete event systems*, John Wiley & Sons Publ., New York et al.

[28] J.-P. Quadrat/ Max-Plus working group (1994), 'Max-Plus algebra and applications to system theory and optimal control', in *Proceedings of the ICM, Zürich.*

[29] G. Cohen and J.-P. Quadrat, editors (1994), 'Proceedings of the 11th Conf. on Analysis and Optimization of Systems: Discrete Event Systems', *Springer Lect. Notes in Control and Inf. Sci.*, **199**.

[30] G. Rote (1985), 'A systolic array algorithm for the algebraic path problem (shortest paths; matrix inversion)', *Computing*, **34**, 191–219.

[31] Y. Robert and D. Tristram (1987), 'An orthogonal systolic array for the algebraic path problem', *Computing*, **39**, 187–199.

[32] P. I. Dudnikov and S. N. Samborskiĭ (1987), *Semimodule endomorphisms over a semiring with an idempotent operation* (Russian), Inst. Math. Ukrainian Acad. Sci., Kiev.

[33] P. I. Dudnikov and S. N. Samborskiĭ (1991), 'Endomorphisms of semimodules over semirings with an idempotent operation', *Izv. Akad. Nauk SSSR, ser. math.*, **55, 1**, English transl. in *Math. USSR Izvestiya*, **38, 1**, 91–105.

[34] V. P. Maslov and S. N. Samborskiĭ (1992), 'Stationary Hamilton-Jacobi and Bellman equations', in [22], 119–133.

[35] S. N. Samborskiĭ and A. A. Tarashchan (1992), 'The Fourier transform and semirings of Pareto sets', in [22], 139–150.

[36] V. P. Maslov and S. N. Samborskiĭ(1992), 'Existence and uniqueness of solutions of the steady-state Hamilton–Jacobi and Bellman equations. A new approach', *Russian Acad. Sci. Dokl. Math.*, **45**, 3, 682–687.

[37] P. I. Dudnikov and S. N. Samborski (1994), 'Networks methods for endomorphisms of semimodules over Min-Plus algebras', in [29], 319–321.

[38] S. N. Samborski (1994), 'Time discrete and continuous control problems convergence of value functions', in [29], 297–301.

[39] V. N. Kolokoltsov and V. P. Maslov (1987), 'The Cauchy problem for the homogeneous Bellman equation', *Soviet Math. Dokl.*, **36**, 2, 326–330.

[40] V. N. Kolokoltsov (1992), 'On linear operators in idempotent analysis', in [22], 87–101.

[41] V. N. Kolokoltsov (1992), 'The stochastic Bellman equation as a non-linear equation in Maslov spaces. Perturbation theory', *Russian Acad. Sci. Dokl. Math.*, **45**, 2, 294–300.

[42] V. N. Kolokoltsov (1996), 'Stochastic Hamilton-Jacobi-Bellman equation and stochastic WKB method', in [63].

[43] P. Del Moral (in press), 'A survey of Maslov optimization theory', in [24], Appendix.

[44] P. Del Moral, J.-Ch. Noyer and G. Salut (1994), 'Maslov optimization theory: stochastic interpretation, particle resolution', in [29], 312–318.

[45] P. Del Moral and G. Salut (in press), 'Maslov optimization theory', *Russian J. Math. Phys.*

[46] P. Del Moral and G. Salut (in press), 'Particle interpretation of non-linear filtering and optimization', *Russian J. Math. Phys.*

[47] I. Simon (1988), 'Recognizable sets with multiplicities in the tropical semiring', *Lecture Notes in Computer Sciences*, **324**, 107–120.

[48] I. Simon (1994), 'On semigroups of matrices over the tropical semiring', *Inform. Theor. Appl.*, **28**, 3–4, 277–294.

[49] A. Jean-Marie and G. J. Olsder (1993), *Analysis of stochastic min-max systems: results and conjectures*, INRIA, Sophia–Antipolis.

[50] E. Wagneur (1991), 'Moduloids and pseudomodules 1. Dimension theory', *Discrete Mathematics*, **98**, 57–73.

[51] E. Wagneur (1994), 'Subdirect sum decomposition of finite dimensional pseudomodules', in [29], 322–328.

[52] M. A. Shubin (1992), 'Algebraic remarks on idempotent semirings and the kernel theorem in spaces of bounded functions', in [22], 151–166.

[53] J. S. Golan (1992), *The theory of semirings with applications in mathematics and theoretical computer science*, Pitman monographs & surveys in pure and applied mathematics, **54**, Longman Sci. & Tech., Harlow & New York.

[54] G. Cohen (1994), 'Dioids and Discrete Event Systems', in [29], 223–236.

[55] G. J. Olsder (1994), 'On structural properties of min-max systems', in [29], 237–246.

[56] M. Akian, J.-P. Quadrat and M. Viot (1994), 'Bellman processes', in [29].

[57] S. Gaubert (1992), *Théorie des systémes linéaires dans les dioïdes*, These, École des Mines, Paris.

[58] S. Gaubert (1994), *Introduction aux systèmes dynamiques à événements discrets*, Notes de cours ENSTA, INRIA, Rocquencourt.

[59] S. Gaubert (1994), 'Rational series over dioids and discrete event systems', in [29].

[60] D. Krob (1992), 'The equality problem for rational series with multiplicities in the tropical semiring is undecidable', *Lect. Notes in Comp. Sci.*, **628**, 101–112.

[61] J. Gunawardena (1993), 'Timing analysis of digital circuits and the theory of min-max functions', in *TAU'93, ACM Int. Workshop on Timing Issues in the Specifications and Synthesis of Digital Systems*.

[62] J. Gunawardena (1994), 'Min-max functions', *Discrete Event Dynamics Systems*, **4**, 377-406 .

[63] J. Gunawardena, editor (1997), *Idempotency*, Publications of the Isaac Newton Institute, CUP, Cambridge (the present volume).

[64] J.-E. Pin (1996), 'Tropical semirings', in [63].

[65] A. V. Finkelstein and M. A. Roytberg (1993), 'Computation of biopolymers: a general approach to different problems', *BioSystems*, **30**, 1–20.

[66] M. A. Roytberg (1994), *Pareto-optimal alignments of symbol sequences*, Inst. of Math. Problems of Biology, Pushchino.

[67] M. D. Bronstein and A. S. Cherevatskiĭ (1991), 'Variational problems for functionals determined by the maximum of a Lagrangian', *Soviet Math. Dokl.*, **43**, 1, 117–121.

[68] M. D. Bronstein and A. S. Cherevatskiĭ (1991), 'A variational method of solving boundary value problems for ordinary differential equations', *Soviet Math. Dokl.*, **43**, 2, 572–575.

[69] M. D. Bronstein (1991), 'Functionals containing idempotent integration, and their applications to estimating the smallest eigenvalue of differential operators', *Soviet Math. Dokl.*, **44**, 2, 417–421.

[70] M. D. Bronstein and S. N. Samborskiĭ (1991), 'A multicriterion variational method for systems of nonlinear partial differential equations', *Soviet Math. Dokl.*, **44**, 2, 603–607.

[71] M. Lorenz (1993), *Object oriented software development: a practical guide*, Prentice Hall, Englewood Cliffs, N.J.

[72] G. L. Litvinov, V.P. Maslov and A.Ya. Rodionov (1995), *Unifying approach to software and hardware design for scientific calculations*, preprint.

[73] S. C. Kleene (1956), 'Representation of events in nerve nets and finite automata', in *Automata Atudies* (J. McCarthy and C. Shannon), Princeton University Press, Princeton, 3–40.

[74] V. N. Kolokoltsov and V. P. Maslov (1995), 'New differential equation for the dynamics of the Pareto sets', in [63].

[75] P. Lancaster (1969), *Theory of matrices*, Academic Press, New York and London.

[76] V. N. Kolokoltsov, V. P. Maslov (1989), 'Idempotent calculus as the apparatus of optimization theory, I,II',*Functional Anal. i Prilozhen.*, **23**, 1, 1–14; **23**, 4, 52–62 (Russian); English transl. in *Functional Anal. Appl.* (1989), **23**.

[77] R. E. Bellman, S. E. Dreyfus (1965), *Dynamic programming and applications*, Dunod, Paris.

[78] V. N. Kolokoltsov, V. P. Maslov (1987), 'The general form of endomorphisms in the space of continuous functions with values in a numerical semirings with idempotent addition', *Soviet Math. Dokl.* **36**, 1, 55– 59.

[79] R. W. Floyd (1962), 'Algorithm 97: shortest path', *Commun. ACM*, **5**, 6, 345.

[80] S. Warshall (1962), 'A theorem on Boolean matrices', *Journal ACM*, **9**, 1, 11–12.

[81] V. V. Voevodin (1991), *Mathematical principles of parallel computations*, Moscow State University Publ., Moscow (Russian).

[82] V. P. Maslov *et al.* (1991), *Mathematics of semirings and its applications*, Technical report. Institute for New Technologies, Moscow (Russian).

[83] H. T. Kung (1985), 'Two-level pipelined systolic arrays for matrix multiplication, polynomial evaluation and discrete Fourier transformation', in *Dynamical Systems and Cellular Automata* (J. Demongeof *et al.*, Eds), Academic Press, New York, 321–330.

[84] S. G. Sedukhin (1992), 'Design and analysis of systolic algorithms for the algebraic path problem', *Computers and Artificial Intelligence*, **11**, 3, 269–292.